Upper Pennsylvanian section on the north wall of Dry Canyon in the Sacramento Mountains, 6 km northeast of Alamogordo, New Mexico. The bottom of the creek is about the base of Missourian beds. The base of Virgilan strata is about the base of the pronounced biohermal level. The well-marked dark thin bed below the main bioherm is composed of peloidal grainstone, foreset-bedded to the left (west into the Oro Grande basin). The bioherms are algal plate mud mounds and are developed at a very gentle shelf margin and on the west flank of a Pennsylvanian anticline whose axis lay to the right edge of the photo. The axis is cut transversely by the canyon. The two mounds at the west end are micritic cores with abundant algal plates. The main mound is composed chiefly of foreset-bedded, detrital, bioclastic flanking strata derived from organisms growing on mounds exposed on the south side of the canyon from where the picture was taken. The ledges above the main mound level are principally limestones capping sedimentary shelf cycles described on Figs. VI-7, VI-12, VII-8. The cycles grade a few kilometers east to red beds and conglomerates of the Pennsylvanian upland.

JAMES LEE WILSON, B.A., M.A., Ph.D., Department of Geological
Sciences, University of Michigan, Ann Arbor, Michigan 48109, USA.

Cover motif from Paul N. McDaniel, ERICO, Woking, England.

ISBN 0-387-07236-5 Springer-Verlag New York Heidelberg Berlin (hard cover)

ISBN 3-540-07236-5 Springer-Verlag Berlin Heidelberg New York (hard cover)

ISBN 0-387-90343-7 Springer-Verlag New York Heidelberg Berlin (soft cover)

ISBN 3-540-90343-7 Springer-Verlag Berlin Heidelberg New York (soft cover)

Library of Congress Cataloging in Publication Data. Wilson James Lee, 1920 — Carbonate
facies in geologic history. Bibliography: p. Includes index. 1. Facies (Geology) 2. Rocks.
Carbonate. I. Title. QE651.W52 551.8'1 75-15667

© by Springer-Verlag Berlin · Heidelberg 1975

9 8 7 6

Printed in the United States of America.

James L. Wilson

Carbonate Facies Geologic History

With 183 Figures and 30 Plates

Springer-Verlag

New York Heidelberg Berlin

To my wife
Della Moore Wilson
and to my parents
Mr. and Mrs. J. Burney Wilson

Preface

Since 1950 geologists have learned more about the origin and lithification of carbonate sediments than in all the previous years of the history of science. This is true in all the diverse fields of carbonate geology: the study of Recent environments, marine zoology and botany, organic geochemistry, trace element and isotope geochemistry, mineralogy, microfacies of depositional environments, and trace-fossil and sedimentary structure investigation. A synthesis of this new knowledge is just beginning to be formulated.

The purpose of this volume is to introduce the advanced student and petroleum explorationist principally to one important aspect of this study: to some of the principles of carbonate geology which may serve to interpret the depositional environments of ancient strata and to better define their sequences and patterns.

Chapter I is a brief review of principles of carbonate sedimentation. (For a full discussion of the mineralogy, geochemistry, and diagenesis of carbonates along with a review of Holocene sediments, one may refer to Bathurst's (1971) and Milliman's (1974) texts.) Chapter II reviews stratigraphic and paleotectonic concepts and discusses a general model for carbonate deposition. Chapter III offers an outline of carbonate petrography, concentrating on lithologic description for the purposes of environmental interpretation. For a further review of this subject and excellent photomicrographs, Horowitz and Potter (1971) and Majewske (1969) may be used.

The remainder of the book (Chapter IV–XII) attempts to synthesize a large amount of descriptive data interpreting depositional patterns from numerous examples of carbonate facies. From this, certain generalizations appear which are mainly set out in Chapters II and XII (Summary). The approach has been mainly inductive: let us look at the record of marine deposition of lime sediment, see what is there and learn how to describe and interpret it. A dual approach is recommended: to focus attention equally on stratigraphic relations and petrographic study. Comparisons between certain patterns are made to facilitate the sorting out of parameters deemed to be most important in their formation.

The inductive approach is always fraught with some difficulty. One is reminded of Mark Twain's remark in *Innocents Abroad*, that despite his perusal of thousands of European cathedrals dedicated to a collection of Saints, he could not honestly admit to an appreciation of such architecture. He felt that he should, perhaps, study another 97,000 or so before making up his mind about their esthetic value. Have we looked at sufficient models for a proper synthesis? The author's answer is obviously "we have made a start at it." The facies patterns described are mostly all from the author's personal experience in European and Middle East regions and from the midcontinent, western Canada, Rocky Mountains, Appalachian and southwestern United States and Mexico. These regions include vast areas of carbonate rock. No personal or up-to-date knowledge of the Great Basin, Australian or Siberian stratigraphy is included. Nevertheless, it appears that sufficient stratigraphy is known to permit some valid generalizations of facies patterns and for them to be used as an effective tool in predicting facies distribution in imperfectly known areas. The middle chapters of the book repeat many patterns with different faunal variations because carbonates tend to follow essentially one basic and fundamental depositional pattern which is superficially modified by tectonic, climatic, and hydrologic factors, thus simplifying our task of generalization.

With the task of synthesis in mind, the book abounds in classifications and outlines attempting to organize and standardize information. This approach includes:

1. Carbonate microfacies (24 SMF types, Standard Microfacies Types).
2. Terminology of carbonate buildups (23 definitions).
3. Types of shelf-margin profiles (3 types).
4. Facies belts along shelf margins (9 in an idealized profile).
5. Sequence in development of an ideal carbonate mound (7 facies).
6. Tectonic settings of carbonate buildups and facies patterns (4 major categories).
7. Carbonate cyclic sequences (5 types).
8. Organism development in carbonate buildups through geologic time (4 stages).

Certain aspects of carbonate sediments are not covered even in the review in Chapter XII. Oceanic sediments, fresh-water and temperate-zone marine carbonates are omitted because practically all common carbonate facies in the geologic record were apparently deposited in shallow, tropical, marine environments. Descriptions of Holocene models of carbonate deposition which are basic to our interpretation of ancient facies patterns have been given by Bathurst (1971) and Milliman (1974), in numerous special publications and memoirs of the AAPG and SEPM, in the Sedimenta Series of the University of Miami, the Persian Gulf volume (Purser, 1973), and numerous guidebooks to the British Honduras, Florida, Bahamas, and Yucatan areas.

The reader may find more serious omissions in the lack of discussion of evaporites, which are an integral part of the carbonate depositional realm. However, a general synthesis of evaporite literature is now available from symposia by major geological societies and several books have appeared on the subject. The related subject, dolomite origin and stratigraphy, is discussed briefly in Chapter X. Tertiary buildups and patterns are not sufficiently known to the author personally to permit accurate description; although they are not discussed, they are equally important to petroleum exploration.

The writer hopes that the labor of reading and studying the book is not quite so formidable a task as was its writing.

July 1975   J. L. WILSON

Acknowledgements

A book of this type would have been impossible without years of cooperative field observation and exchange of ideas with many persons. For these opportunities the author wishes to thank the management of Shell Development and Shell Oil Companies, Shell International Research, and particularly the many fine geologists of these companies who provided stimulating discussion at so many different times and places.

The author is also grateful to the National Science Foundation which, through Grant Number GA10147, made possible the initiation of this compilation, and to William Marsh Rice University for providing facilities for carrying on the research with excellent graduate students and for providing a sabbatical year.

The author would also like to thank Professor Richard Dehm and Dr. Ludwig Happel of the Institutes of Paleontology and General and Applied Geology of the University of Munich for their hospitality during a pleasant year in which much of the writing was accomplished. This was also made possible through a Fulbright-Hayes research fellowship in 1972—1973.

Many persons contributed specifically to the completion of the book. Dr. R. N. Ginsburg and Paul Potter looked at introductory Chapters and gave initial encouragement. Dr. L. L. Sloss undertook the difficult task of reading and criticizing most of the manuscript. C. Kendall, Alan Coogan, Ludwig Happel, G. Shairer, B. H. Purser, and R. N. Ginsburg read and criticized particular Chapters. Valuable suggestions were also made by three graduate students, Steve Schafersman, Edward Thornton, and John Van Wagoner who carefully read the introductory chapters. Schafersman also assisted with the photomicrography. Michael Campbell helped compile a glossary of sedimentary structures.

Special acknowledgement is made to the American Association of Petroleum Geologists and to the Society of Economic Paleontologists and Mineralogists who permitted many illustrations from their journals to be republished. Dr. R. N. Ginsburg of the Miami University Laboratory of Comparative Sedimentology kindly furnished the author eight redrafted Figures by Dr. Bruce H. Purser who gave permission for their inclusion in Chapter X. J. E. Klovan, D. F. Toomey, M. Malek-Aslani, F. R. Allcorn, and C. Kendall assisted greatly by furnishing some of the photomicrographs used on the plates.

The original manuscript editing was largely done by Martha Lou Broussard and completed by Peggy Rainwater. The author is particularly grateful for their conscientious efforts.

Much of the drafting was done by the author's sister, Mrs. Betty Mitchell and the lettering by Michael Carr, John Van Wagoner, and the author's mother, Mrs. J. Burney Wilson.

The author is especially grateful to his wife, Della Moore Wilson, who contributed most importantly; directly by performing the very considerable chore of editing and typing of the manuscript; and indirectly through loving encouragement, unfailing good humor, and pleasant companionship through the two and one half years of the writing.

Contents

Chapter I   Principles of Carbonate Sedimentation   1

The Requisite Marine Environment: Warmth, Light, Water
Movement   1
Carbonate Production Is Basically Organic   4
Principal Hydrographic Controls on Carbonate Accumulation   5
The Local Origin of Carbonate Particles—Textural Inter-
pretation and Classification   7
Carbonate Deposition Is Rapid but Is Easily Inhibited and
Therefore Sporadic during Geologic Time   14
Carbonate Sediments and Rock Are Peculiarly Subject to
Many Stages of Diagenesis   16
Summary   17

Chapter II   The Stratigraphy of Carbonate Deposits   20

Definitions   20
The Basic Facies Pattern   24
Paleotectonic Settings for Carbonate Facies   29
Stratigraphic Sequences, Geometry, and Facies of Carbonate
Shelf Margins and Basins   33
Stratigraphic Sequences in Carbonates of Epeiric Seas on
Shelves and in Shallow Basins   46

Chapter III   Outline of Carbonate Petrography   56

Techniques for Examining Carbonate Rocks   56
Microfacies Interpretations   60
Standard Microfacies Types   63
Diagenetic Changes in Carbonate Sediments   69
Biological Observations   71
Glossary of Sedimentary Structures in Carbonate Rocks . .   75
Environmental Analysis of a Carbonate Thin Section . . .   86
Significance of Color in Carbonate Rocks   89
Clastic Content in Carbonates   90
Porosity and Permeability   93

Chapter IV   The Advent of Framebuilders in the Middle Paleozoic . .   96

The Earliest Buildups   96
Silurian Buildups — Paleotectonic Settings 103
Devonian Buildups 118
Summary   140

Chapter V   The Lower Carboniferous Waulsortian Facies 148

Relation of Waulsortian Facies to Regional Paleostructure . 150
Composition of Typical Waulsortian Facies 162
Theories of Mound Origin 165
Conclusions 167

Chapter VI   Pennsylvanian-Lower Permian Shelf Margin Facies in South-western United States of America 169

Paleotectonic Setting, Geologic History, and Climate . . . 169
Special Organic Communities Forming Carbonate Buildups
in Late Paleozoic Strata 171
Examples of Permo-Pennsylvanian Carbonate Buildups . . 174
Conclusions 198

Chapter VII   Late Paleozoic Terrigenous-Carbonate Shelf Cycles 202

Introduction 202
Yoredale Cycles 203
Pennsylvanian and Wolfcampian Shelf Cyclothems of the
Midcontinent and Southern Rocky Mountains 206

Chapter VIII Permo-Triassic Buildups and Late Triassic Ecologic Reefs 217

Permian Reef Complex 217
The Middle Triassic of the Dolomites 233
Upper Triassic Reef-Lined Banks and Basinal Mounds of
Austria and Bavaria 244
Similarities and Differences between Permian and Triassic
Reef Complexes 254

Chapter IX   Reef Trends and Basin Deposits in Late Jurassic Facies of
Europe and the Middle East 257

Regional Settings 258
Basic Microfacies 262
The Reef Girdle of Central Europe 266
The Solnhofen Facies 274

Chapter X   Shoaling upward Shelf Cycles and Shelf Dolomitization . 281

Oolite-Grainstone Cycles 283
Lime Mud-Sabkha Cycles 297
Platform Cycles with Intense Diagenesis 304
Dolomitization of Carbonate Banks and Interior Shelf Cycles 310
Conclusions 318

Chapter XI   The Rise of Rudists; Middle Cretaceous Facies in Mexico
              and the Middle East 319

              The Rudist Bivalves 319
              Shelf Margin and Plattform-Bank Interior Facies of Middle
              Cretaceous of Mexico and the Gulf Coast 325
              Middle and Lower Cretaceous Facies in the Middle East . 339

Chapter XII   Summary 348

              Stratigraphic Principles 348
              Nine Standard Facies Belts in an Ideal Model of a Carbonate
              Complex 350
              Three Major Types of Shelf Margin Profiles 361
              Carbonate Mounds and Associated Ecologic Reefs 364
              Interrelations of Parameters Controlling and Modifying Car-
              bonate Accumulations 369
              Review of Tectonic Settings for Carbonate Buildups and
              Cycles 375

References   380

Subject Index 399

Plates 411

Chapter I

Principles of Carbonate Sedimentation

The Requisite Marine Environment: Warmth, Light, Water Movement

Most, though not all, carbonate sedimentation results basically from chemical or biochemical processes occurring in a special marine environment: one of clear, warm, shallow water. A world map (Fig. I-1) showing areas of modern carbonate deposition demonstrates clearly a positive correlation between such deposition and the equatorial belt and areas of warm ocean currents. Fairbridge (Chilingar et al., 1967, p. 404) presented a graph showing that neritic carbonates exist chiefly north and south of the equator below latitudes of 30 degrees. Furthermore, whereas the bottoms of deep ocean basins between 40 degress north and south latitudes contain much planktonic carbonate, those in higher latitudes do not, except in the north Atlantic along the Gulf stream. Invertebrates precipitate thicker calcite and aragonite shells in clear, warm waters, many more calcareous algae thrive there, and algal-dependent hermatypic or reef-building corals are restricted to such an environment. Cooler marine waters *do* support swarms of invertebrates whose tests and shells may form local accumulations of shelly lime sand (Lees, 1973; Chave, 1967), but other types of lime sediment, such as ooids, grapestones, peloids, reefy boundstone and lime mud accumulations, are confined to tropical and subtropical waters.

However, tropical water alone is insufficient for calcium carbonate production. The water must be clear. Three great carbonate banks around the Gulf of Mexico are located in areas most protected from major influx of fine clastic sediment. These are far to the southeast of the westward-moving longshore drift of the Mississippi River mud and are protected by deep water from the influx of clay and silt off the large island of Cuba. A reverse situation appears across the Sunda Shelf north of Indonesia where a vast platform, covered by shallow, equatorial water exists. The Sunda Shelf bears only isolated reef accumulations, located along its northern and eastern edges, because large rivers from Sumatra, Java, and Borneo to the south and west muddy the sea and inhibit carbonate formation. Fairbridge (in Chilingar et al.; 1967) pointed out that practically all shallow carbonate shelves in strictly equatorial regions seem to be smothered by fine terrigenous clastics brought by the large tropical rivers.

Given the requisite marine environment, exactly what chemical and biological controls operate to produce abundant carbonate? The complex chemical problems of precipitation of $CaCO_3$ minerals from sea water are beyond the scope of this book. The reader is referred to the discussion of the subject by Bathurst (1971, Chapter 6) and Milliman (1974) who cited pertinent references to modern literature. At the present time, tropical seas are essentially saturated with regard to

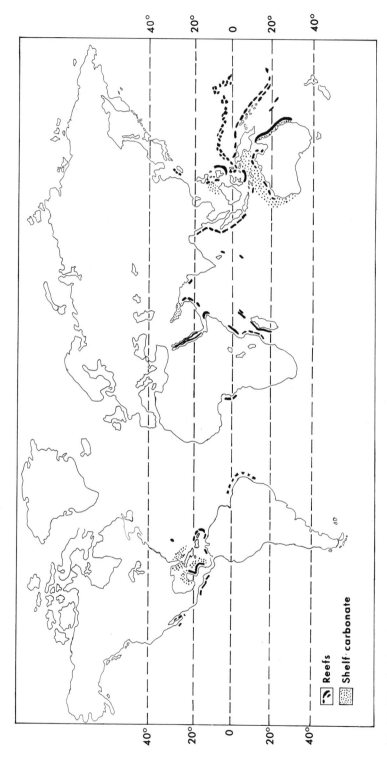

Fig. I-1. Distribution of modern marine carbonate sediments in shallow water

$CaCO_3$. Thus, any process which removes CO_2 from normal water (pH = 8.4), tending to change bicarbonate to carbonate ions, encourages lime precipitation. At least eight mechanisms for this process may be effective: increase of temperature, intense evaporation, influx of supersaturated water to an area where $CaCO_3$ nuclei or catalyzers are present, marine upwelling from an area of high pressure to low pressure, mixing of water high in CO_3 and low in Ca + + with sea water, organic processes in body fluids, bacterial decay to produce ammonia, raising pH and increasing carbonate concentration, and removal of CO_2 by photosynthesis.

The photosynthesis process brought about by metabolism of microplanktonic flora, especially when operating in warm and agitated water, may be of prime importance. If this is so—and biochemical studies indicate more and more that organic amino acids capable of precipitating $CaCO_3$ coat almost all particles in the sea (Mitterer, 1971)—important implications exist for depth control on the rate of carbonate production. Despite the total depth range of tropical marine algae to a 100 m or more, the codiaceans and bluegreen forms are particularly abundant at depths less than 10–15 m. Little green algal growth in deeper water is known except down tropical shelf margins in very clear waters where abundant *Halimeda* grows to a depth of 70 m or so. Generally it appears that a threshold of dominant algal production of $CaCO_3$ may be reached at very shallow depths. Hence any geographic situation resulting in wide areas of water from 10–15 m deep may result in several times more $CaCO_3$ per unit area than in deeper epeiric seas (Fig. I-2).

Not only depth, but also turbidity caused by suspended clay and silt particles in water strongly inhibits $CaCO_3$ production. This occurs in two ways: (1) it cuts down on light to interfere with photosynthesis, discouraging growth of the calcareous algae, the breakdown of whose cortices is a major contributor to aragonitic lime mud. Indeed, if precipitation of lime mud from sea water is chiefly biochemical, the inhibition of phytoplankton by muddy and darker water would essentially eliminate carbonate production: (2) benthonic invertebrates contribute impressive amounts of calcium carbonate particles of all sizes, and most of these animals are inhibited by suspended clay particles which plug up feeding mechanisms.

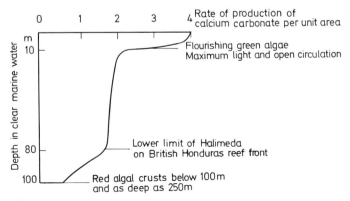

Fig. I-2. Theoretic diagram indicating that production of organic carbonate is probably not a straight-line function of depth. Estimation on rates of production by R. N. Ginsburg

It is noteworthy that most major deposits of limestone or dolomite are remarkably pure carbonate and contain only a few percent of argillaceous or silty insoluble material. Such "contaminants" are obviously detrimental to deposition of calcareous sediment.

Two other processes are important, though probably not necessary for carbonate deposition: agitation of water and strong evaporation. The former process is discussed later. The role of evaporation in concentrating sea water in basins or on tidal flats is well established. Biochemical or wholly inorganic removal of $CaCO_3$ from sea water must occur, before concentration of dissolved salts becomes sufficient for $CaSO_4$ to be precipitated. Interbedded, unfossiliferous, fine grained, homogeneous carbonates, with anhydrite or gypsum in basinal evaporite deposits, attest to this process.

The above factors, which cause formation of marine $CaCO_3$ may be viewed collectively as a system of combined special natural processes. All other sedimentologic processes operate to modify in one way or another the products of this system. Thus the type of available organisms, sea level fluctuation, rate of subsidence, hydrographic factors, and climate, operate on this "carbonate factory" to produce the kinds of limestone and dolomite recognized in the geologic record. See Fig. XII-1 in the Summary Chapter.

Carbonate Production Is Basically Organic

Organisms contribute to detrital deposits of widely varying grain size and, as well, form huge masses of precipitated limestone.

Most Lime Mud Is Organically Derived

Sedimentologists have extensively discussed its origin and modes of accumulation. It is a common product of shallow, tropical water or the upper layers of the open ocean. Cool, marine waters of shallow temperate zones produce only shelly and silty carbonate biogenic debris, and planktonic accumulations in ocean basins are composed mostly of silt-size material formed in the photic zone in lower latitudes.

Fine lime sediment settles to the bottom without appreciable compaction. A few centimeters below the surface the soft sediment consists of about equal portions of $CaCO_3$ and water. The carbonate in modern shallow water lime muds is generally aragonite, but may contain substantial proportions (up to 50%) of high Mg calcite (generally with greater than 10 mol percent Mg in the crystal lattice) and as much as 10–15% low magnesium calcite (generally with less than 5 mol percent Mg).

Lime mud has been demonstrated to derive in several different ways: from death and decay of benthonic organisms (mainly calcareous algae), detritus abraded from larger carbonate particles, accumulation of planktonic biota and possibly from direct precipitation from sea water (perhaps biochemically induced by phytoplankton blooms). Both low and high Mg calcite deposited in sea water are derived from biological or mechanical attrition of shell material, but the origin of aragonite, which constitutes generally more than half and in places up to 95% of lime mud, is much debated. The arguments focus particularly on the extent to which tiny (4 micron) aragonite needles are produced from the breakdown of codiacean algae or to whether they can be inorganically precipitated. Data indicate that whatever its origin, much, if not most or all, fine lime sediment is organically derived.

Most Coarser Lime Particles also Have an Organic Origin

Much sand- to gravel-size carbonate sediment forms from breakdown of shells and tests (bioclasts). For example, mollusks, green algae, modern corals and many foraminifera today contribute aragonite particles; other foraminifera (miliolids and peneroplids), red algae, and echinoderms contribute high Mg calcite. A small amount of low Mg calcite is contributed by breakdown of brachiopod, bryozoans, ostracods, foraminifera, and additionally by trilobites and rugose corals in the Paleozoic. Lime mud may be moulded or aggregated by organisms into fecal pellets and grapestone lumps and transported as very low density sand grains. Organisms also contribute indirectly to construction and modification of sand size particles. Bioclasts are rotted by microboring fungi, algae and sponges, and altered into peloids (rounded homogeneous micritic sand-size particles). Carbonate particles are also formed as larger fragments by collapse of burrows or desiccation of algal mats. Thus, organisms contribute directly or indirectly to construction of practically all major carbonate particle types. This is even true in ooids; coating blue-green algae play a role in the construction of the concentric laminae which are probably formed from precipitated aragonite.

Biologically Precipitated Carbonate Masses Are Abundant in the Geologic Record

In addition to forming both fine and coarse detrital particles of carbonate, many primitive sessile invertebrates, as well as algae, are capable of direct secretion of carbonate in or around their tissues to form various types of massive and rigid frameworks and encrustations. Coelenterates (hydrozoans and anthozoans), sponges, coralline red algae, bryozoans and mollusks are important in this regard. This ability to create rigid boundstone (Dunham, 1962) or biolithite (Folk, 1959) accounts for considerable local buildups of carbonate (organic-ecologic reefs) and is a unique feature of carbonate sedimentation quite different from any mode of terrigenous clastic deposition.

Principal Hydrographic Controls on Carbonate Accumulation

Once carbonate sediment is formed, it is subject to the same processes of marine sedimentation which affect terrigenous clastic sediments. These are especially operative in open sea positions or along the shelf edges of marginal (pericratonic) basins. Currents and waves winnow fine-grain sediment creating sand and gravel lag deposits on the open shallow shelves. These may vary from coastal sand flats in areas with great tidal range (2–3 m) to the broad shelves of the Campeche and west Florida areas now under 50–100 m of water. Waves and currents also tend to pile up carbonate sand and gravel deposits. Coquina beaches, littoral spits and submarine bars due to longshore drift, tidal deltas and bars in tidal passes constitute well-known bodies of mechanically deposited carbonate sediment. Similarly, tidal bars of oolite and peloid grains are known at the edges of major Holocene carbonate banks.

Finer carbonate sediment, winnowed from shelves, tends to accumulate in two preferred locations relative to shelf margins: off the shelf in deeper water marginal to the shore and in protected areas in quiet water behind barriers. In the axis of the Persian Gulf, calcareous muds occur off the northern edge of the Great Pearl Bank, and accumulations of fine calcareous mud sediments are known in the Sigsbee Deep, just north of the Campeche Bank, Gulf of Mexico. Contrasting with these occurrences are shallow lagoons enclosed by coastal spit accretion along the northeast Yucatan coast, which contain 30 feet or more of lime mud, the vast muddy tidal flats on the leeward (western) side of Andros Island, the tidal lagoons along the western side of the Persian Gulf, and the mud-choked Florida Bay. In the latter cases, much fine mud sediment has been brought from outer shelves into lagoons and onto tidal flats by storm and tidal currents and trapped, although *in situ* accumulation is also proceeding in these areas.

Despite the similarity of hydrologic processes in carbonate and terrigenous sedimentation, some additional effects of water movement occur in the deposition of carbonates, because of their predominantly organic origin. These effects can be differentiated into those occurring in marginal basins, with exposure to open sea and those of interior or epeiric seas. Water movements induced by strong currents and crashing waves at coastal margins have an important positive effect on growth rates of carbonate-producing organisms. Removal of CO_2 through wave action and change of pressure, and the bringing of nutrients by fresh marine water to stimulate growth of organisms, encourages $CaCO_3$ precipitation. Modern reefs form best along such areas, particularly where a shelf break occurs. Even in periods of quiet weather or those with onshore winds, upwelling along steep slopes brings fresh nutrients to the shelf margin. Rapid growth to sea level occurs and much debris is shed rapidly from the centers of organic growth. Thus strong water movement itself indirectly creates great volumes of carbonate sediment despite its tendency to erode. Persistently strong to moderate water movement (such as over shallow, tropical shelves) also creates sand size-particles such as ooids, grapestones, and hardened fecal pellets by submarine accretion and cementation. These processes are partly organic and partly physico-chemical, and they not only create hardened sand-size particles, but may also stabilize the sediment through pervasive submarine cementation.

In quiet water areas, behind reef or sand barriers, or across shallow water over wide flat shelves, restricted circulation and climatic factors combine to influence strongly the type of carbonate sedimentation in a different way. These conditions have been exhaustively described by Irwin (1965) and Shaw (1964) and applied to the geologic record by Roehl (1967) and Lucia (1972). Stagnant circulation results in restricted environments for most marine animals. It also results in more variable and extreme salinities when it occurs in an evaporative climate. Pelleting of mud is important in creating some sand and silt in the lagoonal mud sediments where mollusks, certain algae, foraminifera and ostracods dominate the biota and contribute sand size-constituents. In the intertidal zone, a large group of distinctive sedimentary structures forms in the lime mud by intermittent marine flooding and drying due to subaerial exposure. In evaporative climates, calcium sulfate and dolomite commonly form and in tropical rainfall areas, fresh water lenses provide springs and swamps rich in brackish water plants which may precipitate nonmarine, low Mg calcite.

The Local Origin of Carbonate Particles — Textural Interpretation and Classification

Because most carbonate sediment is organically produced, it is fundamentally autochthonous, i.e., it is produced within a given basin and not introduced from without by rivers or streams. Deposits of detrital carbonate sand or silt grains do exist but are extremely rare because of the great solubility of $CaCO_3$ in fresh water, particularly that with dissolved CO_2. Most coarse carbonate grains do not move very far unless spilled over the sides of steep shelf margins or transported by longshore drift parallel to the coast. Most appear to accumulate as detrital particles fallen from growth positions or remain where the producing organism died and decayed, and later moved very little laterally. This is demonstrated in biological studies by Ginsburg (1956) across the Florida Reef tract and McKee et al. (1959), in Kapingamarangi Atoll in the Pacific. In both cases biogenic particles in the bottom sediment, despite extensive bioturbation, reflect rather faithfully the general aspect of present day living biota and hence the environmental conditions (salinity, circulation, temperature, depth, substrate, etc.) of the sea. Even ooids created in tidal bars accumulate more or less where formed. The same constant ebb and flow of water which causes accretion of the particles also constructs the tidal bars in belts or in more or less fan-shaped areas. These build up and remain at select places on the shelf margins, especially where water movement is increased by vertical or lateral constriction. Some tiny particles comprising lime mud are probably transported many miles by storm-wave currents and may accumulate in especially protected localities in a given basin (deep or very shallow water), but can also be produced and accumulated locally in wide lagoons or shallow shelves.

Naturally the autochthonous origin of most carbonate sediment offers a great advantage in its environmental interpretation and increases geological interest and emphasis on identification of particle types, particularly in thin section study. A whole discipline of carbonate petrography has developed beginning with Henry Sorby (1879), and has increased in emphasis tremendously since about the middle 1950's to the present day. Bathurst (1971) has pointed out that the sea floor sediment represents but an insignificant remnant of the teeming variety of organisms and ecological systems which contributed to it. The unravelling of clues left in this residue of debris and sewerage, for purposes of environmental reconstruction, is a fascinating, if difficult, chore. Excellent descriptions and illustrations of typical carbonate particles offering ecologic clues are found in Majewske (1969) and Horowitz and Potter (1971).

The fact that many carbonate particles originate locally and are created in many shapes and sizes, requires quite different classifications and interpretations of textures from those of terrigenous clastic sediments. Much extremely fine-grained material is produced *in situ* as algal or chemically precipitated 2–4 micron aragonite needles, and by microplankton (coccoliths) and tiny detrital carbonate grains. But also the breakdown of calcareous tests of organisms by physical abrasion, organic abrasion or corrosion through ingestion, or simple collapse due to skeletal decay, produces numerous silt, sand, and gravel size particles in a wide variety of shapes. Additionally, these particles vary in mineralogy and in internal structure, which may control the ultimate shape and size of particles much more than the agents of breakdown. Based on the structure of resulting particles, one may recognize several types of skeletons or tests and their resistance to further abrasion. This classification somewhat overlaps but is different in scope from that of Horowitz and Potter (1971, Table 7, p. 36), which is concerned with ultimate shapes of particles for the purpose of identification. The following grouping by R. N. Ginsburg, University of Miami, recognizes six types of skeletons on the basis of their relative resistance to breakdown.

1. *Sheathed and spiculed skeletons* are those in which the mineral matter is small (silt-fine sand size) and loosely held by organic tissue. On death of the organism the organic tissue decays and the particles are released as fine sediment. Examples: *Penicillus*, alcyonarians, corals, sponges, tunicates, holothurians.

2. *Segmented skeletons* consist of mineral particles linked together by organic tissue. Death and decay of the organisms most commonly yields sand-size particles to the sediment. Examples: *Halimeda*, articulated red algae and echinoderms.

3. *Branched skeletons* are composed of well-calcified cylindrical or blade-like projections. The size of fragments found in sediments depends on the original size of the organism, the size and strength of the branches, and the nature and intensity of organic and mechanical breakdown to which they are subjected. Examples: some corals *(Acropora)*, red algae and bryozoans.

4. *Chambered skeletons* include all those that are hollow or partly hollow. Chambers persist after death of the organism, but there are wide variations in the resistance of different types of chambers to breakdown, depending on their absolute size, wall thickness, shape, and microstructure. In general, the arcuate shape successfully resists breakage. Examples: gastropods, serpulid worm tubes, foraminifera, some crustaceans, pelecypods, some echinoderms and brachiopods.

5. *Encrusted skeletons* include all plants and animals that encrust surfaces. The breakdown of the skeletons in most cases depends mainly on the organic breakdown of the encrusted surface or substrate. These are mechanically resistant structures. Examples: some algae, foraminifera, corals, bryozoans, worms, hydrocorallines.

6. *Massive skeletons* are generally large and hemispherical in shape. They are most resistant to breakdown because of their size and in some cases, their microstructure. Examples: corals and some coralline algae.

Throughout the book, the convenient term *bioclast* is employed to designate a fragmental particle derived from breakdown of any sort of calcareous shell, test or skeleton, regardless of whether the breakdown was mechanical or caused by organic agents. The term has been restricted by some authors to skeletal debris resulting only from the latter process. (It was originally defined this way by Grabau, 1920.) Studies of Holocene carbonate sediments show that much breakdown of organic hardparts is caused by various organisms, but the use of the term only for this process is considered too interpretive. The term biogenic or biogenous is of wider application and is used to refer to biological particles formed in any fashion, not merely organic debris, but also fecal pellets, peloids caused by algal rotting of bioclasts, and grapestones agglutinated by organisms.

In contrast to bioclasts, ooids and pelletoids are created within preferred size ranges which probably reflect competence of average currents and waves in very shallow water (1–10 m); most such particles range from about 0.5 to 1.5 mm in diameter.

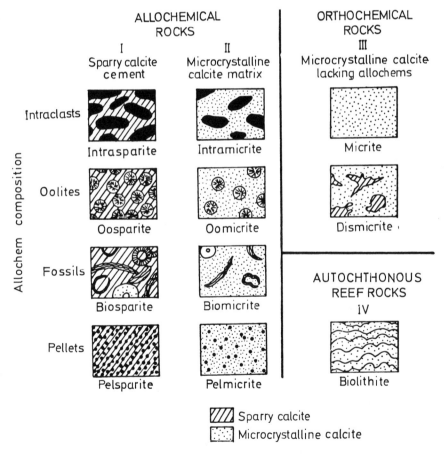

Fig. I-3. Graphic classification table of limestones from Folk (1959, Fig. 4), with permission of American Association of Petroleum Geologists

To repeat: because of the local origin of carbonate particles and the various shapes and dimensions in which they may be produced, it is clear that textural interpretation of detrital carbonates must differ from that of terrigenous clastics. For example, in a mixture of 65% lime mud with 35% sand to gravel size grains of crinoids and bryozoan fragments, is the mud present because it was brought into a site where crinoids and bryozoans were growing? Or, is so much mud present because numerous algae or other potential contributors of fine material were growing in the environment in competition with organisms which contribute coarse particles? Such a question does not arise in terrigenous clastics but must always be considered in carbonates.

The recognition that microcrystalline calcite (micrite) in limestones represents what was originally lime mud and that many carbonate rocks in thin sections could be studied texturally as detrital sediments should be credited to Folk (1959) who proposed a classification and hierarchy of terms for limestones which is now widely used. Many excellent and thoughtful essays on carbonate rock classification, considering the philosophy behind textural groupings, may be found in the American Association of Petroleum Geology Memoir I, edited by Ham (1962). One of the best of these is the brief but significant paper by Dunham. Folk reviewed and expanded his classification also in the AAPG Memoir I. The textural spectrum proposed in this paper embodies extremely useful concepts for making environmental interpretations. Both classifications are given in Figs. I-3 to I-6 and are repeated in many general works on carbonate sedimentation. The author generally uses Dunham's terms in this volume in order to avoid the necessary rigidity involved in applying Folk's particle type prefixes and because the packing concept of Dunham is important and should be designated by a primary name.

Five main principles are significant in categorizing texture in carbonate sediments and are used in these classifications.

1. Presence or Virtual Absence of an Interpreted Very Fine-Grained Carbonate-Lime Mud Matrix

The lithified equivalent is a fine mosaic called microcrystalline calcite (acronym *micrite*) by Folk (particles 4-15 microns). Normally in sediment with *some* lime mud the grain/micrite ratios or percent micrite are so variable as to make its estimate almost meaningless. In nearby areas of the same bed, or even in the same thin section, grain abundance may vary from 10% to 50% or more. Thus, it is more important whether *some* mud exists than to know how much. Dunham's concept (1962) is significant: focus on currents of removal, not delivery.

"The distinction between sediment deposited in calm water and sediment deposited in agitated water is fundamental. Evidence bearing on this problem thus deserves to be incorporated in class names. This can be accomplished in several ways. One is to focus attention on average or predominant size, which erroneously assumes that all sizes in a sample are equally significant hydraulically. Another is to focus attention on the size, abundance, and condition of the coarse material brought to the site of deposition. This emphasis on what might be called currents of delivery has long been successful in dealing with land-derived sediments, but does not work well in lime sediment because of the local origin of many coarse grains. A third way is to focus attention on the fine material that was able to *remain* at the site of deposition. This emphasis on what might be called currents of removal seems advisable if we wish to characterize carbonate sediment systematically in terms of hydraulic environments.

"Inasmuch as calm water is characterized by mud being able to settle to the bottom and remain there, it seems that the muddy rocks deserve to be contrasted with mud-free rocks, regardless of the amount and size of included coarse material" (DUNHAM, 1962).

2. The Possibility of Grain Support Inferred from Estimates of Degree of Particle Packing

Did the sediment consist of a self-supporting framework of grains or do grains float in a micrite matrix which was originally lime mud? The variability of grain to micrite ratios in shallow water carbonate sediments, as well as the extremely variable shapes of particles, makes the packing-framework concept of Dunham fully as useful and about as accurate in application as grain to micrite ratios employed by other classifications. Naturally, the development of grain framework depends not only on the number, but also on the shape of the grains. Solid spheres will form a self-supporting pack with about 60% of the volume being grain bulk. Dense branched twigs and arcuate shells will form a framework with only 20–30% grain bulk (Dunham, 1962, Plate II). Because the grain-supported fabric may lead to important diagenetic modification (e.g., easier early solution or coarser crystallization because of lack of compaction of intergranular mud, and greater grain to grain solution penetration), the concept of grain-supported fabric may be very useful. The carbonate textural spectrum of Folk (Fig. I-4) reflects the packing concept fully as well as does the original Dunham classification (Fig. I-5).

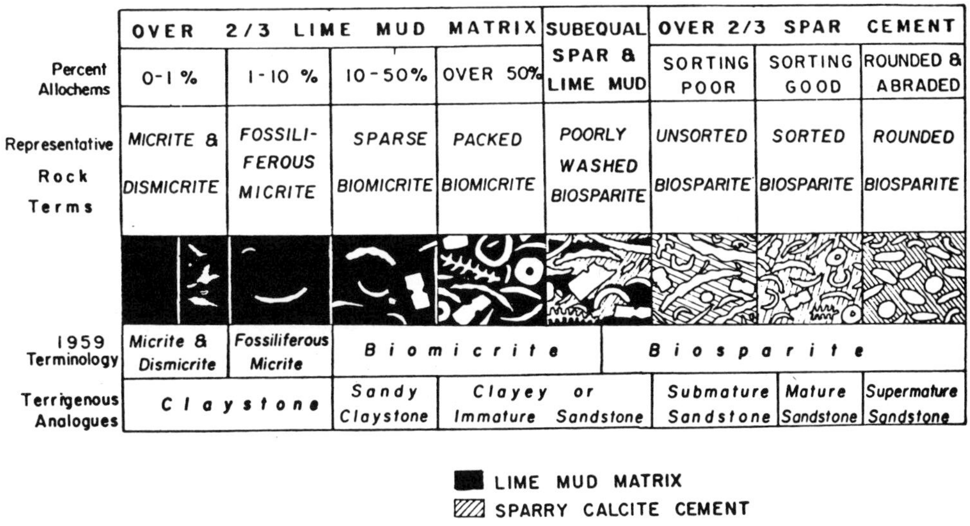

	OVER 2/3 LIME MUD MATRIX				SUBEQUAL SPAR & LIME MUD	OVER 2/3 SPAR CEMENT		
Percent Allochems	0-1 %	1-10 %	10-50%	OVER 50%		SORTING POOR	SORTING GOOD	ROUNDED & ABRADED
Representative Rock Terms	MICRITE & DISMICRITE	FOSSILI-FEROUS MICRITE	SPARSE BIOMICRITE	PACKED BIOMICRITE	POORLY WASHED BIOSPARITE	UNSORTED BIOSPARITE	SORTED BIOSPARITE	ROUNDED BIOSPARITE
1959 Terminology	Micrite & Dismicrite	Fossiliferous Micrite	B i o m i c r i t e			B i o s p a r i t e		
Terrigenous Analogues	C l a y s t o n e		Sandy Claystone	Clayey or Immature Sandstone		Submature Sandstone	Mature Sandstone	Supermature Sandstone

■ LIME MUD MATRIX
▨ SPARRY CALCITE CEMENT

Fig. I-4. Carbonate textural spectrum from Folk (1962, Fig.4), with permission of American Association of Petroleum Geologists

Depositional Texture recognizable					Depositional texture not recognizable
Original components not bound together during depositions				Original components were bound together during deposition... as shown by intergrown skeletal matter, lamination contrary to gravity, or sediment-floored cavities that are roofed over by organic or questionably organic matter and are too large to be interstices.	Crystalline carbonate
Contains mud (particles of clay and fine silt size)		Grain-supported	Lacks mud and is grain-supported		
Mud-supported					
Less than 10% grains	More than 10% grains				(Subdivide according to classifications designed to bear on physical texture or diagenesis.)
Mudstone	Wackstone	Packstone	Grainstone	Boundstone	

Fig. I-5. Classification of Carbonate rocks according to depositional texture from Dunham (1962, Table 1), with permission of American Association of Petroleum Geologists

3. Grain Kind

In addition to purely textural parameters, all modern sedimentary classifications recognize more or less the same basic grain types. These are:

Intraclasts or lithoclasts: Large particles derived by desiccation breakage or burrow disruption of penecontemporaneously deposited carbonate sediment. (Lithoclasts may be also externally derived from older lithified rock and must be specially designated.)

Ooids: Spherical, multiple coated particles in which the laminae are smooth and constitute a relatively thick coating are termed "well-formed ooids" in the volume. Oolitically coated particles (superficial ooids) in which only one or two laminae exist and which commonly retain the original grain shape are also prevalent. Well-formed ooids are products of tidal action.

Bioclasts: Fragmented tests, shells, or skeletons.

Peloids or pelletoids: Fecal pellets and rounded micritic grains of other origin.

Aggregated lumps or grapestones: These are agglutinated lumps of peloids and ooids; they also may be coated.

Onkoids: Algally coated grains, generally over 2 mm in diameter. Coats are generally irregular or crinkly.

Definitions of these basic grain types are given by Powers, Folk, Leighton and Pendexter (see A.A.P.G. Memoir I, 1962), and they are used by most other describers of carbonate rock textures. They are also defined and well illustrated by Horowitz and Potter (1971, pp. 7, 8) and by Milliman (1974). The framework parameter of Dunham cannot be accurately employed without a good understanding of grain type and visualization of their shapes.

In naming carbonate textural types the application of terms of particle kind varies somewhat. Dunham and Leighton and Pendexter preface textural terms with a designation of principle particle kinds: pellet lime wackestone (Dunham,

1962) and pellet micritic limestone (Leighton and Pendexter, 1962). Folk prefers to construct compound abbreviations or acronyms in combining grain kind with textural nouns: pelmicrite and oosparite.

4. Grain Size, Rounding and Coating

From Amadeus Grabau carbonate petrographers inherited some cumbersome but useful words describing grain size in limestone: calcilutite, calcisiltite, calcarenite, and calcirudite. Since most limestones are mixtures of all four of these sizes, and since the names alone do not indicate whether or not lime mud (micrite) is present in the coarse varieties, these terms are usually relegated to a secondary role in modern textural classifications. An exception occurs where such criteria as good sorting, sparry cement between grains, rounding and coating of grains and certain sedimentary structures indicate an origin as a grainy, well-washed lime sand. In such cases, grain size becomes an important attribute, just as it is in a quartz sandstone, although even in clean lime grainstone mixtures of sand and gravel size particles are common. Recently, Embry and Klovan (1971) (Fig. I-6) added a grain size modification to the Dunham classification, based on work on very coarse reef-derived sediments. A framework of coarse gravel with little or no matrix is termed a rudstone by these writers. Cobbles floating in a finer-grained matrix (sand to mud sizes) constitute a floatstone. The matrix of the floatstone may be described separately using the accepted Dunham terms. The well-known

Allochthonous limestones original components not organically bound during deposition						Autochthonous limestones original components organically bound during deposition		
Less than 10% > 2 mm components				Greater than 10% > 2 mm components		By organisms which act as baffles	By organisms which encrust and bind	By organisms which build a rigid framework
Contains lime mud (< .03 mm)			No lime mud					
Mud supported		Grain supported		Matrix supported	> 2 mm component supported			
Less than 10% grains (> .03 mm < 2 mm)	Greater than 10% grains							
Mudstone	Wackestone	Packstone	Grainstone	Floatstone	Rudstone	Bafflestone	Bindstone	Framestone

Fig. I-6. Amplification of original Dunham (1962) classification of limestones according to depositional texture by Embry and Klovan (1971, Fig. 2), courtesy of Canadian Society of Petroleum Geologists

and widely-applied classification of Folk (1959, 1962) also designates class names based partly on size. A biomicrite in "Folklore" or a bioclastic lime wackestone in "Dunhamese" becomes, with coarser grains (e.g., oyster shells entombed in a mud matrix), a biomic*rud*ite or bioclast floatstone.

Grain shape is a useful parameter to describe, provided one recognizes grain type. Grains from gastropods and crinoid stems are formed round and not subject to the same consideration as originally angular mollusk shell fragments which have been rounded and coated in agitated water.

5. Biogenically Precipitated Carbonate Masses

These form an additional textural category treated in all classifications. Folk's term *biolithite* is essentially equivalent to *boundstone* of Dunham whose classification scheme offers three signs of binding which may be used as evidence: visible construction of organic framework, stromatolitic lamination contrary to gravity, and the presence of roofed over, sediment-floored cavities which appear to have been organically constructed. Such cavities may be recognized even when an organic frame is not. They are of improper shape to be solutional and too large to be normal interstices. Embry and Klovan (1971) added to Dunham's boundstone category, attempting a genetic interpretation of the type of binding.

Bafflestone: Sediment with abundant stalk-shaped (dendroid) fossil remains which are interpreted to form baffle for matrix accumulation; matrix is volumetrically important; commonly ill-sorted.

Bindstone: Tabular-lamellar organisms binding and encrusting a large amount of matrix. No self-supporting organic fabric.

Framestone: *In situ* massive fossils which construct a rigid framework; matrix, cement, or void space fills in the framework.

Carbonate Deposition Is Rapid but Is Easily Inhibited and Therefore Sporadic during Geologic Time

It is commonly stated that carbonate deposition is very slow compared to that of terrigenous or evaporite sedimentation. Indeed, when one compares great deltaic sections such as that within the south Louisiana Tertiary depocenter with maximum thickness of carbonate banks in the same marine basin, for example, the Florida Shelf or Great Bahama Bank, one finds about twice as much terrigenous clastic as carbonate sediment from earliest Cretaceous to the present day. But, actually, rates of shallow neritic carbonate sedimentation derived from deposited thicknesses in the last 5000 years are extremely fast. Table I-1 compares these from the Caribbean, Gulf of Mexico, Atlantic and Persian Gulf areas. The Persian Gulf tidal flats are building out so fast that they should be 100–200 km wide in 100000 years (Kinsman, 1969, p. 839), filling in the whole Trucial Coast embayment. Indeed, volumetric calculations show that the whole Persian Gulf, including the 90 m deep axial trough, could well be buried by lagoonal and intertidal

Table I-1. Comparison of modern rates of $CaCO_3$ sedimentation with depositional rates of some thick limestone sections

Reference	Locality	Maximum thickness meters	Time years	Rate meters per 1000 years	Depositional environment
Enos (1974)	Florida reef tract	25	7000	3+	Reef and debris
Turmel-Swanson (1972)	Rodriguez bank	5	Less than 5000	1+	Open sea bank
Stockman et al. (1967)	Florida bay-Crane key	3	3000	1	Lagoon
Shinn et al. (1965)	Andros Island	1.5	2200	0.7	Tidal flats
Bathurst (1971), and Cloud (1962)		3	3800	0.8	
Illing et al. (1965)	Sabkha Faishak	4	4000	1	Sabkha
Kinsman (1969)	Trucial coast	2	4000–5000	0.5	Sabkha-intertidal
Brady (1971)	N.E. Yucatan	5	5000	1	Lagoonal average from bank thickness
Holocene Average rate of shallow water $CaCO_3$ production[a]				1.0	Lagoons, tidal flat, sebkhas, reefs
Goodell-Garman (1969)	Superior well, Andros Island	4600	120×10^6	0.035	Bank sediment
	Suniland field Florida	4000	120×10^6	0.03	Bank-shallow shelf
Coogan et al. (1972)	Golden Lane bank (Albian-Cenomanian)	1500	20×10^6	0.08	Bank sediment
Wilson, J.L.	Persian Gulf Mesozoic-Cenozoic-maximum	6000	200×10^6	0.03	Shallow marine and tidal flat
Ham, W.E.	Arbuckle Group (Lower Ordovician portion)	3000	100×10^6	<0.03	Tidal flat-lagoonal
Maximum rate of $CaCO_3$ production from ancient rocks				0.04	Variety of shallow sediments like those of Holocene

[a] These figures are maximum thickness of unconsolidated mud or reef growth over Late Pleistocene subaerially exposed and hardened sediment. They represent accumulation since the last sea level rise. (Post-Wisconsin glacial maximum).

sediment in a very few million years. Considerable discrepancy exists when such rates obtained from deposition on modern tidal flats and reefs are applied to thicknesses of ancient neritic strata. For example, the Great Bahama Bank should have 35000–50000 m of post-Cretaceous sediment instead of 4500+ (Goodell and Garman, 1969, p. 528). Since these rates do not jibe with the rates of deposition of even thickest known ancient carbonate deposits of comparable environment, we assume that the carbonate producing system operates intermittently and is very sensitive. That is, it may cease abruptly and start again when conditions are favorable. Carbonate sedimentation resembles a powerful Cadillac with a defective carburetor.

This principle, of fast but intermittent deposition, is important in interpreting thickness and stratigraphic relations in carbonate rocks. These are discussed in Chapter II, but some of these effects are listed below:

1. When conditions remain favorable, carbonate production can keep up with almost any amount of tectonic subsidence or eustatic sea level rise. Carbonate sedimentation may be intermittent, is also commonly diachronous and generally regressive.

2. One may expect great irregularities in thickness of carbonate deposits at certain places, particularly along shelf margins where conditions become optimum for production and accumulation and where subsidence is more continuous. Commonly, sudden thickening occurs down depositional dip in limestones which, in thinner beds, have followed for many miles across shelves with no consistent thickness change. Carbonates may build out and up to great "pods" of sediment in miogeosynclines. In the Lower Ordovician of North America, great local thicknesses occur in the Central Appalachian and Anadarko basins (Arbuckle Mountains of Oklahoma) (Fig. II-8). Similar areas of great local thickness are seen in the Middle Cretaceous of the Mexican geosyncline.

3. The vagaries of carbonate sedimentation make isopach maps of carbonate units difficult to interpret unless facies information is available. Thick carbonate accretion can occur over or downflank from positive areas. Very thin carbonate deposits may occur over strongly positive areas. Slowly subsiding basins may contain a great thickness of carbonates, while strongly subsiding troughs contain water too deep for carbonate accumulation and become sediment-starved.

4. Rapid, but intermittent accretion of carbonate banks may be responsible for much pervasive meteoric diagenesis to which such rocks are subjected. When subsidence ceases, carbonate sedimentation can build a wide shelf or bank rapidly to sea level and above; the recently formed sediment is affected by fresh or hypersaline water depending on climate and geography.

Carbonate Sediments and Rock Are Peculiarly Subject to Many Stages of Diagenesis

Puzzling and confusing differences appear when a microscope view of Holocene carbonate sediment is compared with a thin section cut from ancient limestone. Pure lime mud from a modern lagoon is a creamy, stiff gel of organic slime, aragonite needles, tiny detrital bits of high Mg calcite and coccoliths, and is 50% water. The typical limestone micrite apparently equivalent to this is a mosaic of 3–4 microns or larger, more or less equidimensional, low Mg calcite crystals. It is often dense, without measurable porosity, or perhaps chalky with the calcite crystals loosely arranged. Modern lime sands deposited and cemented in entirely marine environments contain interstitial fibrous aragonite or "dogtooth" Mg calcite. Ancient lime sands are generally dense with intergranular space filled in by later generations of calcite mosaic. Usually, ancient carbonate rocks show a reorganization of crystal structure from the original grains and cement, as well as complete mineralogical changes. Boundaries of former grains are recognized by differences in their crystal sizes and shapes, or by color distinctions based on inclusions and trace elements within replacement crystals or overgrowths. To paraphrase St. Paul, "now we see through a glass darkly...", all the millions of years of diagenetic alteration undergone by the typical carbonate rock. This is not only because carbonate particles are very susceptible to solution and mineralogical alteration when exposed to fresh or migrating connate waters, but also because in the sea, prior to final deposition, intense (bio) chemical and organic activity can easily alter or destroy lime particles, or cement them into aggregates.

The importance of diagenetic study of carbonates is signified by the current major literature on the subject. The contributions edited by Pray and Murray (1965), Chilingar et al., (1967), Purdy (1968), and Bricker (1971), are outstanding modern references in this burgeoning research endeavor. Additionally, and inevitably, classifications of carbonate porosity involve diagenesis. The outstanding papers of Murray (1960), Powers (1962), Thomas (1962), and Choquete and Pray (1970), may be cited. An outline of diagenetic events affecting calcium carbonate sediment is appended to Chapter III.

Summary

This Chapter has briefly listed a few special qualities of carbonate sedimentation which distinguish it from sandstone-shale deposition and which constitute important principles to be considered in employing petrography and stratigraphy in the environmental interpretation of carbonate depositional patterns. Useful corollary generalizations by Laporte (1969) have been added.

Sedimentological Controls

1. Most carbonate sediments are formed in a special depositional environment: warm, generally shallow, clear marine water. Despite the presence of shelly calcium carbonate deposits in temperate and northern zones, lime sediment similar to that found in the geologic record is today essentially a product of low latitudes.

2. Carbonate formation is essentially autochthonous. In Laporte's words "intrabasinal factors control facies development."

3. Basin configuration and water energy are the dominant controls on facies genesis and differentiation (Laporte, 1968). Depth control of facies exists because depth controls water energy. Light control is important, confining important biological production to the photic zone. Certain hydrographic controls operate on carbonate sediments in a way different from the obvious mechanical ones of transport, piling, and winnowing. Marine upwelling and water agitation induce considerable *in situ* organic production on shelf margins, and lack of circulation in restricted marine areas results in hydrographic conditions leaving strong imprints on the sediments. Such very shallow marine water and intertidal sediments make up a great volume of carbonate rock.

Biological Influences

4. Carbonate formation is basically biochemical and organisms are all-important in creating and modifying all types of carbonate particles from tiny grains to large precipitated rock masses.

5. A corollary pointed out by Laporte and Ginsburg is that organic abundance and diversity in carbonate sediments reflect original conditions despite a bias caused by differences in preservation.

6. A consequence of dominant organic composition is that biofacies and lithofacies often correlate. Organisms cause certain typical lithofacies (e.g., hard, massive coral reefs) and the same environmental factors which cause inorganic grain types affect the ever present organisms (e.g., oolite commonly possesses thick-shelled gastropod fragments). Substrates in turn control types of organisms inhabiting them.

7. The relative rates of physical and, particularly, biological reworking are ascertainable from sedimentological study and have environmental significance.

Textural Influences

8. Since carbonate particles are produced approximately *in situ* and in a variety of shapes and sizes, special sedimentary principles and textural classifications are necessary to further our understanding of them. Modern classifications are based principally on presence or absence of interpreted original lime mud and on types of grains; the amount of mud matrix is considered a better guide to water energy and circulation than grain size or shapes of particles.

9. As corollary to the above concepts Laporte pointed out that since most carbonate grains accumulate where produced and little net transport occurs, the textures of many carbonate sands are more dependent on the nature of contributing skeletal producers than on external agents.

Rates of Sedimentation

10. Carbonate sedimentation can be extremely rapid when the special marine environment is right. The measured rates of neritic sediment for the last 5000 years are on an order of magnitude too high for the maximum amounts of thick,

shallow water carbonate known to have been produced in the geologic record. Presumably, the production mechanism is sensitive and easily shut off by building to sea level or by turbidity induced by change of climate or source area. This is an important consideration in interpreting time correlation on stratigraphic surfaces in the geologic record.

Diagenetic Susceptibility

11. Carbonate sediments and rocks are very sensitive to diagenesis. Changes begin in the loose grains before and during deposition and continue on the sea floor. Drastic solution and mineralogical changes occur when carbonate sediment is taken from the marine and placed in a subaerial, fresh water environment. Other changes occur when evaporative brines and other types of connate waters migrate through limy strata both before and after deep burial. The changes occur through various stages of lithification. They include altering of marine carbonate minerals to low Mg calcite, organic rotting of particles, solution, cementation of pore space in lime sands, crystal rearrangement (neomorphism), metasomatic replacements (dolomitization) which rearrange and enlarge pore space, and anhydrite replacement which makes the sediment more susceptible to later dissolution.

Chapter II

The Stratigraphy of Carbonate Deposits

The basic processes of carbonate sedimentation result in certain stratigraphic relations and in predictable facies patterns which are widespread in the geologic record. These patterns recur in a variety of tectonic settings. This Chapter considers these interrelations and also treats correlation problems in carbonate strata and the techniques for recognition of approximate "time planes" in shelf and shelf-margin areas. The Chapter contains many of the basic premises from which the author formulates stratigraphic and petrologic interpretations. Later Chapters furnish considerable basic data and offer numerous examples of facies patterns in a variety of tectonic settings, all of which serve to test some of the stratigraphic ideas set forth below. A summary Chapter repeats most of these generalizations.

Definitions

Recent rapid growth in understanding the formative processes of carbonate buildup has resulted in redundancy and confusion of terminology used in describing and interpreting such sedimentary bodies. The number of scholarly papers, including a review of the term "reef" alone, probably exceeds twenty. Therefore, rather than reviewing further the evolution of terms and concepts, only definitions of common expressions are given. The terms below are those now widely employed—even though some of those defining composition and genesis are somewhat overlapping. Several preliminary observations are worthwhile: (1) It is important to separate expressions describing shape of carbonate bodies from those describing their internal composition—the latter necessarily intermingle concepts of genesis. (2) Distinction must be made between names for discrete smaller scale limestone bodies and large regional configurations formed by carbonates (e.g., "Bank" has been used in two ways). (3) The term "reef" has evolved rapidly and geologists have modified and restricted its meaning. This has caused confusion to the extent that the term should be used *always* with a modifier to designate its meaning (Heckel, 1974).

General Definitions

Carbonate buildup: A body of locally formed (laterally restricted) carbonate sediment which possesses topographic relief. This is a general and useful term because it carries no inference about internal composition (Fig. II-1).

CARBONATE MASS

CARBONATE BUILDUPS

Fig. II-1. Definition of carbonate masses and carbonate buildups

Carbonate mass: A carbonate localization developed with only slight relief caused by facies change from compactible argillaceous strata to noncompactible pure limestone (Fig. II-1).

Geologic reef (Dunham, 1969, p. 190) or *stratigraphic reef* (Dunham, 1970), (Heckel, 1974 p. 93): A general term for a carbonate body which encompasses both the above concepts and includes both local mound-like and regional curvilinear trends. No inference as to origin or internal composition is included.

Definitions Based on Configuration of Regional Features

Carbonate ramps: Huge carbonate bodies built away from positive areas and down gentle regional paleoslopes. No striking break in slope exists, and facies patterns are apt to be wide and irregular belts with the highest energy zone relatively close to the shore (Fig. II-2).

Carbonate platform: Huge carbonate bodies built up with a more or less horizontal top and abrupt shelf margins where "high energy" sediments occur. The normal processes of carbonate sedimentation effectively and rapidly turn ramps into platforms and create narrow, steep shelf margin ridges. Slopes on some ramps may be so gentle as to make them commonly indistinguishable from platforms. Thus, these terms are often used interchangeably (Fig. II-2).

Major offshore banks: Complex carbonate buildups of great size and thickness well offshore from the coastal ramps or platforms (Fig. II-3).

Shelf: An area on top of a ramp or platform.

Shelf margin: The edge of the shelf on a platform.

Shelf lagoon: Often used for the shallow neritic shelf seas on a platform.

Definitions for Configurations of Local Carbonate Features

(Mostly implying organic accumulation rather than purely mechanical accumulation.)

Mound: Equidimensional or ellipsoidal buildup.

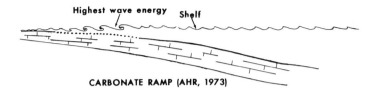

CARBONATE RAMP (AHR, 1973)

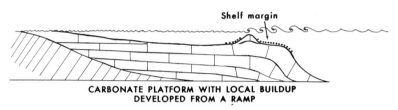

CARBONATE PLATFORM WITH LOCAL BUILDUP
DEVELOPED FROM A RAMP

Fig. II-2. Definition of carbonate ramps and platforms

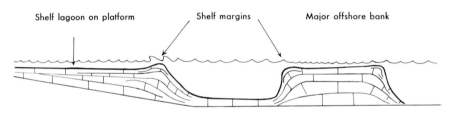

Fig. II-3. Definition of carbonate platforms, shelf margins and offshore banks

Pinnacle: Conical or steep-sided upward tapering mound or reef.

Patch reef: Isolated more or less circular area of organic frame-constructed buildups. In modern seas patch reefs are mainly on shelves and rise into wave base and close to sea level.

Knoll: Isolated more or less circular area of carbonate accumulation in deeper water below wave base. *Knoll reef:* a knoll of organic frame-built growth. In practice, whereas patch reefs commonly denote buildups in shallow shelf positions, reef knolls or pinnacles are used to identify individual buildups of shelf margins or in basins. The German term "Knollenriffe" exemplifies this usage.

Atoll: Ring-like organic accumulation in offshore or oceanic position surrounding a lagoon of variable depth.

Faro: Ring-like organic accumulation with shallow central lagoon located shelfward of a barrier reef trend.

Barrier reef: Curvilinear belt of organic accumulation somewhat offshore and separated from the coast by a lagoon.

Fringe reef: Curvilinear belt of organic accumulation built directly out from the coast.

Compositional Terms

(Often employed both descriptively and with genetic implications.)

Sediment pile: An accumulation of sediment in almost any shape whose internal composition shows it to originate mainly through mechanical piling of moved particles (e.g., dunes, bars, spits, tidal deltas, etc.). The term bank has also been used to mean a sediment pile.

Organic bank: Buildup whose internal composition permits the inference that it is formed mostly of detrital organic sediment accumulated in place by trapping or baffling but also in part through mechanical piling by waves and currents.

Bioherm: Buildup whose internal composition shows it to be largely derived from *in situ* production of organisms *or* as framework or encrusting growth as opposed to mainly mechanical (hydrodynamical) piling.

Lime mud mounds and linear mud accumulations: Lime mud matrix dominates other constituents such as organic boundstone and bioclastic debris. Such buildups are commonly perceived to accumulate both through hydrodynamic processes and *in situ* organic production.

Organic framework reef or ecologic reef (Dunham, 1970): Buildup formed in part by a wave-resistant framework constructed by organisms. An accessory part of such a definition implies that the reef exerts some degree of control over its surrounding environments. Other types of buildups behave similarly. Some geologists have more or less informally used the simple term "reef" for this concept. Others, principally petroleum geologists, have advisedly used "reef" alone for any carbonate buildup. The original term meant a ridge or shoal on which a ship could ground. Because many such features were coral-algal reef, geologists appropriated the term to their own special jargon and with two specialized meanings: (1) organic framebuilding communities and (2) organic buildups. Many earth scientists wish that the simple term "reef" could be returned to navigators and dropped altogether from geological use. It is undoubtedly too late for this. An obvious solution is to modify the term whenever it is used.

Heckel (1974), in an attempt to make compositional terms more objective, proposed the terms in Table II-1 in a long essay on carbonate buildups. Though

Table II-1. Terminology for carbonate buildups

Heckel (1974) Compositional descriptive terminology		Common usage followed in this text; descriptive with genetic implications			Dunham (1970)
Major mixed buildups	Encrusted skeletal buildup	Organic framework reef	Bioherm	Carbonate buildup	Ecologic reef
	Loose skeletal buildup	Organic bank			Stratigraphic reef
	Lime mud buildup	Lime mud accumulation			
	Sorted-abraded skeletal buildup	Sediment pile			

cumbersome, they are purely descriptive and may be accurately employed until geologists are sufficiently informed in specific cases to be able to replace them with the parallel descriptive, but partly genetic, terms. As Heckel pointed out, most major buildups are composites, combining several of the four basic compositional types.

The Basic Facies Pattern

Chapter I has pointed out that the favored realm of carbonate sedimentation is in warm shallow water on or bordering shelves in tectonically stable areas. In such areas sedimentation is mostly autochthonous, the locally produced organic carbonate being mainly accumulated close to its site of origin (Shaw, 1964; Anderson, 1971, 1974). On a very gently sloping shelf there is a tendency for a seaward, low energy zone to develop below wave base and a zone of higher wave energy to be situated somewhat shoreward where waves drag the bottom and where maximum organic productivity occurs. A third, interior or shoreward low-energy zone also develops. These three zones may reach a considerable thickness, commonly forming a prograding or up-building sequence after a period of marine transgression. Thus they normally form a carbonate ramp or platform. Detailed discussion of this process follows later in this Chapter.

The hydrologic, climatic, and organic controls exerted on the *in situ* production of lime sediment elaborate this simple trio of environmental belts (basin, shelf margin, and "backreef") into about nine sub-environments. These are expressed by a surprisingly regular facies sequence which exists in various tectonic settings (see below). Its outer belts encircle basins, exist at the edges of major carbonate banks and form halos around mildly positive areas. It is significant that this pattern is so persistent: it offers essentially a single model for prediction of geographic distribution of rock types. It thus becomes a tool for use in practical field mapping, in designation of rock units for correlation purposes, for depositional interpretations and in the search for petroleum and for metallic ores such as lead, zinc, and silver, whose distribution may be facies-controlled. The basic model is now well known and has been discussed in the several major papers and books cited below.

The first men to point out the similarity between the Bahama Banks sediments and facies in the ancient limestones were Maurice Black (1930) and R. M. Field (1930). About two decades later Thomas Grimsdale, of the Royal Dutch Shell group, recognized the general application of the pattern in the geologic record. The Bahama Bank studies of N. D. Newell and students and the publication in 1953 of the book *The Permian Reef Complex* by Newell et al., did much to bring recognition of the usefulness of the pattern. Shortly thereafter, experienced geologists began to apply the model to Mississippian beds in the Williston basin (Edie, 1958; Shaw, 1964; Irwin, 1965).

Study during the past 20 years has amplified the original three belts very considerably. (Dooge, 1966; Coogan, 1972; Tyrell and other writers in S.E.P.M. Spec. Pub. 14, 1969; Wilson, 1970; Armstrong 1974). A brief summary of the model is presented by Horowitz and Potter (1971).

Standard Facies Belts

In the following discussion the pattern is conventionally portrayed as a profile across a gently sloping shelf atop a platform with an abrupt shelf margin. Methods used in identifying the nine facies and their interpreted environments noted below (Fig. II-4) are reviewed in the next Chapter and their lithologic characteristics are summarized in the final part of the book.

1. *Basin facies—starved or filled basin (fondothem):* Water is too deep and dark for benthonic production of carbonate, and deposition is dependent on the amount of influx of fine argillaceous and siliceous material and the rain of decaying plankton. Euxinic and hypersaline conditions may result.

2. *Shelf facies (deep undathem):* Water with a depth of tens or even a few hundred meters, generally oxygenated and of normal marine salinity. Good current circulation. Deep enough to be below normal wave base but intermittent storms affect the bottom sediments.

3. *Basin margin or deep shelf margin facies (clinothem):* Formed at toe of slope of carbonate producing shelf from material derived from the shelf. Depth, wave base conditions and oxygen level about that of facies 2.

4. *Foreslope facies of carbonate platform (marine talus; clinothem):* Generally the slope is located above the lower limit of oxygenated water and from above to below wave base. Material is debris deposited on an incline commonly as steep as 30 degrees, is unstable and varies greatly in size. Bedding contains slumps, mounds, wedge-shaped foresets, and large blocks.

5. *Organic reef of platform margin:* Ecologic character varies dependent on water energy, steepness of slope, organic productivity, amount of frame construction, binding, or trapping, frequency of subaerial exposures, and consequent cementation. Three types of linear shelf-margin organic buildup profiles may be discerned: Type I is formed by downslope carbonate mud and organic debris accumulations. Type II consists of ramps of knoll reefs, organic framebuilding organisms in isolated clumps or encrusting sheets or organisms growing up to wave base and stabilizing debris accumulations. Type III are frame-constructed reef rims like modern coral-algal assemblages with sessile forms growing up through wave base into the surf zone.

The present author (Wilson, 1974) applied these three types of profiles to the shelf margins discussed within the book. The summary Chapter reviews them and offers additional examples.

6. *Winnowed platform edge sands:* These take the form of shoals, beaches, offshore tidal bars in fans or belts, or dune islands. Depths of such marginal sands range from 5 or 10 m to above sea level. The environment is well oxygenated but not hospitable to marine life because of shifting substrate.

7. *Open marine platform facies (shallow undathem):* Environments are located in straits, open lagoons and bays behind the outer platform edge. Water depth is generally shallow, a few tens of meters deep at most. Salinity varies from essentially normal marine to somewhat variable salinity. Circulation is moderate.

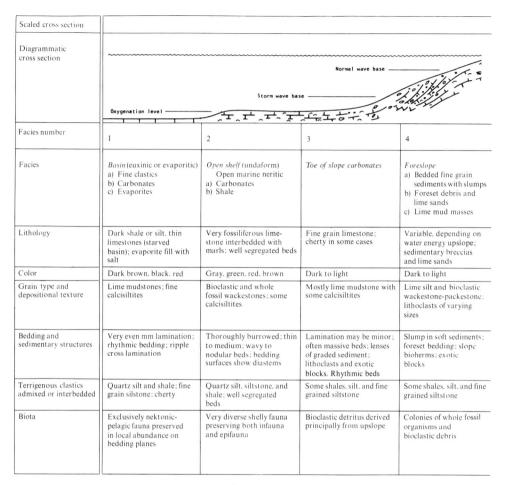

Scaled cross section				
Diagrammatic cross section				
Facies number	1	2	3	4
Facies	*Basin* (euxinic or evaporitic) a) Fine clastics b) Carbonates c) Evaporites	*Open shelf* (undaform) Open marine neritic a) Carbonates b) Shale	*Toe of slope carbonates*	*Foreslope* a) Bedded fine grain sediments with slumps b) Foreset debris and lime sands c) Lime mud masses
Lithology	Dark shale or silt, thin limestones (starved basin); evaporite fill with salt	Very fossiliferous limestone interbedded with marls; well segregated beds	Fine grain limestone; cherty in some cases	Variable, depending on water energy upslope; sedimentary breccias and lime sands
Color	Dark brown, black, red	Gray, green, red, brown	Dark to light	Dark to light
Grain type and depositional texture	Lime mudstones; fine calcisiltites	Bioclastic and whole fossil wackestones; some calcisiltites	Mostly lime mudstone with some calcisiltites	Lime silt and bioclastic wackestone-packestone; lithoclasts of varying sizes
Bedding and sedimentary structures	Very even mm lamination; rhythmic bedding; ripple cross lamination	Thoroughly burrowed; thin to medium; wavy to nodular beds; bedding surfaces show diastems	Lamination may be minor; often massive beds; lenses of graded sediment; lithoclasts and exotic blocks. Rhythmic beds	Slump in soft sediments; foreset bedding; slope bioherms; exotic blocks
Terrigenous clastics admixed or interbedded	Quartz silt and shale; fine grain siltstone; cherty	Quartz silt, siltstone, and shale; well segregated beds	Some shales, silt, and fine grained siltstone	Some shales, silt, and fine grained siltstone
Biota	Exclusively nektonic-pelagic fauna preserved in local abundance on bedding planes	Very diverse shelly fauna preserving both infauna and epifauna	Bioclastic detritus derived principally from upslope	Colonies of whole fossil organisms and bioclastic debris

Fig. II-4. Idealized sequence of Standard Facies Belts from J. L. Wilson (1970, 1974). See also Fig. XII-1. Illustration with permission of American Association of Petroleum Geologists

8. *Facies of restricted circulation on marine platform:* Includes mostly fine sediment in very shallow, cut-off ponds and lagoons, coarser sediment in tidal channels and local beaches, and the whole complex of tidal flat environment. Conditions are extremely variable here and constitute a stress environment for organisms. Fresh, salt, and hypersaline water occur as well as areas of subaerial exposure, both reducing and oxygenated conditions and marine and swamp vegetation. Windblown terrigenous material may contribute significantly. Diagenetic effects are strongly marked in the sediment.

9. *Platform evaporite facies:* Supratidal and inland pond environment of the restricted marine platform developed in an evaporative climate—the areas of sabkha, salinas, salt flats. Intense heat and aridity is common, at least seasonally. Marine flooding is sporadic. Gypsum and anhydrite form from the evaporating sea water both as depositional and diagenetic sediment.

Fig. II-4 (continued)

5	6	7	8	9
Organic (ecologic) reef a) Boundstone mass b) Crust on accumulation of organic debris and lime mud; bindstone c) Bafflestone	*Sands on edge of platform* a) Shoal lime sands b) Islands with dune sands	*Open platform* (normal marine, limited fauna) a) Lime sand bodies b) Wackestone-mudstone areas, bioherms c) Areas of clastics	*Restricted platforms* a) Bioclastic wackestone, lagoons and bays b) Litho-bioclastic sands in tidal channels c) Lime mud-tide flats d) Fine clastic units	*Platform evaporites* a) Nodular anhydrite and dolomite on salt flats b) Laminated evaporite in ponds
Massive limestone-dolomite	Calcarenitic-oolitic lime sand or dolomite	Variable carbonates and clastics	Generally dolomite and dolomitic limestone	Irregularly laminated dolomite and anhydrite, may grade to red beds
Light	Light	Dark to light	Light	Red, yellow, brown
Boundstones and pockets of grainstone; packstone	Grainstones well sorted; rounded	Great variety of textures; grainstone to mudstone	Clotted, pelleted mudstone and grainstone; laminated mudstone; coarse litho-clastic wackestone in channels	
Massive organic structure or open framework with roofed cavities; Lamination contrary to gravity	Medium to large scale crossbedding; festoons common	Burrowing traces very prominent	Birdseye, stromatolites, mm lamination, graded bedding, dolomite crusts on flats. Cross-bedded sand in channels	Anhydrite after gypsum; nodular, rosettes, chickenwire, and blades; irregular lamination; carbonate caliche
None	Only some quartz sand admixed	Clastics and carbonates in well segregated beds	Clastics and carbonates in well segregated beds	Windblown, land derived admixtures; clastics may be important units
Major frame building colonies with ramose forms in pockets; in situ communities dwelling in certain niches	Worn and abraided co-quinas of forms living at or on slope; few indigenous organisms	Open marine fauna lacking (e. g. echinoderms, cephalopods, brachiopods); mollusca, sponges, forams, algae abundant; patch reefs present	Very limited fauna, mainly gastropods, algae, certain foraminifera (e. g. milio-lids) and ostracods	Almost no indigenous fauna, except for stromatolitic algae

Discussion of Idealized Pattern

The pattern of Fig. II-4 results from a combination of effects of slope, geologic age, water energy, and climate, and as these vary, so will the patterns they control. Also, any ingress of terrigenous clastics will affect it. It is therefore obvious that no one example should include all nine facies belts. Clearly, for example, whether belt 1 or 2 occurs is dependent on whether the constructed carbonate bank or ramp rises from a deeper water euxinic basin or whether it rises above a shelf sea with open circulation. Similarly, occurrences of belts 3 and 4 are determined by the steepness of the slope, the depth of water into which it plunges, and the amount of wave energy at its upper margin. The organic reef (belt 5) may alternate along facies strike with lime sand (belt 6), or both may be present, dependent on the combinations of geologic age and water energy. If a deep enough lagoon exists behind the barrier belts in a constantly temperate or tropical climate, it may have good circulation and no restricted marine carbonates or evaporites may occur. The numerous examples organized in the following Chapters demonstrate a range of variations. The different factors controlling these variations are outlined in the last Chapter. Yet it is remarkable how uniform the facies patterns may be in carbonate strata. The Permian Reef Complex stands as an almost ideal model of the complete gamut of facies (Dunham, 1972; Meissner, 1972), and the Cretaceous pattern described by Coogan (1972) is almost complete.

This is not the only facies pattern that has been recognized. Ahr (1973) and Anderson (1974) described a carbonate ramp situation in which a higher energy zone exists along the coast and grades outward across the shelf to fine carbonate mud deposited in open marine conditions. Modern carbonate shelves contain shoreward lime sands but are not geologically typical, their sedimentary patterns resulting from geologically recent inundation and showing only the beginning of a sedimentary cycle of progradation. In the geologic record such facies patterns occur but rarely. Ahr cited two geological examples of his "carbonate ramp facies model": the Smackover Jurassic around the Gulf of Mexico and the Pleistocene around the now-submerged Campeche bank. Both examples consist of comparatively narrow belts at the edge of the continental shelf. The slope at the edges of the belts is relatively steep. Under such conditions, even if considerable sea level

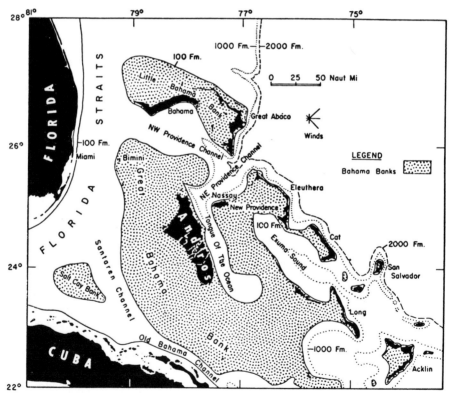

Fig. II-5. Bahamas Banks and environs from Purdy (1961) after Newell (1955), with permission of American Association of Petroleum Geologists

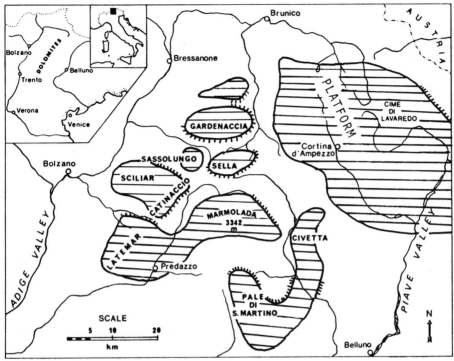

Fig. II-6. Location map of the Dolomites and their major banks after Bosellini and Rossi (1974). Hachured lines indicate foreslope strata marking original edge of bank. See also Fig. VIII-10. Illustration with permission of Society of Economic Paleontologists and Mineralogists

rise occurs, shelf margin barriers cannot easily form. Land is too close and the belt for maximum carbonate production is too narrow. The tectonic setting for such narrow facies patterns is defined later in this Chapter, termed a "fringe or halo" bordering a cratonic edge.

Paleotectonic Settings for Carbonate Facies

Tectonism will, by controlling subsidence, partly influence thickness, facies, and vertical sequence of the nine facies belts. But climatic, hydrographic, and organic controls are so strong that the basic pattern will occur in varied megatectonic provinces. These are outlined below, organized into several major types and fitted to the outline of Krebs and Mountjoy (1972). The examples are all drawn from the geologic record. It is difficult to find analogs from Holocene carbonate sedimentary realms because the sedimentary record covering Tertiary and Recent platforms is so young and thin. This problem is reviewed in the summary Chapter.

1. Basinal Buildups in Areas of Great Regional Subsidence

The largest of these features includes major offshore banks. These occur well within geosynclines or pericratonic basins. They contain some of the thickest carbonate sections known, and

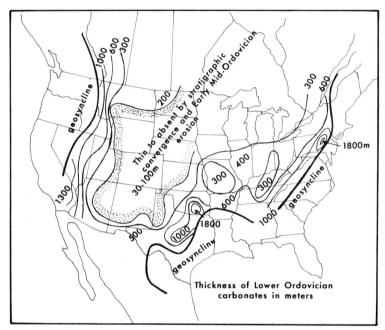

Fig. II-7. Thickness of Lower Ordovician around North American craton. The great "halo" of shallow marine and intertidal carbonate portrayed is mainly depositional and its outward thickening forms platforms on all sides of the central cratonic axis

develop on isolated highs in offshore basins, probably originating at lower sea level stands. Once started, they manage to maintain themselves, and grow upward through rapid accumulation in spite of considerable subsidence. Small examples are often termed pinnacle reefs. These may appear in lineations and rise from platforms developing over irregularities or fault scarps along its surface.

The slopes and flat tops of such banks resemble those of the major platforms. Unlike the platforms adjacent to cratonic blocks, whose facies progression faces the seaward side, the belts completely encircle the great offshore banks, having only minor facies differentiation on the windward side. For example: the Great Bahama Bank (Cretaceous to Recent age; Fig. II-5), the Cretaceous Valles and Golden Lane platforms of central Mexico; the Central Basin platform of West Texas, Middle and Late Permian; the Middle Triassic Dolomites of South Tyrol, northern Italy (Fig. II-6); Pennsylvanian banks (including Jameson Field and the Horseshoe "Atoll") in the Midland Basin (Fig. VI-14); Silurian pinnacle reefs in the Michigan Basin (Fig. IV-11); and the Zama area buildups of northern Alberta, Middle Devonian.

2. Carbonates Developed off Major Cratonic Blocks during Great Regional Subsidence

a) Major platforms and ramps built out from cratonic blocks: These occur at edges of miogeosynclines or pericratonic basins. Shallow carbonate sediments are built out to form large sloping ramps which evolve rapidly into platforms whose outer (seaward) slopes may range from 1 or 2 degrees to as much as 30 degrees. Thicknesses may be on the order of hundreds or even a few thousand meters. Facies belts on the edges of such platforms generally may be only a mile or two wide, whereas the interior facies may be tens of km across, with almost flat surfaces (slopes of 30 cm per km are common). An example is the Lower Ordovician of North

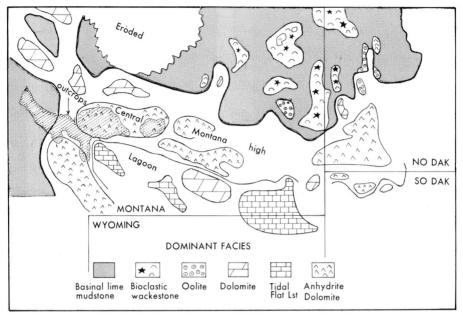

Fig. II-8. Generalized Early Mississippian facies of Williston basin and Montana shelf. The map is of a single interval (Nesson zone) about 30 m thick in a carbonate sequence of about 500 m (Mission Canyon facies of Madison group). Such intervals are subdivided on the basis of electrical log characteristics and cyclic lithofacies. The dominant facies patterns are indicated where the rock type represents more than 50% of the interval. Note the evaporite and dolomite across the central Montana high, the tidal flat-lagoonal limestone and dolomite south of the positive area, and the irregular "basinal" lime mudstone north, with scattered large banks of bioclastic wackestone and oolite sands. This cyclic progression of facies was built northward off the low Wyoming shelf into the gently subsiding Williston basin. The facies belts are wide and somewhat irregular

America, a great shoal water and intertidal wedge of carbonates bordering the Shield and extending into miogeosynclines east, west, and south of it (Fig. II-7). The Cretaceous bordering the Gulf of Mexico and the Mesozoic section east and south of the Arabian shield also typify this deposit. In the latter case, the present Persian Gulf lies in the shallow foredeep formed between the Arabian Shield and the rising Zagros chain.

  b) Narrow fringes or halos at cratonic edges: When tectonic activity is strong and rapid, basin subsidence occurs at borders of old orogenic ridges or cratonic plates and a relatively narrow, but thick, fringe of carbonate facies forms. Such patterns are common around tectonic lands within geosynclines or on steep borders of some platforms. If the slope is sufficiently steep, no barrier and no back reef lagoon develops. An example of the latter would be the halo of high energy Smackover (Late Jurassic) sediments crowded against the roots of the Ouachita orogenic belt and the northern Gulf of Mexico (Bishop, 1968, Fig. X-12).

3. Carbonates Developed Adjacent to Platforms in Areas of Moderate Subsidence

These are built out mainly within shallow intracratonic basins. They are similar in construction to large carbonate ramps and platforms but the strata are thinner and all facies belts may be tens of km wide; the gentle depositional slopes into the basin center may be on the order of only 20–30 cm per km. Somewhat irregular facies patterns occur along platform margins and

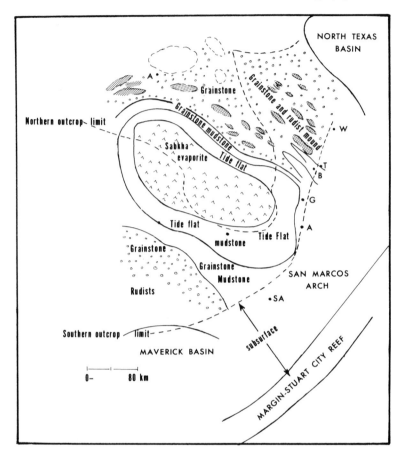

Fig. II-9. Fredericksburg facies over central Texas-San Marcos arch after Fisher and Rodda (1969). The central Texas (Llano uplift) area was a gentle dome on the larger Comanche platform but sufficiently positive to influence the carbonate facies. For additional diagrams and discussion see Chapter XI. This illustration from original with permission of American Association of Petroleum Geologists

within basins and in most cases clearly defined organic buildups are lacking. An example is the Mississippian of the Williston Basin (Fig. II-8). The original facies model of Edie (1958), Shaw (1964), and Irwin (1965), was drawn from these strata. Shoaling upward shelf cycles following the standard carbonate facies patterns are characteristic. No pronounced rim of organic buildups occurs in these strata, but scattered shelf mounds or patch reefs are common. Chapter IV (Middle Paleozoic Buildups) and the summary in Chapter XII describe several types of these shelf buildups. Shallow intracratonic basins within the subtropical belts which produce carbonate sediments are essentially absent on the present surface of the earth.

4. Buildouts from Low Positive Areas on Platforms — Areas of Moderate Subsidence

Because of reduced subsidence at margins of these low structural highs, low depositional relief develops and irregular patterns and wide facies belts exist. Nevertheless, the familiar pattern may be discerned. For example, Mississippian facies across and around the Central

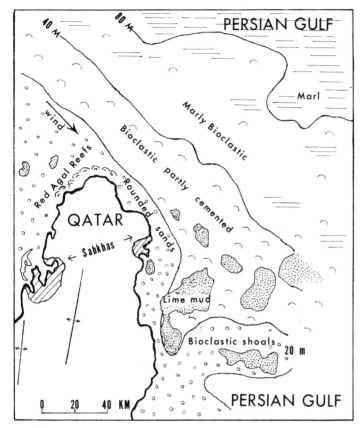

Fig. II-10. Bathymetry and surface sediments of Persian Gulf after Houboult (1957), showing facies change in lime sands in progressively deeper water out from the isolated gentle positive area of the Qatar peninsula. The slope northeast off Qatar is extremely slight ($^1/_2$ m per km) but a facies progression exists which reflects the systematic though very gradual deepening

Montana high; the Albian Cretaceous beds of central Texas (Glen Rose Formation and Fredericksburg Group, of central Texas, Fig. II-9); a modern analog may be found in the facies developed around the northern end of the Qatar Peninsula in the Persian Gulf (Fig. II-10).

Stratigraphic Sequences, Geometry, and Facies of Carbonate Shelf Margins and Basins

The basic carbonate facies model divides logically into two parts based on geometry and stratigraphic relations, degree of slope, and tectonic position: (1) Shelf margin complexes at the edges of major basins or along margins of major offshore banks within such basins, areas of considerable subsidence; and (2) Shelf strata laid down across flat cratonic areas in clear epeiric seas, areas of moderate subsidence.

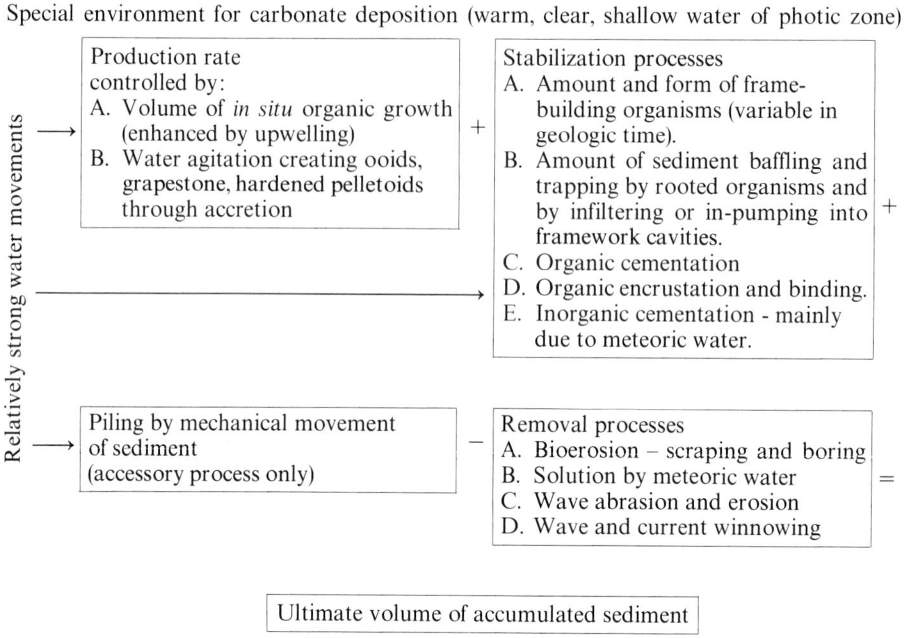

Special environment for carbonate deposition (warm, clear, shallow water of photic zone)

Relatively strong water movements →

| Production rate controlled by: A. Volume of *in situ* organic growth (enhanced by upwelling) B. Water agitation creating ooids, grapestone, hardened pelletoids through accretion | + | Stabilization processes A. Amount and form of frame-building organisms (variable in geologic time). B. Amount of sediment baffling and trapping by rooted organisms and by infiltering or in-pumping into framework cavities. C. Organic cementation D. Organic encrustation and binding. E. Inorganic cementation - mainly due to meteoric water. | + |

Relatively strong water movements →

| Piling by mechanical movement of sediment (accessory process only) | − | Removal processes A. Bioerosion – scraping and boring B. Solution by meteoric water C. Wave abrasion and erosion D. Wave and current winnowing | = |

Ultimate volume of accumulated sediment

Fig. II-11. Processes forming and shaping carbonate accumulations at shelf margins from Wilson (1974), with permission of American Association of Petroleum Geologists

Each of these areas has its own stratigraphic peculiarities and problems of correlation. The shelf margins and major basinal buildups are areas of pronounced depositional topography, steep slope, sharply defined facies, and thick sections.

Natural Construction of Carbonate Ramps-Platforms

The standard sequence of facies is based on the sedimentary construction of a wedge-shaped platform or ramp on a gently subsiding planar surface and the consequent development of submarine topography with a seaward face of variable steepness. Modern barrier and fringing reefs typically have this configuration even though most are built on paleotopography inherited from the Pleistocene. This commonly developed form results from the basic organic origin of carbonate sediment whose production is speeded up over preexisting high areas as well as over those created by local production or sediment piling.

Carbonates are produced most abundantly in marine water which is warm, shallow, clear, sunlit, and free of clayey contamination. The special processes operating to enhance the biological productivity on such sea bottoms are numerous and result from an interplay of hydrologic and biologic factors. The ultimate volume of accumulated sediment is dependent upon the increased production rates plus the stabilization processes and mechanical piling less the normal processes of sediment removal. Fig. II-11 diagrams these processes.

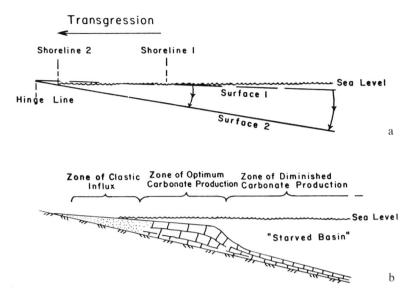

Fig. II-12a and b. Creation of shelf to basin topography from Meissner (1972, Figs. 4 and 5), with permission of West Texas Geological Society. (a) Stratigraphic model showing surfaces of deposition related to a transgression caused by differential subsidence. Rate of subsidence is assumed to be greater than rate of deposition. (b) Stratigraphic model showing the creation of shelf to basin topography caused by differential rates of carbonate sedimentation. Little or no subsidence following a rapid initial transgression is assumed

The belt of maximum carbonate accumulation on a planar surface is apt to occur removed from land, down the paleoslope, and paralleling ancient coast lines, but in water shallow enough to be within the zone of maximum biological productivity. Consequently, the mass of organic carbonate grows most rapidly along preferred positions on the upper part of any gentle seaward slope. Continued submergence of a positive block, plus the fact that organic buildup and sediment accumulation, when once started, is normally very rapid, results in the formation of a line of shelf margin organic buildups (Fig. II-12). Concurrent fill-in of the lagoon, between the shelf margin and land, occurs by restricted marine or evaporite deposits, and eventually a sloping ramp or level platform is constructed surrounding the positive element.

Not much sediment accumulates simultaneously in the area downslope from the shelf margin, and as the platform buildup occurs, and relative sea level rises, the water offshore becomes relatively deeper. The normal result is a basin surrounded by carbonate ramps or platforms which built out from the positive elements rimming the original depression and possess certain special facies down the slope into the basin (Fig. II-13). The basin seaward of the platform remains relatively deep for three reasons: (1) it may be the site of more rapid subsidence; (2) it is commonly starved of sediment, receiving only pelagic, calcareous and siliceous organic debris, and some fine argillaceous matter, either wind blown or brought in from distant rivers; and (3) such argillaceous sediment compacts to a higher degree than pure carbonate shelf or shelf margin deposits. The latter are

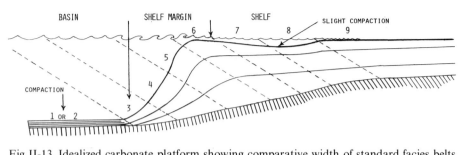

Fig. II-13. Idealized carbonate platform showing comparative width of standard facies belts and areas of slight compaction in shelf lagoon and starved basin. See also Meissner (1972, Fig. 8), with permission of West Texas Geological Society

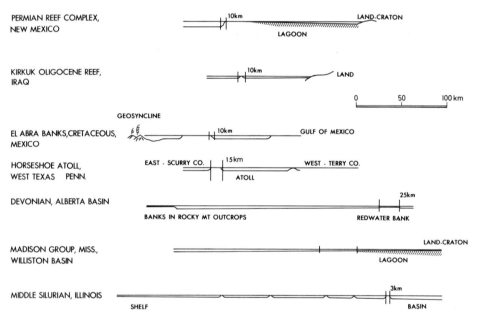

Fig. II-14. Comparison of several major carbonate platforms and offshore banks. Reduced vertical exaggeration. Horizontal scales are all the same. Vertical exaggeration is X 10. Shows relative width of facies belts at shelf margins

commonly cemented early in their diagenetic history and compact only slightly owing to some fresh water solution of individual grains.

A knowledge of this sedimentary topography and its rapidly changing facies is critical to proper correlation, particularly in subsurface mapping. This may not be easy if the basin is filled later with other sediment. The recognition of this depositional topography is credited to John Rich (1951) who christened its shelf, slope and basin components respectively *undaform, clinoform, and fondoform*. Though Rich was concerned principally with regional outbuilding by terrigenous clastic sediments, the principle applies equally to carbonates.

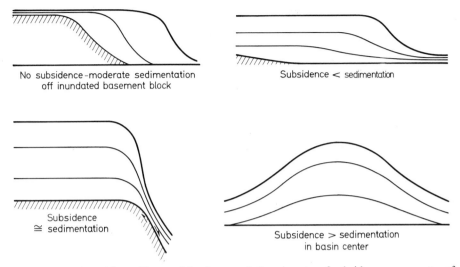

No subsidence-moderate sedimentation off inundated basement block

Subsidence < sedimentation

Subsidence ≅ sedimentation

Subsidence > sedimentation in basin center

Fig. II-15. Stratigraphic profiles resulting from variations in rates of subsidence *versus* rates of sedimentation

Facies Belts Vary in Width and Uniformity along Such Ramps

A typical facies complex described above usually consists of more or less wide-spread and uniform sedimentation across the shelves and within basins, and much narrower belts showing rapid change of contrasting sedimentary facies across the intervening shelf margin. Shelf and basin sediments commonly extend across tens or even hundreds of miles, whereas individual facies bands of shelf margins are only a mile or two wide. Generally, the steeper the slope, the narrower the facies belts (Figs. II-13 and II-14).

The Effect of Differential Subsidence during Deposition

Modification of the above simple model based on linear subsidence may be expected if subsidence rate increases toward the basin during deposition. This may be due either to differential downwarping of the basin or to contemporaneous faulting. The effect of varying subsidence with sedimentation rate is outlined by Meissner (1972, p. 212, see Fig. II-15).

The most common cases seen in the geologic record are those displaying progressive outbuilding as carbonate sediments are produced and accumulated faster than basin subsidence can accomodate them. For example, the well-known Permian Reef Complex can be seen to have prograded basinward more or less continuously over several miles during Late Permian time alone. Shelf margins around the Delaware and Midland basins show a general pattern of several abrupt basinward marine regressions in step-outs, diagrammed nicely by *Van Siclen* (1972, see Fig. II-16).

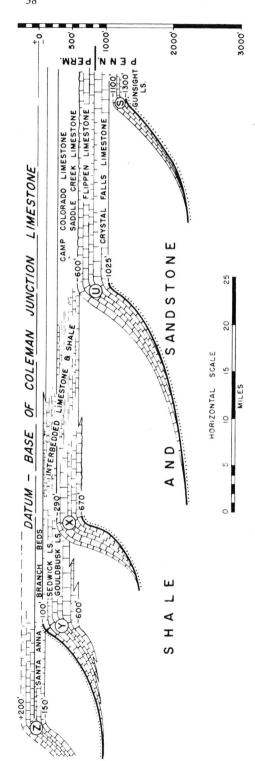

Fig. II-16. Periodic regression by outstepping of carbonate shelf margins resulting from cyclic and reciprocal sedimentation, from Van Siclen (1972, Fig. 4). Vertical exaggeration is 40 times. Original illustration in Bulletin of American Association of Petroleum Geologists (1958)

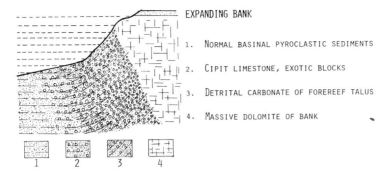

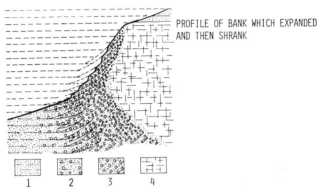

Fig. II-17. Upbuilding, outbuilding, and inbuilding of carbonate banks in Dolomites of South Tyrol, Italy as discerned from foreslope beds, after Leonardi (1967, Figs. 154 and 155)

Some of the greatest carbonate banks have mainly built upward; in areas of geosynclinal subsidence, sedimentation often just manages to keep pace. The El Abra Cretaceous of central Mexico and the Middle Triassic of the Dolomites are notable examples (Fig. II-17).

Rarely is there found a continuously transgressive record of carbonate facies. The best documented is that of the Helderberg Group of New York State (Laporte, 1969) developed in beds marginal to the Appalachian geosyncline where clear evidence exists of westward inundation followed by a seaward progradation back into the geosyncline to the east.

Usually carbonate production keeps up with relative sea-level rise. Transgressive reefs forming a continuous stratigraphic unit are essentially unknown, but the pinnacle form developed on some offshore banks indicates that in some basins subsidence has overcome sedimentation. Generally, some cause other than a too-rapid subsidence or sea-level rise ends the carbonate building process. These causes might include a rapid, but brief, sea-level drop preceding the next marine incursion, terrigenous influx, climatic change, or restriction of the basin to evaporitic conditions.

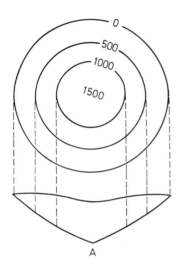 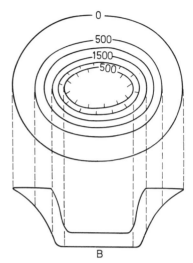

Fig. II-18. Two common isopach patterns for carbonate time-stratigraphic units. Pattern B forms with development of a carbonate platform around the basin and rapid rate of subsidence of the basin center. Gradual and slow subsidence may result in detrital infilling of the basin and the development of the pattern A

The Interpretation of Thickness Maps in Carbonate Strata

Isopach maps are easily and rapidly constructed and may permit some tectonic interpretations of rate of subsidence and location of structural trends and provinces. This is a reliable procedure only if boundaries at the base and top of the isopached unit appear logically to be time-stratigraphic, or if their onlap and offlap relations are clear. Careful correlations within the isopached unit offer a check on this. In carbonate strata, however, even if "time planes" are known, accurate interpretations of an isopach map alone may be difficult. This is because, as pointed out above, there exist dual mechanisms for carbonate deposition: (1) Detrital accumulation washed into or created *in situ* within a shallow subsiding basin; in this case greatest thickness lies at the center of the downwarping basin. (2) Carbonate may be anomalously thick over an intrabasinal topographic high which subsided regionally.

Furthermore, as Fig. II-18 illustrates, different isopach patterns in carbonate strata may result from simple basin filling or from gradual basinward accretion of the shelf margin. Accurate distinction between them at an early stage of stratigraphic information depends on lithofacies analysis as well as thickness information.

Trends of Carbonate Buildups May Be Controlled by Tectonic Movements in the Basement

It is important to relate regional structure to reef configuration and facies, for purposes of predicting distribution of such carbonate bodies. For example, fault movements on basement blocks may offer the primary control for linear carbon-

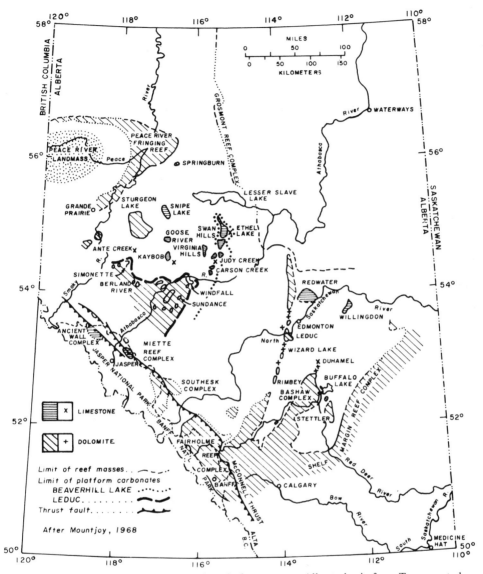

Fig. II-19. Linear trends of Late Devonian platforms across Alberta basin from Toomey et al. (1970, Fig. 1)

ate buildups. Movements both preceding and during carbonate deposition are known to be important. A good illustration of this in the Holocene and late Tertiary is seen in the British Honduras—Yucatan barrier reef around the Caribbean and the several large atolls or banks off shore from it. Often such faults cause multiple lineaments of carbonate bodies which are recognizable in the geologic record. Many of these have underlying positive gravity anomalies (Ball, 1967a) with intervening basins represented by negative anomalies. Examples are: the Bahama Islands which are formed on a north-west trending pre-Cretaceous positive element; Central Basin Platform of the West Texas Permian basin (with more than 100 milligals rise in gravity across it); and the major Alberta reef trends with north-northeast alignment, the Leduc-Rimbey trend of this area having a faint underlying positive gravity anomaly (Fig. II-19). The Late Paleozoic algal plate mounds also display multiple linear trends.

These linear trends, presumably resulting from basement faulting, exist because of the sensitivity of carbonate production to any kind of pre-existing topographic high, whether structural or geomorphic. Carbonate accretion may be expected to occur along any sudden break in slope, perhaps a response to drape on basement faults. Once initiated, growth is apt to be continuous and rapid. Topographic control for Holocene barrier reef development has been advocated by Purdy (1974a and b) who relates it to a late Pleistocene karstic surface as well as underlying Tertiary faulting. In addition, basinal areas commonly contain numerous areas of small circular and elliptical buildups of carbonate, developed as a result of preferential accretion atop slight topographic rises. Note that lines or elongate areas of such pinnacle mounds or reefs are known in several basins discussed in Chapters IV and V. These may be structurally induced along basement faults or shifting blocks. Massive carbonate caps are known to grow on rising salt domes both in the Gulf of Mexico and in the Persian Gulf. In the latter sea only a very few meters of original relief at depths of 30 or more meters is sufficient to cause pure $CaCO_3$ to accrete rapidly over the elevated humps.

Cyclic and Reciprocal Sedimentation

Alternation of basin and shelf sedimentation results when the buildups of shelf margin slopes around a basin are formed during periodic sea level fluctuations, particularly when the climate and lack of hinterland relief encourage the carbonate producing process to operate efficiently.

Fig. II-20 shows that during high stands of sea level the shelf margin is protected from clastic influx, which is caught many miles away near the shoreline. In the clear shelf water, carbonate production and sediment stabilization by organic and inorganic processes are particularly efficient, resulting in rapid accumulation on the shelf and its margin and almost none in the basin. If sea level remains high, the carbonate shelf margin builds out, followed in regressive sequence by its several shelf facies belts (lime sand, lagoonal lime muds, tidal flats and saline lagoons and supratidal sebkhas). At the same time any minor accumulation occurring in the deeper basin consists only of pelagic and windblown material in an euxinic environment. Additionally, perhaps at the foot of the constructed platform, some fine carbonate detritus is deposited from the shelf. Obviously the basin starves during this time, although in many cases a thin, dark limestone or shale unit may be formed completely across its floor.

A second phase of sedimentation occurs when sea level drops. Clastics are reworked by rivers, particularly if tilting of the hinterland has increased gradients. Sands and clays are carried out in channels across the shelf, bypassing the margin and accumulating in the basin, filling it, and reducing the marginal relief created during the high stand carbonate phase. There is a growing awareness among

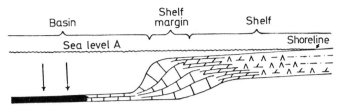

CARBONATE STAGE   High sea level, shoreline far from shelf margin

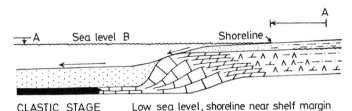

CLASTIC STAGE   Low sea level, shoreline near shelf margin

Fig. II-20. Stratigraphic model of cyclic and reciprocal sedimentation from Meissner (1972, Fig. 7), courtesy of West Texas Geological Society

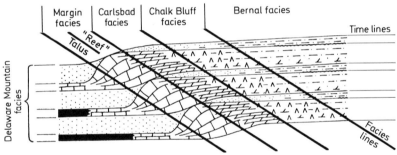

Fig. II-21. Stratigraphic model of cyclic and reciprocal sedimentation in Permian Reef complex, New Mexico from Meissner (1972, Fig. 8). Shows the repetition of several depositional cycles each composed of a "carbonate" and a "clastic" stage. Lithofacies terminology of the Artesia Group has been applied to the basinward prograding diachronous and regressive facies. Illustration courtesy of author and West Texas Geological Society

geologists of how common this type of sedimentation may be. Meissner's models (Figs. II-20, II-21) are derived from the Late Permian of the Delaware basin and Van Siclen's initial presentation (1958) comes from the Pennsylvanian-Wolfcampian, eastern shelf of the Midland basin. Wilson (1967a, 1972) has described this type of sedimentation in detail from Late Pennsylvanian beds in the Oro Grande basin. Additional examples are from the Middle Pennsylvanian evaporite filling of the Paradox basin (Peterson and Ohlen, 1963), Permo-Pennsylvanian beds of the Glass Mountains (Ross, 1967), Devonian shale filling the Alberta basin (Oliver and Cowper, 1963), and the Mississippian beds of the Illinois basin (Lineback, 1969). The recognition of cyclic and reciprocal sedimentation is largely based on thin subsurface correlation markers seen in slope and basinal strata, and on recognition of equivalence in time of both basinal and shelf sandstones.

Correlation Problems along Edges of Shelf Margins

Usually bed-by-bed correlation becomes difficult along a carbonate shelf margin. For one thing, the even bedding planes of the shelf facies give way to widely-spaced irregular "breaks" or to massive unbedded strata characteristic of some particular organic accumulations at the shelf margins. In addition, considerable changes of thickness coincide with the more closely-spaced changes in facies at the shelf margin, making rock-unit correlation particularly difficult. The normal condition is for limestone strata to thicken gradually to the shelf break and suddenly thin at the same place where they dip over into the basin. Such steep margins, where thickness and facies changes are rapid, are hard to evaluate in the geologic record. This is true even on outcrops where major faults may commonly occur at the line of the slope break and obscure the significant correlative horizons. For example, despite many years of study, argument still prevails concerning the cause of the great topographic relief on the margin of the Cuesta El Abra (outcrop) and the Faja del Oro (Golden Lane) of Mexico. The prevailing view is that it is depositional, but an interpretation of downfaulting along this trend and a disbelief in the fact of great depositional relief on such banks is still maintained by some experienced geologists (Coogan et al., 1972).

In many respects, subsurface data are more convincing than outcrop studies for correlations along normal depositional topography at shelf margins. Four techniques are used:

1. Tracing of thin carbonate horizons basinward: Units can be followed down the flanks of thickened carbonate masses at the break in the slope. Such key beds may be seen to "drop" basinward for hundreds of feet and may be traceable well into the basin itself. They are best recognized on electrical or radioactivity logs as a series of special patterns or signatures and generally considered to represent "time lines" or depositional surfaces. Fig. II-22 is an excellent demonstration of the use of such markers in the subsurface correlation across a Late Paleozoic shelf margin. Here middle Wolfcampian beds show a 3 degree depositional slope.

2. Sandstone-siltstone correlations: Where slopes are very steep, bed by bed correlation may not be possible but another technique is usable if cyclic and reciprocal sedimentation has occurred. Sandstone beds on the shelf may have correlatives in basinal sandstones even though the two series are disconnected across the shelf. At lower sea level stands, sand may migrate as regressive sheets across the shelf and spill over into the basin, either bypassing the exposed carbonate edge in channels or being windblown. Comparative sections with multiple sandstone may be used to effect good correlations. For a good diagram of this see Meissner (1972, p. 218, Fig. 11 and Plate II) where the Bell Canyon sands are correlated with 6 sands of the Yates Formation shelf across a depositional topographic relief of 400 m in the Permian Reef Complex. See also the Bolliger and Burri (1970) correlation across the Swiss Jura reef front.

3. Paleontologic zonation: Whenever possible paleontology is critical to the certification of such correlations. It is significant that large amounts of topographic relief at the edges of subsurface basins were first recognized in the Late

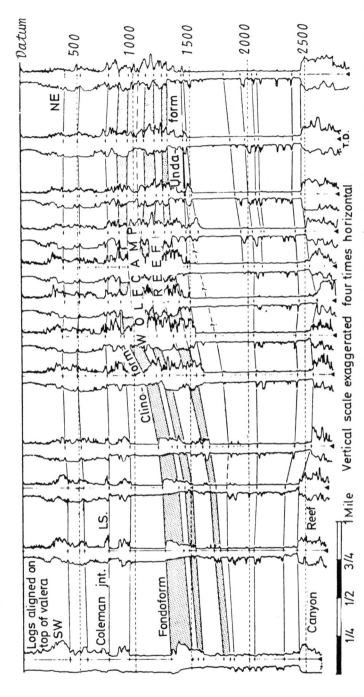

Fig. II-22. Subsurface correlation of electric logs across Middle Wolfcampian shelf margin off the Kelly-Snyder field, Scurry County, Texas from Van Siclen (1972, Fig. 2). Shows tracing of clinoform down depositional topography off the buildups. Canyon reef is close to top of the Upper Pennsylvanian. Depositional dip on top of Wolfcamp reef is less than three degrees. Basin was about 150 m deep at end of Wolfcamp carbonate buildup. Original illustration in Bulletin of American Association of Petroleum Geologists (1958)

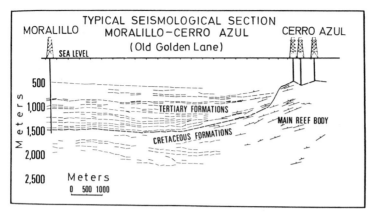

Fig. II-23. Seismic profile interpretation across west side of Golden Lane bank south of Tampico, Mexico; slightly modified after Guzman (1967, Fig. 5)

Paleozoic beds of West Texas where close biostratigraphic control was made possible through the use of large robust foraminifera, the fusulinids. Twenty or more years ago, much difference of opinion and discussion among petroleum geologists concerned the evaluation of fusulinid determinations in West Texas and New Mexico. Fusulinids are predominantly shelf and upper slope organisms but commonly form detrital particles, and readily drift downslope to be reworked into toe-of-slope limestones. Such beds at the basin margin obviously cannot be older than the fusulinids found in them, but can always be younger than their faunas indicate if reworked forms are present. Normally, resolution of seemingly anomalous fusulinid occurrences is made if biostratigraphic data are plotted on carefully correlated electric log cross sections.

4. Seismology: Seismic profiles probably offer the clearest demonstration of subsurface topographic relief and the ability to distinguish such from faulting and structural downwarping, Guzman (1967), Fig. II-23) offers an illustration of this for the Golden Lane in Tamaulipas, Mexico.

Stratigraphic Sequences in Carbonates of Epeiric Seas on Shelves and in Shallow Basins

Sequences of both uniform and cyclically alternating, geographically widespread thin rock units are common behind shelf margins. Such sequences include open and restricted marine and evaporite environments of the carbonate facies spectrum (belts 7–9). They also occur in shallow cratonic basins where a wider range of facies belts from the basin center to the shelf may be present but wide and irregular. Petroleum geologists have long termed such sequences "onionskin" or "layer-cake" stratigraphy. The surfaces over which such strata are deposited are extremely flat, generally dipping only 20–30 cm per km (less than the slope of the lower Gulf of Mexico coastal plain!). This is true for surfaces of the interior of present day carbonate banks and for modern shelves built from the continental blocks.

By analogy with present shelf depths, whose Holocene sediments much resemble ancient limestones, we can safely assume that in the geologic past, shelves and platforms hundreds of miles wide were covered with water only a few tens of meters deep. Again, by analogy with actualistic models, it is hardly conceivable that tidal currents and wave action in such widespread and shallow seas could have been very effective. Yet, many deposits across the North American craton contain uniform sequences of rock types for hundreds of miles and such deposits commonly include beds of clean quartz sandstone and pelletoid or oolitic lime grainstones! These are clearly the result of wave and/or current activity and yet are widely distributed over thousands of square miles. On such flat-bottomed, shallow seas not much wave energy can be generated and tidal effects are severely restricted (Keulegan and Krumbein, 1950). It is thus probable that such deposits were formed by seaward progradation of shorelines and offshore oolite bars.

Studies of the mechanism of accretion of lime muds in Florida Bay, on the tidal flats in the Bahamas, and along the Persian Gulf, reinforce this conclusion. Shelf areas of the earth are now covered by wide expanses of shallow marine, tropical water as a result of the post-Wisconsin sea level rise. Fine lime sediment is produced abundantly in some of these waters, perhaps mostly by breakdown of algae and attrition of calcareous tests of a wide variety of organisms. This fine sediment is carried continuously landward on flood and wind tides and trapped effectively against the shore on tidal flats which build progressively seaward. A regressive sequence, about 4000 years old is well documented by coring on parts of Andros Island and in the Persian Gulf. As pointed out in Chapter I, this sedimentation has been extraordinarily rapid since the beginning of the Holocene. Thus progradation over a very flat surface would result in quiet water, open marine muds overlain by sheets of grainstone, which in turn are overlain by wide belts of shallow water carbonate muds and evaporites bearing sedimentary structures which form today in intertidal areas. Unless such deposits were formed in tides which daily transgressed inland hundreds of miles, we must conclude that these facies result from environments migrating over thousands of years. The reader is referred to an extensive explanation of the origin of carbonate facies over these surfaces and their diachronic mode of deposition by Shaw (1964) and Irwin (1965). Stratigraphers recognize two types of such strata: uniform single lithosomes and sedimentary cycles.

Widespread Uniform Single Lithosomes

The midcontinent area of North America, including the Williston basin, typically contains "layer cake" strata as does also the Arabian shield and Russian platform. Some of these lithic units cover remarkably wide areas. Krumbein and Sloss (1963, p. 374–378) considered some of them "time-parallel rock units". Consideration of facies within these strata shows that it is unlikely that some of them were deposited in contemporaneous layers. Certain of the beds, such as cross-bedded

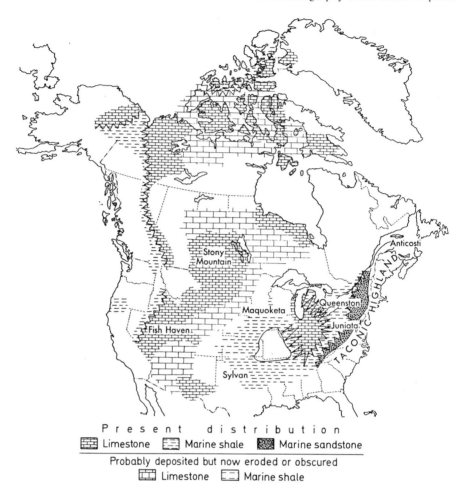

Present distribution
▦ Limestone   ▤ Marine shale   ▨ Marine sandstone
Probably deposited but now eroded or obscured
▢ Limestone   ▢ Marine shale

Fig. II-24. Facies map of Upper Ordovician strata in North America. After Clark and Stearn (1968) illustrating the widespread single argillaceous carbonate sheet across the Midcontinent and Cordilleran miogeosyncline. In the Midcontinent and western regions this unit directly overlies an equally widespread and lithologically similar pure carbonate unit—the Red River, Bighorn, Viola-Montoya. Fine brick pattern indicates the carbonate facies. Coarser brick marks, eroded areas where the unit formerly extended. Note the Appalachian Queenston delta along eastern border of craton. Illustration from Geological Evolution of North America (2nd Edition) with permission of authors and The Ronald Press, New York City

marine orthoquartzites and glauconitic oolitic limestones, represent residual (or lag) strata formed by repeated marine transgressions and regressions across the shelves and during times of relatively slow sedimentation.

Other single, uniform units may represent widespread stable marine conditions and consist of members deposited more or less at the same time. The Middle and Upper Ordovician west of the Mississippi River contain two lithosomes which fall in this category. The Red River-Viola-Montoya shallow marine carbonate unit and the overlying shale, Stoney Mountain-Maquoqueta-Sylvan, are faunally and lithologically similar over the vast area stretching from Hudson Bay and Manitoba in the north to Sonora, Mexico in the south, and from Missouri to Colorado in an east-west direction (Krumbein and Sloss, 1963, Fig. 10–15; see Fig. II-24). The lower, carbonate portion of the couplet is almost everywhere a characteristic brown color, contains large, dolomitized burrow mottling and a characteristic fauna of large fasciculate corals, *Receptaculites*, and nautiloids. Its thickness variation (100–200 m) is slight•considering its extremely broad distribution.

Remnants of the upper portion of the lower Devonian (Camden chert) stretching from Tennessee through the southwest USA to Chihuahua, constitute an equally impressive unit of distinctive rock types: light novaculitic and cherty limestone, everywhere bearing the same brachiopod and trilobite faunas.

The Washita beds of the Texas Cretaceous (Late Albian-Cenomanian) constitute another widespread transgressive carbonate unit of rather uniform, though argillaceous, character. According to rules of sedimentation in clear-water, epeiric seas, as set forth by Irwin and Shaw, one would expect such widespread units to consist of highly restricted carbonates and evaporites, to show gradual facies differentiation, and be naturally diachronous. Indeed, they do change facies somewhat into the mildly subsiding basins across the vast midcontinent area of North America, become more laminated and cherty there, but appear always to have the same faunas. One may conclude that such limestone sheets must have been deposited slowly, with many breaks in sedimentation, perhaps under water a few hundred feet deep, under hydrologic and climatic condition which encouraged some degree of circulation. No modern model for such shelf sedimentation is known.

Carbonate Shelf Cycles

Commonly, shelf sediments consist of depositional cycles in which repetitive sequences of rock types are spread over vast areas. Individual beds in such cycles may be traced for many miles. The earliest studied and best known cyclothems of the geologic record are those in the Carboniferous (Duff et al., 1967, p. 4) where complex combinations of clastics and carbonates make for easy recognition. One of the most interesting stratigraphic results of carbonate study in the last 20 years has been the recognition that *most* thick limestone and dolomite sections consist of cyclically repeated strata and are not homogeneous units. This is particularly true of shelf deposits. (See Chapters VII and X.).

One may commonly recognize cyclicity in strata after measuring and describing a section in some detail; some basic rock types may be repeated in a regular order. Computer programs are available to aid in this recognition. A further question is: between which rock types are the stratigraphic contacts most abrupt? The evaluation of such boundaries should consider evidence of subaerial exposure or depositional still-stand under marine conditions (see outline of sedimentary structures in Chapter III). At this stage interpretations of either truly cyclic (ABCBA) patterns or rhythmic (hemicyclic-ABCABC) patterns may be possible. Once cycles are recognized, further work on the section may involve resampling

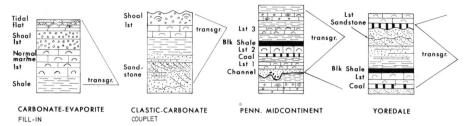

Fig. II-25. Common patterns of cyclic sedimentation. These sequences are caused basically by the timing of terrigenous influx and climatic variations with rising or falling of sea level relative to land. Since the mechanisms are multiple and somewhat interrelated, results could be theoretically complex. The observation that in many places patterns such as these are discernible and repeated through long sequences, means that many times only one or two mechanisms operate and result in fairly simple cyclic patterns

and closer study of microfacies and sedimentary structure for more accurate environmental interpretation.

Despite the time and effort required, several advantages exist in the use of cyclothems in sedimentological analysis of strata. (1) Their recognition may aid in additional environmental interpretations. Certain rock types are more clearly interpretable than others. The establishment of an orderly and consistent sequence of rock types permits comparison of depositional conditions with beds above and below any bed in question. For example, a pellet calcarenite lying between beds of bioclastic micrite more probably represents an organically pelleted lime mud than an originally current-laid lime sand. (2) Recognition of sedimentary cycles also offers aid to time-stratigraphic correlation within a given depositional basin. The use of boundaries between cycles or the points of major marine invasion or "deepest deposition" in correlation, has been discussed by Krumbein and Sloss (1963, p. 383–386). (See later discussion of correlation.) (3) Discernment of cyclicity in strata aids in predicting facies distribution. Walther's Law (1893, p. 979) states that facies vary in analogous sequence both vertically and horizontally. This premise forms the first half of Shaw's *Time and Stratigraphy* (1964); see also Coogan (1972). Cycles will generally change facies in a regular way, some within a distance of a few miles, some gradually for several hundreds of miles. Recognition and description of regular vertical repetition in strata is a valuable first step in construction of lithofacies maps.

Three types of cyclic patterns are typically seen in shelf deposits. These are diagrammed in Fig. II-25. It is convenient to organize the vertical sequence into three main environmental subdivisions: lower terrigenous phase, middle normal marine, upper shoaling, and final exposure. The categories of cycles are: (1) Upward shoaling (fill-in) carbonate or carbonate-evaporite hemicycles with essentially a regressive record (Chapter X); (2) simple terrigenous clastic-carbonate couplets or, more complex and complete clastic-carbonate cycles like those of the Late Paleozoic shelves of the mid continent (Chapter VII); and (3) cycles with a transgressive lower carbonate and upper regressive terrigenous phase.

Many other examples of cyclic sedimentation may be found in extensive references: The Kansas Geological Survey Bulletin 169 (Merriam, 1964) and Duff et al.

(1967). Microfacies and environmental interpretation of several carbonate shelf cycles are discussed in Chapter X. As Coogan (1972, p. 7) has observed, most shallow shelf or platform cycles are asymmetric or rhythmic, containing a thin transgressive record, often in sharp contact below with beds representing the top of the underlying cycle. The balance of the sequence is gradational upward and is regressive, culminating in a shoal-water phase.

Causes of Cyclicity in Shelf Strata

A tectonically balanced wide platform built just to sea level is widely affected by repeated marine transgression and regression. But what are the mechanisms and underlying causes of cyclic sedimentation over such platforms? The repetition of rather complex sedimentary sequences argues strongly for some systematic causes. These have been discussed in so many scientific papers and textbooks that they are not treated in detail here (see references above). The various hypotheses which may apply to the common patterns observed include:

1. Steady subsidence plus an independent external (out-of-basin, world wide) mechanism for causing repeated sea level fluctuations. Subsidence added to independent eustatic sea level rise results in relatively rapid transgression.
   (a) Eustatic sea level changes due to glacial periods such as in the Pleistocene and Late Paleozoic.
   (b) Eustatic sea level changes caused by periodic large-scale movements of the earth's tectonic plates.
2. Steady subsidence and stable world-wide sea level, plus an external mechanism for causing repeated periods of sedimentary fill-in with a built-in mechanism for stopping sedimentation usually by completely filling up the basin.
   (a) Alternately widening and narrowing of marine source area for lime mud by outbuilding of carbonate sheets.
   (b) Periodic climatic changes controlling development of reefs or carbonate banks causing restricted circulation and more evaporite conditions in the basin.
   (c) Periodic climatic changes acting on land source areas for terrigenous clastics, and/or periodic shifting of deltaic distributaries to furnish alternately distant and nearby sources for clastics.
   (d) Tectonism in distant land masses to change stream gradient and furnish clastics to the basins.
3. Episodic subsidence of shelf with gradual fill in of steadily produced carbonate-evaporite sediments or a combination of episodic subsidence and episodic production. Tilting down of the shelf and up of the hinterland, "seesaw action" along a hingeline.
4. Episodic "bobbing up and down" of shelf with a total effect of subsidence (to preserve the record) and a more or less continuous sedimentation.

These processes are not necessarily mutually exclusive. This lies at the heart of the complexity of the phenomenon and the difficulties in its explanation. Prevailing views on the causes of shelf cycles usually favor an eustatic mechanism instead of purely local tectonic causes. One cogent reason for this is that shelf cycles occur in areas of variable tectonic activity. They are found over interiors of large carbonate banks situated in active geosynclines (e.g., Lofer facies of Dachsteinkalk) and across shelves and shallow basins a thousand miles from the nearest orogenic belts and well within the stable interior of cratonic blocks (e.g., cyclical Ordovician to Mississippian in the Williston basin). Cycles may pass from one tectonic unit to another, changing in facies but still maintaining a cyclic character. The Mississippian Madison Group shows this in passing from the Williston basin on to the Central Montana high, as do the Pennsylvanian cyclothems traced from the Pedernal uplift to the Oro Grande basin of New Mexico. Very possibly some world-wide mechanisms operate to cause sea level fluctuations and in addition more local causes, such as tectonic uplift and weather changes, add to and compli-cate the pattern (Wanless, 1972, p. 41).

The favored eustatic mechanism for cyclicity, particularly in the Late Paleo-zoic, has long been continental glaciation and many papers discuss this (Wanless and Shepherd, 1936; Wanless, 1972). This is a very attractive mechanism for Late Paleozoic strata, which are strikingly cyclic in the northern Hemisphere and which in general correlate in time with a long period of southern Hemisphere glaciation. Cyclicity of Late Paleozoic strata, however, is so clearly manifest and so easy to recognize mainly because they commonly consist of interbedded sandstones, shales, and limestones. Even at a distance, most Pennsylvanian beds may be recognized by their "ledgy" character. The terrigenous influx is undoubtedly a response to the widespread Late Paleozoic orogeny.

It is important to consider that pure carbonate shelf cycles are prevalent in *all* parts of the geologic column. They are, for example, particularly well-developed in Cretaceous and Jurassic beds in both Europe and North America, as well as in Devonian strata. If glacially induced eustatic sea-level movements are generally responsible for sedimentary cycles, a more or less continuous waxing and waning of polar ice caps is called for, at least throughout Phanerozoic time.

Perhaps other causes for eustatic sea-level movements exist, e.g., megatectonic (Wells, 1960), but evidence shows that sinking of oceanic plates or uplift of the mid-oceanic rises accounts for only a small amount of sea level displacement. Duff et al., (1967, p. 246) quoted figures from Menard of a sea level rise of 0.2—0.3 cm per 1000 years based on megatectonic adjustments in the Pacific since the beginning of the Tertiary. Mathews (1974) concluded that, except for the most tectonically active island arcs in the northwestern Pacific, a tectonic displacement rate of 1 m per 1000 years is average. See Mathews' review of Christensen's study of the San Joaquin Valley of California (1974, p. 68). Hallam studied upper Liassic geologic history in western Europe and estimated that sea level may have risen about 15 m in 3 million years (during 3 Toaracian ammonite zones formed during a presumably nonglacial period in earth history). This is a rate of only 0.5 cm per 1000 years. These data are not as accurate as are needed but they indicate clearly that (1) megatectonic forces are presumably continuously operating to cause sea level variations and (2) that regionally none of these rates approach the most recent glacially induced eustatic fluctuations of the Late Pleistocene. During the last 120000 years, sea level dropped in spurts of several meters per 1000 years at an over all rate of 0.8 m per 1000 years. More recently, during the last 15000 years, sea level rose at almost 8 m per 1000 years!

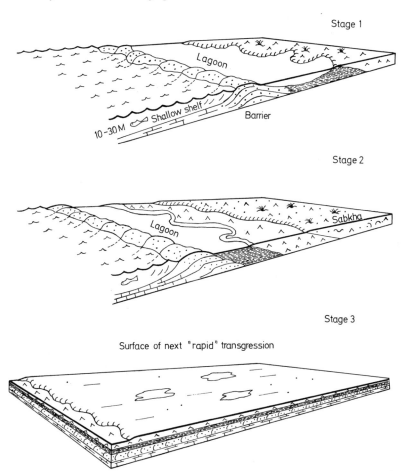

Fig. II-26. Construction of single sedimentary sequence (upward shoaling cycle) by progradation or seaward outbuilding of carbonate-evaporite facies

    In any event, cyclic sedimentation and sea level fluctuation independent of local tectonics are both the rule in geologic history. The causes of both may be sought in normal uniformitarian processes and not in any special conditions in time and space.

Use of Cycles in Chronostratigraphic Correlation

Rock unit correlation is facilitated by grouping beds into cyclothems, but many geologists have also employed cyclic sedimentary patterns as a valuable tool in chronostratigraphic correlation. (For an extended criticism of this method, see Shaw, 1964, *Time and Stratigraphy*.) Following principles stated by Coogan (1972, p. 5–16) and Krumbein and Sloss (1963), the succeeding arguments may be cited as support for the proper use of cyclical units as "time markers" (Fig. II-26).

1. Marine transgressions must have occurred over extremely level surfaces formed by a preceding regression associated with construction and seaward migration of a flat coastal plain.

2. Any kind of independent sea-level rise would "rapidly" flood wide areas of such surfaces, especially when they are tectonically subsiding. But, although marine transgression occurs very "rapidly," wave action to erode the substrate is dampened by the very low relief over such wide areas. Marine transgression naturally cuts off sediment supply by causing upstream aggradation, decreases sediment influx, and so gradually increases its own progress. Transgression is apt to *appear* more rapid than it actually is. Thin transgressive units are the rule, but open marine shelf strata may be thin due to slower deposition and, as pointed out by Fischer (1961), some transgressions cannibalize their own stratigraphic record. Offshore waves beyond barrier bars and island complexes may in some instances erode the previously deposited record as they move across the swampy inner coastal sediments.

3. Once a typical shelf is flooded, regressive outbuilding occurs if carbonate production is established. This happens because shallow water carbonate sedimentation is an extremely rapid process and may fill in large areas in a geologically brief span of years (note figures for Holocene shallow sedimentation given in Chapter I).

4. The rate of regressive outbuilding is, for any one cycle, generally so rapid that it is beyond the resolution of our biostratigraphic calendar. Sedimentary cycles form faster than the evolution of practically any organisms upon which paleontological zonation is based. The sharp boundaries between such cycles and the thin transgressive record at the base represent a duration of only a few thousand years (see below).

Thus, although each hemicycle or upward shoaling sequence so created follows Walther's Law, and consists in the main of regressive, diachronously deposited facies, the boundaries between these sedimentary rhythms may closely approximate "time markers" and are more useful as such than the diachronous facies within each cycle. Geologists commonly use either the contact between cycles where inundation begins or the maximum inundation phase of an individual cycle as "approximate time lines." The principles underlying the use of such boundaries in "time" correlation are described by Krumbein and Sloss (1963) who term the procedure "correlation by position in the bathymetric cycle." Additionally, there are certain types of strata commonly associated with transgressive-regressive cycles which may be assumed to have formed only along flat planar surfaces and to have been preserved by relatively rapid transgression (e.g., coals and the top of sabkha evaporites). When sequences of such key beds lie parallel to each other, or when their intervening beds thin or thicken at regular rates, most geologists assume that such strata represent, for practical purposes, time-stratigraphic markers within a given basin. Other key beds which may be used as time-stratigraphic markers in beds on platforms or shallow basins are bentonites, radioactive silt zones (often causing gamma ray deflections in thick carbonate strata) and certain calcareous concretion zones in shales. These key units may be used to check cyclothem boundaries considered as "time markers."

It is obvious that such chronostratigraphic correlation is essentially intrabasinal, although certain thin, lithic units have been carried great distances across several tectonic provinces (see Wanless, 1972, on coals).

Arguments for the time stratigraphic use of these cycle boundaries and thin markers are mainly deductive and they should be tested wherever possible against as much detailed paleontology as is available. Shaw's system (1964) of statistical treatment of sequence of faunal zones, related to a composite standard, offers a hope of providing the kind of paleontology zonation needed for this. However, the requisite biostratigraphic detail with which to check our cyclic correlations is generally missing. For example: The Late Devonian beds of the Williston Basin contain perhaps between 25–30 cycles, unusually regular and widespread. At best, our paleontologic zonation provides only a half dozen zones, whether based on goniatites, conodonts, or brachiopods, for these Frasnian and Fammenian beds (estimated time, 15–20 million years). The same duration can be estimated for Late Paleozoic cycles in New Mexico. The Late Pennsylvanian and lowest Permian of the Sacramento Mountains, (narrow east shelf of the Oro Grande basin) consist of approximately 50 cycles through beds in which the best fusulinid experts can at present find only about 5 or 6 subdivisions of the zones *Triticites* and *Schwagerina*. Time involved in this case must be about 15 million years. Estimated times (Duff et al., 1967, p. 246) for Late Paleozoic cyclothems range from 30000 to 300000 years each. Fischer (1964) and Zankl (1971) stated that the duration of the 300 Triassic Dachstein cyclothems was from 20000 to 100000 years each, deposited within a period of 15 million years.

Therefore, whereas at present the most detailed correlatable paleontologic zones embrace about one to two million years or so (Hay, 1972), the sedimentary cycles discussed above were probably deposited in as little as 1/10 to 1/20 of this time!

Chapter III

Outline of Carbonate Petrography

In the history of carbonate geology, progress in understanding limestone and dolomite as sediments accelerated tremendously as petrographic data were added to careful field study. There exist various levels of petrographic observation useful in depositional interpretation. Some require more sophisticated approaches and more complex techniques than others; the list below progresses from field mapping and section measuring to microscopic and geochemical study in the laboratory.

1. General lithology.
2. Stratigraphic relations indicating submarine topography, cyclic sequences, facies changes.
3. Paleontology including paleoecological observations relating organic associations and sedimentary features.
4. Sedimentary and organic structures, including trace fossils, bedding, directional properties.
5. Color variations.
6. Microfacies—texture.
7. Microfacies-particle kind identification, mainly biological but including identification of non-carbonate, acid-insoluble particles.
8. Diagenetic observations; interpretations including consideration of porosity and permeability measurements.
9. Mineralogical and geochemical laboratory data as needed.

   This Chapter reviews most of the above classes of data in outline form, referring to more comprehensive papers on the subject and the stratigraphic relations discussed in other Chapters.

Techniques for Examining Carbonate Rocks

Some of the "tricks of the trade," employed in carbonate petrography are briefly reviewed below. See also Chilingar et al., (1967), Horowitz and Potter (1971), and Milliman (1974).

Sawing and Polishing

One reason that petrographic study of carbonates lagged far behind that of sandstones (excepting Sorby, 1879; Cayeux, 1935; Sander, 1936, transl. 1951), was that geologists simply could not see enough of limestone fabrics in the field. The sawing and polishing of slabs and wafers (plaquettes), as well as thin-section study, is a necessary basic procedure for satisfactory observation and description. A normal rock saw and a progressive polish with 400–600–1000 carborundum grit is required. A good procedure is to cut a plaquette (about 3–5 cm square and a few mm thick) across the bedding and polish one side. A weak acid (acetic or formic) in 1:5 water solution is used to etch half of the already polished surface. This improves the polish in a pure limestone and etches out dolomite rhombs, clay, and such insoluble grains as quartz, silt and sand. Such preparations may be routinely made and studied under a stereo-binocular scope with reflected light. A light mineral oil on the surface permits fine observation at high magnification. The closer the refractive index of the oil to that of calcite the better the visibility. Except for the rock saw, no cumbersome or powered equipment is necessary. If the saw can be mounted with an independent power source, the above procedure is essentially adaptable to the field.

Thin Sections

Thin sections have long been the standard means of carbonate study. They are expensive if purchased commercially costing several dollars each, depending on the size. They are slow to manufacture without proper equipment. In many respects, however, no substitute for them exists. For most purposes in carbonate study, they should be ground just slightly thicker than the usual 30 microns to get deeper color contrast under plane light. They should never have a permanent cover-glass. (They can always be thinned if necessary and a temporary cover glass mounted with immersion oil if needed.) Uncovered sections may be oiled, scratched, plucked, stained, and acid etched under the microscope, using polarized light to aid in determining the character of the residue. Because carbonate minerals are optically difficult to distinguish from each other, the above simple techniques of identification are useful.

Acid Etching

Use of progressive acid etching on a thin section or slab enables identification and exact placement of dolomite, anhydrite, quartz, chert, and clay in the calcite fabric. One may first use formic or acetic acid (about 1 part concentrated acid to 5 parts H_2O) to dissolve calcite and preserve dolomite and later 1/10 HCl to dissolve the dolomite. The organic acid tends to coat the dolomite rhombs with a fine residue of partly dissolved carbonate and makes them stand out clearly. Progressive acid etching of calcite, when observed under a microscope, also gives useful information about crystal structure and orientation.

Acid solution of carbonates may also be carried out on small chips of approximately equal volume, placed in a transparent glass finger plate. Progressive acidization of residue, with projection of polarized light through it, will determine both kind and quantity of terrigenous impurities and dolomite-anhydrite in the limestone. Water flushing and blotting with absorbent cotton is necessary during application of multiple acids.

Staining

A very useful review of stains for carbonate mineral identification is presented by Friedman (1959) and the reader may refer also to Wolf et al. (in Chilingar et al., 1967) for thorough discussion of techniques. The following are the most useful stains.

1. Titian yellow in 30% NaOH. Stains Mg calcite and dolomite orange-red in weak (5%) NaOH.
2. Clayton yellow in NaOH and EDTA. Stains Mg calcite orange-red (Winland, 1971).
3. Alizarin Red S in 30% NaOH. Stains Mg calcite and dolomite purple.
4. Alizarin Red S in 2% HCl. Stains calcite and aragonite red. Stains ferroan dolomite and calcite purple.
5. $K_3Fe(CN_6)$ in dilute HCl—stains all carbonates dark blue if they contain iron. Because most dolomite is iron-rich, this is a useful stain to distinguish dolomite from calcite.
6. $K_3F(CN_6)$ and Alizarin Red S in 2% HCl (Evamy, 1963). Stains calcite red and ferroan calcite purple.
7. Feigel's solution (Ag_2SO_4 in dilute solution of $MnSO_4$). Stains aragonite black, leaving other minerals untouched.

Staining is more usefully employed on thin sections than on plaquettes, on light-colored rock than on dark, and on coarse rather than on finely crystalline fabric. Indeed, in the latter texture, crystalline boundaries tend to soak up the stain and obscure its effect. Microscopic porosity (e.g., chalk texture) essentially prohibits staining. The distinction between dolomite and calcite by stains is really one of iron content as well as solubility. The above variations in effectiveness indicate that considerable caution should be exercised in the use of stains for mineralogical identification.

Acetate Surface Replicas—Peels

The use of cellulose acetate to make delicate replicas of etched surfaces of carbonate rock was begun by paleontologists 30 years ago. It is now widely used to help in the quick interpretation of carbonate textures and to view at high magnification cements, micrite, and internal structure of bioclasts (Beales, 1960). Peels are made by pouring a solution on a carbonate surface which has been polished with 1000 grit carborundum and lightly etched in weak acid. Depth of etching controls the quality of the replica but varies considerably with the type of rock. The following procedure and formula for peel manufacture is useful.

Make a creamy solution of cellulose acetate dissolved in amyl acetate or acetone incorporating drying inhibitors such as a little xylene and chloroform. Such a solution, if tightly bottled, may be kept for a long time. After acid etching the rock surface, carefully wash it by immersion in still water, and dry it in air. Flooding with acetone also quickly removes the water. Level the rock slab in a sand box and pour acetate solution in the center of the slab. Do not let it run over the edges. Let it dry overnight and peel off with a dissecting knife or razor blade. Dip in HCl to remove plucked carbonate grains, wash, dry, label with india ink and mount peel between glass slides.

Peels may also be made more quickly by flooding the etched surface with acetone and rolling on a thin sheet of acetate. Commercial acetate film (0.005 inches) is recommended. Place the roll of film in the middle of the slab and spread it on evenly to prevent entrapment of air bubbles. Let it dry 15 min after the application and then peel it off. Such sheets may also be mounted between glass slides to prevent wrinkling or may be previously glued to glass slides with a silicone rubber cement. The etched and acetone soaked rock is then pressed onto the slide. Plexiglass pieces may also be used. The "dry peel" technique is quicker than using peel solution and is better for more porous rocks because the wet peel solution tends to run down and harden in minute openings. This causes the peel to tear when removed from the rock. Pore space may be outlined on polished rock prior to acid etching by buffing with chrome oxide. This fine powder is transferred on the peel. There are thus several advantages to the manufacture of dry peels although generally better results (and less air bubbles) are obtained by use of the solution. Peels are excellent for high magnification work where crystals of a few microns need to be distinguished; the 30 micron thickness of a "thin section" obscures small crystals. Peels also make good photographs. The petrographer misses the color contrast of a thin section and, of course, no mineral identification is possible through use of polarized light. Peels can be used effectively with transmitted light through any type of microscope if the light is well diffused and slightly oblique.

Cathode Luminescense (Sipple and Glover, 1965)

It is known that luminescence is produced when a broad beam of low energy electrons is focused on a thin section or plaquette of a carbonate rock in a vacuum chamber. This color glow outlines in great detail structures within calcite and dolomite crystals which are not seen in polarized or white light. The very characteristic orange luminescence of calcite is due to the presence of divalent Mn. This luminescence is quenched by the presence of iron, cobalt, and nickel. Dolomite shows a similar characteristic but with shifts in the spectrum to reds and yellows. The trace elements are concentrated on surfaces of slow crystal growth. Thus some idea of the trace element content of the water which precipitated the carbonate cement may be obtained. Advantages of this technique to the petrographer include the better recognition of growth stages in void-filling calcite cement and in similar dolomite cements or replacements, the delineation of vein calcite which is seemingly in optical continuity with the host rock and the better discernment of fossils in altered carbonates (Sipple and Glover, 1965). The instrument for attachment to the microscope, complete with vacuum chamber and pump, is commercially available for about $3000.00 USA.

Scanning Electron Microscope

The Scanning Electron Microscope (SEM) is one of the latest developments in electrical optical instrumentation. The image is derived from scanning the gold-plated surface of a specimen with an electron beam and detecting secondary and back-scattered electrons. Depth of field is at least 20 times that of a light microscope at comparable magnifications. Magnifications of X500–10000–20000 are commonly employed with SEM. This instrument, along with the microprobe, furnishes abundant new information about such things as the constituents of lime mud, e.g., the character of calcitic plates and aragonite needles and the abundance of submicroscopic algae, coccolithophores, comparative studies of the diagenesis of lime mud and its alteration to limestone, investigation of trace-element chemistry and crystal structure of carbonate cements, studies of microscopic structural deformation of fabric and micron-size surface coatings of carbonate particles.

Microfacies Interpretations

Microscopic study is the most important of the various levels of observation possible in the broad field of carbonate petrograpphy. But, despite the advantages that identifiable carbonate particles offer toward environmental interpretation, detailed petrographic study of limestones and dolomites may be difficult because of their susceptibility to diagenetic alteration. Prevailing neomorphic crystallization, cementation, and mineralogical replacement of the original carbonate sediment result in a mat of dense crystal fabric. Ordinarily, forms seen in a thin section or peel are caused by color and impurities (dust lines) and by different crystal shapes and sizes rather than the original particles and fine-grained matrix. Carbonate rocks are in a sense metamorphic—at least metasomatic, the replacement occurring through the passage of time and water rather than through the agencies of heat and pressure.

Nonetheless, petrographic study lies at the heart of the considerable success geologists have had in depositional interpretation of carbonate strata, particularly when petrography is combined with detailed stratigraphic control and when Holocene sediments are studied to offer depositional models. Many authors have treated the subject of depositional interpretation in part, particularly relying on interpretations of organic particles and on large numbers of photomicrographs. See especially Horowitz and Potter (1971), the International Sedimentary Petrography Series of Cuvillier and Schürmann (1951–1969), and Carozzi et al. (1972).

The following check list of observations and pertinent questions should aid in the study of limestones as they are observed in polished and oiled plaquettes, peels, and thin sections. Several other such lists are available (e.g., Klovan's list given in Horowitz and Potter, 1971, p. 10). The present list is adapted particularly for depositional interpretation. It is followed by an outline on diagenesis and an example of actual thin-section interpretation.

1. What are the relative amounts of the major components?
   a) Carbonate grains
   b) Silt and quartz sand grains
   c) Micrite matrix (lime mudstone)
   d) Cement (sparry calcite)
   e) Clay
   f) Authigenic minerals including dolomite

2. Character and amounts of different types of bioclastic grains.
   a) Are many different organisms represented or only a few monotonous types? Diversity indicates generally open marine conditions.
   b) Are open marine types abundant or is the biota of restricted marine type? E.g., brachiopods, cephalopods, echinoderm, red algae indicate open marine conditions whereas oysters, clams, snails, many foraminifera, ostracods, and types of green algae indicate more restricted circulation.
   c) What types of algae are present? See Ginsburg et al., (1971) for environmental description. These are reviewed below in some detail (Fig. III-3).
3. Preservation of grains.
   a) How sharp or ragged are the boundaries of bioclasts?
   b) Are rinds on them micritized (shell structure altered to microcrystalline calcite)?
   c) Are internal pores filled with mud or spar-cemented?
   d) Is there pervasive internal micritization or does one observe well-preserved internal shell structure?
   e) Are original aragonite grains preferentially dissolved?
   f) Are the grains rounded or angular?
   g) How much rounding can be attributed to micritization?
4. Non-bioclastic grains.
   a) Are ooids present or absent?
      (1) Is oolite pure or are ooids mixed with other particles?
      (2) Is there mere superficial coating of particles or complete ooids?
      (3) What is the relative size range of the nuclei?
      (4) Is the grain size of ooids uniform despite disparity in size of nuclei?
      (5) How regular are the oolite coatings? Do they show extensive boring by filamentous blue-green algae?
      (6) What types of particles make up the nuclei?
      (7) What is the packing of the ooids?
      (8) If overpacked could this be caused by early solution compaction rather than by later stylolitization?
   b) Peloids (fecal pellets) and indeterminate pelletoids (rounded homogeneous micritic grains).
      (1) Is there any size variation in grains?
      (2) Do isolated nests of large pellets occur?
      (3) Are there many small well-sorted round ones?
      (4) Do those in the shells have well-preserved form and are those in the matrix outside of the shells squashed?
      (5) Is there evidence of agglutination of peloid particles?
      (6) Does peloidal micrite include silt grains as well as finer particles?
      (7) Is there evidence in some grains of transition from rotten bioclasts to structureless microcrystalline rounded grains, i.e., making peloids by micritizing other grains? This is a process occurring only in very shallow water.
   c) Lithoclasts are derived locally and penecontemporaneously (intraclasts) or from outside the basin. Allochthonous lithoclasts are eroded fragments of previously lithified rock derived from some distance away from the site of deposition, i.e., pebbles in a conglomerate.
      (1) What is their external shape and size?
      (2) Are they flat chips or equidimensional?
      (3) Are they notably larger than associated peloids?
      (4) Do they possess rinds?
      (5) Is there a color difference relative to micrite matrix?
      (6) What is their internal composition—similar or different from the local matrix?
      (7) Is the arrangement of clasts parallel to bedding or crosswise?
   d) Onkoids are large particles formed by coating of algae and other organisms.
      (1) What is their size range?
      (2) What type of matrix, calcarenite or lime mudstone?
      (3) Are their laminae crinkled or even?

(4) Do they show more or less equal concentric growth or periodic growth stages on different sides?

(5) Internally do they contain algal filaments (microtubules of *Girvanella*) or other encrusting organisms such as foraminifera, sponges, spirorbids, or stromatoporoids?

(6) Were the onkoids hard or soft when emplaced in the sediment?

(7) Are they deformed in any way?

e) Grapestone lumps are a compound of agglutinated or aggregated particles.

(1) How much larger are such particles than peloids or lithoclasts?

(2) Are they coated or not?

(3) Does intergrain material show any organic structure?

(4) Does fibrous druse exist between grains?

(5) Are grains on the edge of the lump truncated or whole?

(6) Are there internal algal filaments?

(7) What kinds of particles are agglutinated?

(8) Are they exclusively peloids?

5. Textural considerations.

a) Is there evidence of any systematic size-sorting of grains?

b) Is there evidence of any systematic shape-sorting of grains?

c) Are elongate axes of grains and structures parallel with bedding or is fabric homogenized by bioturbation (burrowing)?

d) Is there evidence of textural inversion, i.e., grains of a given size and shape indicative of higher energy environment deposited in a micritic matrix?

e) Is there infiltering of mud between grains? Are there bridging and umbrella effects, i.e., grains caught above flat fragments or flat fragments acting as protectors for underlying pore space which became filled with spar? This would indicate original deposition of a grain-supported fabric. Use Dunham or Folk textural criteria to determine whether the fabric is grain or mud-supported as a background to further observations on grain packing.

6. Compaction history.

a) Is there evidence of early (syndepositional) solution compaction? Examine types of grain contacts: points, facial, or sutured contacts. In sands with spherical grains, more than 0.7 point contacts per grain constitutes overpacking.

b) Is there any evidence of multiple (horsetail) microstylolites or large individual ones? Stylolites are always very late diagenetic features.

7. Types of cement (see this chapter's outline on diagenesis and Horowitz and Potter (1971, Fig. 3).

a) What is the form of cement? Isopachous rims, dogtooth spar, fibrous druse, palisade crystals, micrite?

b) Are there one, two or three generations of cement? Does any blocky calcite cement appear to have existed from the beginning of cementation or is it purely a late feature?

c) Is the blocky calcite mosaic of evenly uniform texture or irregular in size? Are crystals filling void space equidimensional or do they enlarge centripetally?

d) Is the later generation of cement ferroan compared to earlier cement?

e) Is earliest cement cloudy and later cement clear, perhaps indicating respectively marine and meteoric origin?

f) Are there overgrowths on echinoderm grains? On other grains as well?

g) Are there enfacial angles (Bathurst, 1971)?

h) Are there enclaves of irregular-sized calcite crystals?

i) Is the cement inside hollow grains like that outside?

j) What is the age relation of cement types to compaction?

8. Dolomite content.

a) Are the rhombs of uniform size?

b) Do the rhombs possess clear rims and cloudy centers?

c) Are the rhombs located in such positions to infer that early permeability or fluid content controlled replacement? E.g., are they preferentially in micrite or mud pellets? Do rhombs avoid what were originally dense calcitic bioclasts?

    d) Are shells which were presumably originally aragonite preferentially replaced by do-
lomite or dissolved?

    e) Are there coarse dolomite veins or patches?

    f) Are dolomite rhombs particularly iron rich?

    g) Is dolomitization pervasive and fine-grained and apparently in no way controlled by
original fabric of sediment?

9. Geopetal structure.

    a) Does micrite exist as internal sediment inside of shells or cavities?

    b) Is the internal sediment laminated, pure or silty?

    c) Are pellet forms better developed in and under shells?

    d) Is dolomitization better developed in and under shells? (This may be due to lack of
early compaction under the protection of the shells and hence more permeability, fluid
flow, and dolomitization during subsequent diagenesis.)

    e) Are the levels of internal sediment horizontal or tilted affording a level bubble for
original depositional dip?

    f) Was mud and silt which infiltered the original cavities deposited horizontally or is it
micro cross-laminated or slumped?

    g) What is the crystal form of cement in the upper part of the geopetal cavity?

10. Micrite matrix.

    a) Is it pure lime mudstone?

    b) Is its grain size that of true micrite (4–5 microns) or is it of microspar range (10–
20 microns)?

    c) Is the matrix vaguely pelleted or clotted, grumelous? This type of matrix commonly
occurs in grain-supported packstone or in interstices of boundstone.

    d) Is the matrix entire or brecciated?

    e) Does it have a fenestral or birdseye fabric with enclosed geopetals?

    f) Is the matrix laminated or homogeneous?

    g) Does the microcrystalline calcite consist purely of rhombs or platy crystals or are
remains of nannoplankton abundant?

11. Burrows.

    a) What biological interpretation is possible for the burrows? What organisms caused
them?

    b) Is there a mottling effect which outlines burrows, i.e., a color difference between
burrow fill and matrix sediment? Such differences may be caused by a different micro-
chemical environment due to organic decay of material in the burrow.

    c) Is there a grain-size difference in burrow sediment and matrix? This might indicate a
later filling of the burrow.

    d) Was the sediment hard, soft, or viscous when burrowed? Was matrix sediment soft
enough to be distorted by compaction after burrows were formed?

    e) Is there evidence that lithoclasts could have been formed by collapse of burrows?

    f) Can one see discoloration and sharp contacts at edges of the burrows, i.e., were
burrows lined by mineral matter or mucous?

    g) Are most burrows vertical and straight?

Standard Microfacies Types

This Section lists 24 standard microfacies types which are considered of prime
sedimentological significance. The concept of depositional interpretation of microfacies may be credited to the French micropaleontologist J.Cuvillier of the
Sorbonne in Paris in the early 1950's. See Fairbridge (1954) for an early review of
the importance of the concept. Many of the basic types have been categorized
by Erik Flügel (1972) who added sedimentological criteria to the basic paleontological approach used by many European researchers. In addition, some of the

basic microfacies have been illustrated by Horowitz and Potter (1971) disguised by a few highly imaginative names such as "Satisfactory Succotash" and "Pleasant Potpourri." Plates I to XVIII illustrate most of these sedimentary types.

The designated classes are attempts to interpret with the microscope what Bathurst has aptly termed the "insignificant remnant of sea floor life and ecology" furnished us by carbonate rock. Considering the range of sedimentological parameters which control deposition in the marine environment (depth, latitude, salinity, water movement, light penetration, etc.) the organization of microfacies into a limited number of categories is clearly an oversimplification. The suggested grouping of them into standard facies belts of a generalized model as outlined below (see Fig. II-5) contains some overlapping and inconsistencies and omits many variations. But the grouping is applicable to enough known geological facies complexes to demonstrate its general accuracy and to show the usefulness of reduction to a limited number of types.

These types do not employ specific faunal and floral identifications but they may be added as necessary when dealing with rocks of various geological ages. Of course, in addition to sedimentological variations, biological changes through geologic time strongly influence microfacies from System to System and complicate the petrographer's task of interpretation.

In summary, despite obvious difficulties, a combination of the general paleoecological observations of Flügel with Dunham's or Folk's textural classes results in very useful general categories. These are employed throughout the book, symbolized in the general legend (Fig. III-1, SMF-1 to 24), are keyed into the facies-environment schema of Fig. II-5 and in Fig. XII-3 in the final chapter.

Basin and Lower Slope Environments (Facies Belts 1 and 3)

SMF-1   Spiculite (Plate II)
It is a dark, organic rich, and argillaceous lime mudstone or wackestone; siliceous spiculitic calcisiltite. Spicules are usually oriented, generally siliceous monaxons, commonly replaced by calcite.

SMF-2   Microbioclastic calcisiltite (Plate II)
This is a mixture of fine bioclasts and peloids with a very fine grainstone or packstone texture. Fine ripple cross-lamination is common.

SMF-3   Pelagic lime mudstone (Plates III, XXIX)
Its micrite matrix contains scattered fine sand or silt grains composed of pelagic microfossils (e.g., radiolarians or globigerinids) or megafauna such as graptolites or thin-shelled bivalves like *Halobia*.

Slope Environments (Facies Belts 3 and 4)

SMF-4   Microbreccia or bioclastic-lithoclastic packstone
Grains are commonly worn and of originally robust character. They may consist of both locally derived bioclasts and previously cemented lithoclasts; commonly they are graded. Grains may be either polymictic in origin or of uniform composition. Quartz and chert grains, as well as carbonate fragments, may be present. This rock type includes both fine talus and coarser debris resulting from turbidites. The term "allodapic limestone" of Meischner (1965) encompasses this microfacies.

SMF-5   Bioclastic grainstone-packstone; floatstone (Embry and Klovan, 1971) if clasts are
        of gravel size with finer matrix supporting fabric (Plates I, IV)
        This is a common reef flank facies composed mainly of organic debris from organ-
        isms inhabiting reef top and flanks. Geopetal fillings and umbrella effects from
        infiltered finer sediment are common.
SMF-6   Reef rudstone (Embry and Klovan, 1971)
        Coarse gravel of biogenic pieces derived from reef top or reef flank organisms with
        no matrix material. The facies occurs commonly within organic buildups formed in
        zones of high wave energy.

Organic Buildup Environments (Facies Belt 5)

(Plates IV, V, XVII, XVIII, XX, XXIV–XXVI)

SMF-7   Boundstone (Dunham, 1962), *in situ* organic growth. Three subtypes proposed by
        Embry and Klovan are useful: (a) Massive upright and robust forms constitute
        Framestone, (b) encrusting lamellar mats enclosing and constructing cavities and
        encrusting micrite layers are termed Bindstone, and (c) when delicate, complex,
        frond-like forms are abundant in a fine matrix, the sediment may be interpreted as
        trapped by growth of the organisms and the term Bafflestone is applied. (Commonly
        the micrite is clotted or vaguely pelleted.)

Shelf Facies-Open Circulation (Facies Belts 2 and 7)

SMF-8   Whole fossils wackestone (Flügel, 1972) (Plates V, XVII B)
        This is defined by sessile organisms rooted in micrite which contains only a few
        scattered bioclasts. The sediment is formed in quiet water below normal wave base
        and contains preserved infauna and epifauna.
SMF-9   Bioclastic wackestone (Dunham, 1962) or bioclastic micrite (Flügel, 1972) (Plate VI)
        Almost invariably the sediment contains fragments of diverse organisms jumbled
        and homogenized through burrowing. It is formed in shallow neritic water of open
        circulation at or just below wave base. Bioclasts may be micritized.
SMF-10 Coated and worn bioclasts in micrite; packstone-wackestone (Flügel, 1972)
        (Plates XVA, XXIIIA)
        This sediment shows textural inversion and formed in swales in proximity to shoals.
        Dominant particles are of high energy environment and have moved down local
        slopes to be deposited in quiet water.

Shoal Environment in Agitated Water (Facies Belt 6)

SMF-11 Coated bioclasts in sparite, grainstones (Flügel, 1972). Bioclasts may be micrizited
        (Plate VII)
        This sediment formed in areas of constant wave action, at or above wave base so
        that lime mud is removed.
SMF-12 Coquina, bioclastic grainstone or rudstone, shell hash (Flügel, 1972) (Plate VIII)
        Sediment formed in an environment of constant wave or current action with mud
        removed by winnowing. Concentrations of special types of organic debris may be
        significant; e.g., dasycladacean grainstones accumulate in very shallow water. En-
        crinites are a special microfacies of SMF-12, requiring winnowing but less strong
        water movement for their formation. This type of concentration is a common slope
        and shelf edge sediment.

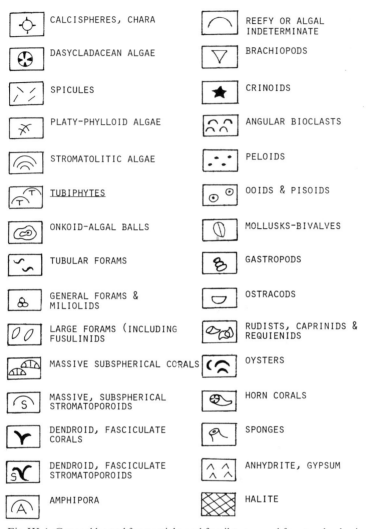

CALCISPHERES, CHARA		REEFY OR ALGAL INDETERMINATE	
DASYCLADACEAN ALGAE		BRACHIOPODS	
SPICULES		CRINOIDS	
PLATY-PHYLLOID ALGAE		ANGULAR BIOCLASTS	
STROMATOLITIC ALGAE		PELOIDS	
TUBIPHYTES		OOIDS & PISOIDS	
ONKOID-ALGAL BALLS		MOLLUSKS-BIVALVES	
TUBULAR FORAMS		GASTROPODS	
GENERAL FORAMS & MILIOLIDS		OSTRACODS	
LARGE FORAMS (INCLUDING FUSULINIDS		RUDISTS, CAPRINIDS & REQUIENIDS	
MASSIVE SUBSPHERICAL CORALS		OYSTERS	
MASSIVE, SUBSPHERICAL STROMATOPOROIDS		HORN CORALS	
DENDROID, FASCICULATE CORALS		SPONGES	
DENDROID, FASCICULATE STROMATOPOROIDS		ANHYDRITE, GYPSUM	
AMPHIPORA		HALITE	

Fig. III-1. General legend for particle and fossil types and for standard microfacies. The brick limestone pattern always indicates a micrite matrix and a large arc above a fossil symbol means boundstone or "reefy" tendency

SMF-13   Onkoid biosparite grainstone (Flügel, 1972) (Plate IX B)
       Algally coated particles formed in a moderately high energy, very shallow water environment.

SMF-14   Lag (Plate IX C)
       Coated and worn particles, in places mixed with ooids and peloids which are blackened and iron-stained, with phosphate. Allochthonous lithoclasts may be present. These lags are characteristically thin deposits, representing slow accumulation of coarse material in zone of winnowing.

SMF-15   Oolite (Flügel, 1972), ooid grainstone (Plate X)
       Well-sorted, well-formed, multiple-coated ooids ranging commonly from 0.5 to 1.5 mm in diameter; commonly the fabric is overpacked. The sediment is always crossbedded. It originates through water movement on oolite shoals, beaches, and tidal bars. The best-formed ooids are typically produced on tidal bars.

	SMF-1	BASINAL SPICULITE, BLACK LIME MUDSTONE
	SMF-2	MICROBIOCLASTIC AND PELOID
	SMF-3	PELAGIC FORAMS, NANNOPLANKTON
	SMF-4	TALUS, LITHOCLASTIC CONGLOMERATE
	SMF-4	LITHOCLASTIC, BIOCLASTIC MICROBRECCIA
	SMF-5	FORESLOPE CALCARENITE, BIOCLASTIC GRAINSTONE, PACKSTONE-FLOATSTONE
	SMF-6	REEF RUDSTONE
	SMF-7	BINDSTONE, LAMELLAR, ENCRUSTING
	SMF-7	PLATY ALGAL BAFFELSTONE
	SMF-7	TUBIPHYTES BINDSTONE
	SMF-7	DENDROID CORAL, STROMATOPOROID BOUNDSTONE
	SMF-7	MASSIVE CORAL, STROMATOPOROID BOUNDSTONE
	SMF-8	WHOLE FOSSIL WACKESTONE
	SMF-9	FORESLOPE, NERITIC OPEN MARINE BIOCLASTIC WACKESTONE
	SMF-10	ROUNDED BIOCLASTS IN MICRITE, TEXTURAL INVERSION

Fig. III-1 (continued)

Restricted Marine Shoals (Facies Belt 7 and 8)

SMF-16 Pelsparite (Flügel, 1972) or peloidal grainstone (Plates IX A, XI B, XIII A)
This consists of what are probably hardened fecal pellets, in places admixed with concentrated ostracod tests or foraminifera. The peloids are derived from organic pelleting of mud and may represent only very slight water movement. Such sediment may grade into peloidal wackestone (Type 19) and is common on tidal flats and natural levees where it contains cm thick graded laminae and fenestral fabric. This is typical Loferite sediment of Fischer (1964).

SMF-17 Grapestone pelsparite or grainstone (Plates XIII A, XXX C)
This is a mixed facies of isolated peloids, agglutinated peloids, some coated particles and lumps which are in part small intraclasts. These are the bahamite type particles of Beales (1958). The facies is formed in very warm, shallow water with only moderate circulation.

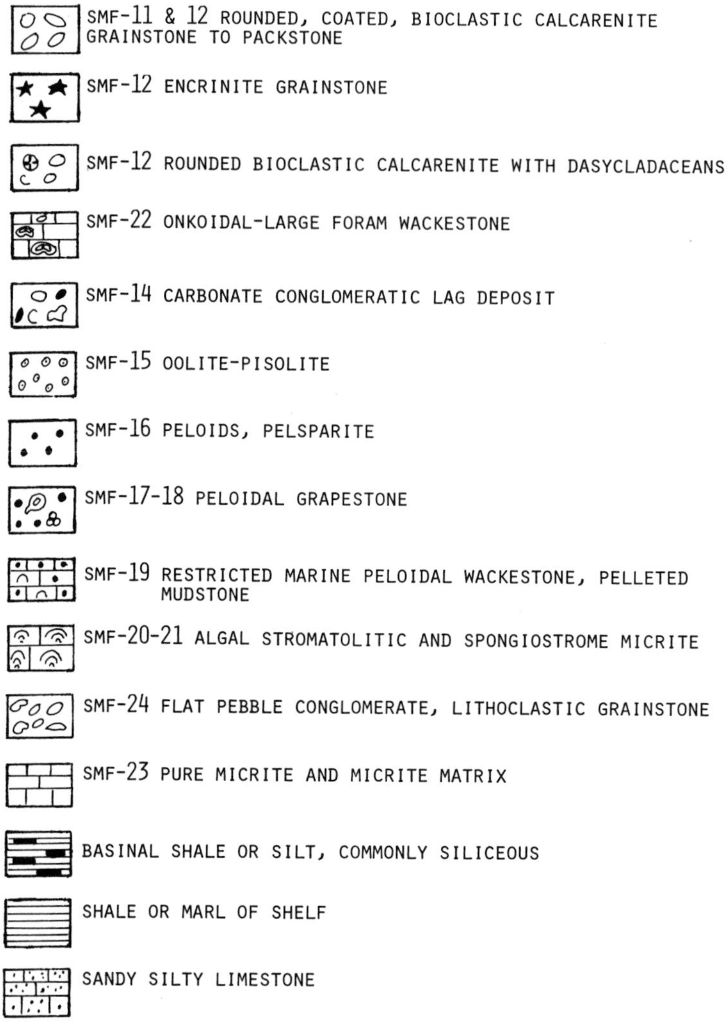

SMF-11 & 12 ROUNDED, COATED, BIOCLASTIC CALCARENITE GRAINSTONE TO PACKSTONE

SMF-12 ENCRINITE GRAINSTONE

SMF-12 ROUNDED BIOCLASTIC CALCARENITE WITH DASYCLADACEANS

SMF-22 ONKOIDAL-LARGE FORAM WACKESTONE

SMF-14 CARBONATE CONGLOMERATIC LAG DEPOSIT

SMF-15 OOLITE-PISOLITE

SMF-16 PELOIDS, PELSPARITE

SMF-17-18 PELOIDAL GRAPESTONE

SMF-19 RESTRICTED MARINE PELOIDAL WACKESTONE, PELLETED MUDSTONE

SMF-20-21 ALGAL STROMATOLITIC AND SPONGIOSTROME MICRITE

SMF-24 FLAT PEBBLE CONGLOMERATE, LITHOCLASTIC GRAINSTONE

SMF-23 PURE MICRITE AND MICRITE MATRIX

BASINAL SHALE OR SILT, COMMONLY SILICEOUS

SHALE OR MARL OF SHELF

SANDY SILTY LIMESTONE

Fig. III-1 (continued)

SMF-18 Foraminifera or dasycladacean grainstones (Plates VI, XIII B, XXX B–D)
        This sediment occurs as concentration of tests with peloids commonly in tidal bars
        and channels of lagoons.

Restricted Marine Shelf Lagoons, Protected Environment (Facies Belts 7 and 8)

SMF-19 Laminated to bioturbated pelleted lime mudstone-wackestone grading occasionally
        into pelsparite with fenestral fabric, Loferite (Plate XII B)
        An ostracod-peloid assemblage is common within these mudstones. Micrite with
        scattered foraminifera, gastropods and algae also occur. This biota represents depo-
        sition in very restricted bays and ponds.
SMF-20 Algal stromatolite mudstone (Flügel, 1972) (Plate XIV)
        Dense and closely-spaced growth laminations swelling over protuberances charac-
        terize this sediment. Fine lime mud is preferentially trapped on the highest areas

resulting in a lamination contrary to gravity. Such stromatolitic structure is commonest in the intertidal zone.

SMF-21 Spongiostrome mudstone fabric (Plates XI A, XIV)
Tuffeted algal fabric is preserved in fine lime mud sediment in tidal ponds.

SMF-22 Micrite with large onkoids (Plate XV A)
Wackestone or floatstone texture. This is a quiet-water sediment with algal balls composed of light organic matter which is later calcified or which traps fine detrital lime mud. They are typical of shallow water backreef environment, found typically on edges of ponds or channels.

SMF-23 Unlaminated homogenous unfossiliferous pure micrite (Plate XV C)
This lime mudstone is mainly deposited in somewhat saline or evaporative tidal ponds. Selenite crystals may form at random or in rosettes in this sediment.

SMF-24 Coarse lithoclastic-bioclastic rudstone or floatstone (Plate XVI)
The clasts are generally of unfossiliferous micrite or calcisiltite, the matrix variable but sparse. Crossbedding and edgewise arrangement of pebbles may occur. The sediment is normally termed intraformational limestone pebble conglomerate. It is formed as a lag deposit in tidal channels.

Diagenetic Changes in Carbonate Sediments

Chapter I mentions the critical importance of diagenesis in interpreting carbonate facies and the extensive research being conducted in this field. In view of Bathurst's recent excellent synthesis (1971) of this important subject, there follows only an outline of the many diagenetic effects which can be recognized. Since many of these effects can be confused with original depositional grains and fabric, any petrographic study of carbonate in thin sections cannot avoid consideration of diagenesis. The alteration of carbonate sedimentary particles and matrix is a continuing process. It begins during deposition and continues long after burial and the first stages of lithification.

1. Diagenesis in shallow marine water.
   a) In lime muds:
   In Holocene fine-grain carbonate sediments there is little or no carbonate mineral alteration but only slight compaction and fabric rearrangement as the soupy mud settles on the sea bottom. The change in density is from about 80% water saturation in the upper 20 cm to about 50% water saturated pore space after the bottom ooze has settled. As far down as measurements have been taken in such mud (7–8 m) no additional compaction has been observed. Decay of entombed organic matter in relatively impermeable water-saturated mud results in a reducing environment. Apparently little more has happened to carbonate muds produced and left in the marine environment during the last 5000 years. Thin sections of ancient limestones show that no additional compaction normally occurs after burial in the subsurface.
   b) In lime sands:
   (1) Void-filling cement evenly surrounding sand grains may be precipitated as fibrous aragonite or high Mg calcite grain-skin or isopachous cement in the form of tiny palisadelike crystals.
   (2) Beach rock is formed as lime sands in the splash zone are cemented by aragonite and Mg calcite, the same minerals present in wholly marine cements.
   (3) Aragonite crystal growth hardens pellets and cements grapestone lumps in areas of moderate water circulation. Even ooid formation may be viewed as a process of submarine cementation by the plastering of aragonite needles, tangential to a nucleus moving in marine water.
   (4) Particle surface alteration (micrite rinds). The process is one of infilling with micritic carbonate, of tiny algal or fungi borings on lime sand grains. The process goes on

centripetally within the grain and may completely convert it to micrite. The process is not fully understood; it is probably biochemical and both aragonite and high Mg calcite minerals are known to participate in it.

c) Reef cavities contain linings and fillings of both aragonite and high Mg calcite down to depths of at least 70 m. Layers of coarse druse as well as micritic linings are known. To what extent such carbonate is organic, inorganic or fully marine is now debated.

d) Cementation of sands just below sediment-marine water interface and creation of hard grounds has been extensively described both from ancient and recent deposits (Purser, 1969). Many of these instances must be completely marine and represent stillstands in deposition.

In the Recent Persian Gulf marine hard grounds (Shinn, 1969) the processes of aragonite and Mg calcite cementation, mutual mineralogical replacements, and detrital infilling by mud may proceed from the surface to at least half a meter below the surface.

2. Diagenesis in deep marine water.

a) Bathurst lists (1971, Table XVI, p. 376) fifteen occurrences of modern to Tertiary cemented globigerinid oozes from the deep sea (from 90 to 3300 m). In most of these samples the mineral is high Mg calcite but some low Mg calcite is also present.

b) Red nodular limestone with encrusting Mn and Fe minerals. A process of aragonite and/or calcite subsolution on deep sea bottoms is proposed for creation of these nodules (Garrison and Fischer, 1969). They are common sediment deposited as bathyal deposits on geosynclinal swells in many Mediterranean-Alpine Jurassic strata. Partial aragonitic solution and recementation in deep water has also been described from the Holocene.

3. Diagenesis by meteoric water in the vadose and phreatic (water table) zones.

a) Solution and void-creating mechanisms are as follows:

(1) Solution of aragonite grains above the water table or solution of anhydrite replacements of carbonate grains occurs.

(2) Collapse brecciation occurs if much solution takes place.

(3) In lime muds some flowage of material occurs to form openings such as stromatactoid structures.

(4) In peloidal muds fenestral fabric (birdseye structure) occurs through gas escaping from organic decay concomitant with desiccation.

(5) Solution compaction of lime sands occurs when fresh water percolates through and dissolves grains in uncemented, loose calcarenites.

b) Void-filling cement types (all low Mg calcite) are varied:

(1) In lime sands: needle fiber (jackstraw) cement, meniscus cement, pendent or microstalactitic cement, coarse to fine blocky (pervasive in phreatic zone), geopetal fillings of vadose silt, rim cement on echinoderms.

(2) In lime muds: alteration to low Mg calcite and growth of blocky calcite crystals from aragonite needles. Preservation of some of the original porosity may occur. There is generally a decrease from 50% (normal original packing porosity) to 25 or 30% when chalky textured rock is formed.

(3) Calcite veins, fissure fillings following joint patterns or lines of weakness in collapse breccias.

c) Calichification: development of fine granular perhaps microsparry micrite, often vaguely pelleted or grumelous (clotted) and with wavy lamellar structure or with concretionary pisoids. Micritized grains and faint remnants of incorporated bioclasts are seen. Circumgranular cracking forms breccias of various sizes. Also coarse blocky calcite patches or poikilitic crystals occur. Color splotches exist around root hairs and casts and needle-fibre cement perhaps follows outlines of fungi.

d) Mineralogical changes and metasomatic replacements:

(1) The exsolution of Mg from calcite lattices.

(2) Formation of siliceous nodules at old water tables.

(3) Dolomite crusts formed by replacement of aragonitic mud at subaerial exposure surfaces through processes of evaporation and capillary attraction.

(4) Dedolomitization by $CaSO_4$-enriched meteoric water which percolates through sediment above water table.

4. Diagenesis by brines in near-surface evaporitic areas.
   a) Processes which preserve void space.
      (1) Creation of chalky texture by lack of cementation during lithification. The process probably involves the transformation, over a long time period, of aragonite to blocky calcite in brines with a low calcium content. Such brines might occur directly below or down hydrologic gradient from evaporite deposits.
      (2) Enlargement of void space through dissolution and the consequent reorientation and preservation of porosity by solution and reprecipitation of local CO_3 during dolomitization. This results in the well known sucrose texture of dolomite (Murray, 1960).
   b) Void-filling by gypsum-anhydrite and metasomatic replacement of calcite by sulfate minerals.
   c) Dolomitization in slightly lithified aragonitic sediments due to shallow reflux.
5. Deep subsurface connate waters (poorly understood).
   a) Pressure solution after deep burial. Stylolites, grain collapse under load due to solution (deformed ooids); possible anhydrite solution.
   b) Cementation processes, difficult to distinguish from those of phreatic zone diagenesis.
      (1) Rim cementation on echinoderm particles, a final filling of intergranular voids (second generation cements).
      (2) Final cementation of lime muds.
      (3) Some vein-filling cementation.
   c) Perhaps also extensive post-lithification dolomitization. In such fabrics no evidence occurs that permeability control on dolomitization existed.
   d) Neomorphic calcite-microspar development.
   e) Replacements by anhydrite and subsequent solution.

Biological Observations

Morphology of Sessile Benthos

In addition to the observations noted during description of basic microfacies, there are some purely biological considerations which aid in environmental interpretation. One of these is careful attention to morphology of rooted bottom dwelling organisms whose life styles and shapes are adapted closely to current activity and sediment accumulation. Such organisms may develop extraordinarily robust forms in rough water. In slight or moderate currents their architecture is adapted for presenting maximum surface area to the water for feeding and respiratory efficiency. Other forms have adopted shapes for preventing accumulation of sediment on the feeding surface.

The following are common growth forms found among such organisms as corals, stromatoporoids, bryozoans, and sponges. Some suggested ecologic interpretations are given.

1. Massive and irregular in areas of high wave energy: Some stromatoporoids, corals such as *Microsolena*.
2. Wavy, erect to encrusting; in areas of high wave energy, but perhaps in more protected places: *Halysites, Agaricia, Millepora*.
3. Branched palmate, elongate in downcurrent direction in areas of high wave energy: *Acropora*.
4. Tabular or irregularly lamellar, in places encrusting and binding fine sediment; in deeper quieter water, wide surface area presented for maximum exposure to water and light: Tabular stromatoporoids, tubular foraminifera, *Tubiphytes, Alveolites*, sheety *Montastrea*.

5. Bulbous-cabbage heads. Rounded shape insures strong architecture but also in turbid water prevents sediment from accumulating on the surface; generally found in quiet water either deeper (below wave base) or in shallow protected back reef areas: Bulbous stromatoporoids, *Diploria, Montastrea.*
6. Basket-shaped; presents double surface area in moderate to slight current: Sponges, *Favosites.*
7. Elongate, erect, tall, and slender. Rapid upward growth as a preventative to swamping by strong sedimentation: Rudists, some deep water modern corals, *Dendrophyllia.*
8. Flexible dendroid, segmented forms. Found in areas of moderate currents and wavy motion. The fenestrate form prevents sediment accumulation and presents large surface area. Associated with bulbous "cabbage head" forms: Crinoids, alcyonarian sea fans, gorgonians, fenestrate bryozoans, sphinctozoan sponges.
9. Rigid dendroid and fasciculate. Found in protected areas in all parts of reef areas. Common in quieter water and on muddy bottoms except for some forms which, although brittle, may be fast-growing: *Acropora cervicornis, Stachyoides* (Devonian), *Thecosmilia, Oculina* (Tertiary-Recent), *Montlivaltia* (Mesozoic), some sponges, disphyllids, *Syringopora* (Late Paleozoic).

Table III-1. Environments for major groups of calcareous marine algae (from Ginsburg et al., 1971)

Algal type and growth habit	Salinity, temperature	Less than average depth	Maximum depth
Coralline Red Algae:			
Massive, and dendroid, Rigid and articulated. Nodules on seafloor. Reef builder and reef fringe dweller	Normal marine. Open marine shelf and bays	<25 m. Very shallow, range variable	200–250 m
Codiacean:			
Encrusting or with leaves erect and articulated. Only calcified cortex preserved.	Warm, shallow marine water of slightly variable salinity. Do not require strong circulation	<10 m	100 m
Dasycladacean			
Erect, articulated segments. Radiating pores which produce calcispheres (fruiting cases).	Warm, shallow, variable salinity, up to 50–60‰	3–5 m just below low tide level	12–15 m
Chara oogonia			
Calcareous fruiting cases. Large calcispheres	Fresh water, but may wash into coastal marine and brackish water	Very shallow	< 10 m
Blue green-cyanophytes			
Irregular tiny tubular bundles forming mats, massive erect domes, dendroid forms, tufts depending on the water energy	Highly variable. Fresh to hypersaline	Intertidal	45 m?

Figure III-2 illustrates the effect of water depth and wave energy on the morphology of corals and hydrozoans in three parts of the geologic record. Several assumptions are made: (1) Morphology has the same environmental significance in hexacorals, rugose corals, and stromatoporoids. (2) Morphology is basically controlled by water movement because this in turn controls such vital functions as food gathering, mud cleaning, respiration, and lime secretion. (3) Water movement is basically depth dependent. The diagram could be an oversimplification because it does not emphasize various other interrelated controls on morphology, e.g., currents, degree of light penetration, symbiosis, amount of fine sediment influx. Even if the analysis is too simple the general sequence of morphologic types seen in the geologic record permits a useful generalization.

Calcareous Algae

Another series of biological observations useful to environmental interpretation can be made on calcareous algae. Although these forms constitute only 5–6% of

FORM AND ENVIRONMENT	HOLOCENE	LATE JURASSIC	LATE DEVONIAN	
REEF FLAT	LITHOTHAMNION CRUST	PTYCHOCHAETETES RED ALGA CRUSTS	SOLENOPORA, PARACHAETETES RED ALGA CRUSTS, NODULES	
WAVES CRASHING. LESS THAN 1 METER; PALMATE ORIENTED BRANCHES, BATTLEMENTS	ACROPORA PALMATA AGARACIA MILLEPORA (DEEPER)	↑	↑	
WAVE BASE TO 10 METER; MASSIVE AND IRREGULAR, INTERSTITIAL LIME SAND AND SOME MUD	↑	ENCRUSTING HYDROZOANS AND MICROSOLENA STYLINA	LARGE STROMATOPOROIDS AND FINGER-LIKE FORMS STACHYOIDES ALVEOLITES (TABULATE) THAMNOPORA (TABULATE FINGER CORAL) PHILLIPSASTREA (MASSIVE RUGOSE) MUCH FORESLOPE AND FLANK INTERSTITIAL DEBRIS	
MORE THAN 10 METERS; DEPTH APPLIES ONLY TO WINDWARD SIDE OF BARRIER. MASSIVE KNOBBY AND DENDROID, FINE SAND AND INTERNAL MUD FILLINGS BETWEEN HEADS	SEAFAN DIPLORIA, MONTASTREA KNOBBY PORITES THICKETS OF ACROPORA CERVICORNIS	CALAMOPHYLLIA IN MICRITE 5 METER HIGH LATOMEANDRA	DISPHYLLIDS IN MICRITE	TABULAR STROMATOPOROIDS IN MICRITE
BELOW 30 METERS; DARK WATER; SHEETY FORMS. DEPTH APPLIES ONLY WINDWARD	SHEETY MONTASTREA	THAMNASTERIA GONIOCORA CLADOPHYLLIA	2 TO 3 METERS HIGH	

Fig. III-2. Comparative morphology in genera of corals, red algae, and stromatoporoids controlled by depth and water energy. This admittedly over-simplified sequence may be modified by varying amounts of suspended fine clay and organic particles and by biological evolution

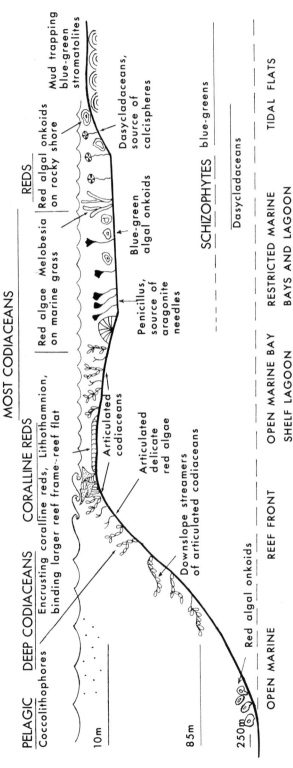

Fig. III-3. Ecology of calcareous marine algae; depositional environments along an idealized profile of a carbonate shelf margin

all algal genera, they have great importance in the carbonate regime of clear tropical marine water. Simpler forms are probably essential to carbonate production from seawater. Furthermore, algae which secrete preservable calcareous skeletons can be divided into several major groups, each with its own ecological character. A series of descriptive work resulting from a lifetime of study by Johnson (1957, 1961, 1967) is available in the American literature and furnishes pertinent information on the taxonomy of calcareous algae and their environmental interpretation. See also Horowitz and Potter (1971, p.75–80) for key to identification and for references in European literature to works by Pia, Ott, Elliot and Maslov. The most comprehensive review of the geological application of these forms is by Ginsburg et al. (1971) from whose "*Notes for a Short Course*" the Fig. III-3 and Table III-1 are derived.

Glossary of Sedimentary Structures in Carbonate Rocks

About 80 commonly recognized sedimentary and organic structures in carbonate strata are described below. They are grouped according to the standard facies belts which occur along a typical shelf-margin profile, but numbered consecutively. A few of the smaller structures are discussed in this Chapter also under microfacies. The references appended to most of the categories are not complete but were chosen for availability and for good illustrations.

Slope and Basin (Standard Facies Belts 1, 3, and 4)

1. Flysch-bedding
Rhythmically monotonous marine sequences composed of thin (10 cm–50 cm) resistant limestones interbedded with somewhat thinner marls or calcareous shales. The limestones are poorly fossiliferous, and graded when composed of calcisilt or fine calcarenite (Thomson and Thomasson, 1969, p.69; Wilson, 1969, p.8; and AGI Glossary, 1973, p.270) (Plate XXVII B).

2. Bouma sequences
Characteristic and ideal successions of five intervals making up one complete sequence of a turbidite. One or more of the intervals may be missing. The intervals from the top: (E) pelitic; (D) upper parallel laminations; (C) current ripple laminations; (B) lower parallel laminations; and (A) graded coarse sediment (Bouma, 1962; Thomson and Thomasson, 1969).

3. Mudflows with exotic blocks — debris flows
Boulders and blocks up to 10 feet and larger in diameter surrounded by mud flow material, i.e., much finer matrix (Thomson and Thomasson, 1969, p.65).

4. Flute casts
A sole mark, a raised subconical structure, the up-current end of which is rounded or bulbous, the opposite end flaring out and merging with the bedding plane. Formed by filling of an erosional scour or flute (Pettijohn and Potter, 1964, p.307).

5. Load casts

A sole mark, bulbous, mammillary, or papilli forms which are downward protrusions of sand produced by load deformation in underlying hydroplastic mud; due to yielding under unequal load (Pettijohn and Potter, 1964, p. 318; Kuenen, 1953, p. 1048 and 1058). Termed "flow cast" by Shrock (1948, p. 156).

6. Groove casts and tool markings

Sole marks, rounded or sharp-crested rectilinear ridges produced by filling of grooves (Pettijohn and Potter, 1964, p. 311; Shrock, 1948, p. 162).

The grooves may be produced by engraving tools such as shells, sand grains, pebbles, and logs swept over firm lutite bottoms by currents. The marks are preserved as casts on bases of overlying beds (Dzulymski and Sanders, 1959; Pettijohn and Potter, 1964, p. 348).

7. Conglomerate channel fills

Bodies of carbonate, chert, and sandstone pebble clasts which have flattish tops and shallow convex bases. These may interrupt the normal thin and planar bedded basinal and slope strata (Young, 1970, p. 2304–2305).

8. Glide surfaces

Observed within uniform and thinly bedded lime mudstones and calcisiltites as major discontinuities in sequence, formed probably as a result of large-scale slippage or slumping of strata without much internal deformation. Such structures may be of great lateral extent, masses of several hundred feet long apparently having slid across the stable part of an exactly similar section. Contacts are sharp. J. Harms (personal communication) believes that some of these could be shallow channels cut in fine-grained sediment by dense brines and penecontemporaneously filled in by the fine sediment (Garrison and Fischer, 1969, p. 38; Wilson, 1969, p. 9–11).

9. Soft sediment slumps

These are rarer in carbonates than in clastic sediments (Plate XXVII). *Convolute bedding:* wavy or contorted laminations that diminish upward and downward within a given sedimentational unit (Kuenen, 1952, p. 31 and 1953, p. 1056; Pettijohn and Potter, 1964, p. 292). *Flame structures:* the mud plumes separating the downward bulging load pockets or load casts of sand and sand-shale interface. Also described as "streaked-out ripples" (Pettijohn and Potter, 1964, p. 305).

10. Mn-Fe crusts and nodules

Carbonate lithification on the present-day sea floor mainly of globigerinid oozes, commonly with an admixture of benthonic skeletal matter occurs at depths ranging from about 200–3500 m. In places there exists a superficial crust overlying essentially unconsolidated sediments. Some of these crusts, nodules, and coatings are rich in manganese and iron oxide resulting from continuous solution and reprecipitation processes within the chemical gradient that exists near the sediment-water interface (Price, 1967; Garrison and Fischer, 1969; Morgenstein, 1973).

11. Subsolution forming *in situ* conglomerates (Plate XXIX B)

A solution rubble of limestone crusts formed *in situ* on the sea floor and perhaps in places moved slightly downslope. Some such blocks or pebbles are bounded with a ferruginous rind while others are not; matrix contains higher proportion of microscopic skeletal material (ostracods, foraminifers, etc.). Stylolites commonly occur separating pebbles (Hollman, 1962, Garrison and Fischer, 1969, p. 26–27).

12. Resedimented clasts and retextured sediments

Beds in which slumped or otherwise dislocated complexes of rock occur. Original stratification, or traces of it, may still be recognizable; locally, a slumped mass may consist of a chaotic melange of large tilted blocks with no internal deformation (see No. 3 in this list). In other cases, beds are deformed plastically and pass laterally into retextured sediments. The interstices between blocks and slump folds are filled with slump rubble (Bernoulli and Jenkyns, 1970, p. 512).

13. Current ripple mark or small scale cross-lamination

A common synonym for transverse asymmetrical ripple mark (Pettijohn and Potter, 1964, p. 297). Ripple mark formed with ridges transverse to current. Periodic undulations of primary origin at the interface between water and granular material on the sea floor cause such ripples. They are usually on a small scale and are basically current-formed, below wave base (Shrock, 1948, p. 99–113; Pettijohn and Potter, 1964, p. 333).

14. Mud mounds on slope

Mounds of skeletal lime wackestone or relatively pure lime mudstone usually enclosed by beds of dark laminated sandy beds, or shales (Dunham, 1972, p. III-19, III-20).

15. Planar-bedded lime mudstone with even mm laminae (Plate XII A)

Common within thin beds of lime mudstone and reflected as color banding caused by concentrations of organic matter, iron, or minor amounts of clay (Wilson, 1969, p. 7). Where interbedded with $CaSO_4$ (see below) perhaps bacterial reduction of the SO_4 augmented by associated lime precipitation will result in dark fetid lime muds (Friedman, 1972).

16. Evaporites in basins with mm laminites of carbonate

Dark finely and planar laminate sediment, commonly interstratified carbonate and anhydrite or even pure anhydrite resulting from thorough replacement of the carbonate. May be minutely interstratified with thin graded carbonate beds (Davies and Ludlam, 1973).

Shelf Margin and Shallow Shelf in front and behind Shelf Margin
(Standard Facies Belts 2, 5–7)

17. Massive to medium-bedded strata, irregularly interbedded

Beds of varying thickness from homogenous thick massive strata to thinner beds (less than 50 cm), commonly with shaly intervals. Common in shelf strata (McKee and Weir, 1953; Pettijohn and Potter, 1964, p. 321).

18. Flat lens-shaped beds with shale partings

Subplanar beds which taper gradually over distances of many tens of meters, as opposed to regular planar beds. Also typical of shelf deposits.

19. Nodular to wavy bedding

Layers which consist of undulose beds grading to loosely-packed nodular bodies of rock in matrix of like or unlike character (commonly shale); also termed "lumpy bedding" (Pettijohn and Potter, 1964, p. 324, pl. 15b, pl. 16b and c).

20. Ball and flow, sedimentary boudinage, or flaser bedding

Differential compaction of originally patchily distributed shale and carbonate resulting in irregular, closely spaced, almost nodular structure caused by a disruption of layers by stretching and flowage. Original inhomogenities in the fabric may result from burrowing, patches of shells, or pebbles on the sea floor. The argillaceous material compacts regularly while the calcareous material cements early and resists compaction (McCrossan, 1958, p. 3161; Pettijohn and Potter, 1964, p. 287, pl. 16a). Similar to pull-apart structures (Natland and Kuenen, 1951, p. 89).

21. Local patch reefs

Recognized as small subequidimensional or irregularly shaped, flat-topped organic buildups, less extensive than "platform reefs". Mostly dominated by frame-building organisms. These occur locally as small, thick, generally unbedded lenses of very fossiliferous carbonate, more or less isolated and surrounded by rocks of unlike facies (Henson, 1950, p. 222; AGI Glossary, 1973).

22. Foreset bedding or flank beds on slopes of knoll or patch reefs

Loosely used for any inclined bedding off flanks of carbonate buildups. Some may be steeply inclined beds (about 25° but up to 45°) developed on seaward slopes of major shelf margins (Stricklin et al., 1971).

23. Bioturbation and burrows

Thorough homogenization of sediment by burrowing organisms. Sinuous or straight burrow forms commonly preserved as traces through color mottling and textural differences, or outlined by dolomitization. Tubular openings made and refilled by worms, mollusks, echinoderms or other organisms (Pettijohn and Potter, 1964, p. 288; Seilacher, 1964 and 1967; Shinn, 1968b, p. 879; R.W. Frey in Perkins, 1971, p. 91–124).

24. Trace fossils

Tracks, trails, grazing traces, nests, etc., on bedding planes and within the substrate. For references see No. 23 above.

25. Ripple mark

This structure is varied in the shallow marine environment. In general ripples are either symmetric or asymmetric. (1) Symmetric marks have symmetric profile in cross section, and crest in plan view is predominantly straight. (2) Oscillation ripple marks are straight with sharp, narrow, relatively straight crests between broadly rounded troughs, formed by the orbital or to-and-fro motion of water agitated by oscillatory waves on sandy bottoms above wave base. (3) Asymmetric ripple marks are characterized by short and steep slopes facing down-current and more gentle slopes facing up-current. (4) Current ripple marks are asymmetric and formed by currents moving in a uniform direction over a sandy surface. They are common in shallow water but may also form in deeper water (see above) (AGI Glossary, 1973, p. 4, 44, 503, and 717 respectively).

26. Festoon, medium-scale cross bedding

Cross beds in sets up to $^1/_2$ m thick; caused by scour and subsequent fill by megaripples formed in moderately strong currents. Beds exhibiting this structure are generally very lenticular and often truncate lower beds. The structures are common in outer, seaward parts of shelves, downslope from beaches where long-shore currents exist (Thomson and Thomasson, 1969, p.62; Stricklin and Smith, 1973).

27. Onkoids in lime sand or in lime mud matrix (Plates IX B, XV A)

Balls with concentric lamellae, commonly irregular and crinkled; formed by coating of particles by algae, foraminifera, serpulids; where occurring in micritic matrix they are seen in widely distributed beds. Such particles must have formed in areas of moderate circulation but were deposited over broad areas of shallow, standing water. When seen in sandy matrix the balls may have been formed and deposited in channels or near shoals where water movement was swifter than normal (SMF-13 and SMF-22).

28. Local small lime mud mounds

Small, rounded, well-defined, algal-foraminiferal micrite accumulations (not more than 1–2 m in diameter). These occur generally in well-bedded limestone deposited in shallow marine water on shelves or on gentle slopes off shelf margins.

29. Calcareous concretion zones (commonly in marls)

Prevalent in shallow basinal shales equivalent to shelf carbonates. They are considered to be syngenetic or early diagenetic in origin, are formed within the substrata and represent a slowdown in the normal argillaceous sedimentation (Weeks, 1957, p.98; Waage, 1964, p.550).

Lime Mud Mounds on Shelves and Shelf Margin Slopes
(Standard Facies Belts 4 and 7)

Several of these sedimentary structures are discussed in more detail in later Chapters.

30. Collapse breccia and vein patterns within micrite core (Plate XXII B)

Thorough syngenetic brecciation, mud flowage, with superimposed later brecciation and development of vein calcite is common in lime mud buildups. It is probably caused by early dissolution, collapse, and slumping of the mixture of lime mud and fragile shelly components. This may have occurred through gravity-slumping under marine conditions or upon exposure to meteoric water when the mass was at or near the sea surface (Heckel and Cocke, 1969; Choquette and Traut, 1963).

31. Stromatactoid structure (Plate XVIII B)

Originally open-space sedimentary structures characterized by horizontal or nearly flat bottoms and by irregular, digitate upper surfaces. The original cavities are generally filled by internal micritic sediment or by sparry calcite in patterns indicating void-filling precipitation. The structure occurs only within lime mud buildups. The origin of the cavities is much discussed in the literature (see Chapter V) and may be due to slumping of partly lithified lime mud, settling of rigid platy shells in the mud, rotting of the form of some unknown organism (Philcox, 1963, p. 905; Heckel, 1972a, p. 12, 13).

32. Injection dikes

Beds cutting vertically through massive or normally bedded strata and filled by material squeezed up from below through loading of the substrate by the carbonate mass. May be somewhat irregular and anastomozing or straight and sharp-sided.

33. Filled fissures, Neptunian dikes

Discordant sediment filling fissures which cut across massive to normally bedded strata; sediment filled from above and generally composed of sand or coarser material. The fissures may have been opened by slumping or tectonic activity at an early stage in the geologic history of the mound (Shrock, 1948, p. 212; Fischer, 1964).

34. Geopetal sediment fillings in mound cavities (Plate XXV B)

Structures that indicate the relations of top to bottom during, or shortly after, sedimentation. Various types of sediment fillings may occur in bottoms of cavities. Color changes, textural differences, and cross lamination generally mark these internal sediments which may also occur in several generations. Spar-filling in upper parts of cavities, only partly filled with sediment, indicates top. Vadose silt of Dunham (1969) is a special type of internal sediment resulting from early breakdown and deposition of partly lithified lime mud, presumably under conditions of meteoric water flow (Sander, 1936, p. 31, 1951, p. 2; Shrock, 1948, p. 4).

Disconformities—Surfaces of Non-Deposition or Very Slow Deposition Marking "Stratigraphic Breaks" (Standard Facies Belts 5–8)

35. Planar corrasion zones

Abrasion and truncation of surfaces by marine currents, meteoric water solution of organic activity. The structure may be indistinguishable from some "hard grounds". The processes of formation are partly the same.

36. Hard grounds or bored surfaces

Organically bored marine or littoral surfaces encrusted commonly with fossils in growth position. The sediment was syngenetically lithified by marine micritization or void space cementation. Such surfaces are encountered at tops of regressive carbonate sequences (Hallam, 1969, p. 240; Purser, 1969, p. 217; Shinn, 1969, p. 109).

37. Oyster plasters and other encrusters

Oyster-encrusted surfaces are commonly associated with "hard grounds", typically within Mesozoic and younger sediments (Klüpfel, 1917, pl. III).

38. Oxidation films

Red-stained zones due to thorough oxidation of iron in the top of the substrate beneath a disconformity surface. Commonly but not exclusively due to meteoric water (Krumbein, 1942).

39. Small serpulid mounds

Commonly cup-shaped patches with raised rims, actually colonies of serpulid worms. Size may vary from that of a bread loaf up to several m across. Serpulids may thrive in very shallow marine or hypersaline, very warm water, covering wide shallow areas on hard bottoms of very shallow subtidal and intertidal zones.

40. Blackened pebbles

Dark-colored, brecciated limestone and isolated blackened particles may indicate subaerial exposure in salina areas filled periodically with hypersaline water. Algal mats flourish here during high-water periods but are intermittently exposed to times of extensive decay on the edges of ponds which dry out completely. The black color of the bed rock is derived from decaying organic matter, chiefly from microfilamentous boring algae which thrive in the mats and residual "stewy" water and thoroughly impregnate the underlying substrate. The reworking of this brecciated material into overlying beds is common (Ward et al., 1970, p.549; Rose, 1972, p.98).

41. Lag deposition, glauconite and phosphatic bone beds (Plate IX C)

The accumulation of chemically and physically resistent coarser particles in thin beds formed through long periods of non-deposition and persistent winnowing of the sediment on the sea floor. Lithoclasts and blackened coated particles may also mark such deposits along with fish bones and teeth, phosphatized organic remains, glauconite and quartz sand grains (Krumbein, 1942; see discussion of SMF-14).

42. Brecciation of lithified bed rock

Polygonal cracking and shrinkage causing brecciation of hardened or semi-lithified bed rock owing to temperature changes and force of crystal growth within sediment (Ward et al., 1970).

43. Karstic collapse

Collapse of limestone in sink holes, fissures, and irregular caverns may be common as much as tens of meters beneath unconformities. These represent fillings of caves and various underground drainages formed by solution. Dolomitization is known to follow some of these pathways (Shrock, 1948, p.66–69).

Intertidal Environments, Tide Flats, Channels, Levees, Ponds, Beaches
(Standard Facies Belt 8)

These many subenvironments have been recognized and closely studied. Manifold sedimentary structures are described in the Symposium Tidal Flat Deposits, 1973, Comparative Sedimentology Laboratory, Miami University.

44. Homogeneous and unusually thick storm layers

Such widespread, commonly unfossiliferous and well-marked layers result from major storms and are deposited when rising tides and high winds mobilize mud on bottoms of normally quiescent shallow ponds and lakes and deposit single thick layers on shoal areas (Shinn et al., 1967, p.583; Shinn et al., 1969, Fig.22).

45. Algal mat mm laminites forming stromatolites (Plate XIV)

Mats consisting of various assemblages of mostly blue-green algae and diatoms provide a semiresistant open network into which lime mud and sand grains infiltrate and are trapped. The mucilaginous secretions of the algae and diatoms in association with fine filaments act as binding agents. No trace of the calcareous internal organic structure remains except "*Girvanella* tubules", but fine lamination is preserved reflecting the intermittent mud accumulation (Davies in Logan et al., 1970, p. 169–205; Kendall and Skipwith, 1968; Neumann et al., 1970, p. 285).

46. Spongiostrome fabric in stromatolites (Plate XI A)

Certain modern algal growths (e.g., *Scytonema*) resulting in tiny bush-tufted mats whose fabric is calcified and preserved as irregularly anastomosing filamentous structure (Johnson, 1961, Pl. 103; Davies in Logan et al., 1970, p. 190).

47. Bulbous algal stromatolites (Plate XIV A)

Includes the typical "*Collenia* or *Cryptozoan*" forms of stromatolites, laterally linked hemispheroids. The bulbous form results from sustained energetic but intermittent algal growth, probably over original protrusions of substrate such as arched mud-cracked polygons. Tidal ranges above half a meter or so are necessary (Logan et al., 1964; Kendall and Skipwith, 1968; Davies in Logan et al., 1970, Fig. 20 and Fig. 22).

48. Fenestral laminate fabric (Loferites, birdseye limestone) (Plates XII B, XIII A, XIX A)

Laminate lime mudstone or dolomite riddled with parallel aligned pores; larger ones with stromatactoid shapes; in places with geopetal fills of internal sediment. Sediment is commonly pelleted. The structure results from desiccation and is found best developed in areas of tidal flats where persistent inundation is combined with maximum exposure (Fischer, 1964, p. 124; Tebbutt et al., 1965).

49. Small scale (cm thick) layers of micrograded pelleted sediment (Plates II B, XI, XII)

Characterized by a clastic texture and normal "fining-upward" size grading of tiny clasts and peloids. Such cm graded beds commonly occur on natural levees and slightly higher areas of better drainage on the flats. The cm grading of sand-size grains may be capped by a micrite layer which could represent deposited lime mud or coarser sediment whose texture was obliterated by micritization or caliche processes (Shinn et al., 1969; Wood and Wolfe, 1969).

50. Mud balls and soft pebbles

These are semi-hardened globs formed by erosion of pre-existing lime mud or balled-up sediment by digging arthropods. They may be rounded through rolling by gentle current action and may form armored mud balls when shells or twigs adhere to them (AGI Glossary, 1973, p. 130).

51. Edgewise conglomerate (Plate XVI)

Flat pebble conglomerate, imbricate lithoclasts accumulated in places in fan-like, swirly patterns due to strong currents. They fill channels and pot holes. The clasts are mainly derived from the dried-out lime mud on levees and algal mats, the chips being abraded, moulded and redeposited in areas of current on floors of tidal channels (Shinn et al., 1969).

52. Lenses of cross-bedded lithoclastic-bioclastic calcarenite

Occurring isolated in finer-grained, planar-bedded sediments and representing channels cut down in tidal flats and filled with coarse sediment. Such sediment occurs often in intrachannel bars (Shinn et al., 1969).

53. Rill marks

Narrow, deep grooves formed when small dendritic channels bifurcate upstream; commonly found in subaerial portions of beaches, sand bars, and sand flats (Shrock, 1948, p. 128; Pettijohn and Potter, 1964, p. 332).

54. Raindrop prints

Small, shallow, circular to elliptical, and vertical or slanting depressions or crater-like pits surrounded by a slightly raised rim, formed in soft and relatively fine sediment on tidal flats by impact of falling raindrops (Shrock, 1948, p. 141–146).

55. Crinkly laminae from plant rootlets

Rootlets penetrating both vertically and horizontally, shoving aside and lifting fine layers of sediment and imparting a characteristic wispy, wrinkled, fabric. The disturbance of normally laminated sediment by plant growth on tidal flats is common. In carbonate sediment the roots and roothairs may be preserved by precipitated sheaths (Shinn et al., 1969, Fig. 21).

56. Preserved roots and peat seams

Organic tubes, both calcified and carbonized, in carbonate tidal flat and marsh sediment. Coal and peats occur as remnants of extensive swamps at the shoreward edge of tidal flats in tropical climates (Shinn et al., 1969, Fig. 21 B).

57. Algal mat blisters from rain showers

Sudden rain soaking rubbery, algal mats causes them to become impermeable. Gases released within the mat through organic decay raises immediate blisters following rain. The blisters may be calcified and preserved *in situ* or dry up and peel off to form small circular clasts (Ginsburg, personal communication).

58. Peeled and curled mud chips

Thin films of lime mud and raised layers of algal mats when dried out are curled into flakes. These may be buried more or less *in situ* by rapidly deposited sediment or they may be redeposited in nearby channels and rills as intraclasts (lithoclasts) (Shinn et al., 1969).

59. Mud cracks

Positive expression of desiccation mud cracks appearing as polygonal forms on bedding planes, or cracks filled with later sediment and preserved as raised ridges (casts) on undersides of beds. In cross section, desiccation cracks are commonly seen as vertical v-shaped cracks (Pettijohn and Potter, 1964, p. 323).

60. Dolomite and limestone crusts

Hardened crusts present within subdivisions of the supratidal-intertidal zone where most frequent intermittent wetting and drying occurs; related to sedimentary highs such as levees and beach ridges, or encircling edges of ponds. Formation is due to evaporation and precipitation of carbonate minerals in concentrated brines in capillary zone at the surface. This is a favored place for dolomitization in tropical tidal flats (Shinn et al., 1965, p.118, 119; Shinn et al., 1969, p.1217).

61. Minor beach deposits

Thin beds of fine sand and shell hash with low angle cross bedding formed on beaches separating the edges of flats from large lagoons. In many places these beaches are exclusively of cerithid snail shells (Shinn et al., 1969, p.1212, Fig.15).

62. Burrows

Filled traces of organic burrowers, dominantly vertical, rather long, distinct, and isolated in sediment that is intertidal and partly lithified as contrasted with subtidal sediment that is thoroughly homogenized, horizontally burrowed with shallower penetration (Frey in Perkins, 1971).

63. Tracks and trails on bedding planes

Organic markings of land or amphibious animals on bedding planes usually indicate intermittent exposure in lime muds which are readily hardened only by subaerial exposure. Arthropods, snails, mollusks, and vertebrates all commonly leave trails and tracks on such surfaces (Perkins, 1971).

Supratidal (Sabkha) Evaporites and Related Carbonates of the Intertidal Zone (Standard Facies Belts 8 and 9 under a Rigid Evaporitic Climate)

64. Sheet cracks in limestone or dolomite

Planar cracks concordant to bedding attributed to shrinkage of sediment, perhaps due to drying out or to de-watering in an aqueous environment with varying brine concentration (Fischer, 1964, p.148).

65. Zebra limestone or dolomite

Rock banded by parallel sheet cracks filled with calcite or dolomite druse. Bands are at an angle to normal bedding. It is not known whether the steep angle represents traces of bulbous algal mats coating lime mud mounds or failure by slumping of piles of lime mud along lines normal to stress. Perhaps a shrinkage process causes the structure. Such oblique bands are known in sediment within fissures (Fischer, 1964, p.135).

66. Druse veins

Concordant and discordant veins of sparry calcite in the form of coarse druse or palisade crystals filling an irregular cavity or opening. The walls of the fissure are lined with projecting crystals usually of the same minerals as those of the host rock (Fischer, 1964; Dunham, 1972, p.II-3, II-47).

67. Tepee structures

Cross sections of very large desiccation polygons with prism cracks. In a cross section they appear as disharmonic folds, resembling chevrons or inverted, depressed Vs. The structures are obviously caused by expansion of the rock, presumably during hydration of anhydrite interbeds or failure of the limestone during expansion caused by crystal growth or extreme variations in temperature. Parallel beds below and above such features clearly show them to be syngenetic (Newell et al., 1953, p.126; Kendall, 1969, Fig.9; Assereto and Kendall, 1971; Smith, 1974c).

68. Caliche breccia and wavy laminites

Calcareous micritic crusts up to 1 or 2 m thick formed on surfaces of subaerial exposure by continuous solution and reprecipitation of $CaCO_3$ over long periods of time, usually in semi-arid climates where there are repeated long dry periods alternating with a wet season. The wavy laminae are convex downward, developing progressively at the base of the crust. The continuous wetting, drying-out, and calcite precipitation results in brecciation (Dunham, 1972, Fig. II-39. See also under Diagenesis, this Chapter).

69. Vadose pisolite (Plate XV B)

Large (up to several cm in diameter) concentrically laminated concretionary growths interpreted to form in limestones in the vadose zone. The balls are commonly bound together by sparry cement or encompassing laminae. Other evidence for *in situ* growth includes a tendency for the pisoids to form in downward elongation and to fit together. They commonly show reverse grading. Pisolites may cut across normal bedding but may also follow it. They apparently originate as fresh-water vadose caliche, or as "cave pearls", or perhaps also in the littoral splash zone (Dunham, 1969a; Dunham, 1972, II-27, II-40).

70. Evaporite solution breccia

Breccia with abundant vein calcite usually in widely distributed and thin beds with a sharp base and irregular top. Caused by collapse of carbonate layers due to solution of interlayered anhydrite (Blount and Moore, 1969; Dunham, 1972, p. II-51, 52).

71. Salt hoppers and selenite bladed crystals (Plate XV C)

Isolated molds and casts of the typical crystal forms of halite and gypsum may appear scattered within carbonate beds. The scattered slash-mark of selenite blades is most common. These mark solution voids or carbonate replacements of the soluble original minerals which earlier replaced carbonate sediment and grew in its void spaces on sabkhas (salt flats).

72. Ptygmatic or enterolithic structure in gypsum-anhydrite

Contorted or thin beds of calcium sulfate probably caused by *in situ* crystal growth and consequent volume increase under vertical confining pressure. The structure has also been considered to form by coalescence of nodules perhaps during compaction. That it is commonly an early diagenetic structure is evidenced by its presence in the upper meter of sediment in modern sabkhas of the Persian Gulf. Tiny nodules and fine enterolithic structure are also described in laminated evaporites considered basinal by Davies and Ludlam (1973, p. 3533). (For illustrations of evaporites considered intertidal, see Maiklem et al., 1969, Shearman and Fuller, 1969, Fig. 4).

73. Nodular, pearl, flaser-chicken wire anhydrite-gypsum

Nodular structures formed generally by compaction of various sizes and amounts of calcium sulfate nodules in sediment of varying permeability. Such structures are seen today forming on modern sabkhas but also result from compaction following burial (Shearman and Fuller, 1969, Fig. 6, pl. 2; Maiklem et al., 1969; Bebout and Maiklem, 1973).

Eolianites—Carbonate Dunes

General references are Ball, 1967b, Ward, 1970

74. Steep cross beds, dipping shoreward in large sets

Dips from 25 degrees to 45 degrees common, progressive migration of dunes downwind resulting in pronounced crossbedding directions shoreward but within an arc of 180 degrees. The preservation of large "sets" of cross beds, up to 15–20 m is possible.

75. Preserved dune forms

Upward convex forms formed by spillover lobes preserved in eolianites because of rapid cementation in the vadose zone in some climates and because of protection by an impermeable caliche crust.

76. "Ribbing" caused by differential cementation of layers

Alternating fine and coarse layers showing on weathering a characteristic fine ribbing. This is due to differential cementation because generally the finer sand sizes hold water by capillary attraction long enough for $CaCO_3$ cementation, whereas water drains from the coarser layers which remain loose and therefore weather deeper.

77. Red zones

Old soil zones represented by limonitic red or orange-weathering, brecciated, powdery and chalky beds containing chunks of cemented eolianite.

78. Snails and calcified insect cocoons

These occur in old soil zones. The snails are the only body fossils commonly preserved. Tiny egg-shaped cocoons occur rarely in these zones.

79. Root casts and root hairs

Rhizocretions are calcareous lined tubes or solid cylindrical forms resulting from calichification and cementation along roots. Even root hairs may be so preserved as tiny calcareous sheaths. The characteristic dense network and downward decrease in size of branches distinguish these from burrows which they superficially resemble. These masses occur commonly below old soil zones.

Environmental Analysis of a Carbonate Thin Section

A great amount of environmental information is available from just a single thin section, as illustrated by study of Plate I. The field is about 1.0 by 1.4 cm and magnification is X16, plane light. The sample is from the top of a $3^1/_2$ m thick bed of gently foreset strata forming a capping bed above strongly foreset dipping strata constituting biohermal flanking beds (Fig. XII-6). It is a variety of SMF-5. It is of Virgilian (Late Pennsylvanian) age and is from the Sacramento Mountains, north wall of Dry Canyon above New Mexico State highway 52, 3 miles northeast of Alamogordo, New Mexico. This single thin section indicates the water salinity, wave energy, coherence and oxidation of substrate sediment, how the sediment accumulated, and to some extent its later diagenetic history. Proper vertical orientation of the sample is also seen from its fabric.

Physico-Chemistry of Water Mass during Deposition
(as Derived from Ecological Interpretation of Biota)

The biota indicate warm tropical water of normal marine or perhaps slightly higher salinity for the following reasons: Echinoid spines and echinoderms indicate generally normal marine salinity; Foraminifera are abundant and varied; *Triticites* (fusulinid), *Paleotextularia*, *Globovalvulina* are large foraminifera and are presumably characteristic of warm open marine water; *Tuberatina*, *Calcitornella*, and *Archeodiscus* are all encrusting forms and appear to have grown on plant life or soft-bodied creatures—as well as on the abundant platy algae; platy codiacean algae today flourish in warm, tropical, sun-lit water generally less than 15 m deep; common encrusting blue-green algae, in corsortium with foraminifera, also indicate shallower photic zone water; ostracods—many scattered throughout, colonies of bryozoans in the form of hollow spheres embedded in micrite and introduced in the flanking beds from a quieter environment; worn, broken, gastropod fragments, encrusted inside, obviously introduced from outside the immediate environment.

Note certain characteristic bioclasts that are lacking: brachiopods, common bryozoans, crinoids, red algae, corals, common mollusks. The lack of, or rarity of, these forms indicates that the bioclastic material does not derive from organisms living in fully open marine conditions, but probably from water on a shelf, water somewhat warmer than open marine, nutrient depleted, and even slightly more saline than in the open sea.

Water Energy

Note the texture: a packstone with a wide range of bioclasts of all sizes, and lithoclasts of several shapes and sizes and obviously of different origins. The unsorted grains constitute about 50% of sediment volume. The matrix consists of clotted silt and fine sand-size carbonate with many tiny patches of coarser sparry calcite. Larger fragments such as algal plates and echinoid spines are rounded, coated, and micritized indicating considerable movement of the grains in water before burial. Note the recrystallization and the cement infilling the interior of algal plates. No cortical structure is preserved.

Smaller bioclasts are less damaged, protected from mechanical abrasion by surface tension effects. Lithoclasts comprise additional coarse fragments. The left center and center field of Plate I contains two rounded lumps about 3 mm across in actual size. Their matrix is micrite as indicated by the denser, darker color than the general matrix of the sediment. The lithoclasts contain bioclastic and pelletoidal material. These lumps may have been formed by crabs or other arthropods which "ball-up" excavated material. The larger lump contains a bryozoan colony in the shape of a hollow sphere. The lump may have been shaped around the bryozoan colony. The central lump was still soft when deposited. It is pressed flat against the algal plate above it. This is not a stylolitic contact formed through later pressure solution. About twenty dark micritic peloids 300–400 microns in diameter are scattered about within the lumps and in the general matrix. These are of previously well-consolidated lime mud and may be either fecal pellets or have formed as steinkerns (internal moulds) of ostracod tests. (Note the ostracod tests in the lower right corner.) However formed, these large clasts were probably torn out of the sediment and drifted around before redeposition, indicating some water movement.

Fabric Packing, Diagenesis, and Orientation

The coarse fragments are jumbled together and form a self-supporting pack. The pack may have been deposited rapidly with fine sand and silty-muddy sediment or more probably the finer matrix material may have drifted in later. In any event the finer matrix has altered to microspar. A large calcite area which was originally void space stands out in the lower right-hand corner. The centripetally enlarging calcite of this area and the rather frequent occurrence of enfacial angles (Bathurst, 1971) at contacts of calcite crystals argue strongly for a cement infilling. Various origins are possible for the void: it may be a cavity protected by the framework of algal plates, it may be a burrow, or the space earlier occupied by a soft-bodied organism which rotted away. Whatever its origin, sediment fell or was washed to its bottom, larger pieces following the smaller. This sediment probably came from roof collapse; its larger pieces are angular and the roof of the cavity is more jagged than the base. That this partial sediment in-fill occurred early in the lithification history is evidenced by isolated, loose foraminifera tests as pieces of the in-fill.

Several other lines of evidence indicate that micritic sediment filled in a packstone which was supported by the larger grains. For one, a settling effect is noticed. Preferentially one side of the larger, elongate grains contain calcite mosaic spar-filling which enlarges centripetally. This is void-filling calcite which was deposited in space caused by the finer intergranular sediment settling down in protected areas. That this type of early diagenesis occurs on a smaller scale and throughout the fabric is indicated by the light color of the irregular, almost clotted character of the general matrix which is a coarse microspar in size. The grains separated by microspar and constituting the general matrix are fine sand to silt size (50–100 microns).

This fabric is quite common in packstones with an appreciable amount of varied sized framework grains and affords a means of orienting the rock sample. In addition to the above described settling effect under grains, a cavity just left of the center of the photo contains floored sediment of the "vadose" crystal silt described by Dunham (1969b). The jumbled large grains show no preferred lineation but diagenetic crystal growth has been controlled by the original fabric, a fact that shows early cementation. It is not possible to ascertain whether such cementation was under the influence of meteoric water or was submarine. The surface of the thick ledge from which the sample comes shows some oxidation (red staining) and is at least a surface of nondeposition if not subaerial exposure.

To sum up: This sediment represents bioclastic trash which drifted around on the sea floor for some time in very shallow tropical water of less than 50 feet depth and of normal marine or just slightly restricted marine salinity. The fabric shows deposition probably below storm wave base in a zone where finer particles (i.e., lime mud) were being winnowed away. The area was a shoal above and on the edge of an earlier formed algal plate mud mound. Settling of finer, mostly silt-size sediment, occurred as it was deposited with the framework of larger grains, perhaps during a single storm. Lithification began early with finer grain silt sediment settling down through the pack, some moving to fill cavity floors. Calcite cement filled pores of all sizes forming large areas of calcite in-fills or in places with more lime mud, a generally clotted, grumelous matrix.

Significance of Color in Carbonate Rocks

In outcrop and core study much time can be occupied with a standard color chart describing precise tone and hue, for color is an easily observable though variable attribute of rock and is useful for environmental interpretation (Weller, 1960, p. 129–141; Krumbein and Sloss, 1963, p. 123). However, variations due to grain size of matrix, crystallinity, pigment content, and weathering are great, and detailed color description is generally useless. For carbonate rocks only three basic colors are important indications of environment: light, dark, and reddish hues. In some fine-grained rock of even texture, only a trace of pigment is necessary to create pronounced coloration. Absolutely black limestone of unusually fine crystalline texture (e.g., Marbre Noire of Belgium) contains on the average only 1–2% insoluble matter of which but 0.2% is organic carbon. Light colored limestone therefore is almost completely pure calcium carbonate lacking any trace of pigment. Its color is commonly enhanced by oxidation and weathering, traces of Fe_2O_3 giving usually a buff or faded yellow or cream color. In quarries of shelf limestone deposits, the water-saturated rock is normally gray, particularly if fine-grained.

Shallows bays and lagoons are surfaced with lime mud which is generally oxidized during turn-over by burrowing and perhaps by oxygen produced by photosynthesis of blue-green filamentous algae. Burial of this material under a few cm of sediment and removal from the O_2 of sea water results in a rapidly acquired reducing condition with production of H_2S, a gray color, and blackening of some organic rich particles. Decay of buried organic slime in the mud causes the reducing environment which presumably contains active anaerobic and sulfate-reducing bacteria. Such shallow water muds, however, are deposited with only a few tenths of a percent of organic carbon. Their color is basically medium gray. It is obvious that a truly dark gray, dark brown, or black carbonate must result from much more reducing conditions and somewhat higher content of organic carbon than normally observed in shallow shelf lagoons. Shelf limestones traced into miogeosynclines or basins generally darken and become almost black due to preservation of trace amounts of organic matter and ferrous sulfides thoroughly disseminated in the rock matrix. Only 1 or 2% is necessary for thorough pigmentation. The preservation is probably due to more rapid burial and more uniform marine conditions with no chance for periodic oxidation due to storm wave action. Limestone found with siliceous or argillaceous basinal deposits is generally (but not always) dark.

Not only is an environmental distinction between light and truly dark carbonate sediment useful, but attention to reddish-purple colors is also important. It is well documented that the more or less resistant ferric oxides are the main pigmentation of terrigenous clastics and are oxidation products either inherited from a previous environment or of the direct depositional environment of the sediment. Carbonates in shelf redbed sequences are usually light-colored unless externally stained on weathering. On the other hand, reddish-purple carbonates, particularly encrinites, are not uncommon in basinal and geosynclinal strata. The iron pigment may have formed in oxidizing conditions on submarine swells and washed down into geosynclinal troughs. Deeper water carbonates may be red or pink or purple through preservation of Fe and Mn oxide pigments, just as occurs during formation of radiolarites and siliceous shales. Preservation of hydrous Fe_2O_3 in deep basins with slow sedimentation is probably a function of the rate of burial of decaying organic matter (Fischer in Mesolella et al., 1974, p. 54). Limonite is produced widely in trace amounts in sea water. Normal sedimentation entombs enough organic matter to reduce the Fe to pyrite within the substrate, but in areas of very slow sedimentation, common in deep starved basins, bacterial decay oxidizes the organic matter before its burial. Mocassin Ordovician limestone of Tennessee, Hierlatzkalk (Jurassic) of Northern Calcareous Alps, Chapel (Mississippian) limestone of the Llano Uplift of Texas, and the Cretaceous-Jurassic boundary beds of the Oman geosyncline of Arabia are all examples of pink basinal crinoidal beds. The Ammonitico Rosso and Adneterkalk (Jurassic) are conglomerate beds of red color produced on swells in the Alpine geosyncline.

Clastic Content in Carbonates

The small amount of terrigenous or biogenic siliceous clastic material in carbonate sediment may be very significant in environmental interpretation. It may be disseminated in the carbonate matrix, or present as grains, or concentrated in thin discrete layers. The normal techniques of multiple acid etching on thin sections and manufacture of insoluble residues in transparent finger plates for examination with a stereomicroscope have been previously described. X-ray defraction mineralogical study is also a standard technique which may be employed although not as a substitute for petrographic study. Since most carbonate rocks are relatively pure, an increase in trace amounts of insolubles, particularly in the coarser fraction, may indicate the landward direction. A specialized study of carbonate insoluble residues was developed in the 1930's by McQueen (1931) of the Missouri Bureau of Geology and Mines and reviewed by Ireland (1947). This technique was used chiefly to discern correlation markers in Cambro-Ordovician and Mississippian strata in the Midcontinent and Texas. Types of authigenic chert and quartz and detrital sand grains were described in detail according to texture, color, and included impressions of fossils or dolomite rhombs. A complete roster of descriptive terms for these is found in Ireland's paper. The causes of the wide distribution of various chert types and recognition of the depositional facies in relationship to chronostratigraphy were never developed and study of chert residues along with that of heavy minerals in terrigenous sediments declined somewhat in interest, although both techniques are still useful in subdividing thick carbonate strata.

Two types of detrital siliceous grains occur extensively though in trace amounts in major carbonate strata. Zones of rounded, frosted, rather large, quartz grains are known from many sections and probably represent coastal dune and beach sand reworked in the shallow marine environment. Such zones are known widely in the Lower Ordovician platform composed of shallow shelf and intertidal deposits surrounding the Canadian shield of North America. Some of these zones of quartz grains occur now in areas hundreds of km from the original sandy coastlines of the great carbonate producing sea which lay across the shield. The southern shelf of the modern Persian Gulf has up to 10% of such grains distributed from large sand dunes into the sea along the coast south of the Qatar Peninsula (Shinn, in Purser, 1973). Quartz sand is so resistant that it may be widely distributed in the carbonate environment without appreciable abrasion or solution.

The other siliceous detrital grains common in carbonates are silt-size angular quartz and feldspars which are also widely distributed, probably windblown. Again, the Persian Gulf Holocene sedimentary regime serves as a model (Kukal and Saadallah, in Purser, 1973). Vast dust storms in the Gulf are capable of blowing silt and clay size material completely across the sea (a distance of 400 km). Probably a large part of the fine material of the axial part of the Persian Gulf east of the Qatar Peninsula (both carbonate and terrigenous) is wind-blown. In the geologic record such silts may occur as concentrates in intertidal beds, which often also contain rounded, frosted quartz sand grains and traces of detrital thorium. Such beds cause pronounced high gamma readings. Wind distribution of radioactive detrital material may be indicated in the cyclic Late Devonian Duperow strata of the Williston basin of North Dakota, which contain argillaceous silty dolomite beds in sabkha evaporite strata (Chapter X). These beds contain radioactive silt grains whose distribution is indicated on Fig. III-4. Stratigraphic evidence indicates a sandy source area to the south and the intensity of the gamma ray deflections increases in this direction.

The same strong gamma ray deflections or "kicks" are common in thick, and otherwise pure, carbonates in Ordovician, Silurian, and Mississippian strata of the Williston basin (Porter and Fuller, 1959; Krumbein and Sloss, 1963, p. 382). Similar thin zones of high gamma ray response are known in Siluro-Devonian strata in the West Texas basin. In all these examples such markers are widely traceable across the basins, and the ubiquitous, silty, terrigenous material is probably windblown.

Beds of altered volcanic ash (bentonites or metabentonites) may be important units in carbonate strata, particularly useful as key beds ("time markers") in correlation. These have been widely traced in the Middle Ordovician of the Appalachians (Dott and Batten, 1971, Fig. 10–18). Kay (1953) used these bentonites to effect correlations between the eastern black shale and western limestone facies of the Trenton group in New York state. Volcanic activity in the mobile

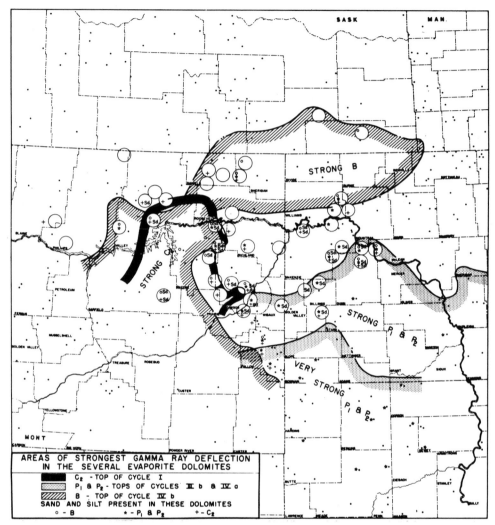

Fig. III-4. Areas of strongest gamma ray deflections and sand-silt distribution in several evaporative dolomites in tops of cycles of lower Duperow Formation of Williston basin, Montana, North Dakota, and Saskatchewan. The markers C_2, P_1, P_2, and B are very thin, silty zones in gray-green dolomite micrite associated with anhydrites. Patterns plus occurrence of sands indicates source of fine terrigenous (wind-blown?) material to south. From Wilson (1967b, Fig. 13). See also Chapter X for discussion and maps of Duperow beds

land areas east of the Ordovician Appalachian geosyncline and winds from the northeast low-latitude trades of Ordovician time distributed ash as far west as Ohio. Bentonites have been used in subsurface correlations within Middle Permian (Word) strata in the San Andres Formation in west Texas, east of the Central Basin platform. Bentonites are also recognized in Upper Cretaceous (Senonian) strata in the Boquillas and San Felipe formations of West Texas and northern Mexico. The even thin-bedding, sharp segregation of carbonate and bentonitic marl, and pelagic fauna indicate deposition in a rather deep water, offshore environment over which westerly winds distributed ash. The same events were taking place along the whole Cretaceous Rocky Mountain orogenic belt. Bentonites of the same age are widely traceable east from this front in terrigenous strata in Wyoming and Montana.

Clay seams are of consequence in the petrography of carbonates, although clay mineralogy itself is of debatable use in environmental interpretation because of the detrital sources and the existence of diagenetic alteration in given environments (Grim, 1958, Weaver, 1958). Krumbein and Sloss (1963) furnish a table giving, in abridged form, possible environmental interpretation from major clay minerals (kaolinite, montmorillonite, illite, chlorite).

In addition, the mere presence of clay "contaminants" in dominantly carbonate rock is significant, commonly indicating quiet-water deposition in deeper water conditions. In many places basinal, dark limestones thin to dark argillaceous units in the basin center. In such strata, clay layers may be well-segregated from the carbonates, both rock types being rather pure, although regularly interstratified. In shallower water deposits, currents, wave action, and burrowing tend to intermix carbonate and argillaceous material, part of which was originally deposited in segregated layers. Burial of such partially mixed, partially segregated material results in compacted clay seams and noncompacted limestone nodules or lenses—ball and flow structure or sedimentary boudinage. Seams of argillaceous carbonate are subjected to stylolitization and contacts between boudins, or for that matter, between bioclastic grains (e.g., crinoid pieces), are often sharply defined by stylolites which concentrate organic matter, clay, or dolomite rhombs. There is some reason to believe that clay seams in carbonates control position of stylolites which are commonly parallel to bedding. It is of interest that clay content of limestone generally is small, only from 5 to 10% being sufficient on outcrop to cause weathering profiles generally characteristic of those of pure clay and silt. As pointed out earlier, the general presence of argillaceous particles in the sea inhibits organic production of $CaCO_3$ and carbonate is easily swamped by clay content. Many marls (carbonate rich clay and silt) contains more than 50% $CaCO_3$ but are in no sense limestones.

Porosity and Permeability

Pore space is an important attribute of some carbonates and has naturally been much studied by petrographers to ascertain its origin and relationship to permeability and fluid transmissibility. A discussion by Harbaugh (1967, in Chilingar et al., 1967) concerns techniques for porosity and permeability determination. These include bulk density—grain density determinations, mercury capillary pressure injection, plastic impregnation of porosity moulds, buffing of highly polished surfaces of rock with jeweler's rouge or chrome oxide to outline pores and thin section study under polarized light.

Porosity is notoriously erratic in carbonates compared to that of sandstones and is generally slight. Producing reservoirs with 5 to 10% pore space are not

BASIC POROSITY TYPES

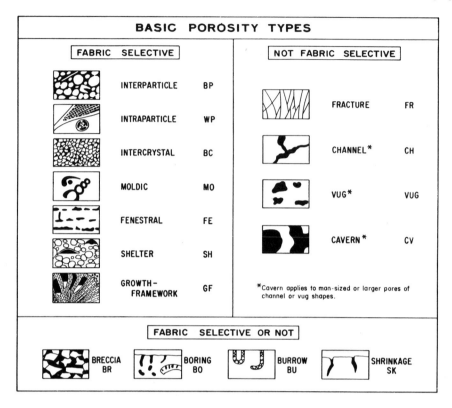

FABRIC SELECTIVE

	INTERPARTICLE	BP
	INTRAPARTICLE	WP
	INTERCRYSTAL	BC
	MOLDIC	MO
	FENESTRAL	FE
	SHELTER	SH
	GROWTH-FRAMEWORK	GF

NOT FABRIC SELECTIVE

	FRACTURE	FR
	CHANNEL*	CH
	VUG*	VUG
	CAVERN*	CV

*Cavern applies to man-sized or larger pores of channel or vug shapes.

FABRIC SELECTIVE OR NOT

BRECCIA BR   BORING BO   BURROW BU   SHRINKAGE SK

MODIFYING TERMS

GENETIC MODIFIERS

PROCESS

SOLUTION	s
CEMENTATION	c
INTERNAL SEDIMENT	i

DIRECTION OR STAGE

ENLARGED	x
REDUCED	r
FILLED	f

TIME OF FORMATION

PRIMARY	P
pre-depositional	Pp
depositional	Pd
SECONDARY	S
eogenetic	Se
mesogenetic	Sm
telogenetic	St

Genetic modifiers are combined as follows:

PROCESS + DIRECTION + TIME

EXAMPLES:

solution-enlarged	sx
cement-reduced primary	crP
sediment-filled eogenetic	ifSe

SIZE* MODIFIERS

CLASSES

			mm[†]
			—256—
MEGAPORE mg	large	lmg	—32—
	small	smg	
			—4—
MESOPORE ms	large	lms	
	small	sms	—1/2—
			—1/16—
MICROPORE mc			

Use size prefixes with basic porosity types:

mesovug	msVUG
small mesomold	smsMO
microinterparticle	mcBP

*For regular-shaped pores smaller than cavern size.

[†] Measures refer to average pore diameter of a single pore or the range in size of a pore assemblage. For tubular pores use average cross-section. For platy pores use width and note shape.

ABUNDANCE MODIFIERS

percent porosity	(15%)
or	
ratio of porosity types	(1:2)
or	
ratio and percent	(1:2)(15%)

uncommon. The reader is referred to several studies on this subject (Murray, 1960; Thomas, 1962; Harbaugh in Chilingar et al., 1967; Choquette and Pray, 1970, Fig. III-5 of this volume). Several types of porosity can be recognized and classified genetically:

1. Secondary solution along fractures due presumably to meteoric water.
2. Secondary solution in the form of vugs or cavities presumably due to meteoric water.
3. Primary vugs and cavities in reef framework.
4. Primary interparticle porosity in sands and gravels preserved from cementation. Primary interparticle porosity in pelagic microfossil silts (chalky).
5. Chalky (very fine porosity) caused by lack of early cementation and by replacement of original lime mud fabric by blocky calcite crystals. Preservation of up to 30% porosity, very little permeability.
6. Sucrose dolomite porosity, which forms during early dolomitization of nonlithified calcareous mud. This process preserves original porosity, using up carbonate ion as Mg replaces pure Ca in the original mud and growing dolomite crystals to form a noncompactible framework.

Fig. III-5 is recommended as a key for describing porosity in the following order: genetic modifier plus size plus type plus percentage of bulk volume of rock. E.g., secondary eogenetic intercrystalline and moldic dolomitization porosity— 10% (Choquette and Pray, 1970).

◄ Fig. III-5. Geologic classification of pores and pore systems in carbonate rocks. From Choquette and Pray (1970, Fig. 2), with permission of American Association of Petroleum Geologists

Chapter IV

The Advent of Framebuilders in the Middle Paleozoic

Whereas the preceding Chapters discussed the fundamentals of carbonate sedimentology, stratigraphy, and petrography, this second portion of the book describes a variety of carbonate facies complexes given more or less in order of geologic age. These examples serve as documentation for many of the ideas expressed in the first three Chapters. The facies occur in many diverse tectonic settings and are found in shelf cycles, mounds and patch reefs on shelves, buildups at shelf margins, basinal banks, and pinnacles. Well-studied outcrop areas as well as subsurface petroleum reservoirs are included. The examples cited may serve as models for interpretation of less well-known facies patterns.

Considerable changes in the biota of carbonate accumulations occurred at the beginning of the Middle Ordovician, just prior to the Carboniferous, in the Late Triassic, Cretaceous, and early Tertiary. All of these changes are significant in terms of size and amount of framebuilders, prediction of biofacies patterns, trend and shape of bodies of sediment, and susceptibility to diagenesis. Similarities in growth form appear from System to System in the geological record. "Carbonate buildups are like Shakespeare; the plays go on—only the actors change" (Ginsburg).

The Earliest Buildups

The first of the critical biologic changes affecting carbonate reefs and mounds occurred in the Middle Ordovician. This was the development of lime-precipitating coelenterates which were capable of encrusting each other, branching, and forming a framework. Before this time, from middle Precambrian through earliest Ordovician, carbonate buildups were formed mainly by calcareous algae in the form of tiny bushes or wiers or by stromatolitic laminae forming bulbous or lamellar-undulose layers. These minute baffles or gummy layers acted as traps and permitted accretion of considerable quantities of carbonate mud which was carried to them by gentle tides or currents. This process has been well studied in modern tidal flat areas. The only other sessile marine organisms known to participate in the buildups formed by such trapping of mud are sponges and related forms whose fossils appeared first in Cambrian strata. Low banks of lime mud were formed by these archeocyathids in Lower and Middle Cambrian time.

Regional studies, principally by the Canadians Aitkin (1966) and Hoffman (1974), have demonstrated that the limited and simple algal flora was capable of stabilizing massive carbonate accumulations at shelf margins. The process was rapid enough to form considerable relief on tectonic hinge-lines (Fig. IV-1). Black shale basins, such as the Burgess (Middle Cambrian of Alberta) and Middle

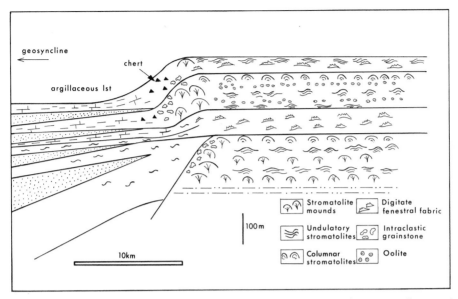

Fig. IV-1. Facies of Middle Precambrian carbonate platform, Coronation geosyncline, northwest Canadian Shield, N.W.T., Canada. After Hoffman (1974), with permission of American Association of Petroleum Geologists

Precambrian turbidites in the Great Slave Lake area (Hoffman, 1974) occurring on the seaward side of such platforms, are evidence for these steep margins.

The best-known early buildups of North America, however, are small mounds of lime mud located toward the edges of the wide shelves covered by Cambro-Ordovician seas. Algae from these masses have been described from the Late Cambrian of Texas by Ahr (1971) (Fig. IV-2). In Lower Ordovician strata the possible dasycladacean *Calathium*, the lithistid sponge *Archeoscyphia*, a spinose stromatolitic organism *Pulchrilamina*, and tiny branching, cylindrical, and encrusting algae (?) *Nuia*, *Epiphyton*, and *Renalcis* have been described by Toomey (1970), Toomey and Ham (1967), and Riding and Toomey (1972) from mounds in the southwestern U.S.A. The sponges are small basket-shaped, thick-walled forms (Plate V A). The algae are encrusting in habit and *Pulchrilamina* forms stromatolite structure on the mound tops. Where not dolomitized, the Cambro-Ordovician strata preserve internal structures beautifully and permit detailed microscopic study of these primitive organisms.

The mounds are composed of massive micrite apparently representing lime mud trapped by the organisms. They are seldom more than 10 m thick and most are considerably smaller. Their micrite cores may intergrade with bedded bioclastic sediment. On the same buildup, however, bioclastic strata may overlap parts of the mound showing that in places considerable relief had developed over the seafloor (Fig. IV-3). Many of the mounds show two stages of formation: (1) An early massive lime mud accumulation replete with algae which trapped the sediment. (2) This may be followed by a mound-capping phase of stromatolite. Such sequences may form composite masses (Fig. IV-2). The mounds may be cut by

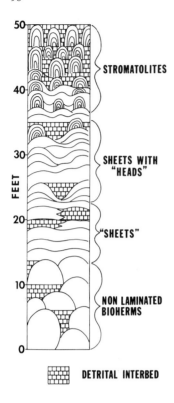

STROMATOLITES

SHEETS WITH "HEADS"

FEET

"SHEETS"

NON LAMINATED BIOHERMS

DETRITAL INTERBED

Fig. IV-2. Idealized Late Cambrian algal mound section in lower Point Peak Member, Wilberns Formation, southwest Llano uplift, Texas. Upward progression indicated schematically from nonlaminated mounds of algal bafflestone to algal sheets and capping stromatolites. From Ahr (1971, Fig. 7), with permission of Society of Economic Paleontologists and Mineralogists

tidal channels and fissures caused by slumping. These are filled with coarse bioclasts of echinoderms, brachiopods, and trilobites, and lithoclasts (pebble conglomerates of reworked desiccated tidal flat sediment).

It is probably paleogeographically significant that even such small mounds are known mainly around the edges of the North American craton, in the limestone facies of the Arbuckles and Wichita Mountains of Oklahoma, as exotic blocks in the Marathon Formation, and in the El Paso limestone of West Texas. Such mounds occur as well in the upper Pogonip of the Cordilleran geosyncline in Utah and Nevada where Church (1974) described the same sponge-algal assemblage. Ross and Cornwall (1961) noted carbonate masses almost 100 m thick in southern Nevada in the Miekeljohn Peak area. These consist of laminated micrite. The rare megafauna includes only a few brachiopod layers, sponges, and cephalopods. These large mounds also lie toward the cratonic margin. Lower Ordovician mounds are rarer and smaller in the more interior platform dolomitic facies, such as in the Ellenburger Formation of Texas, the type Ozark uplift strata, and the Oneota and Manitou Dolomites of Wisconsin and Colorado. Presumably the continental interior contained too restricted an environment and too shallow water for the biological assemblage to flourish.

The above algal-sponge buildups prevailed at shelf margins and, in the form of small mounds, over shallow shelves for a very long period in earth history. More than one billion years passed from the time of the Great Slave Lake deposition

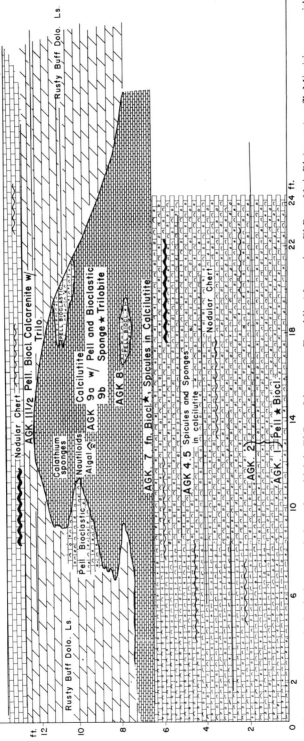

Fig. IV-3. Sponge-algal micrite mound about 2 m thick in El Paso Limestone, Franklin Mountains, El Paso Unit Bl, lower "reef". Micritic mass with nautiloids, sponges and *Calathium* has pockets and flank beds of peloidal bioclastic grainstone and packstone. AGK-numbers indicate sample locations

described by Hoffman to the beginning of the Middle Ordovician, about 450 million years ago. At this time, in North America and around the Baltic Shield, the first real "reefy" faunas began to develop. About 100 families are found as fossils in Cambrian strata. Twice this many occur in the Middle Ordovician despite the loss of two-thirds of the Cambrian trilobite families which constituted the majority of the Cambrian life forms. Many representatives of these new families appear first in small carbonate buildups. Bryozoans, rugose corals, stromatoporoids, and tabulate corals constitute the important new Ordovician sessile benthos capable of carbonate construction. Accessory organisms such as cephalopods, trilobites, and brachiopods, and above all the crinoids and blastoids, become significant on and around the buildups.

The change from Lower Ordovician to Middle Ordovician faunas occurred during formation of a regional unconformity in North America. It was an event almost of equal evolutionary consequence to the first appearance of shelly calcareous tests in the Cambrian 150 million years earlier and the biological changes ending the Paleozoic 200 million years later. The ability of these varied organisms to form a definite ecological association, to accrete carbonate in their tissues at a rapid rate, and to proliferate and diversify, was a significant event in earth history.

Large discrete masses of calcium carbonate were built up at continental margins and as smaller mounds around and within shallow shelves and basins. Circulation became more restricted, encouraging evaporite deposition. At the same time basin interiors within, and marginal to the craton, became starved of terrigenous sediment. With regional subsidence the basins became more sharply differentiated from the shelves which, themselves, were built up by increased carbonate accumulation. Up to this time, mainly physical processes of tidal flat construction had added carbonate sediments to these shelves. Now effective organic construction became an important process. These early "reefy" faunas and floras offer significant ecologic comparisons with the larger ones of the Mesozoic and the present day. It is also of interest to observe the burst of evolution in this biota during the Middle Paleozoic. It begins in the Middle Ordovician and declines during Fammenian time in the Late Devonian.

The earliest carbonate buildups of the Middle Paleozoic have been well described by Pitcher (1964) and Toomey and Finks (1969), who studied them in the New England Chazyan beds along Lake Champlain, an area which extends across the Canadian border north of Montreal in Quebec province. This shelf lay west of the Appalachian geosyncline and north of the Adirondack-Frontenac axis. It accumulated 300–1000 m of well-bedded carbonate, beginning in the earliest part of the Middle Ordovician. Micritic mounds are distributed in these sediments, but observation of regional trends or evidence for localization of particular mounds is not permitted by the scattered outcrops. It is evident that the mounds formed on a shallow shelf in conditions of varying water agitation. All mounds are small, from 2 to 8 m high at most. They are the same shape and size as the alga-sponge mounds existing in the Lower Ordovician and Upper Cambrian, but faunally they are much more complex.

Distinct biological assemblages are recognized in these mounds but a clearcut vertical succession of faunas is rare in the oldest Middle Ordovician. Figure IV-4 is a schematic diagram of these mounds and faunas from Pitcher (1964).

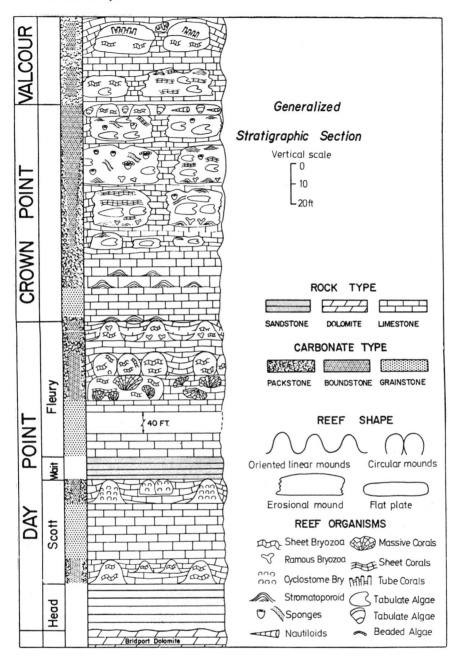

Fig. IV-4. Generalized stratigraphic section showing reef types and organic growth forms in Middle Ordovician Chazyan strata around Lake Champlain. From Pitcher (1964, Fig. 8)

Table IV-1. Growth habits of the Chazyan mound organisms

Encrusting (binders-stabilizers of fine sediment)	Knobby (binders)	Upright (frame-tubular)
Bryozoans: *Batostoma*, *Cheiloporella*		*Atactotechus*
Encrusting stromatoporoids: *Cystostroma*	*Pseudostylodictyon* Coral *Billingsaria*	Coral *Eofletcheria*
Encrusting coral: *Billingsaria*	*Lamottia* *Billingsaria* (rare)	
Encrusting sponge: *Zittelella* (rare) (lamellar form)		*Zittelella* (most common)
Algae: *Solenopora* *Girvanella*-anastomosing, penetrating tubules *Rothpletzella* (tiny beaded chain algae) Stromatolites *Sphaerocodium?* beads	*Solenopora*	

The earliest Chazyan mounds are among the smallest and are linear ridges of micrite with flat or bulbous encrusting bryozoans (of the genus *Batostoma*). They are surrounded by washed bioclastic and oolitic packstone-grainstone. Later, in Chazyan time (Crown Point Formation) when less agitation and less terrigenous influx existed, many diverse organic assemblages constructed mounds. These also remained small.

Much faunal diversity occurs: (1) An assemblage of *Solenopora?* and an encrusting stromatoporoid, (2) mounds dominated by the lithistid sponge, *Zittelella*, (3) masses exclusively with the tabulate coral *Billingsaria*, and (4) mounds composed of three assemblages, (a) stromatolitic algae and small heads of *Solenopora?* with scattered abundant orthoconic cephalopods, (b) the encrusting bryozoan *Batostoma* with the tubular tabulate coral *Eofletcheria*, and (c) *Batostoma* with *Solenopora?*.

Some vertical succession is indicated in small mounds of this age in Quebec. Here the lowest beds are bioclastic packstones overlain by the encrusting bryozoans in micrite and the encrusting coral *Billingsaria* occurs gradually higher. The mounds are capped by the tubular *Eofletcheria* or other Tabulates. In the Crown Point beds of Lake Champlain the conical sponge *Zittelella* is prominent down the flanks, overlain by bryozoan micrite. The two-meter high mound is capped by lamellar stromatoporoids. The mounds are cut by channels filled with calcarenite much like those of the Lower Ordovician and Upper Cambrian mounds. Growth habits of the varied Middle Ordovician forms are given in Table IV-1.

A food chain (trophic) analysis is suggested by Toomey and Finks (1969). The sponges, stromatoporoids, and bryozoans are regarded as suspension feeders, filtering organic matter from moderately quiet water. The algae grew among them. The only sessile benthos resembling modern macrophagous carnivorous coelenterates would be the tabulate corals. Accessory dwellers include the trilobite *Glaphurus*, a detritus feeder, the very abundant suspension feeding pelmatozoans, and the brachiopods living in channels and perhaps on upper mound surfaces. Grazers common on the mounds included the large gastropod *Maclurites*. Numerous cephalopods may have been the only carnivorous predators. The same gastropod and nautiloid association is seen in the Lower Ordovician algal-sponge buildups.

Middle Ordovician mounds of the same size range have been described by Alberstadt and Walker (1973) in the Carters and Holston formations of Tennessee. Holston mounds are faunally monotonous. They begin with a stabilization phase of pelmatozoan debris and small branching bryozoans and change upward through a dominance of branching to encrusting bryozoans. The Carters organic buildups are more complex. Sediment with pelmatozoan debris was apparently stabilized by upright branching *Receptaculites* forms (dasycladaceans), encrusting bryozoans, and pelmatozoans. This earlier assemblage was colonized by laminated stromatoporoids which were followed by a diversified assemblage of domical corals, laminated stromatoporoids and upright branching dasycladaceans, pelatozoans, and ramose bryozoans. Finally, laminated stromatoporoids dominated the top of the mound as did encrusting bryozoans in the Holston Formation.

As with all other older Paleozoic mounds on shallow shelves, hydrologic and depth controls are deemed less important than the progressive creation of sites for growth concurrent with an increase in diversity of biological communities (Alberstadt and Walker, 1973; Alberstadt et al., 1974). The sequence is outlined as follows:

Table IV-2. Biological evolution in organic buildups

Encrusting and laminate forms	No diversity, one form dominant
Diversification phase. Upright and encrusting forms dominating	Very high diversity
Colonization phase	Diversity increasing
Stabilization phase Upright forms in tangled masses	Less diversity
Pelmatozoans acting as baffles for lime mud accumulation	

Silurian Buildups-Paleotectonic Settings

Silurian coral-stromatoporoid micritic mounds and ecologic reefs are among the best-known in the world due principally to careful and extensive study in two localities: the Gotland Island exposures off the Swedish Baltic coast (A. Hadding,

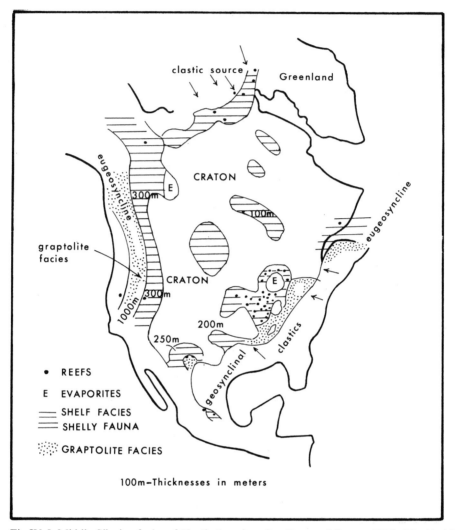

Fig. IV-5. Middle Silurian facies of North America after Dott and Batten (1971, Fig. 11–11) showing peripheral distribution of mounds and reefs

V. Hede, V. Jux, and A. A. Manten) and the northern United States outcrops in Illinois and Indiana (H. A. Lowenstam, E. R. Cummings and R. R. Shrock, A. V. Carozzi, D. A. Textoris, and J. J. C. Ingels).

Regionally, the Gotland occurrence of small Silurian mounds lies along a shallow shelf bordering the Baltic shield east of the Caledonian geosyncline. The related Wenlock mounds of Britain (Colter, 1957; Scoffin, 1971) lie within the geosyncline itself but presumably on a shallow platform near its border.

Silurian outcrops in North America are widely distributed though scattered across the craton and it is possible to obtain a clearer picture of the regional setting than in Europe. Silurian strata were extensively removed by erosion during Devonian time but sufficient remnants exist to assure that by Middle Silurian time practically the whole continent was inundated by shallow seas producing marine carbonate (Fig. IV-5). Patches of mounds and ecologic reefs of algae, corals, and stromatoporoids are seen generally to rim the great central craton and to lap over the eastern side of it in a wide belt mostly now exposed in Indiana, Michigan, and Illinois. In addition, three cratonic basins are bordered by Silurian reefy carbonate buildups: the Michigan basin, the Illinois (or Bainbridge basin), and the West Texas-New Mexico (Tobosa) basin.

The Cordilleran geosyncline also contains a shelf margin of Silurian age in central Nevada and in the northern Canadian Rockies. In none of these areas has a barrier reef rim of any sort been demonstrated but only a relatively abrupt facies change from fossiliferous carbonate into deeper water, graptolite-bearing, dark or reddish argillaceous sediment. The Nevada slope facies has been well described (Winterer and Murphy, 1960). Clear evidence of turbidites and slump deposits exists here indicating steep slopes, but the shelf margin is so thoroughly dolomitized that individual organic buildups are not observable and the real character of the shelf margin is unknown. In most places a regional belt of limestone borders the dolomitized craton but faunas are the same in this belt as within the interior.

Little is actually known about major Silurian shelf margins. Detailed description of Silurian buildups, therefore, comes almost solely from shelf areas, where several different kinds of carbonate masses have been exhaustively studied for at least thirty years, first biologically and more recently petrographically. Extensive dolomitization, so characteristic of the North American Silurian, has hampered the latter effort.

Illinois and Indiana Silurian Mounds and Reefs

The stratigraphic section in the northern middle western states, consists of about 100–200 m of marl, dolomite, and limestone. This covers an extensive shelf between the Michigan and Illinois basins west of the Cincinnati-Findlay arch. Only the northern edge of the shelf is exposed; the larger part lies in the subsurface of Illinois and Iowa. Fig. IV-6 shows the location of about 160 known carbonate buildups in Niagaran (Wenlockian and Ludlovian) strata on this shelf and around the Michigan basin. Shaver (1974) points out that many additional buildups remain undiscovered.

Lowenstam (1950, 1957) divided the shelf into areas of pure carbonate, low terrigenous, and high terrigenous sediments. These belts represent progressive changes in deposition along a shallow shelf sloping into the Illinois basin. The clastic free belt contains some evidence of tidal flat and restricted marine conditions but normal marine waters covered much of the area. Water depth is probably indicated by average height of the buildups above sea floor which is only a few meters in the northern outcrops and as much as 100 m in the subsurface of Illinois. The buildups occur in irregular clusters in the non clastic and low clastic belts, but no linear trends are obvious, except where they surround the Michigan basin.

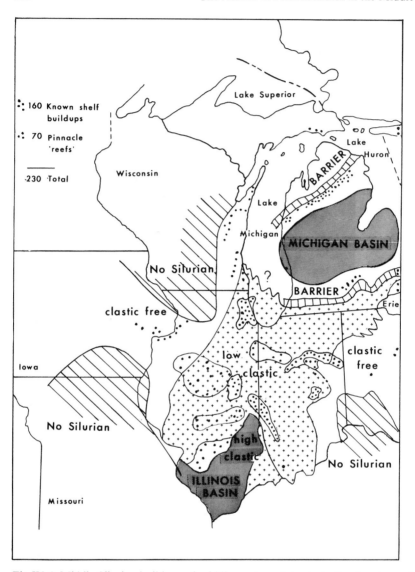

Fig. IV-6. Middle Silurian buildups of middle western states including barrier reef belt surrounding Michigan basin. After Lowenstam (1950), Sangree (1960) and Mesolella et al. (1974)

Shaver (1974) has related the Silurian buildups of Indiana to a biostratigraphic framework and pointed out some interesting faunal generalizations. Almost 400 species of reef and interreef organisms are known. These faunas evolved continuously through the Silurian, increasing in diversity. The reef dwellers (flank and top and niche environments) became more diverse than the frame-builders; the non-reef forms show little change. This fact and the increased upward expansion of the reefs show a tendency in time for growth into wave base and expansion of the Middle Silurian reefs to restrict the circulation significantly. Shaver pointed

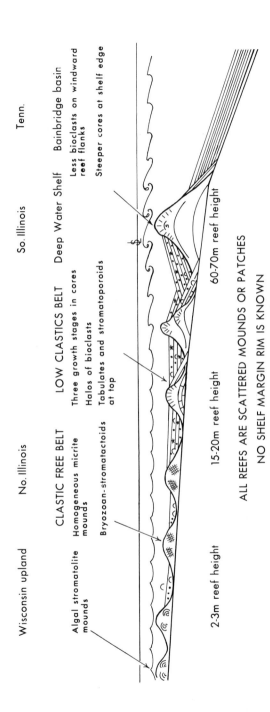

Fig. IV-7. Idealized stratigraphic model north to south across Middle Silurian shelf in midwestern U.S.A. over distance of several hundred kilometers

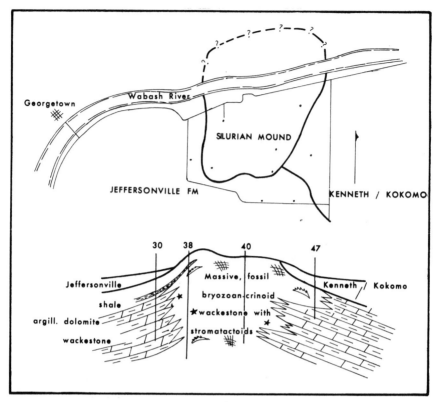

Fig. IV-8. Simple Middle Silurian mound near Georgetown, Indiana in clastics-free belt of Lowenstam. Original relief low, composed merely of micrite with bryozoans and crinoids. Lower diagram not to scale. Mound is about 2 km across and 40 m thick. Core holes outline mound through glacial drift and partly under thin cover of Devonian Jeffersonville and Kenneth-Kokomo Formations. Petrography by O. P. Majewske

out that reef construction is later in time on the Indiana shelf than previously realized. It is equivalent in part to the Salina evaporites of the Michigan basin. Shelf buildups are varied, types being controlled by location on the shelf which sloped south into the Illinois basin (Fig. IV-7). Five types are described below.

1. Algal spongiostrome stromatolitic mounds in Ohio and Wisconsin and northern Indiana (Textoris, 1966; Soderman and Carozzi, 1963; Textoris and Carozzi, 1966)

These mounds, which are small (some less than 3 m high and a few tens of meters across), developed in a restricted marine environment far up on the shelf on the perimeter of the Michigan basin. The algal stromatolite beds are brecciated in places, owing to intermittent desiccation and wave action. At some localities, the stromatolite caps a mud mound of the type described below.

2. Bryozoan-mud mounds with abundant stromatactoid structures

These correspond to "reefs" in the semirough water stage of Lowenstam (1950, 1957) and to stage 2, microfacies 3, of Textoris and Carozzi (1964). They formed below active wave base and grew to heights of some tens of meters, a figure partly determined by water depth (wave base).

This type of buildup forms the base of stromatoporoid-tabulate ecologic reefs, but may occur alone. The Georgetown reef, cut by the Wabash River, was extensively cored and a three dimensional picture can be ascertained (Fig. IV-8, Georgetown mound).

Some lime mud mounds may be as much as one mile in diameter and are simple, low, conical forms consisting of pure limestone or dolomite, surrounded by vuggy, argillaceous, somewhat cherty, dolomite-wackestone with less than 10% bioclasts. The mound rock is massive wackestone-packstone with a bioclastic content varying from 5 to 50%.

A diverse fauna contains ostracods, trilobites, brachiopods, corals, stromatoporoids, cephalopods, gastropods, bivalves, and sponges. This is a very similar biota to that of the normal, bedded Silurian limestone; no vertical or lateral differentiation of fauna exists in these mounds. The dominant particle types are crinoids and cryptostome bryozoans including many fenestrates and small stick-like forms. Corals are mostly small solitary rugose and colonial tabulates. Stromatoporoids are small, cabbage-head types and not framebuilders. Algae are rare—only a few *Solenopora*. Stromatactoid structures occur throughout. Petrographic study indicates that these sparry calcite-filled cavities are caused by collapse and separation of mud away from bryozoan fronds and as well are fillings of cavities formed by protective umbrellas of bryozoans. They are not considered dissolved organic frame builders as interpreted by Lowenstam, Lees, Lecompte, Bathurst, and other researchers (see Chapter V).

3. Crinoidal-tabulate coral mud mounds in Iowa (Gower Formation, Philcox 1970, Fig. XII-4)

These are essentially masses of lime mud, replete with crinoids and scattered colonies of *Favosites, Halysites*, and rarer plate-like stromatoporoids. The accumulations are from 5 to 30 m high. They built up close to wave base where extensive flank beds developed in a marginal (windward side) zone.

Early in the mound history, crinoids may have induced mud accumulations which were differentiated on the windward, north, and west facing sides, but these gradually fused together leeward and upward into a main mound. On the steeper, windward sides of the mound, a blanket of tabulate corals developed. At a late stage, the amalgamated mass developed into a platform near wave base. Then colonies of organisms developed as wedge-shaped biostromes on the windward side of the platform. These consist of algal stromatolites, beds of rhynchonelled brachiopods and thickets of the interwoven dendroid coral, *Amplexus*, with scattered colonial tabulates. This organic veneer grades down-slope into a laminated dense dolomite, a quiet-water lithofacies which is also developed in the lee of the mound complex.

4. Full development of Silurian ecologic reefs

The development into wave-resistant framework-built bodies is described thoroughly by Lowenstam (1950, 1957) for buildups in the low clastic belt of Illinois and Indiana. His studies included the subsurface as well as outcrops. Later petrographic study of outcropping reefs in this belt was conducted by Textoris and Carozzi (1964) and a detailed investigation, with some imagination, was made by Ingels on the Thornton reef (1963). Despite the different approaches, and the cumbersome separation of paleontology from petrography, a coherent picture emerges. These buildups tended to form farther down the paleoslope and in water of greater depth than mounds previously described. Some of them grew to the appreciable height of more than 30–70 m above the sea floor.

Several growth stages are recognizable. The buildups began with crinoidal wackestone accumulations below wave base, much like that of the Gower Formation of Iowa. Faunally, this corresponds to Lowenstam's pioneer population of *Syringopora, Favosites*, dendroid tabulates, and few stromatoporoids. Much sedimentary infill of argillaceous lime mud occurs and also networks of stromatactoid structure. Crinoids consist mainly of *Pisocrinus*. Several genera of ornate trilobites and four genera of sponges are recognized.

Stromatactoid mudstone-wackestone and calcisilt replete with bryozoans, forms the next important stage and is identical to the total mound development seen at Georgetown and most of that at Wabash and Mattock (stages 2 and 3 of Textoris and Carozzi, 1964). A gradual increase occurs in the varied reef fauna which culminates in the true framework of wave-resistant organisms at the top and includes an impressive list of reef-associated (niche-

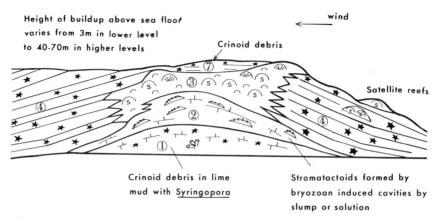

Fig. IV-9. Idealized fully developed Middle Silurian ecologic reef common in the low clastics belt of Lowenstam (1950). Boundstone of stromatoporoids developed on top of an accumulation of crinoid-bryozoan debris in a micrite matrix. Mound facies follow pattern outlined in Figs. XII-5 and VI-25: 1. Basal bioclastic pile of initial accumulation; 2. Micrite bafflestone core-mud mound buildup below wave base; 3. Crestal boundstone, framebuilt organic reef; 4. Flank beds of crinoids derived from reef top dwellers, not framebuilders; 5. and 6. Talus and organic veneer not recognized; 7. Capping grainstone of crinoidal sand

dwelling) forms. Lowenstam noted increasing abundance of stromatoporoids but that the corals *Favosites, Halysites, Heliolites, and Syringopora* form merely low bosses. Camerate crinoids are abundant and brachiopods such as *Chonchidium*, rhynchonellids, atrypids, and spirifers occur. Seven diverse trilobite genera are present and also many mollusks.

In the frame building (rough-water stage of Lowenstam, 1950, stage 4 of Textoris and Carozzi, 1964) the buildup reached above wave base and the typical reef fauna flourished. More and larger tabulate corals occurred with the stromatoporoids. New corals added include *Arachnophyllum, Thecia, Alveolites, Fletcheria, and Coenites.* Sponges disappeared. Trilobites remained varied and are represented by eleven genera, although the convex smooth-shelled *Bumastus* is by far the most abundant. Brachiopods are represented also by many genera, including the heavy-shelled *Trimerellacea* and pentamerids. *Conchidium* occurred on outer edges of buildups. The echinoderms were also important: blastoids and cystoids (especially *Caryocrinites*) and dominant camerate crinoids occurred. Reef grazers or carnivores are represented by 22 nautiloid genera! The lower reef stages representing quieter water have orthocones but coiled and breviconic forms characterize the later stage of the buildups, especially in the shelfward, low clastic belt. Gastropods occur in varied shapes from flat to turreted. The large probable dasycladacean *Ischadites* occurs also in this crestal stage.

The final and complete development of a typical Silurian shelf reef (stages 5 and 6 of Textoris and Carozzi, 1964) was reached when the upward organic growth reached into wave base and stopped. Lateral accretion of extensive flank beds forms a platform and to a limited extent, outward core growth over the flank beds occurs on the windward side. Faunal differentiation also marks the windward side and occasionally a capping bioclastic-oolitic deposit developed. The flank beds, which are mainly crinoidal, may constitute much more than half of the bulk of the buildup showing that bioclastic production by niche dwellers was prolific on the reef top and lasted a long time. The very abundance of organic production at this stage argues for a water depth measurable in meters across the top of the platform. Fig. IV-9 is a summary diagram of growth history.

Study of the Thornton quarry south of Chicago by Ingels (1963), resulted in reconstruction of the final stage of growth of the reef crest and its relation to flank development on windward and leeward sides. Exposures of this buildup permit only a limited view of the core facies which is but a minor part of the rock volume, most of it being steeply dipping flank

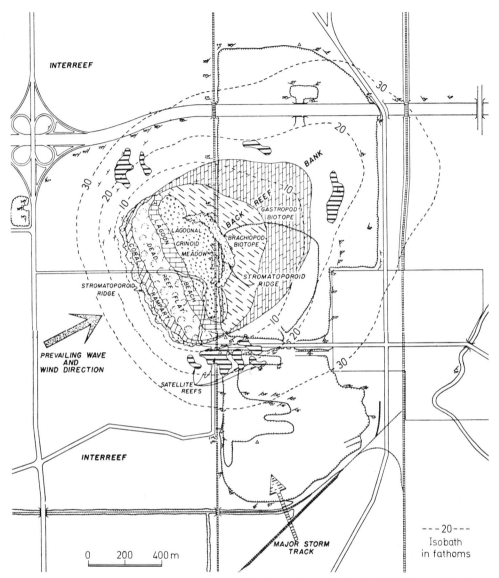

Fig. IV-10. Paleoecologic map of Thornton Silurian reef complex in South Chicago, Illinois, at a level 100 m above its base. From Ingels (1963, Fig. 16). Map reconstructs the key biofacies for the buildup as conceived to result from growth of ecologic reef into wave base. Illustration with permission of author and American Association of Petroleum Geologists

beds (35° off the central core area). The asymmetry of the flat conical elliptical mass is clearly shown, the flanking strata tailing off in the leeward direction.

Small satellite mounds are mapped in flank positions surrounding the main core which is composite and grew outward and upward in both windward and leeward directions. Wave-resistant frame-built core on the windward side of the reef is indicated by coral-stromatoporoid ridges around the center. The belt of such boundstone is four times as wide on the southwest (windward) side. A zone where the flattish convex trilobite *Bumastus* is common is interpreted as a reef flat bordered by a beach on which orthoconic cephalopods accumulated. A crinoid meadow with scattered large patches of tabulate corals is considered to have filled in the lagoon leeward of the main southwest facing coral-stromatoporoid ridge. Coarser crinoid debris occurs on the southwest (windward) side. The presence of echinoderms and cephalopods indicates water of normal marine salinity and of some depth over the top of the reef—not an intertidal environment. Relatively fewer bioclastic beds exist down the windward flank giving a northeast elongation to the mass. Suggested reef-derived breccia is reported along the southwest quadrant of the mass, an unusual feature in Silurian reefs. Most of the flanking material consists of thin, rhythmic beds, alternating dolomite mudstone, and bioclastic beds. Fig. IV-10 is an environmental interpretation by Ingels of a restored Thornton reef mass.

5. *The Michigan Basin "Pinnacle Reefs"*:

Superimposed and interlaced reefs of the above types are known to form a Middle Silurian barrier complex 150–200 m thick around the Michigan basin which was sediment-starved and accumulated only 30 m of sediment during this time (Fig. IV-11). Sequences of cored wells passing through this barrier reef encounter the following sequence above basinal carbonate muds: (1) coarse carbonate sands and gravels with cross-bedding grading up into massive sands (15–100 m thick), (2) floatstone with large heads of stromatoporoids with the Tabulate corals *Coenites* and *Favosites* in a very coarse sand matrix (15–80 m), (3) coarse sands with no fossils grading up through green shale, to (4) mottled and laminated backreef lime muds about 10 m thick. This regressive sequence is Wenlockian in age but on the basinward side of the barrier younger Silurian (Ludlovian) with stromatoporoids and algae is present.

Within the basin numerous isolated buildups parallel both the north and south flanks of the barrier. Accumulation on the slopes into the basin managed to keep pace with sedimentation. Briggs and Gill (1971) and Mesolella et al (1974) described the ideal sequence developed in these Wenlockian pinnacle reefs (Fig. IV-12). The sequence begins with crinoidal bryozoan micrite which formed in water of some depth. Piles of crinoidal calcarenite occur in this early phase. Tabulate corals (the genera *Coenites*, *Favosites*, and *Halysites*) with some stromatoporoids developed. They are encrusted by algae. Evidence from cross-bedded detrital sediment shows a subsequent fall of sea level. Much of the mass may have been subjected to vadose diagenesis. During this time of lower sea level (Ludlovian) evaporites began to form in the basin (A-1 unit). A sea-level rise then permitted laminar stromatoporoids and algal stromatolites to form. Another period of exposure and desiccation is indicated by flat pebble conglomerates at the top of this unit. During this time shallow-water evaporites were precipitated from brine ponds and sabkhas in the basin (A-2 evaporites). The pinnacles were later buried by accumulation of the evaporites which form the remainder of Salina Formation. Very similar Middle Devonian stromatoporoid reefs, buried by saline deposits, are also known in the Zama areas of western Canada where they produce considerable oil and gas. Zechstein algal mounds in Germany lie peripheral to a basin of Permian age and were also buried by evaporites. This common association is of consequence. The evaporites were deposited during sea-level lowering and basin filling during cyclic reciprocal sedimentation. During these low sea-level stands the mounds were exposed, leached, and dolomitized. Porosity and permeability is thereby increased in the organic carbonate buildups intermittently during the evaporite fill-in of the basin. The same evaporites also act as seals for the reservoirs created during their formation.

Evidence Determining Trends, Paleocurrents, and Wind Directions of Silurian Buildups

In the mounds and ecologic reefs of the Silurian shelf of Indiana and Illinois, wind and wave direction was generally from southwest to west. The evidence for this offers an illustration of what kind of data may be used for directional interpreta-

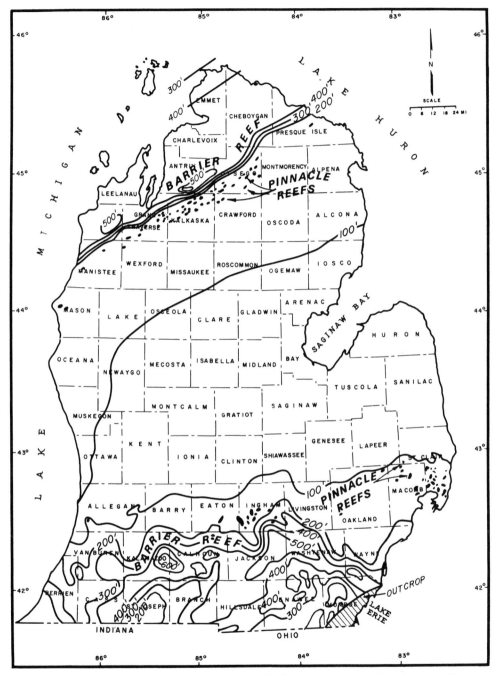

Fig. IV-11. Thickness variations within Middle Silurian carbonate of Michigan basin. From Mesolella et al. (1974, Fig. 2). Contour interval is 100 feet. Thick dolomitized carbonates associated with barrier-reef complex at basin margins grade basinward into thinner non-reef limestones. The facies transition is characterized by many hydrocarbon-bearing pinnacle reefs, i.e., offshore mounds and small banks. Illustration courtesy of author and American Association of Petroleum Geologists

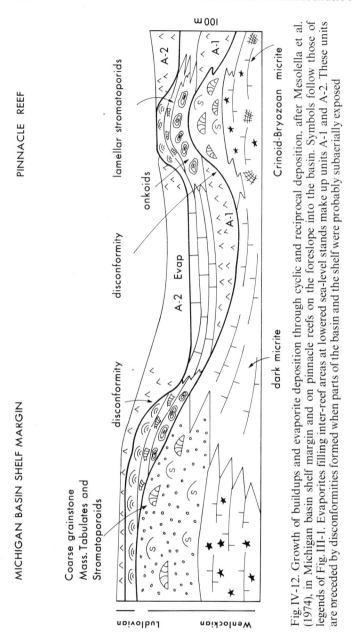

Fig. IV-12. Growth of buildups and evaporite deposition through cyclic and reciprocal deposition, after Mesolella et al. (1974), in Michigan basin shelf margin and on pinnacle reefs on the foreslope into the basin. Symbols follow those of legends of Fig. III-1. Evaporites filling inter-reef areas at lowered sea-level stands make up units A-1 and A-2. These units are preceded by disconformities formed when parts of the basin and the shelf were probably subaerially exposed

tions even in a stable shelf environment. Such differentiation and sedimentary features may occur even where the framework of organisms did not grow far into wave base or caused surf, but only extended *to* wave base.

1. Form. Since reef frameworks grow windward while their debris forms tails to leeward, owing to wave refraction, an asymmetrical form results. A "tail" or double "horn" of reef debris protrudes leeward. Flank beds not only extend farther away from the reef on this side, but their slope is less than on the steep windward side.

2. When sea level is stabilized, core or framework tends to grow over the flanking beds on all sides of a patch reef, but more on the windward side. Satellite reefs grow preferentially on windward sides.

3. Faunal differentiation occurs, forming parallel, semi-encircling belts which are narrower and more numerous on the windward side. Special organic communities tend to grow downflank on this side. These include coral thickets, crinoidal and brachiopod beds.

4. Crinoidal bioclastic debris from thickets growing on flanks surrounds the reef core. These calcarenites are thicker on the windward sides. Coarser, less disarticulated fragments occur here, despite the more widely spread sediment of this type on leeward flanks.

5. Breccia beds and coarse detritus are rarely present on these Silurian reef flanks but may occur on the windward side.

6. More argillaceous, quieter water beds interfinger with reef calcarenite on leeward flanks, wave action preventing clay deposition in the windward quadrant.

Silurian Buildups in Gotland

Silurian carbonate buildups and masses of several types have been extensively studied in the Baltic islands of Gotland and off the coast of Estonia. An extensive literature including studies by Hede (through the years 1917–1960), Hadding (through the years 1927–1958), Jux (1957), and Manten (1971) is available. The latter book discusses the Gotlandian strata in great detail and has a good bibliography. The Baltic island outcrops are part of a belt of limestone and marl trending northeast-southwest along the axis of the present Baltic Sea. These and older Paleozoic beds were deposited in shallow basins or shelf environments south of the Baltic-Scandinavian shield and east of the Caledonian geosyncline whose thick argillaceous graptolite-bearing strata crop out in southern Sweden (Scania) (Fig. IV-13) and are known in the subsurface south of the Baltic. The composite of Silurian strata exposed on Gotland is about 600 m; biostratigraphically it embraces practically the whole system. The strata dip gently southeast and are exposed in low cliffs in outcrops striking oblique to the long axis of the island, which extends for about 100 km north-south. The elaborately subdivided section consists of alternating limestones with shales and marls. The section varies lithologically because of successive marine shallowing and deepening and clay influx. It is a typical neritic section and is highly fossiliferous. Several horizons bear carbonate buildups. These occur in both regressive and transgressive sequences. According to Manten the buildups vary systematically from base to top of section.

The oldest and most northern Visby beds contain small subspherical to inverted conical steepsided knolls only a few meters in diameter. These are embedded in marls and lack flank beds. They bear a limited number of species, main reef builders being coral colonies of bell and tower shapes, with rare flattened stromatoporoids and, according to Manten, almost no algae. Crinoids are similarly absent. The matrix is lime mud. Such buildups are believed to have formed in muddy water some tens of meters deep. The Visby beds are of Late Llandoverian age. Wenlockian Silurian strata in Estonia contain superficially similar small buildups but of a different organic composition, consisting of bryozoans and calcareous algae. Hadding (1959) described a similar but *algal* bioherm at Smöjen in northeastern Gotland.

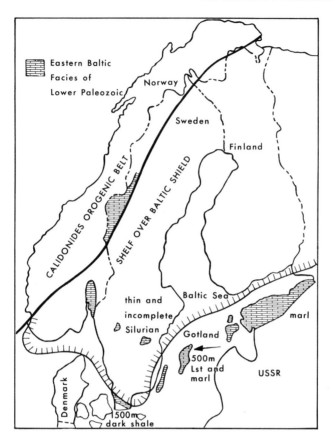

Fig. IV-13. Silurian outcrops over Baltic shield showing marl and carbonate in Gotland belt and across the shield and thicker dark shale section south of the neritic area

The first large buildups occur in Wenlockian strata on Gotland (the Holburgen type of Manten). Some of these cover 100 square m and are of flattened lens-shape. They comprise complexes of massive, highly fossiliferous, micritic "reefy" limestone and are associated with horizontal to steeply dipping flank beds. Depositional relief on the order of 5–15 m can be ascertained from field relations but is generally less. In some places the unbedded masses grade laterally into horizontal thin-bedded strata lying between them. Elsewhere the masses are onlapped by flanking beds. Such strata, which may dip up to 25 degrees, are halos consisting mainly of crinoidal debris. Marine erosion of the tops of some buildups can be observed.

These buildups resemble some of those in the middle stage of development in North America but are smaller. The masses are linear, bread-loaf shaped, trend northeast with long axes parallel to the belt of outcrops. The buildups apparently trended with the Silurian coastline, which at times shifted to a more east-west direction. The general belt containing these masses may be some kilometers wide. The masses themselves contain a highly varied fauna: some 15 genera of

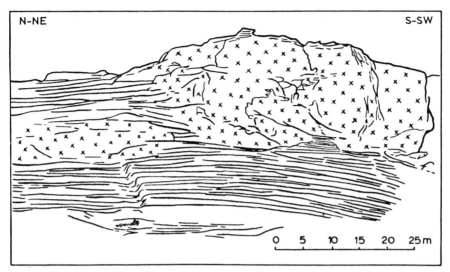

Fig. IV-14. Silurian reef on Gotland, north of Lundsklint in Högklint beds. Hoburgen reef type of Manten. From Manten (1971, Fig. 49), with permission of author and Elsevier Publishing Company, Amsterdam

Tabulates and Heliolitids and 25 genera of Tetracorals are recognized, but volumetrically stromatoporoids replace the corals as major framework builders. Algae are also common. Altogether 260 species of organisms have been recognized from these Wenlockian buildups and their flank beds. The environment is interpreted as shallow water, photic zone, but not far into active wave base (Fig. IV-14).

The highest carbonate buildups in the Gotlandian Silurian (Holmhällar type of Manten) are of Ludlovian age and are large, flattened, irregular masses built chiefly of large stromatoporoids and algae with almost no corals. The fauna is very limited; much crinoid reef flank debris occurs. These masses, not more than a few tens of meters thick, occur in pure limestone strata and may be greater than a square kilometer. They are wide and flattish, and in some places with a cuspate shape. The convex arch of the crescent faces in the windward direction away from the coastline. Their environment is considered to have been shallow, very clear and perhaps slightly restricted marine water. Shallowness is indicated by many interruptions in growth, giant fissure fillings, and coarse blocks of talus on the flanks. In addition, major circular reefs occur on the Karlsö Islands with large corals, overlain by stromatoporoids. "Flank reefs" or satellite buildups are also known here.

In some respects, the Gotlandian shelf reefs differ from those of the American Middle West. Over all, the shelf must have had better circulation and more normal marine water. Argillaceous influx was greater in Gotland. Depth may have been shallower or subsidence slower for the buildups are smaller. Flanking beds in Gotlandian reefs show less consistent wave and current direction. Yet, coarser crinoid debris occurs closer to cores and there is some evidence of a southeast wave direction based on the arcuate shape of some reefs.

No vertical biological sequence is observed within most of the individual cores of the buildups. Perhaps they existed in water so shallow and quiet that differences of relief were insufficient to cause biological zonation. In larger reefs, on the Karlsö Islands, large corals at the base give way to more stromatoporoids and Hadding (1959) suggested that here a sequence of mud-trapping globular stromatoporoids and tabulates formed a base for micrite accumulation. Such cores built up into fragmental limestone with abundant algae but hardly into the surf zone. All in all, biological zonation is not prominently displayed. In this and in the small size of most buildups, Gotlandian masses more resemble the early to middle growth stages of North American Silurian mounds and reefs. Occasionally, smooth stromatoporoids, compact corals, and patches of derived conglomerates in flanks and interruptions of growth indicate development into active wave base, but this type of reef is not very common. Most flank sediment is fine-grained. Many times Hadding (1959) stressed the role of algae in binding other sessile organisms and debris in some buildups. *Solenopora* nodules and encrusting spongy stromatolites occur. Marly intermound beds are noted for their abundant algal balls (*Sphaerocodium* and *Spongiostroma*) so that algae are of general importance in these strata.

Strangely, no stromatactoid structure is reported from the Gotlandian buildups.

Environmental conclusion: these buildups must have formed in very quiet but normal marine water with not much consistent wave and current direction. The area was far removed from the basin and close to a fine clastic source area. Buildups began on soft substrates. Water was of moderate depth, a few tens of meters. In places buildups grew to wave base in later Silurian times. Some accumulated up into water 5 m or so deep and were much like Illinois-Indiana reef masses.

Silurian buildups, both in North America and in the Baltic regions, are typically stable shelf deposits much like those described from the Ordovician. They formed as mud mounds, replete with bryozoans and often with a stromatolitic algal cap; apparently they were capable of beginning on soft substrate in quiet water of moderate depth. Growth into wave base resulted in cores of corals and stromatoporoid boundstone and extensive flank beds consisting of debris or organisms living on the tops of the buildups. Probably framework organisms grew only into wave base and remained covered by water some meters deep. Larger banks and shelf margin buildups, if they existed at all, were composites of individual cores or patch reefs. The Silurian features typify a sequence of growth stages seen in many other mounds, e.g., in the Permo-Pennsylvanian and Jurassic. The typical sequence is reviewed in this Chapter and Chapter XII.

Devonian Buildups

Older Devonian buildups consist of coral and stromatoporoid patch reefs much like those in the Silurian. No continuous shelf-margin developments are known in strata this old. In the Middle and Late Devonian, however, a bloom of coral and

stromatoporoid evolution occurred over the whole world, establishing the culmination of Middle Paleozoic reef biota. Colonial rugose tetracorals were added as
reef forms to the dominating tabulates. This assemblage took many forms and
existed in many tectonic situations. Almost all types of carbonate buildups were
represented from stromatoporoid-coral capped micrite mounds in basins and
geosynclinal troughs to micrite mounds, faro-like patch reefs on shelves, and
gigantic sprawling banks across marginal cratonic basins. Linear barrier reef
trends were also represented in Devonian time.

   Carbonate buildups of latest Silurian and earliest Devonian age are very rare,
despite extensive Middle Paleozoic exposure in the northern hemisphere. However, reefs whose age is bracketed between Ludlovian and Emsian Series are known
in central Asia. Stauffer (1968) described a narrow Siluro-Devonian reef trend
25 km long near Nowshera in northeastern Pakistan, close to the Indus River
near Kashmir and Afganistan. This is probably part of an east-west trending
platform edge although regional relationships are not clear. It consists of individual elongate mound-like cores tens of meters thick and hundreds of meters long.
Flanking facies consist of crinoidal beds and reef-derived breccia, both lithoclastic
and bioclastic. The latter consist of crinoids and pieces of dendroid favositid
corals. An interesting faunal sequence is described from top to base of the reef
cores, indicative of growth into wave base by the communities:

   Top layer 4. Tabular massive stromatoporoids more than 30%, abundant orthoconic
nautiloids. No algae. Other coelenterates not common. Common gastropods.
   Layer 3: Tabular massive stromatoporoids about 30%, rugose corals, massive *Favosites*,
crinoids and gastropods common. No algae.
   Layer 2: Abundant *Thamnopora* colonies and delicate dendroid favositids. Massive tabular stromatoporoids 25%, rugose corals and crinoids common.
   Layer 1: *Thamnopora* colonies and dendroid favositids, less than 20% massive tabular
stromatoporoids. Common rugose corals. Abundant brachiopods.

   This sequence rests on a sparsely fossiliferous micritic limestone with some
brachiopods, crinoids, and scattered stromatoporoids.

General Summary of Later Devonian Facies

Pioneer ecological studies of Late Devonian buildups derive from paleontologic
and stratigraphic work by Marius Lecompte in the Dinant Basin, Belgium. Vast
petroleum reservoirs were discovered in dolomitized and partly altered limestone
strata in the Late Devonian of the subsurface of western Canada about 1950. This
spurred the expansion to North America of the detailed biological and petrographical knowledge amassed by Lecompte. The discovery and description of a
beautifully exposed reef trend of the same Late Devonian age in western Australia
later added immensely to our knowledge of such strata. The Moscow basin also
contains Devonian petroleum reservoirs. No other reefy strata have been so
thoroughly analyzed in so many different parts of the world as the Givetian-
Frasnian beds have been from 1950–1970.

   In all the buildups discussed below, the biological assemblage is basically
similar. A generalized review of microfacies and lithofacies is given which combines petrography, organic content, and sedimentary structures. World-wide ex-

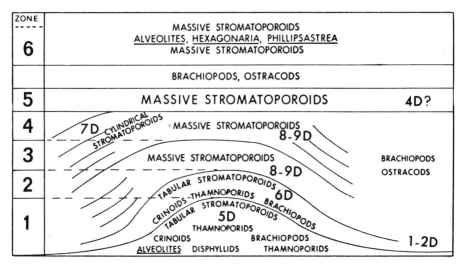

Fig. IV-15. Biofacies in carbonate buildups in Mercy Bay Member at Manning River, N.W.T., Canada. From Embry and Klovan (1971, Fig. 4). Numbers 1 D to 9 D refer to Devonian microfacies types described in Chapter IV. Illustration courtesy of authors and Canadian Society of Petroleum Geologists

tent and varied geographical settings of the Middle and Late Devonian buildups make it impossible to list all types of their microfacies. Surprisingly, however, a group of about a dozen types recur in areas where careful study has been made. These are listed below in order from basin to shelf with comments on special characteristics and environmental interpretations. The list includes numbers 1-D to 13-D, the letter desgination for Devonian to distinguish these microfacies from the standard types (SMF-). The types are chosen with particular attention to examples from western Canada where the most detailed petrographic study has been done. Photomicrographs of most of these facies are found in Klovan (1964), Fischbuch (1968), and Jenik and Lerbekmo (1968). From study of almost perfectly exposed beds in the Arctic Banks Island, Northwest Territories, Canada, Embry and Klovan (1971) were able to position some of these facies in terms of absolute water depth along slopes of organic carbonate accumulations assumed to have built up in a situation of stable sea level. Their water depth estimates are given below. Some of the microfacies are "keyed in" to the idealized diagram of Arctic Banks Island buildup described by Embry and Klovan (Fig. IV-15). A classification of Devonian faunal assemblages based simply on wave relationships was given by Lecompte (1970) and is also noted in the outline.

Basinal facies
  1D. Brown-black organic rich shale with *Styliolina*, *Buchiola*, *Tentaculites*, conodonts, goniatites. Deposits around buildups in deep, less aeriated waters. Typical in Duvernay of Alberta, and interreef deposits of Canning Basin, Australia, and Dinant Basin, Belgium. Waters as deep as a few hundred meters are estimated from the thickness of the section onlapping against the buildups. Lecompte's "deep zone".
  2D. Gray-green shale with ostracods, *Tentaculites*, *Buchiola* and some foraminifera. The Ireton shale blanket which fills in around banks in Alberta, Canada is characteristic, perhaps in water more than 100 m deep.

Shelf or platforms of open circulation

3 D. Normal marine lime wackestone-packstone. Often brown to cream color in Alberta. Varied fauna dominated by crinoids, brachiopods (excluding rhynchonellids), and with scattered rugose corals and bryozoans. Homogenized by burrowing, many lithoclasts, bioclastic debris rather macerated. Lecompte's "quiescent zone". SMF-9. Jenik and Lerbekmo 1968, Plate 2, Fig. 1.

4 D. Globular stromatoporoid-bearing shelf wackestone-packstone in biostromal layers. A few corals and red algae are included. Matrix contains organisms typical of normal marine limestone. The Cairn Formation of the Canadian Rockies is typical. They may have formed just below wave base in very shallow shelf water. Special variety related to SMF-9. Fischbuch 1968, Plates 12, 13 (Plate XVII).

Gentle slopes on banks

5 D. Coral beds. Large heads of dendroid and fasciculate Rugosa such as *Disphyllum* and massive forms such as *Phillipsastraea*, dendroid stromatoporoids like *Stachyoides*. Matrix is bioclastic wackestone; particles have no micritized rinds. Relatively low on slope in an argillaceous environment. Lecompte's "below turbulence zone". Termed coral bafflestone by Embry and·Klovan (1971) who give examples from the Arctic. Water depth estimated as 20–30 m.

6 D. Tabular stromatoporoids (25–50%) with much *Stachyoides* (dendroid stromatoporoids), sorted packstone-wackestone texture. Also includes angular nonsorted bioclasts without micritized rinds. Biota is normal marine and varied, with brachiopods and crinoids. Midslope "pure" limestone facies, nonargillaceous environment—associated with coral beds in some examples, Lecompte's subturbulent zone. Termed coral-tabular stromatoporoid "bindstone" by Embry and Klovan (1971). Depth estimate from 10–20 m. Fischbuch, 1968, Plates 16, 17 (Plate XVIII).

7 D. *Stachyoides* thickets. Boundstone with many irregular dendroid forms and some tabular stromatoporoids. Brachiopods, crinoids, and red algae. Some grainstone-packstone down middle of slope. Fischbuch, 1968, p. 19.

Bank margin facies

8 D. Massive-irregular stromatoporoids. Middle of marginal slope with grainstone-packstone matrix; much *Stachyoides*, red algae and normal marine bioclasts with micritized rims. Fairly agitated water. Lecompte's "Zone of Turbulence," probable water depth of few meters (Plate XVII).

9 D. Massive-irregular stromatoporoid with wackestone interstitial matrix. Grades bankward to backreef facies. Top of marginal slope. No normal marine bioclasts, more spherical stromatoporoids. Includes much *Stachyoides*. Embry and Klovan's depth estimate is from 10 m to sea level. Jenik and Lerbekmo, 1968, Plate 6, Figs. 1, 2.

10 D. Bioclastic calcarenite, wackestone-packstone, not much stromatoporoid. Normal marine fauna, locally with onkoids; alternating with stromatoporoid beds all along slope and front of bank; includes also *Amphipora*, a backreef organism which is probably allochthonous in the marginal environment. SMF-9. Fischbuch 1968, Plate 10; Jenik and Lerbekmo, 1968, Plate 4, Fig. 5.

Bank interior facies

11 D. Biostromes of *Amphipora*, a tiny spaghetti-like stromatoporoid in a pellet mudstone-wackestone matrix. Some pellet calcarenite packstone. Microfauna of radiosphaerid calcispheres and the foraminifera *Parathurammina*. Special biostrome with a variety of SMF-6. Fischbuch, 1968, Plate 6, Fig. 1 (Plate XIX).

12 D. Pellet calcarenite with calcispheres and foraminifera and no *Amphipora*. SMF-18. Fischbuch, 1968, Plate 4, Fig. 2, Plate 9 (Plate XIX).

13 D. Laminites of fenestral algal mats. Some *Amphipora*. SMF-19. Fischbuch, 1968, Plate 5, Fig. 1.

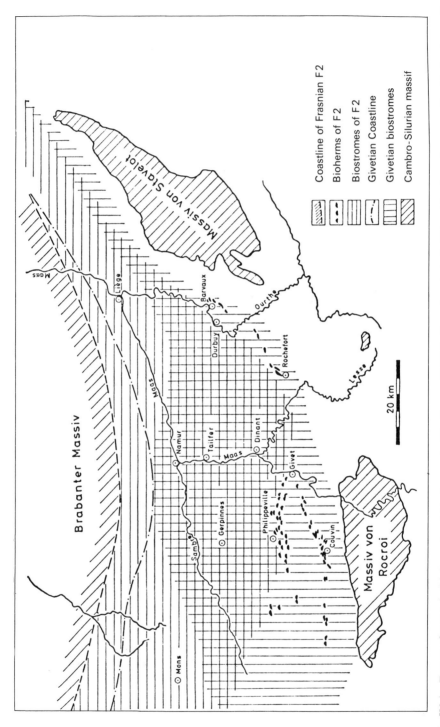

Fig. IV-16. Extent of mounds (bioherms) and biostromes in Givetian and Middle Frasnian strata in Dinant basin and surrounding Brabant massif, Belgium. From Lecompte (1970, Fig. 5). Frasnian bioherms are in a shale basin south of shelf and overlying Givetian bedded limestones

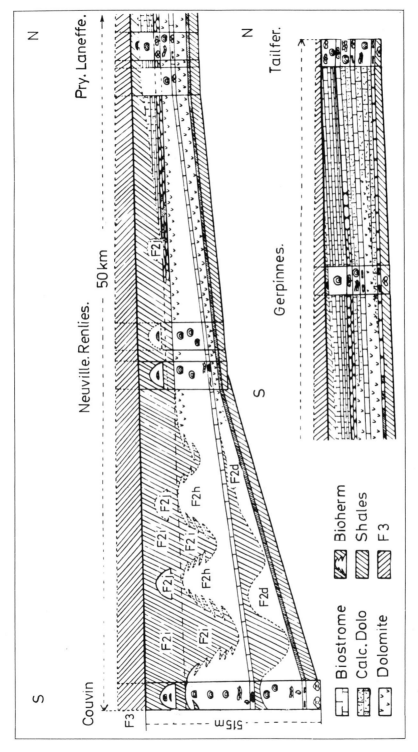

Fig. IV-17. Restored south-north section, stratigraphic model of Middle Frasnian (Upper Devonian) facies across the Dinant basin. From Lecompte (1956)

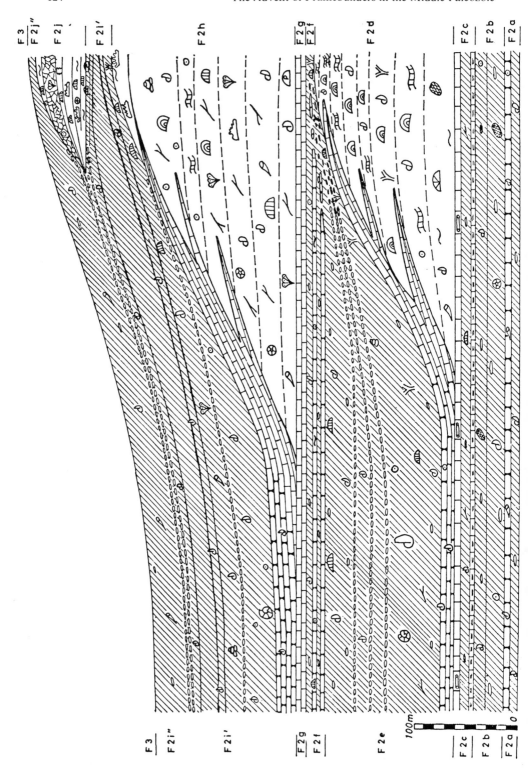

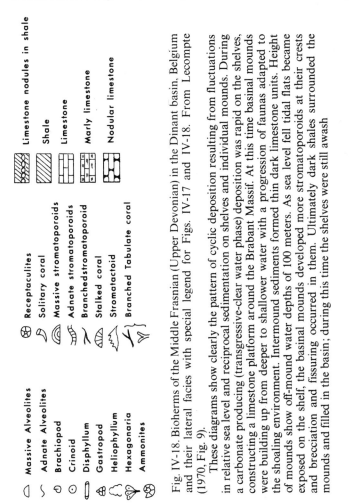

Fig. IV-18. Bioherms of the Middle Frasnian (Upper Devonian) in the Dinant basin, Belgium and their lateral facies with special legend for Figs. IV-17 and IV-18. From Lecompte (1970, Fig. 9).

These diagrams show clearly the pattern of cyclic deposition resulting from fluctuations in relative sea level and reciprocal sedimentation on shelves and individual mounds. During a carbonate producing (transgressive-clear water phase) deposition was rapid on the shelves, constructing a limestone platform around the Brabant Massif. At this time basinal mounds were building up from deeper to shallower water with a progression of faunas adapted to the shoaling environment. Intermound sediments formed thin dark limestone units. Height of mounds show off-mound water depths of 100 meters. As sea level fell tidal flats became exposed on the shelf, the basinal mounds developed more stromatoporoids at their crests and brecciation and fissuring occurred in them. Ultimately dark shales surrounded the mounds and filled in the basin; during this time the shelves were still awash

Dinant Basin of Belgium: Stromatoporoid and Coral Pinnacle Buildups and Shelf Equivalents

Figures IV-16 and IV-17 outline the settings of the well-known Belgian section described over the period of 1954–1971 by Lecompte who provided much of the basis for facies interpretation.

The beds are now tightly infolded with Carboniferous strata owing to Late Paleozoic orogeny, and outcrops are confined largely to quarries in hilly grass-covered terrain. Years of biostratigraphic study, however, have made the area the type for world wide stages of the Devonian. Couvin, Givet, Frasnes, and Famennes, are all villages in this area. Devonian deposition occurred south of the Brabant massif, the southern shore of the Old Red Sandstone continent, on a narrow shelf and in a narrow shale basin of the Variscan geosyncline. The basin was bordered on the south by swells within the geosyncline, the Stavelot and Rocroi massifs. The northern Brabant massif provided reddish terrigenous clastics to the trough. Thickness of the Late Devonian in the basin is 510 m, about twice the amount on the shelf. Marine regressions and transgressions, coupled with periodic influx of clastics, resulted in cyclic sedimentation in the area throughout Late Devonian time. During periods of transgression, carbonate sedimentation prevailed on the shelf. At this time shelf biostromes of stromatoporoids develop while mounds of carbonate mud began to accumulate downslope in the subsiding basin. In periods of marine regression or stillstand, clastics crossed the shelf and flooded into the basin. In some basinal positions continued carbonate accumulation overcame both subsidence and clastic influx and isolated carbonate mounds built steadily upward to heights of tens of meters. As this occurred a faunal sequence adapted to shoaling formed progressively higher in the buildups.

The mounds are surrounded by dark shale and overlie a Middle Devonian carbonate platform. They built upward through coral limestones to a stromatoporoid cap whose constituents became massive and irregular as wave base was approached. The progression through dark goniatite-conodont shales to brachiopod-bearing limestone to coral limestone, to tabular stromatoporoids, to massive irregular forms, is essentially that found in the Canadian buildups. Lecompte envisaged this sequence as representing progressive shoaling zones. It is now accepted that this scheme is too simple and that argillaceous influx and substrate character played a role. Figure IV-18 is Lecompte's classic presentation of this faunal succession.

The mounds contain much micrite matrix and stromatactoid structure. The "organism" *Stromatactis* was originally described from these beds. It denotes an area of calcite spar with a flat base and digitate top. Much of the sparry calcite and internal micritic sediment of stromatactoid structure obviously filled cavities protected by irregular tabular organism (e.g., bryozoans, *Alveolites*). For discussion of the origin of these see Chapter V. Major fissures in the mounds occur in places, presumably caused by large-scale slumping. The reddish color of much late Frasnian sediment outlines these fissures and the internal fillings of the stromatactoid structures.'

Flank beds in the Belgian mounds consist principally of bioclastic medium-bedded limestone with brachiopods, solitary corals, and conodonts. Extensive encrinites are not present in them. The beds intertongue with parts of the massive mound rock. The relationships indicate mound height of a few tens of meters above the sea floor.

The Dinant basin mounds offer a good illustration of how effective carbonate accumulation may be even in conditions of rapid subsidence and the biologically unfavorable terrigenous mud environment. Even though they began accumulating below wave base and in muddy water, and were originally not very far above the sea floor, the upper portions managed to build up into a zone of moderately active turbulence. Since no clear windward and leeward flank beds and no coarse foreslope breccias are known, one may assume that the tops of the bioherms remained under some meters of water. Lack of encrusting forms and a few green or blue-green stromatolitic algae on the tops support this view.

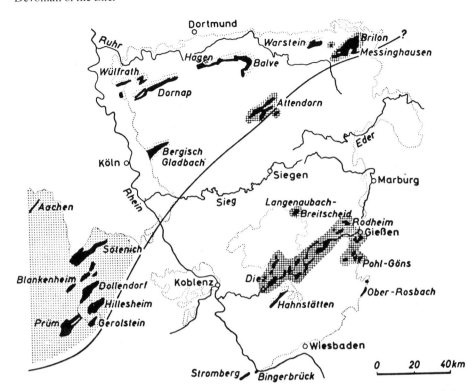

Fig. IV-19. Types of buildups and their tectonic positions in the Middle Devonian of the
Rhenish Schiefergebirge. From Krebs (1974, Fig. 12). Black areas are outcrops. Black line
separates trough to east from shelf to west; cross-hatch in lower right indicates basinal
buildups over volcanic highs in Variscan geosynclinal trough (Type 1b); cross-hatch in upper
right marks buildups at shelf margin, edge of cratonic block (Type 2b); light dot in upper left
marks buildups in interior of shelf (Type 3c); light dot in lower left indicates biostromes,
continuous reefy beds on shelf (also Type 3c). The buildup types given here are those of the
tectonic classification of Krebs and Montjoy (1972) and the author. Krebs (1971) termed them
as follows (Types A, B1, B2, and C). Illustration with permission of author and Society of
Economic Paleontologists and Mineralogists

Devonian of the Eifel

An important eastward continuation of the classical Belgian Devonian exists in
the Rheinish Shale Mountains in the Eifel district of West Germany. The tectonic
setting here is similar to that of the Brabant massif and its southern narrow shelf
but it is more complex and was subjected to much greater subsidence. A strong
northeast-southwest linearity suggests basement faulting as a control of buildups.
Studies by Jux (1960) and Krebs (1971) show a shelf area bordering the northern
massif (Fig. IV-19). East of the Rhine this area subsided as a marginal miogeosynclinal shelf. The normal interreef facies (Schwelm) consists of well-bedded dark
limestone and shales with some terrigenous clastics. Isolated elongate NE-SW
carbonate masses and buildups occur along this shelf (Dorp Facies). Buildups

with considerable relief above the bottom, existed toward the shelf margin (Type 2b on Fig. IV-19). Total thickness of Dorp facies limestone here is close to 1000 m. Individual limestone banks with less sea floor relief and whose "reefy" facies grades to interreef strata, exist in the northern, most interior parts of the shelf (Type 3c). All of these banks are flat and wide, may be hundreds of meters thick and up to 100 km. square. A shallow shelf area like that of the northern Dinant basin, existed West of the Rhine and contains biostromes of corals and stromatoporoids long termed meadow (rasen) reefs, ruben (carrot) reefs, knobby (knollen) reefs and blocky reefs by the Germans, depending on growth forms of organisms composing them (Fig. IV-19).

Farther south in the Eifel Variscan trough, reefy buildups also exist which differ from the steep-sided bioherms of the Dinant shale basin. These are larger and flatter mounds on volcanic rises capped by atoll-like stromatoporoid build-ups (Type 1b).

Despite the geosynclinal tectonic setting, the familiar facies pattern of Middle-Late Devonian buildups occurs in the Eifel and can be favorably compared with that of both Belgium and the Alberta banks (Krebs and Mountjoy, 1972). The facies sequence characteristic of the Dorp reef complex in the Eifel region is as follows:

Basin: dark shales, black, thin-bedded limestone, some turbidites.

Lower Slope: fossiliferous calcarenite and local breccias, brachiopods, bryozoan, crinoid faunas.

Upper Slope: reef flank, coarse breccia with drusy lined cavities, dendroid and tabular stromatoporoids.

Reef: massive, irregular-tabular stromatoporoids with some rugose and tabular corals. Algae not important. Some *Solenopora*.

Backreef: mainly reef detritus. *Stachyoides*, *Amphipora*, tabular corals, echinoderm peloid micrite facies.

The buildups and banks are framed by a relatively narrow reef band with *in situ* stromatoporoids, interstitial fragments, and coarse drusy linings. Unlike the Alberta reefs, echinoderms and corals are present in the bank margins and back-reef areas. This indicates considerably less restriction on the external shelf. Some of the Eifel banks do not contain an interior lagoonal facies but only a thick cap of stromatoporoids and corals. Parts of the shelf area possess vertically alternating beds with the usual backreef *Amphipora*, fenestral laminites, globular stromatoporoids, and micritic limestones.

Alberta Basin Banks

The Middle and Late Devonian typical reefoid facies exists along edges of huge, flat banks, and wide shelves surrounding the Devonian Alberta basin and along narrow banks and linear trends extending into it. Oil discoveries, beginning in 1950, in the vast plains and foothills area of western Canada resulted in extensive study of the subsurface Devonian to the east, in front of the magnificent Devonian outcrops lying westward in the Canadian Rockies. Since carbonate banks are

Table IV-3. Correlation of early Late Devonian in Alberta

Rocky Mountains outcrops	Alberta Basin subsurface	European stage
Sassenach	Alexo	Famennian
Mt. Hawk / Southesk / Mt. Hawk / Cairn / Perdrix / Maligne / Flume (Fairholme group)	Nisku / Ireton / gray-green shale / Leduc dolomite / Duvernay brown limestone shale / Cooking L. platform lst. (Woodbend group) — Waterways / Swan Hills limestone / shale-limestone (Beaverhill Lake)	Frasnian

exposed in these strata of the Cordilleran miogeosyncline, the mountain sections offer valuable clues to the Alberta basin subsurface.

Three Devonian basins were discovered during petroleum exploration of the prairie provinces of Canada. They lie between the Precambrian Canadian Shield and the western Cordilleran geosyncline. From north to south, these are: the Hay River, Alberta, and Williston. The basins were dotted with stromatoporoid build-ups in Middle Devonian time, and their edges were lined by the facies. During a regression toward the end of the Givetian, the basins were filled with evaporites. Marine transgressions from the northwest occurred episodically but with ever-increasing effect throughout the Frasnian until a major regression began in Famennian time. The Frasnian transgressions are represented in the Alberta Basin by two major groups of formations: Beaverhill Lake and Woodbend. During each of these depositional intervals, tectonic positive elements in the basement must have formed the bases for extensive carbonate platforms, rimmed by massive irregular stromatoporoid accumulations. Some of these appear as great halos built out from major elements. Carbonate buildups also developed on multiple linear and roughly parallel structural trends as well as on isolated equidimensional ones, presumably along basement faults, and over horst blocks. Positive gravity anomalies are known along some of these. The map on Fig. II-19 indicates the rather clear NE-SW tectonic grain displayed by the linear buildups and trends of shelf margins. The buildups are chiefly flat lens-shaped bodies; their lower beds grading into the surrounding shale. Marginal slopes are from 3/4 degree to 5 degrees and are generally (not always) steeper on the eastern or northeastern sides, where more organically constructed boundstone is seen. Thickness of build-

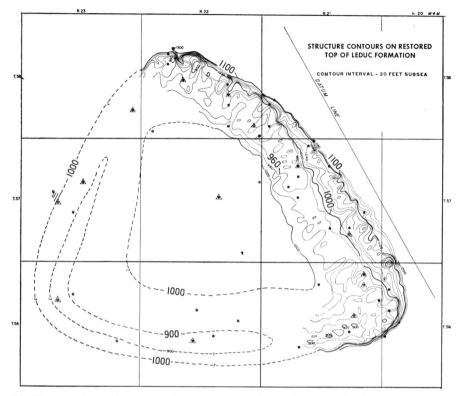

Fig. IV-20. Surface of Redwater offshore carbonate bank, Alberta basin, Canada. From Klo-van (1964, Fig. 5). Note wide central depression which may be an artifact of later compaction and not depositional. Contours are in feet. The grid marks townships which are 6 miles square. Illustration with permission of author and Canadian Society of Petroleum Geologists

ups may be hundreds of meters but platform-bank relief is variable from a few meters to 200 or so. Many buildups rise from previously formed, even wider platforms indicating that they may be residual organic accumulations surviving regional deepening of the sea and/or clastic influx by keeping "heads above water." The presence of multiple trends across the Alberta basin indicates no major response to prevailing winds and currents. The organic community was able to thrive almost as well within a semi-protected basin on its open sea side.

The intervening basinal areas of Alberta were areas of slightly deeper and considerably muddier water than the carbonate producing banks. The first transgression after the Elk Point evaporite period, resulted in widespread deposition of a fossiliferous gray shale and limestone (Waterways Formation); at this time scattered carbonate bank development became extensive. The best known carbonate complex of this earliest time is embodied in the Swan Hills Formation. Both of these units are subdivisions of the Beaverhill Lake Group of early Frasnian age. These early Frasnian buildups are regionally north and west of the maximum carbonate development of later Frasnian time. Before the major stage of buildups in Woodbend time and after construction of the Cooking Lake Plat-

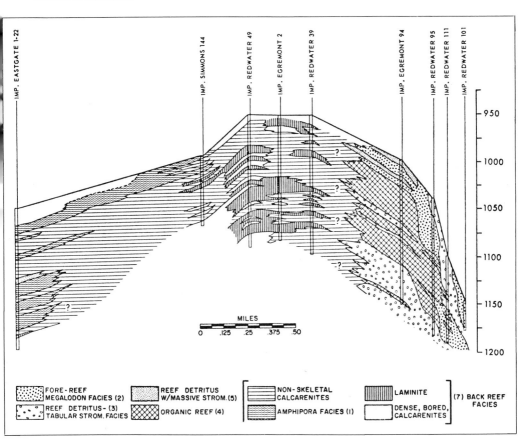

Fig. IV-21. Composite cross section of Upper Redwater reef complex showing facies across eastern edge of the large offshore bank in the Alberta Basin, Canada. From Klovan (1964, Fig. 11). Great vertical exaggeration; the true dip of eastern slope is only about 2 degrees. The interior westerly dip between Imperial Eastgate and Imperial Simmons wells is possibly an artifact of compaction. Illustration courtesy of author and Canadian Society of Petroleum Geologists

form, thin deposits of black-brown calcareous shale (Duvernay) accumulated in the basins. This time of basin starvation coincided with considerable upward growth of the surrounding banks. A gray-green shale (Ireton) filled in the reef-lined basins which may have been 200 m deep by this time (see Table IV-3).

Most of the buildups are dolomitized, a process responsible for much of the porosity, permeability, and good reservoir quality. Several of the limestone banks have been studied in great detail both paleontologically and petrographically. The microfacies types listed above in this chapter derive mainly from such study. Most of the Swan Hills banks have been described (Fischbuch 1968; Murray, 1966; Jenik and Lerbekmo, 1968) and complementary studies have been made on Woodbend strata in the Canadian Rockies (e.g., Dooge, 1966; Dolphin and Klovan, 1970; Noble, 1970). A classical study is by Klovan (1964) of the Redwater

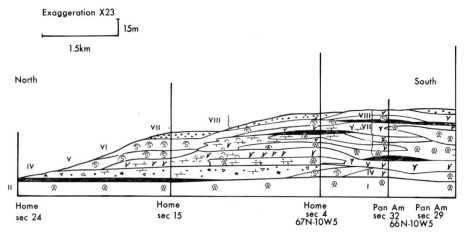

Fig. IV-22. Upper Devonian carbonate bank, Swan Hills oil field, Alberta. After Fischbuch (1968, Fig. 11 a). Bank is a buildup from normal marine limestone to a shoal of *Amphipora* Ⓐ and dendroid stromatoporoid (branching symbol). This was covered by layers of massive encrusting stromatoporoids Ⓢ and capped by grainstone as the bank accumulated into wave base. The Roman numerals indicate stages of growth in time-stratigraphic units. These are defined by thin shaly units traceable across the buildup. The upward facies progression is repeated laterally from bank edge to bank interior in most of the time-stratigraphic units. For legend see Fig. III-1. Illustration courtesy of author and Canadian Society of Petroleum Geologists

bank of Woodbend age lying east of one of the linear trends which crosses the Alberta basin (Rimbey-Leduk-Meadowbrook). Figs. IV-20 and IV-21 indicate the profile and paramount sedimentary facies of this bank. Figure IV-22 is based on Fischbuch's reconstruction of an older and flatter Swan Hills bank (1968) but plotted with a reduced vertical exaggeration.

Despite the differences in age, size, and shape, the Late Devonian Alberta banks are all generally of low profile, are flattish, isolated buildups, with essentially level tops containing deposits of very shallow lagoons. In places they developed on wider platforms, sticking up as flattened "pinnacle reefs." Marine circulation was restricted and winds, waves, and currents very moderate. Oolite is essentially lacking, indicating quiet water with not much tidal action. Multiple rows of buildups also indicate reduced effects of hydrographic controls. Evaporite deposition occurred only on the interior of the broadest shelves to the south. Tidal flats are indicated in several places by the fenestral laminite facies. The sediments also share the same upward changing sequence of organisms described by Lecompte for the smaller bioherms of the Dinant basin, Belgium.

Deposits of Steep Bank Edges Found in the Outcrops of the Canadian Rockies

Several major carbonate banks have been studied in outcrops of Late Devonian strata in the Canadian Rockies. These range from banks with very little relief (Dooge, 1966; Cook, 1972) and gradational facies over considerable distance, to much steeper shelf margins with more than 100 m relief and 2–10 degree slopes (e.g., on the southeast sides of Wapiabi Creek of Miette Bank (Noble, 1970), and at the Ancient Wall complex). These banks consist mostly of interior facies, light-colored, well-bedded limestones and dolomites with restricted marine faunas (*Amphipora* beds and peloid mudstones with radiosphaerid calcispheres). Their margins of massive light-colored, coarse, sucrose dolomite must represent organic growths of algae and stromatoporids. In the mountains these facies are termed Southesk Formation. The slope deposits are of more interest. Within a few hundred meters the obvious bank facies gives way to a variety of calcarenitic beds with onkoids, coarse bioclastic fragments and in some places, coral heads. Crinoidal beds and *Thamnopora* debris are known in such strata at Wapiabi Creek; at Miette the detritus is lithoclastic with calcirudite pieces composed of cemented lime sand in a calcarenite matrix whose grains were originally uncemented. This interstitial calcarenite was later much compacted compared to the fragments of coarser debris included within it as clasts. Such beds may represent sand flows down the slope. Spectacular megabreccias are described at Ancient Wall resulting from debris mud flows which moved large blocks up to some tens of meters in diameter. These blocks occur chaotically in a micritic matrix and are derived from various facies on the bank and its margin. No grading is obvious in the megabreccias but some finer beds composed of locally derived shelf grains are graded (allodapic limestone of Meischner). The megabreccias occur in beds from 3 to 25 m thick, some with channelled but generally planar bases (see illustration in Cook et al., 1972; Mountjoy et al., 1972).

The megabreccias extend a few km into the basin where they are interbedded with dark shale and siltstone. Slopes no greater than 2 degrees were necessary for the processes of slumping and flowage which created these wholly submarine deposits. Presumably they were caused by storm wave or tsunami phenomenon or possibly by earthquakes, although regionally the environment appears to have been tectonically stable. Such coarse sedimentary deposits are unknown in the low-relief banks drilled in the Alberta subsurface.

Canning Basin, Western Australia

The Middle and Late Devonian (Givetian and Frasnian stages) of semiarid northwestern Australia, contain some of the finest exposures of ancient carbonate buildups in the world. They are arranged along a Devonian to Late Paleozoic fault system which trends within and borders the narrow Lennard shelf on the northeastern side of the Fitzroy trough. This shelf separated the Kimberley Precambrian block of the Australian continent from a deep water basin, later filled by Permian terrigenous and glacial sediments and now buried in the subsurface. We owe our present clear understanding of this paleogeography and the related reefs to an excellent publication of the Geological Survey of Western Australia (Playford and Lowry, 1966, see Fig. IV-23).

The exposed Lennard shelf trends about 400 km in a southeasterly direction, inland from Derby, Western Australia. It contains about 500 m of Devonian strata with classical interreef, forereef, reef, and backreef facies. The buildups along the shelf take many forms: a barrier with steep slopes directly into the deep

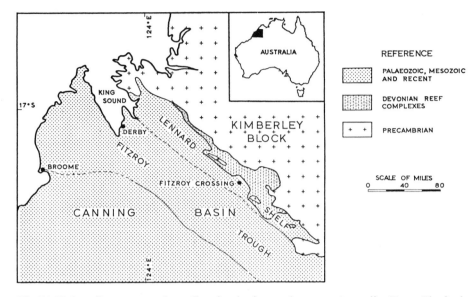

Fig. IV-23. Locality map, northern Canning basin, northwestern Australia. From Playford and Lowry (1966, Fig. 1). Middle and Late Devonian reef trend lies along the Lennard shelf margin

Fitzroy clastic trough (Napier Range), fringing reefs with lower relief lying within the shelf, carbonate ramps, bordering islands and peninsulas of Precambrian "basement" (Oscar Range), and shallow lagoon atolls or faros in the southeastern-most shelf exposures in Emanuel Range (Bugle Gap). Despite the steep shelf margin, immense amount of forereef debris, and proximity to an eroding land mass—all of which are differences from the Alberta buildups, the biota developed in the two areas is very similar. It is instructive to compare two such widely separated areas and fortunately researchers such as P. E. Playford, E. Mountjoy, J. L. Wray and A. J. Wells have been able to do field work in both areas.

The following descriptions derive principally from Playford's work on the steeper barrier reef along the Napier Range. Later some comments are made on the atolls and low relief banks known farther southeastward. The stratigraphic subdivision given on Fig. IV-24 is for the southeastern ranges but serves as well for the northwestern end where the total forereef facies is not subdivided and is included in the Napier Formation.

The forereef and the interreef areas were deposited in water ranging to more than 200 m in depth. Spectacular depositional dipping strata of 25–30 degrees have been mapped over horizontal distances of some kilometers. These have extensively built out basinward from the linear reef trend (Fig. IV-25). Geometric reconstructions from such dips and subsurface indications from a few wells are that the Devonian floor of the basin could have been hundreds of meters deep. This floor is now buried beneath 4000 m of Late Paleozoic sediment filling the Fitzroy trough. The forereef sediments contain blocks of reef limestone as much as 10–100 m across. These megabreccias occur in lithoclastic limestone of float-

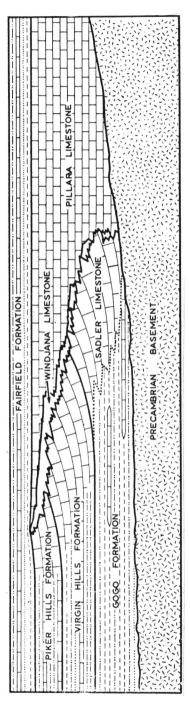

Fig. IV-24. Diagrammatic cross section illustrating stratigraphic relations of Middle and Late Devonian reef trend in the Hull-Emanuel Range area, Canning basin, northwestern Australia. From Playford and Lowry (1966, Fig. 28)

Fig. IV-25. Detail of the most spectacular exposure of the massive barrier reef in Windjana Gorge, northwestern Australia, flanked by detrital forereef and biostromal back-reef beds. Exposure slightly more than 100 m thick. Photo courtesy of A. J. Wells and Royal Shell Exploration and Production Research Laboratory, Netherlands

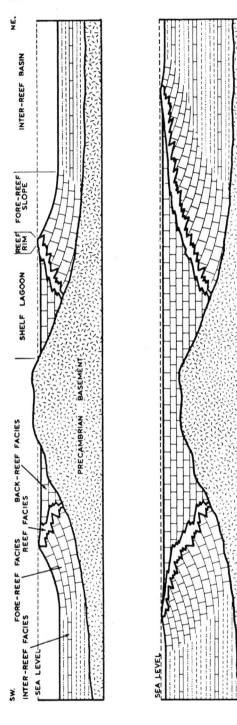

Fig. IV-26. Diagrammatic cross sections illustrating development of the Oscar Range Frasnian reef complex around and over an island of Precambrian basement. Note the shallowing of interreef basins which is believed to be responsible for extinction of the complex. From Playford and Lowry (1966, Fig. 44)

stone texture. The sediments represent gravity debris-flows down steep slopes. The forereef strata are extremely fossiliferous and contain many silicified forms. Fragments of fossils and of eroded and reworked reef rock and debris from upslope are common, e.g., corals and stromatoporoids. Even more common are fossils of indigenous slope dwellers: brachiopods, living a few tens of meters downslope in some atoll areas, sponges, crinoids and bryozoans, and abundant nautiloids. Forereef encrinites are known. Mixed with this benthonic fauna are accumulations of nektonic open-sea forms, such as goniatites and conodonts. The dipping forereef limestone grades basinward to the off-reef equivalent siltstones which also include fine-grained lime wackestone-packstone strata. The fauna here very largely consists of goniatites and conodonts (Gogo beds). Higher up the reddish Virgin Hills siltstone, a shale and lime wackestone-mudstone sequence, displays prominent stromatactoid structures which are interpreted by Playford as calcite filling and matrix replacements of openings along thin organic layers, perhaps under bulbous layers of *Renalcis*. The structure is not associated here with bryozoans as in the Silurian strata. Coarse calcite drusy cement is common in some fragmented and brecciated layers. Bulbous stacked hemispheroidal stromatolites, are known in these slope deposits. *If* sea level remained stable during forereef deposition, such stromatolites must have been deposited in water about 45 m deep. These algal bioherms are capped by the reef-encrusting form *Renalcis* (Plate XX B) which must have lived in much shallower, more agitated water. An unsolved problem in these Devonian strata is how far down the foreslope sea level may have dropped from time to time. Coarse calcite filling of stromatactoids and interbreccia spaces may also indicate deposition from much shallower water than indicated by assuming stable sea level, and projecting down the preserved depositional dip.

Interestingly, as in forereef breccias known in the Permo-Triassic beds of West Texas-New Mexico and the Dolomites in Italy, very little dolomitization has occurred. In all these examples, however, dolomite is prominent in the top and backreef of the buildup. Clearly, such dolomitization was not syndepositional.

Exposures of the relatively narrow reef front proper are marked by an abrupt change from strongly dipping foreslope to massive unbedded strata. The contact is almost vertical in places. This reef front belt is hardly more than 100–200 m wide and has gaps within it just as modern barrier reefs do. These may have been a 100 m or so wide. In many places it can be seen that the barrier was a freestanding, wave-resistant vertical wall. The narrow belt of growth furnished an immense pile of debris both lithoclastic and bioclastic. No detailed biological zonation is seen in the reef but certain important generalizations can be made. The reef crest consists on an average of only 60% organic framework. Detritus can actually range up to 90% of the volume. It is mostly fine bioclastic debris. The most striking organism is massive irregular forms of stromatoporoids, capped by the supposed alga or foraminifera *Renalcis* which was microporous, looks white in hand specimen and forms irregular encrusting chambers. A related form *Chabakovia* is also present. These algae (?) are considered to be the major binders of the reef. They much resemble the tubular encrusting foraminifera of the Late Paleozoic. Many *Solenopora* and other types of algae are seen in the reef. The large dasycladacean-like form *Receptaculites* is common. Sponges are present. The niche dwellers include the stick-like stromatoporoid *Stachyoides* which seemed to flourish widely on slope, reef front, and immediate backreef. Corals are rare in the reef (particularly in Frasnian parts) and totally absent where the reef is wide. They do occur commonly in the backreef. Slightly down the seaward slope from the reef crest several other biological assemblages occur, including brachiopods and other open marine organisms. The matrix becomes micritic and stromatactoid structures appear.

In the backreef strata (Pillara limestone) there are both similarities and differences from the well-known Canadian buildups. These strata are medium, regularly, and horizontally bedded, a striking contrast to the unbedded reef margin. Sedimentation in this environment is more or less independent of the reef itself. Little debris was provided from the barrier to the lagoon. Several hundred meters of such strata exist. The accumulation is typical of a shelf environment in its rapid alternation of rock types most of whose source material is indigenous to quiet, clear, tropical water. Several distinctive organic sedimentary types alternate in upward shoaling cyclic sequences discussed comprehensively by Read (1973). The sediment types are listed in sequence from base upward.

a) Interbedded with the several special kinds of beds are beds of bioclastic lime packstone to wackestone. These lie generally at the base of a sedimentary cycle.

b) Coral biostromes with tabular stromatoporoids, are common in the Givetian and early Frasnian. The coral heads are ball-like to lens-shaped and broken and disarranged in a micrite matrix; floatstone texture of Embry and Klovan (1971). *Hexagonaria, Disphyllum, Temnophyllum,* and *Donia* are known coral genera. The tabulates *Alveolites* and *Thamnopora* occur.

c) Biostromes of small (up to 0.5 m) globular stromatoporoids—mainly of the genus *Actinostroma.* These beds occur as close as 100 m to the reef. They are commonly micritic.

d) Cyclindrical stromatoporoids of the genus *Stachyoides* may replace or occur above the globular stromatoporoids.

e) Biostromes of *Amphipora,* a thin, twisted stromatoporoid forming "spaghetti rock." Matrix of such units varies from micrite to sparry calcite. Calcispheres are common. Such beds may be capped by globular stromatoporoids.

f) Onkoidal beds with irregular algal encrustations on snails or crinoids; matrix is commonly lithoclastic calcarenite with finely coated particles. The large clam *Eumegalodon* is present here; also thick-shelled gastropods.

g) Fenestral fabric in peloidal grainstone-packstone and wackestone with varying amounts of micrite matrix; some mud cracks; occurs commonly at top of sedimentary cycles.

h) A so-called oolite subfacies (Playford and Lowry, 1966, Fig. 24) is very probably synonymous with vadose pisolite of Dunham (1969a) and Thomas (1965) and is not the typical well-sorted, crossbedded, homogeneous oolite. Interpretations of this particular rock type, made in the Permo-Triassic beds, as of vadose concretionary origin would indicate more periodic sea level fluctuation in the Pillara limestone.

The immediate backreef zone is determined where the reefward limit of *Amphipora* and the shelfward limit of *Renalcis* approach each other or overlap slightly. Here globular stromatoporoids existed but not the typical irregular forms. The bedding begins to become thin and regular where the back reef begins.

Shelfward, this whole section grades rapidly to coarse terrigenous conglomerate whose boulders and gravels derive from the nearby Precambrian land mass along which the reefs formed a barrier. In many places the lagoon was narrow. Places are known where fanglomerates, washed out from the land with much finer terrigenous clastics, all but smothered the reef.

All in all, the backreef Pillara limestone is more normal marine and of more open circulation than in the Alberta banks whose interior facies is a rather monotonous repetition of the special rock types "e", "f", and "g", listed above. The narrowness of the shelf is evidenced by the proximity of clastic source. A changeable environment with fluctuating sea levels is indicated by the cycles. What type of reef protected this lagoon? It was relatively narrow and somewhat incomplete. In many places it probably "broke water" inasmuch as it provided an immense amount of detritus, some of it very large. Its profile is steep and sharp and, therefore, in many respects it more resembled a modern barrier reef than other Paleozoic and Mesozoic marginal buildups (profile Type III).

A few tens of km farther southeastward along the shelf, the belt of buildups broadens and shelf platforms or ramps grew over and around highs in the Precambrian basement. Perfectly circular faros with quaquaversal dips can be seen on maps and air photos. Various stages of development around, over, and outward from positive elements can be observed. The reef development is upward or outward depending on the ratio of production compared with subsidence. Channels between platforms filled with interreef sediments; depositional dips off the platform are not so steep into the protected lagoons as on the outer, southwest shelf edge, nor are flank beds so extensive. Some north-south orientation of the platforms is seen as if wind or currents might have shaped their trends. The buildups might have grown on bars of detrital sediment (e.g., crinoidal sand). Satellite algal bioherms up to a few meters high occur on the west and southwest (windward?) sides of the platforms. Brachiopods are extremely abundant downslope of these shelf platforms and faros and are represented by at least 10 common genera.

Playford described and pictured an almost perfectly circular faro (shallow atoll) with protected lagoon, stromatoporoid rim and steeply dipping flank beds. Fig. IV-26 shows a cross section across one of these shallow platforms.

Morocco Devonian

Platform-like bioherms and circular faros very similar to those of Australia, are known in western Morocco and the northern part of the province of Spanish Sahara. These range in age from Eifelian to Frasnian and are also spectacularly exposed. They lie east of a Devonian north-south positive axis trending along the Moroccan coast, are up to 100 m high and 1.5 km in maximum diameter (Elloy, 1972). Dips are quaquaversal and from 30 to 40 degrees. Some of the bioherms are micrite mounds capped by stromatoporoids. The micrite cores with abundant stromatactoid structure, have 10–15% each of bryozoan and crinoid remains and traces of corals, stromatoporoids, brachiopods, and trilobites. Tabular bodies contain more stromatoporoids and corals. An upward faunal progression is seen from disphyllid-*Thamnopora* to tabulate corals and *Hexagonaria* or from the disphyllid-*Thamnopora* assemblage through lamellar corals and stromatoporoids to massive forms of both stromatoporoids and corals. Such bioherms are known in Late Devonian strata for hundreds of kilometers to the west in southern Morocco and Algeria.

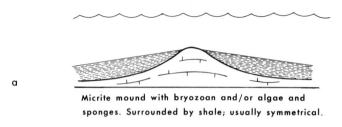

a

Micrite mound with bryozoan and/or algae and
sponges. Surrounded by shale; usually symmetrical.

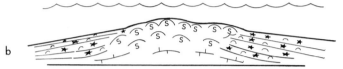

b

Patch reef with stromatoporoid cap; flank beds of crinoid debris.
This may develop into a complex of multiple cores and flank beds.
Windward and leeward sides may be discernible. Shapes may be
elongate and bodies trend parallel to depth contours.

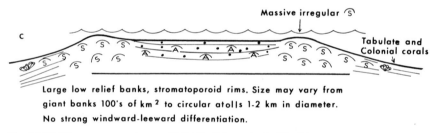

c

Large low relief banks, stromatoporoid rims. Size may vary from
giant banks 100's of km² to circular atolls 1-2 km in diameter.
No strong windward-leeward differentiation.

Fig. IV-27. Three common types of Middle Paleozoic shelf buildups

Summary

Tectonic Settings of Middle Paleozoic Carbonate Buildups

These accumulations exist in a wide variety of forms and tectonic settings. The
growth potential, particularly of the Devonian biota, while not capable of withstanding the strongest surf, was vigorous and could keep up with considerable
subsidence.

Practically all these carbonate accumulations built to wave base but not necessarily to sea-surface. The tectonic classification in Chapter XII includes many
examples of Middle Paleozoic buildups and illustrates their wide variety of tectonic settings. They are found as basinal micritic mounds and organic framework
pinnacle reefs ranging in size from 1 square km to major offshore banks. The
coral-stromatoporoid assemblage also existed in the Variscan geosyncline in Europe. More commonly the biota formed large banks and platforms bordering
large basins in all parts of the world. Such include linear shelf margins and areas
of isolated banks trending parallel to and along the external borders of platforms.

Even in shelf interiors, patches and individual mounds occur. These include simple micritic masses of algae, sponges, and bryozoans, mounds capped with corals and stromatoporoids, and circular faros and atolls of boundstone with crinoidal flanking beds. Fig. IV-27 illustrates these common forms of shelf buildups.

Latitudinal Range of Middle Paleozoic Buildups

It is worth noting how great the latitudinal range of Middle Paleozoic reefoid assemblage really is, based on the present equator. Banks Island, Northwest Territories of Canada, with the reefs described by Embry and Klovan (1971), is within the Arctic Circle (74°), and in North America the southern range is to Arizona, at almost 30° latitude. In Europe the same wide range is demonstrable from Novaya Zemlya (75°) to 25° in Spanish Sahara. About the same range is demonstrable in Silurian buildups which in North America are known from northwestern Greenland (about 80°) to about 32° in West Texas. Comparable ranges for Silurian reefs exist in the eastern hemisphere. Gotland is about 57° North and Nowshera in northern Pakistan is 34°. Evidence suggests a drastic relocation of the Middle Paleozoic equator, particularly in North America. This is because the spread of 45° is much greater than present tropical conditions permit reefoid biota to flourish, and because the north-south extension of such faunas may occur within the same cratonic plates.

Paleoecology of Middle Paleozoic Reef Associated Organisms

1. Tabulate corals: Growth forms in Ordovician and Silurian times were small, nodular to platy colonies in chains or in clusters, particularly when developed in marly sediments. Devonian colonies were hardly ever more than one meter across and usually smaller. The Devonian genus *Alveolites* had many growth forms; on lower slopes it was small and lamellar, thin and encrusting on other organisms, and it occurs generally with micrite matrix. In higher parts of buildups it may be large, irregularly massive to spherical. The Devonian small, stick-like genus, *Thamnopora*, forms biostromes ("Rasenriffe") or is intergrown with larger supporting organisms. It may also form globose colonies. The ubiquitous genus *Favosites* varies from small, flattish, dish-like forms in quiet water or on marly substrates to large, irregular, or globular masses in buildups.

Since many marly beds contain biostromes of *Tabulata* they seem to have adapted to life in somewhat turbid water. Particularly the domical-globular to occasionally elongate forms are considered to result from convex and/or upward growth to avoid being smothered by a persistent rain of fine sediment. Tabulates occur in buildups commonly downslope from stromatoporoids but also occur with them higher in the buildups in "cleaner" water.

2. Rugose colonial corals: There seems little doubt that the elongate curved, or twisted cones of the large solitary tetracorals were an adaptation to life on a soft substrate. Tsien (1971) has shown how some of these forms tended to become adnate on local hard spots on these bottoms. Solitary tetracorals are most common in well-bedded interreef limestones. Colonial *Rugosa* have more complex adaptations. Their Siluro-Devonian growth forms include nodular to massive lumps, e.g., *Hexagonaria*. Tsien has demonstrated how in the Belgium Devonian this genus varies from platy to globular proceeding upslope on buildups and how the calyx of individual corallites becomes deeper in lagoonal forms. The largest colonial tetracorals are the spectacular dendroid-fasciculate Devonian forms such as *Disphyllum*, whose colonies were as much as 2 m in diamter. Presumably the *Rugosa*, like the *Tabulata*, were more capable than stromatoporoids of withstanding a turbid environment. During dominance of the *Tabulata* in the Silurian and earlier Devonian and where conditions of marine salinity were normal,. and in water with moderate turbidity, there formed a general vertical sequence in buildups consisting upwards of *Rugosa* through *Tabulata* to a stromatoporoid cap. There is considerable overlap of forms in this sequence.

Colonial *Rugosa* also occur as isolated clumps in muddy marly beds in biostromes and downslope on edges of banks or reefs in somewhat deeper muddier water. The dendroid-fasciculate form is quite characteristic of the latter environment and has a morphological equivalent in the Jurassic-Triassic "Thecosmilia."

Tsien's analysis (1971) of growth forms for these follows. His characterizations have been somewhat modified based on experience in other areas and by comparisons with modern forms of Hexacorals.
a) Fasciculate forms: It is easy to over-generalize on these forms, many of which dwelt in especially protected niches.
   (1) Small and delicate: far downslope in argillaceous substrate.
   (2) Large and abundant: higher up slope, shallow water of moderate agitation, like modern *Acropora cervicornis*.
   (3) Bushy-knobby: agitated and muddy water (growth form like some modern *Porites*).
b) Massive growth forms:
   (1) Large and globular: Shallow water with frequent sporadic agitation alternating with quieter water, muddy conditions. "Knollenriffe" of German authors. Common in biostromes, not in real buildups, and equivalent to the biostromes of globular (cabbagehead) stromatoporoids.
   (2) Large and irregular: Wave base zone, clear-water reefy boundstone. "Blockriffe" of German authors. Comparable to ledge-flat areas of modern coral growth on Bermuda.
   (3) Large dish-shaped: Buildups, but somewhat downslope.
   (4) Small, thin, dish-shaped: Far downslope of buildups.
       (Note similarity of these shapes to that of modern genus *Montastrea* at depths of 30–50 m.)

3. The stromatoporoids: These probable Hydrozoans or sponges are characterized by very tiny pores and a low-lying, encrusting growth form (excepting some of the later dendroid types). Such morphology prevented successful competition with corals where water was muddy. On the other hand, in the Middle Paleozoic, stromatoporoids are more common than corals across vast platforms or banks of pure carbonate. Probably they were *more* tolerant than corals of warm, nutrient-depleted water of variable salinity, but could also exist with corals where water was generally clear. There is good documentation of stromatoporoids replacing both *Tabulata* and *Rugosa* toward the top of many reefs which show other independent evidence of having grown into wave base. Several basic growth forms of stromatoporoids are ecologically significant and their positions relative to slopes and tops of carbonate buildups have been tested by studies in Canada, Belgium, and Sweden.

a) Massive-irregular and encrusting: Highest water energy at or just into wave base. Matrix calcarenitic, often of broken dendroid stromatoporid fragments.

b) Tabular-lamellar: Developed below wave base encrusting or binding micrite matrix. Found consistently downslope.of buildups and commonly with corals.

c) Globular (cabbage-head) forms: Found in biostromes commonly with micrite matrix. A platform or shelf-adapted form. Known in this environment in Devonian beds from Australia, USA, Canada, and Eifel region of Germany where such strata are termed "Knollenriffe." The subspherical shape probably represents an adaptation to prevent fine sediment accumulation on the colony. Similar shapes are seen in *Favosites* and some colonial *Rugosa* as well as in modern brain corals which live in areas of moderate water agitation.

d) delicate irregular stick-like growth. This is represented exclusively by the genus *Amphipora*, which formed mats of "spaghetti rock" in quiet restricted marine backreef conditions ("Rasenriffe" of German authors).

e) The stubby dendroid form of *Stachyoides*, the same form taken by several Devonian Tabulata (e.g., *Thamnopora*), had a wider bathymetric distribution. Abundant fragments of such stromatoporoids accumulated down the slopes of banks, occur in ancient bank margins, and grade into the bank-interior facies which consists of *Amphipora*, calcispheres, and peloids. These forms probably dwelt in protected areas in turbulent water, existed over a wide range of depths, and perhaps formed thickets just below wave base on the forereef slope. The elongate stick-like form is also considered an adaptation to prevent smothering by mud suspended in water.

4. Algae: The role of algae as binders and encrusters on Middle Paleozoic reef tops was not appreciated until petrographic study was thoroughly accomplished. The algae are present in different forms (Machielse, 1972).

a) *Renalcis* and *Chabakovia*: These are tiny-chambered irregular encrusting organisms with relatively thick and porous wall structure which are generally considered algal, but are possibly foraminifera. They are structured much like the encrusting tubular foraminifera of Late Paleozoic. The irregularly plumose fabrics characterize boundstones from Cambrian to Devonian age, much as do the cornuspirids of the Permo-Pennsylvanian (Plate XX). True reefy boundstone is only well developed by these forms on Devonian shelf margins with steep slopes and detrital breccias in Canada and Australia.

b) Red algae, such *Parachaetetes* and *Solenopora* are common as nodules in buildups from all Devonian areas particularly on shelf edges. Neither in the Silurian nor Devonian do these form impressive encrusting boundstones but they are ubiquitous. They indicate normal marine water. In Chazyan mounds probable *Solenopora* combines with nodules of blue-green filamentous algae and bryozoans to form small buildups.

c) Blue-green filamentous algae coat particles and form nodules and balls (onkoids), and occur just as commonly as in all other parts of the geological record. They occur both in buildups and interreef sediments.

d) Mud-trapping stromatolitic algae make shallow-water buildups of varying size from Precambrian through Silurian.

e) Giant dasycladaceans, *Nidulites* and *Receptaculites* which have been traditionally considered sponge-like in affinity, possess structure much more like that of large dasycladacean algae. They are common in light-colored shelf limestones and in carbonate buildups from Middle Ordovician through Devonian. In the Ordovician they are associated with stromatoporoids and heads of colonial corals, the large gastropod *Maclurites* and nautiloids, an assemblage probably indicating very shallow, warm and somewhat agitated water on shoals and banks.

5. The flanking communities: A wide variety exists.

a) Pelmatozoans. Normally early and Middle Paleozoic buildups of all sizes must have been bordered by thickets and meadows of sessil echinoderms. Even in early Ordovician mounds, flank beds are rich in such debris. Crinoids, blastoids, and cystoids contributed most of the surrounding debris in the Silurian shelf mounds and continue sporadically in the Devonian and Mississippian to play a major role. These lime sands are generally winnowed and foreset-bedded and may dip quaquaversally off mounds or downslopes of linear buildups.

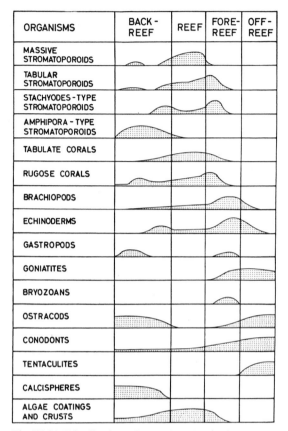

ORGANISMS	BACK-REEF	REEF	FORE-REEF	OFF-REEF
MASSIVE STROMATOPOROIDS				
TABULAR STROMATOPOROIDS				
STACHYODES-TYPE STROMATOPOROIDS				
AMPHIPORA-TYPE STROMATOPOROIDS				
TABULATE CORALS				
RUGOSE CORALS				
BRACHIOPODS				
ECHINODERMS				
GASTROPODS				
GONIATITES				
BRYOZOANS				
OSTRACODS				
CONODONTS				
TENTACULITES				
CALCISPHERES				
ALGAE COATINGS AND CRUSTS				

Fig. IV-28. Distribution of organisms in Devonian reef and off-reef areas in central Europe. From Krebs (1974, Fig. 14), with permission of Society of Economic Paleontologists and Mineralogists

Studies have been made on size and degree of disarticulation of crinoidal particles in the two areas: on Silurian buildups in Gotland by Manten (1971), and by Ingels (1963) in Illinois. In both cases size diminishes and breaking apart increases away from the core. On this basis the windward front of the buildup may be discerned. A leeward crinoid meadow filling a lagoon and a leeward expansion of such bioclastic debris is outlined by Ingels for the Thornton buildup.

b) Brachiopods were also capable of forming substantial meadows on flanking slopes of buildups. These are commonly cited as accessory flank communities. One of the most spectacular of these is the large terebratuloid *Stringocephalus* of the Middle Devonian. Pentamerids are known in similar flank positions in Silurian strata. Philcox (1970) notes flanking communities of rhynchonellid brachiopods in the Silurian of Iowa. Brachiopods in some places are known to inhabit similar environments in later buildups (*Composita* in Upper Carboniferous, rhynchonellids in certain Jurassic and Triassic reefs).

c) Convex, smooth-shelled trilobites such as the Silurian genus *Bumastus*, probably inhabited upper slopes of the buildups. These may be found in shell accumulations on tops and upper parts of the flanks.

d) Nautiloids. These forms are generally oriented downslope or accumulated in fissures in the sides of cores or flanking beds. They appear even within the micritic cores of Lower Ordovician buildups. Orthoconic forms have been reported in Chazyan strata by Pitcher

(1964), at the tops of Silurian reefs by Ingels (1963), and in Siluro-Devonian buildups by Stauffer (1968). Probably nautiloids were predators living on the buildups and their shells were washed down the sides or caught in tidal pools at the tops.

e) Coral mats such as *Amplexus* and Tabulates like *Alveolites* and *Thamnopora* formed down the flanks of some buildups.

f) Small satellite bioherms are known in flank beds of Silurian mounds in Illinois-Indiana shelf areas and on Australian reef slopes. These are similar in position and size to some known in later strata of Pennsylvanian and Triassic age. They are commonly micritic accumulation below wave-base and probably caused by algal growth and lime mud entrapment by very small and delicate dendroid framework organisms.

Figure IV-28 is a chart of the distribution of the biota in Devonian reefs and banks of Eifel district in West Germany; it serves as well for a general summary for the Silurian.

Faunal Evolution in Buildups of the Middle Paleozoic

Givetian and Frasnian strata in the Devonian mark the climax of the coelenterate reefoid faunas of the Middle Paleozoic. In North America most of these corals and stromatoporoids appear fairly abruptly in the Middle Ordovician, with associated bryozoans and sponges and gradually evolve in size and diversity until latest Devonian, Famennian, time when they decrease markedly. Carboniferous buildups are not coralline nor rich in stromatoporoids.

1. Among the first sediment binding, trapping, and framebuilding organisms which appear with the coelenterates are the sponges. Lithistid sponges in the Middle Ordovician replace the Archeocyathids and *Calathium* of the Cambro-Ordovician as reef builders. Sponges in the Chazyan reefs appear to cover from between 25–50% of the surface areas of the lowest mounds and are gradually replaced by lime-secreting corals and stromatoporoids.

2. Bryozoans of both ramose and encrusting types are abundant in micritic early stages of Ordovician and Silurian mounds in the Appalachian miogeosyncline from New York to Virginia. Beginning in the Middle Ordovician the bryozoans become associated with a surprisingly varied coelenterate fauna which gradually replaces them. From a biological point of view, the diverse biota of the Crown Point Chazyan buildups appear to constitute the first fundamentally integrated communities. These possess all the usual reefoid niche-filling forms, including the predators. In addition, the earliest nonalgal shelf margin buildup (Holston Formation of Tennessee is a morphologically diverse but solely bryozoan community. The bryozoan micrite mound assemblages of the Ordovician and Silurian appear to be replaced by corals in the Devonian but reappear in the Early Carboniferous when fenestrate bryozoans become important and corals decline.

3. Stromatoporoids are larger and more numerous and have more diverse forms in the Devonian than earlier. They appear to dominate the tops of mounds even in the Ordovician. Their characteristic growth forms are much more distinctly zoned ecologically in the Devonian.

4. Tabulate corals, although less important than rugose corals in the Devonian, are the first carnivorous coelenterates (*Lamottia* in the Chazyan) and evolve in Late Ordovician and Silurian times to a variety of large globose and dish-shaped forms which are gradually replaced by *Rugosa* in the Devonian. Even though the *Tabulata* decline somewhat in the Devonian they may still be quite varied. The genus *Alveolites* varies from dendroid to thin, platy encrustations of micrite to massive, irregular forms depending on wave base relations.

5. Rugose tetracorals evolved steadily throughout the Middle Paleozoic, gradually replacing the *Tabulata* in importance. Silurian *Rugosa* are smaller than those of the Devonian. The large dendroid-fasciculate forms such as *Disphyllum* are not known in the Silurian.

6. Mounds formed by stromatolitic mud-trapping algae appear less commonly in Devonian than in Silurian and Ordovician strata. Encrusting boundstone of possible algae occur

in Devonian reefs (*Renalcis* and *Chabakovia*) but are not seen in the Siluro-Devonian. This organic bindstone is much like that of the Late Carboniferous and Permian.

7. Crinoidal flank beds are not so abundant in Devonian as in Silurian strata but reappear importantly in the Early Carboniferous.

Generalizations about Water Depths and Turbulence on Tops of Middle Paleozoic Buildups

Several indications exist that these carbonate accumulations may have generally grown to wave base but not necessarily close enough to the sea surface to cause a surf zone. Presumably, except for certain Devonian buildups, the favored position for the assemblage was in water about 5 or 10 m deep.

1. The extensive flank beds developed around shelf buildups are for the most part purely bioclastic, composed of *organisms* which inhabited flank or top of buildups. Many of these (crinoids, blastoids, bryozoans, and long, thin, straight nautiloids) were of construction so delicate that they could not withstand extensive wave battering on the tops of the buildups.

2. Breccia beds with reworked, previously cemented, clasts eroded from the reef front are rare, though present in Australia and Canada in buildups with steeper slopes of 2–30 degrees. More typically, flank debris is composed of organisms growing at the wave base.

3. Middle Paleozoic organisms with frame-building capabilities are small compared to Mesozoic and Holocene forms. Most large colonies are of dendroid-fasciculate forms living rather well down slopes. The most impressive assemblages of encrusting boundstone is the low-lying massive irregular stromatoporoids and algal? *Renalcis* from the crest of Western Australian Devonian reefs.

4. The general absence of sabkha evaporite deposits or tidal flat deposits in the interior of most of the faros (shallow atolls) and great stromatoporoid-rimmed banks in Australia and Canada is significant, particularly when one considers that the Devonian was a time of common evaporite deposition. The characteristic *Amphipora* pellet-calcisphere wackestone-packstone assemblage represents a restricted lagoonal sediment and indicates that sea water had some access over the barrier rim to the bank interior. Some backreef or shelf areas behind reefs, e.g., Lennard Shelf, Australia, and external shelf of Eifel region, Germany, even contain corals and echinoderm beds.

Diagenesis in Middle Paleozoic Buildups

1. Many buildups of coral and stromatoporoids contain large vugs and cavities inherited from the original constructed megaporous reef fabric, extensively bored and rotted. Smaller pore space results from construction by *Renalcis* boundstone. Most cementation of these Middle Paleozoic fabrics seems to have been from organisms and within a totally marine environment. The exact origin of the coarse drusy linings and sparry in-fillings of cavities within boundstone fabric remains, however, a mystery. It is possible that it formed in the marine splash zone like the aragonite druse seen today on arid coastlines of the Trucial Coast. But, because it is also seen filling stromatactoid cavities in mud mounds, it must have originated in quiet-water conditions—either submarine or within buried sediment. It has even been considered a cave deposit from meteoric water. It is not, therefore, a particularly safe indicator for high water energy in the marine environment. This problem is also discussed in later chapters.

2. Stromatactoid structures, patches of spar with flat bases and digitate tops, are common within buildups on shelf areas and in basins but only where cores and flank beds contain much lime mudstone. The structure is not known in normally bedded lime mudstones. A

discussion of it and conjectures concerning its origin will be found in Chapter V. The original cavity is probably basically of slump origin and indicates an early porosity and permeability of micritic core rock.

3. Evidence of subaerial exposure occurs on the edges and in places within some bank interiors. Klovan describes an island rim of tidal flat fenestral laminites on the eastward side of Redwater complex. Petrographic study of the bank margin and bank interior rocks shows some cementation as well as solution under conditions of meteoric water circulation.

4. The dominance of stromatoporoids in the Middle Paleozoic offers improved chances for porosity and permeability in such carbonate strata. That the original loose network of pillars and laminae in stromatoporoids may have been aragonite is inferred by comparing their preservation with that of rugose and tabulate corals. Stromatoporoids are more susceptible to leaching and alteration to coarse calcite and to dolomitization than are corals.

Chapter V

The Lower Carboniferous Waulsortian Facies

Massive lime mudstone containing scattered crinoid and bryozoan fragments and forming lens-like buildups and mounds, constitutes a distinctive and ubiquitous facies in Lower Carboniferous (Tournaisian-Visean) strata throughout the northern hemisphere. The rock of such buildups takes the name Waulsortian, from a village in the Dinant basin, south of Namur in Belgium. Mounds and massive sheet limestone of the same age and type are well known in Pembroke, Derbyshire, and the Pennines of England and in Ireland; they also occur in central France. In the regional paleotectonic patterns of all these areas, the Waulsortian mounds and lenses appear chiefly as an intermediate (shelf margin) facies between geosynclinal basins and shelf deposits which were formed in conditions of open marine circulation (Fig. V-1).

Similar facies are known in the Osagian Series (early Mississippian) of North America. Micritic mounds, surrounded by halos of crinoids, occur in the Big Snowy Mountains, west of the apex of the Central Montana high, in the Bridger Range of Montana (Cotter, 1965; Smith, 1972), in the Boone Formation of northeastern Oklahoma around the Ozark dome (Harbaugh, 1957; Troell, 1962), and in the subsurface of north central Texas. This latter area has an arcuate belt stretching from Comanche to Montague counties and includes circular masses as

Fig. V-1. Early Carboniferous (Mississippian) facies in western Europe showing limestone shelves bordering Hercynian uplifts (dotted pattern). Culm facies is argillaceous, black siliceous limestone and siltstone in intervening basins

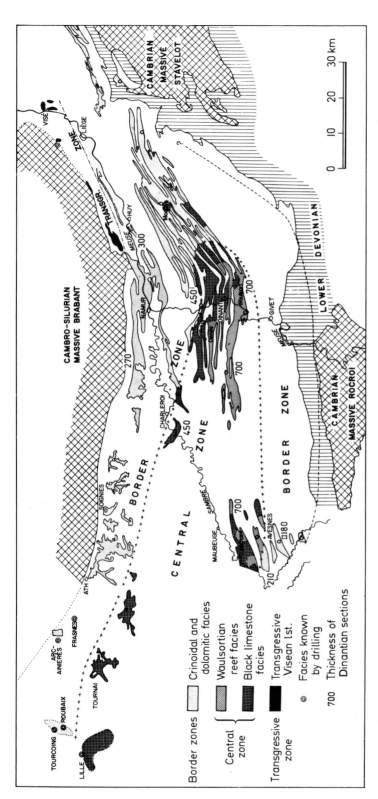

Fig. V-2. Facies of the Dinantian limestone in the Franco-Belgian basin. Compiled by and courtesy of Royal Shell Exploration and Production Research Laboratory, Rijswijk, Nederlands. Note presence of Waulsortian facies at southern edge of basin and the Marbre Noir (Black Limestone) lagoonal facies north of it separated from the Brabant massif by the shoreward crinoidal dolomitic belt. Thicknesses in meters

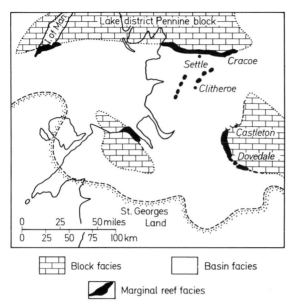

Fig. V-3. Distribution of facies in Upper Dinantian of northern England from Parkinson (1957, Fig. 1) after Hudson and Cotton (1945). Shows marginal Waulsortian facies around the north English basin and interior mounds at Clitheroe. Illustration with permission of first author and American Association of Petroleum Geologists

much as one mile across, rising 500 feet above a platform of Chapel limestone (Turner, 1957, p. 61; Pray, 1958). The most spectacular and best studied North American outcrops of the facies are in the Sacramento Mountains of New Mexico. Irregular bodies of limestone, which may represent the same sedimentary processes, are known also in Alberta in the Upper Banff Formation on Grotto Mountain.

Relation of Waulsortian Facies to Regional Paleostructure

Despite the complexities of geosynclinal structure and sedimentation, European Carboniferous paleogeography has been well worked out and facies can be readily related to tectonics. In the Dinant basin of southern Belgium, a progression of facies can be discerned south from the positive Brabant massif into a Hercynian trough, bearing a fine-grained clastic turbidite (Culm). All these strata were isoclinally folded late in the Paleozoic and it is only owing to many years of the most painstaking stratigraphic work that exact facies relationships are ascertainable within a time-stratigraphic framework (e.g., Mamet, 1962, Dupont, 1969). This has involved the correlation of biostratigraphic zones based on corals and foraminifera in shelf facies with those based on ammonites and conodonts in the troughs.

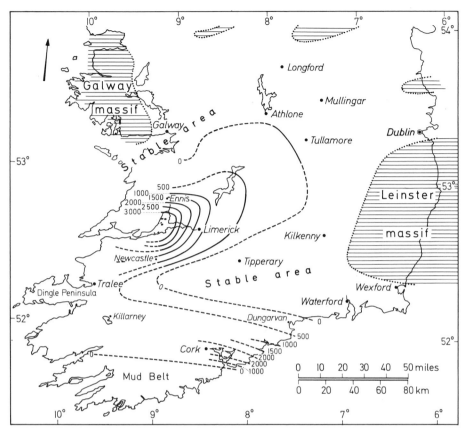

Fig. V-4. Isopach of the Waulsortian mudbank complex in southern Ireland. Thicknesses in feet. Land areas lined. From Lees (1961, Fig. 3)

The massive, type Waulsortian facies encompasses late Tournaisian to early Visean stages and locally varies greatly in thickness (more than 100 m in $1^1/_2$ km), indicating mound-like forms. Facies change is very rapid from the massive Waulsortian into lithoclastic talus and crinoidal limestone, particularly south of an irregular main belt of mounds trending east from Avesnes in France (Fig. V-2). They separate a shallow lagoon in the Namur-Dinant basins to the north, from a deeper trough which extends southward and includes the geosynclinal Culm facies. The Ardennes, a questionable land mass to the south, complicates the paleogeography. Intermound and off-mound strata consist normally of dark crinoidal wackestone-packstone (Yvoir limestone) with chert and are overlain by light-colored, more micritic and less crinoidal limestone (Leffe Formation). The facies transition into the massive Waulsortian strata is highly dolomitized. The uppermost beds of the lagoon, behind the mounds, are of early Visean age and overlap the mound topography. They are black, laminated, thin-bedded peloidal-foraminiferal wackestone and mudstones with some packstones; they represent

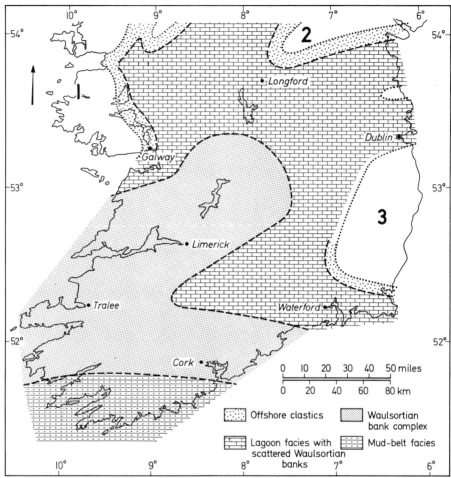

Fig. V-5. Paleogeography and lithofacies of the Waulsortian in southern Ireland. From Lees (1961, Fig. 1). The Waulsortian mudbank facies is shown at its maximum extent occupying a wide belt between the basinal mud belt (Culm) and the lagoonal facies. The latter also contains numerous scattered banks which are not indicated. The land masses surrounded by dotted lines are as follows: 1. Galway-Mayo, 2. Longforddown, and 3. Leinster

generally a subtidal environment. They are the lower part of a famous quarry rock, the Marbre Noir.

A similar shelf margin Waulsortian facies is present in the western Midlands of the British Isles (Lancashire and northwest Yorkshire). It also occurs in central Ireland, surrounding positive elements in the Hercynian troughs and forming large mounds rising more than 100 m within shale basins. Parkinson (1957) offers a good review of these British "tufa mounds" and the map of Fig. V-3 is derived from this work and from that of Hudson and Cotton (1945). The Irish outcrops display relationships very similar to those of Britain (Fig. V-4, V-5). The best modern petrographic studies of the former are by Lees (1961, 1964), Schwarzsacher (1961), and Philcox (1963, 1967).

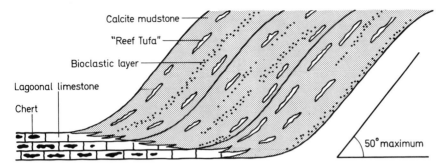

Fig. V-6. Diagrammatic cross sections of Waulsortian banks showing growth of layers of lime mudstone with inclined beds of stromatactoid structure and bioclastic debris parallel with the inclination. Vertical thickness usually only a few meters. From Lees (1961, Fig. 5)

Major Facies and Paleotectonic Patterns in Europe

1. The Waulsortian developed in clear, open marine water: The massive Waulsortian micritic facies lies some distance off from any land mass, and is a type of shelf margin facies. Land masses, which were mainly isolated blocks standing as islands in the shelf, contributed some terrigenous sediments to the shelves but this sediment did not reach the Waulsortian areas.

2. The basinal facies, termed Culm, consists of shale, sandstone, and calcareous mudstone: The terrigenous sediments were apparently derived from somewhere in southern Europe and were partly deposited in deep water. The bedding is flysch-like.

3. The shelf facies behind the Waulsortian mounds are highly varied: These shelves generally contain strata deposited in conditions of open circulation. Black (but not organic) peloidal and laminated limestone of shelf lagoonal facies is present in Belgium. The shelf above the Craven fault in England (Great Scar Limestone in Fig. V-10) contains crinoidal-bryozoan lime wackestones and thin, cross-bedded grainstone with shell banks of *Gigantoproductus*, and some coral beds. These strata are not very fossiliferous. Sheif strata equivalent to the highest mounds were influenced by incursion of deltaic clastics from the north of England and Scotland and form the strikingly cyclic Yoredale beds. In Ireland some shallow water oolitic and algal limestone was apparently deposited around islands but tidal flat deposits are rare. Shelves landward of the major Waulsortian development are spotted with isolated mounds and intermound facies which are clearly of open marine origin.

4. Evidence exists of structural control of Waulsortian facies: The Waulsortian major facies belt subsided rapidly. In Ireland the composite mass is almost 1000 m thick. Basin-prograding growth stages in these massive sheets of strata may be traced by careful attention to depositional dips, indicated by bedding, and oriented sparry calcite fillings of void space (stromatactoids) (Figs. V-6, V-7). Regionally, the Waulsortian facies grades over some distance to the shelf facies

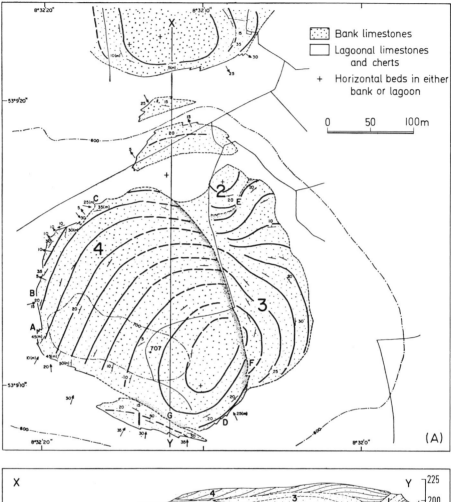

Fig. V-7. Hill in Galway County, Ireland mapped to show concentric growth lines of Waulsortian mudbank accumulation. From Lees (1964, Fig. 10), with permission of author and Royal Society of London

but passes to basin strata within a short distance. It thus appears to form a ramp built out from blocks in northern and western Ireland.

Waulsortian mounds also developed within basins. In Britain (Yorkshire and Derbyshire), the Waulsortian is developed in large or elliptical to circular mounds within the trough which existed between the Pennines and St. George's Land.

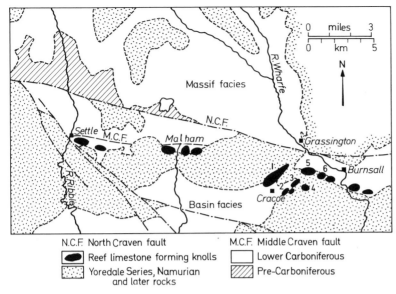

Fig. V-8. Craven fault belt of Waulsortian lime mudstone mounds, West Riding of Yorkshire. Mounds occur on downthrown side of Middle Craven fault. From Parkinson (1957, Fig. 11), with permission of American Association of Petroleum Geologists

Some mounds are between 100 and 150 m high above the old sea floors and as much as several km in diameter (Figs. V-8 and V-9). They are surrounded by dark, fine-grained, terrigenous deposits. Submarine relief is indicated by the steep sides of the sheet-like masses and the mounds which are from 30 to 50 degrees. Some flanking beds of crinoidal grainstones occur, but are not common. Slump and slide masses, breccias, and conglomerates derived syndepositionally from the mounds, are also present but relatively rare compared to other types of mounds and reefs known in the geological record. Evidence of local fault control on mound deposition exists in the North Pennine block in England which is bounded on the south by the Craven fault zones along which Waulsortian mounds occur (Fig. V-8). The faults are of Early Carboniferous age but have been rejuvenated by Tertiary movement and fine exposures occur along them (Fig. V-10). Shelf deposits over the block consist of the Great Scar limestone and cyclic Yoredale beds, the micritic Waulsortian mounds occurring only on the downthrown side of the fault zone. These are surrounded by thin and irregularly bedded black, cherty, limestone with some beds of bioclastic material, including encrinites. Black shale interbeds may be very fossiliferous. The mounds thus accumulated bathymetrically below the shelf on the downthrown sides of growth faults and in a deeper water environment forming a Type I shelf margin profile (Chapter XII and Wilson, 1974). Figures V-8 and V-9 show two adjacent areas in the English Midlands. The original mound topography along the Craven faults is believed by Hudson to have been modified by pre-Namurian (Pennsylvanian) erosion.

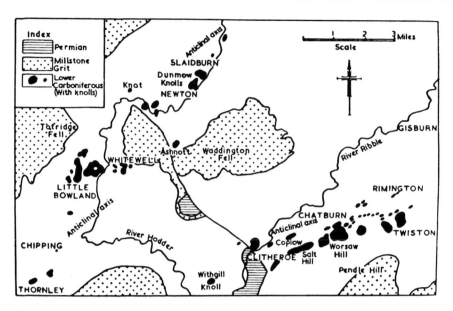

Fig.V-9. Region of "reef knolls" or lime mudstone mounds showing details of two linear trends near Clitheroe and Bowland in north English basin. Note occurrence of mounds off anticlinal axes but parallel to them. See also Fig.V-3. From Parkinson (1957, Fig.5), with permission of American Association of Petroleum Geologists

North American Examples of Waulsortian Mounds in Mississippian Facies

Despite excellent exposures in North America, regional facies relations of mound-bearing strata are less clear because of disposition of the outcrops. In the Sacramento Mountains of New Mexico, Mississippian strata (Lake Valley crinoidal limestones) are exposed along its western scarp in essentially a north-south line for about 35 km and have been carefully mapped (Pray, 1961; Meyers, 1974). In the northern part of this narrow belt, mounds rise from a uniform basal limestone and are surrounded by encrinite (Fig. V-11). They are numerous, lens-like, and somewhat irregular, with elongate axes trending north-northwest. More extensive shelf conditions probably prevailed regionally to the north and east of the present outcrops, but these cannot be seen and must be inferred from broad, regional patterns (Fig. V-12).

The southern end of the Sacramento Mountain belt, on the other hand, contains more isolated, larger, and more conical shaped mounds. Large ones, such as Mule Shoe mound (Pray, 1958), rose more than 100 m from the sea floor (Fig. V-13). The mound-bearing strata essentially disappear southward along the mountain front. Where Mississippian strata are seen in the next mountains southward (Franklin and Huecos), early Mississippian strata are thin, or essentially absent, and a presumed starved basin south of New Mexico (Armstrong, 1962; Wilson, 1969) was filled with Middle Mississippian siliceous platy limestone, the Rancheria Formation.

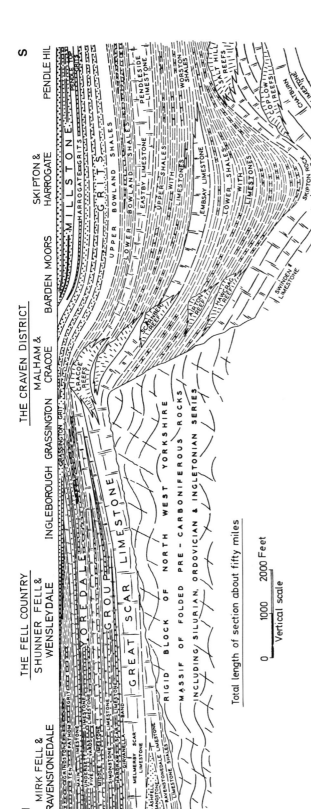

Fig. V-10. Geological section (north-south) showing the relations of Carboniferous limestones, Yoredale Cyclothems, Bowland Shale, and Millstone Grit in the Fell Country and Craven fault zone district. Note the presence of downslope Waulsortian mounds ("reefs") along what is now the downthrown sides of Craven faults. After Hudson (1930)

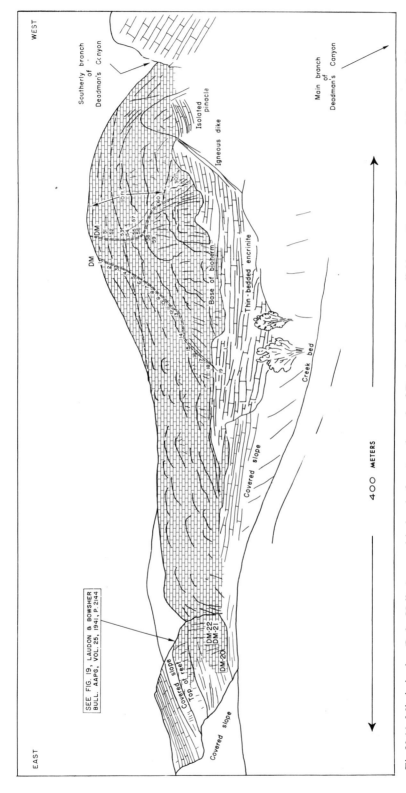

Fig. V-11. Mississippian (Lake Valley) mound in head of eastern branch of Deadman's Canyon off Alamo Canyon, Sacramento Mountains, New Mexico. DM-numbers are sample locations. Natural scale, drawn from a photograph

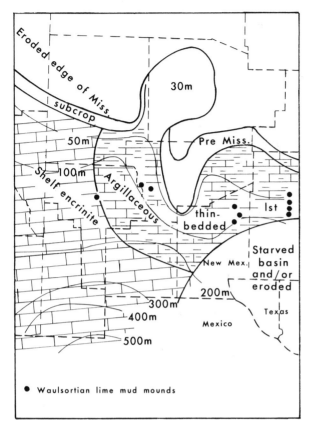

Fig. V-12. Early Mississippian facies in southern New Mexico, after Armstrong (1962), showing position of known Waulsortian mounds

The Lake Valley may have formed on a shelf of moderate depth occupying an intermediate position between shelf and starved basin. The basin was bordered on the east by the Diablo Platform and its northward extension, the Pedernal uplift, but early Mississippian beds are largely removed from this area. Armstrong's (1962) map (Fig. V-12) shows that the west side of the suspected early Mississippian starved basin was bordered by a shelf of thick crinoidal limestones (Escabrosa Formation) whose facies is almost an exact duplicate of the Great Scar Limestone bordering the north Pennine block in England. No bioherms occur in this formation.

Regional facies of early Mississippian strata (Lodgepole Formation) in central Montana have been studied by Smith (1972). The paleogeography is complicated by a Devono-Mississippian horst block (Central Montana high) trending east-west across Montana as a bridge between the Williston Basin and the Cordilleran miogeosyncline. The apex of this Paleozoic positive area lies east of the present structural crest of the Central Montana high, the Big Snowy Mountains. This is indicated by pre-Mississippian erosion of the Devonian.

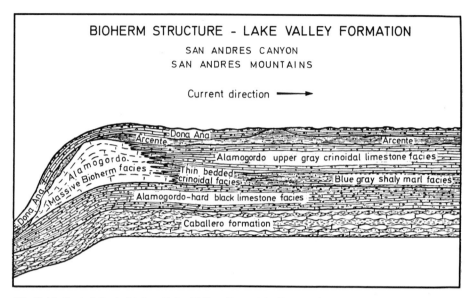

Fig. V-13. Early Mississippian (Lake Valley Group) in San Andres Mountains, New Mexico showing relationship of Waulsortian mound to surrounding strata. From Laudon and Bowsher (1941, Fig. 29), courtesy of Geological Society of America

There are subtle, but real facies changes across the positive block which have been worked out by Smith from study of the Lodgepole cyclic sediments extending across it (Chapter X). The lowest of these cycles embraces the Paine Member of the Lodgepole, a thin and rhythmically bedded argillaceous, siliceous, lime mudstone and wackestone with bryozoans and crinoid fragments (termed "deeper water limestone" by Wilson, 1969). Waulsortian type mounds of moderate size exist in the Paine Member and have been investigated by Cotter (1965) in the Big Snowy Mountains and by Stone (1971) in the Bridger Range, as well as by Smith (1972).

Cotter described the buildups in the Big Snowy Mountains as follows: there are three individual accumulations in the Paine Member in West Swimming Woman Canyon, each about 300 m long and about 70 m high. No oriented trends of the buildups can be ascertained from their outcrops.

The intermound rock is dark argillaceous, cherty, horizontally bedded limestone. The bodies themselves are massive but show vague compositional (mainly textural) layering. The sediment is all wackestone. On an average the rock is 60% micrite and 25% crinoid ossicles and fenestrate bryozoans (about half and half). Less than 5% of other types of bioclasts occur. Around 10–15% consists of sparry calcite, including somewhat irregular stromatactoid structure. The fauna of the banks is somewhat different and more varied than that from the more common basinal Lodgepole. In addition to crinoids, fenestrates, and brachiopods, many mollusk, including nautiloids, occur along with trilobites, calcispheres, ostracods, foraminifera and the solitary coral *Amplexus* and colonial *Syringopora*. No calcareous algae have been reported.

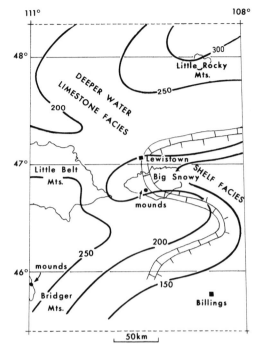

Fig. V-14. Isopach (in meters) and generalized facies map of Paine Member, Lodgepole Formation of central Montana after Smith (1972). Shows offshelf position of Waulsortian mounds in earliest Lodgepole cycle. The Paine member is characteristically "deeper water" limestone

The layering within the mounds dips from 20 to 35 degrees showing deposition on somewhat steep slopes. Erosion has truncated the mounds obliquely so that layering does not outline simple conical bodies. This complexity probably results as well from different growth centers with different directions of accumulation, inferring lack of persistent currents. In addition to complex layering, several dike-like fissures, filled from above with horizontally stratified crinoidal packstone, are noted. These are early postdepositional features resulting from dilation of the mound; They are wider below than above. Irregular, crinoid-filled cavities several meters across also occur. These were either caused by solution or early slumping within the mass. The mounds are best interpreted as biogenically accumulated banks formed below wave base in water deeper than 50 m.

Regional relationships give some evidence to show paleogeographic control of mound deposition. In the Big Snowy Mountains the buildups lie in an intermediate position between the apex of the Devono-Mississippian high, lying to the east, and the offshelf deeper water facies north and west. In the Bridger Range they lie off the Wyoming shelf (Fig. V-14). In facies and faunal content these buildups are identical to those well-known in Europe and in New Mexico. However, the facies changes are more gradual and subtle and the belts of the facies much wider in the Montana-Wyoming areas. The mounds formed as local accumulations, on a sea floor of some moderate depth.

Composition of Typical Waulsortian Facies

The Waulsortian facies thus occupied a shelf marginal position, occurring as massive sheets of limestone, or was deposited in basinal areas where it developed as large individual micrite mounds. Figure V-10 diagrams these two forms. Dips toward the basin on mound flanks may range from 30 degrees to 50 degrees, which is considered to be somewhat beyond the angle of repose of normal lime mud and suggests stabilization of the mud by some organic film or frond-like organisms.

Petrographically Waulsortian facies are remarkably similar both in North America and Europe. The most detailed study has been done in Britain by Bathurst, Lees, Philcox, and Schwarzsacher, and in the United States by Pray and Meyers, in studies of the New Mexico mounds. Only four basic rock types make up the facies although they may occur in very unequal portions.

Lime Mudstone or Wackestone (Plate XXI)

Between 50 and 80% of a mound is clotted, vaguely peloidal lime mudstone. Pray (1958) estimates that about two-thirds of the core rock of the Sacramento mounds is micrite and subsequent experience confirms this. This micrite may occur in several generations: as originally deposited lime mud and as later successive infillings of cavities. Most of the remainder of the rock mass is equally divided between a meshwork of fenestrate bryozoan fronds, both intact and fragmentary, and masses of stromatactoid sparry calcite, whose origin is discussed below. Much of the micrite is finely bioclastic. Stringers of shelly debris and pockets of whole shells include crinoids, dendroid and lacy bryozoans, orthoconic nautiloids, small horn corals (less common than in shelf limestone), brachiopods, trilobites, and ostracods. Many shells in these pockets have both valves intact. Algae of all sorts are notably absent.

Sparry Calcite and Stromatactoid Structure (Plate XVIII B)

The origin of the sparry calcite component of the Waulsortian is much debated. It commonly has stromatactoid form, i.e., extended sparry patches several cm long with a flat base and irregular, digitate top and centripetally arranged coarse crystals. The name *Stromatactis* was first proposed in 1881 by Dupont for what was considered a calcite-filled fossil form whose growth encrusted and paralleled the sides of Devonian mounds in Belgium. Lees (1964, p. 509) lists eight different names applied to it, or very similar sparry patches, by various authors. In the past it has been interpreted as a deposit from meteoric water (hence the British name "reef tufa"). It has also been considered to result simply from aggrading recrystallization of lime mud. There seems now, however, general agreement that the feature results from calcite filling of some type of preexisting hole. The centripetal crystal arrangement and the prevalence of lime mud lying like internal sediment at the bottom of the area argue for this.

The origin of the cavities through decay of some softbodied organism has been accepted by many researchers, particularly in Europe where the structure is so abundant and regular. But the traditional idea that a particular organism formed the structure is open to question. The form is too peculiar and its top is too irregular. Its shape precludes an origin as burrows. Neither is it positioned in most mounds at any certain level, only occurring where micrite matrix abounds. It holds no consistent relationship to the vertical biological sequence which a few mounds display. The multiple layers of it and the tendency for their orientation to be parallel to the outer surface or en echelon and at an angle to it are hard to explain if the structure is organic.

Other possibilities exist for the cavity such as leaching or solution of lime mud and subsequent filling with fresh water tufa, or *in situ* mud collapse and slumping as the lime mud dewatered (Heckel, 1972a). Some additional facts and arguments pertain to the problem. Typically, the base of a stromatactoid structure is smooth and filled in places with calcite mudstone of a different color or texture than that of the host micrite. This supposed internal sediment is laminated in places and may even be cross-stratified. Typical stromatactoids are commonly multiple, oriented parallel to each other in layers a few cm apart, and parallel to bedding or at least to the outer slope of the mound. Significantly, they much resemble, although on a very large scale, fenestral fabric seen in places in intertidal muds. Lees has used them and their inclinations to map growth stages of the massive sheet mounds in Ireland (Fig. V-6). If the structure is organic, the sheet-like organism was often laid down parallel to the outwardly arched surface of the mound. Sheer failure, however, of the collapsing host mud might also account for the en echelon series of oblique crust of spar-like sheet cracks or "zebra rock" seen in many places in the geologic column.

However, configuration of calcite infills in the mounds may be more complex and irregular than the parallel and multiple stromatactoid cavities. Philcox has pointed out (1963) that cavities in the mud, formed by loops of fenestrate bryozoans, may be variously filled with sparry calcite or internal lime mud sediment. Progressive gentle collapse of mud from around such partly supported cavities gives irregular shapes to the holes. The peculiar digitate tops of stromatactoids could be caused by gravity fall of semilithified sediment off the roof of a hole created by down-settling of rigid skeletal debris (e.g., bryozoan fronds) or by slumping of mud which suddenly becomes thixotropic. Later, secondary cracks at high angles to bedding may also be filled with calcite spar, mimicking the early sheet crack spar. Similar irregular stromatactoid forms are common in the brecciated algal plate mound rock of the late Paleozoic discussed in Chapter VI.

However created, the cementation of stromatactoids raises a considerable problem in carbonate geology. All British workers have emphasized the early formation of most of these cavities. Schwarzsacher (1961) pointed out their presence in reworked boulders at the foot of the mounds. Lees and Philcox noted the close association in time and space between collapsed mud matrix (M 1), cavity formation, internal sediment (M 2), and spar filling. Lees offered a quite reasonable explanation for the generation of internal sediment as having been held originally in tissues of sponges or algae and released upon decay. His interpretation accepts that the spar precipitation was submarine and ended the progressive gentle collapse of mud and decay in organisms by causing a rigid cemented framework. No modern analog to such cementation in calcite muds has been found. Indeed, marine cementation seen to date either requires an inordinately long time (deep sea crusts) or is observed in areas where rapidly moving water permeates host rock (e.g., framebuilt reefs or shallow-shelf lime sands). How could moving water in sufficient quantity have cemented cavities in mud without removing the host sediment? The coarse drusy calcite filling of Permo-Triassic mud accumulations in the Dolomites and Permian Reef Complex present the same problem (Chapter VIII). Is the cementation early marine or late vadose?

The most logical explanation for stromatactoids appears to be the filling of cavities formed in mud through settling of plates and fronds, and through slumping or brecciation of the mud mass. But the time of origin of the calcite infill and its diagenetic environment (in marine or connate water) presents a difficult problem.

Bryozoan Fronds and Crinoid Ossicles in Mounds

Not only regional stratigraphy and petrography, but also paleontology bears on the origin of the peculiar Waulsortian facies. After thin section studies were undertaken by several researchers in the 1950's, it was realized that fenestrate bryozoans were an important faunal element in the mud mounds as well as scattered echinoderm plates. Typically these calcified bryozoan fronds, along with the ubiquitous stalked crinoids, show adaptation for food gathering in moderately moving currents below active wave base. The common association of the fenestrates and crinoids leads to the assumption that perhaps the long stems of the echinoderms were rooted in mud and that the bryozoan fronds were growing attached to them as prevention against smothering in the soft muddy substrate. Estimates of bryozoan content range from 4 to 20% of rock volume. Pray (1958) notes that in the micrite of the mound cores bryozoans are more abundant than crinoids. Probably the light and lacy fronds easily floated in to collect with the lime mud. The bryozoan fronds possibly trapped lime mud and stabilized it on the steep slopes.

Flanking Encrinites (Plate XXI)

Large mounds in New Mexico and Montana are commonly flanked by steeply dipping beds of coarse encrinite made of disarticulated stems, arm, and calyx plates. Similar beds occur in Ireland and Britain but are not so common as in the North American mounds. Battered fenestrate bryozoan remains also are abundant with the crinoidal fragments. These flank beds are practically free of lime mud, almost purely bioclastic and petrographically monotonous. Only in a few places are lithoclasts important components of flanking beds despite the size of the mounds and the steepness of their slopes. Apparently the mounds were difficult to erode or they were located below active wave base. Probably Paleozoic crinoid segments like all echinoderm plates were so thoroughly impregnated with organic matter when alive that during decay and before Mg calcite inversion and calcite infill, they should have been very light particles, even floating in sea water. Thus the clean-washed, steeply dipping, crinoidal flank beds do not necessarily indicate very strong current action.

Theories of Mound Origin

Since many large Waulsortian buildups are in the form of spectacular conical mounds, consist mainly of fine lime mud, and lack any abundant framebuilding organisms, their origin has been puzzling to geologists. Any explanations of origin for these accumulations must consider (1) that the Waulsortian beds are principally micrite with scattered minor bioclastic debris in which bryozoans, a few crinoids and stromatactoid coarse spar-filling occur, (2) that paleogeographically they tend to occur below the edges of shelves in belts separating shelf and basin, (3) that the large basinal mounds are roughly equidimensional and without much marked orientation, unless perhaps in crudely defined belts, and (4) that they are layered accumulations in some places. Layering is parallel to their convex upper surface.

The mounds are not erosional remnants but are accretional in origin because of: (1) their regular smooth form, (2) steep depositional slopes (up to 50 degrees) with sheets of stromatactoid and rare bioclastic layers paralleling the external configuration, and (3) essential lack of detrital fans or breccias around most (not all) mounds but common presence of halos of winnowed crinoidal debris.

Possible ways in which such large micrite masses could have come into being include (1) hydrologic accumulations shaped by currents, (2) entrapment and/or precipitation by algae to create local piles of carbonate mud, (3) entrapment of lime mud by baffling action of bryozoans and crinoid, or (4) a combination of baffling and trapping by thickets of these organisms with accumulation in the lee of such thickets induced by gentle currents.

That the mounds are not purely current-piled sediment is indicated by the thorough mixture of the lime mud and sand-size bioclasts which indicates a lack of sorting during accumulation. The smooth, rounded, and almost perfectly conical form of many mounds also argues for accumulation in very quiet water.

The first workers to describe the Mississippian mounds of the Sacramento Mountains (Laudon and Bowsher, 1941), proposed that they originated through disintegration of non-calcareous algae which trapped lime mud or precipitated it through photosynthesis. Because of petrographic studies made during the last ten years, this hypothesis seems untenable. Very few calcareous algae appear in these mounds despite occurrence of abundant forms of these in other North American Mississippian beds. Alleged stromatolitic algae are present in one mound in Belgium at Colebi, (Dupont, 1969). On the other hand many micritic mounds *with* preserved calcareous algae occur in shallow water Paleozoic strata in North America. Why not in these Lower Carboniferous buildups? It is perhaps significant that, in addition to the absence of calcareous algae in the Lower Carboniferous buildups, no workers have described a consistent vertical sequence of faunal

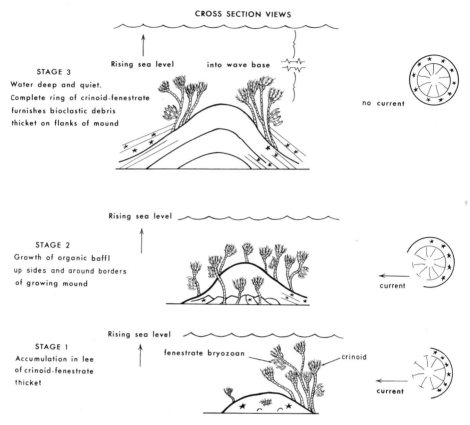

Fig. V-15. Theoretical diagram illustrating hypotheses of development of equidimensional Waulsortian mounds through progressive colonization of current-induced pile of fine sediment in lee of crinoid-fenestrate bryozoan thicket. Rising sea level accounts for mud accumulation in center of thicket growing on all sides of mound. Flanking beds develop coarse debris derived from outer edges of thicket where finest lime-mud sediment is winnowed

or textural types in them. Most geologists, familiar with the Lower Carboniferous bioherms, now consider them to be accumulations formed below wave base, perhaps in water deep enough to be below the photic zone.

The third hypothesis of origin, that of accumulation of mud under and around thriving masses of stalked crinoids and bryozoans, is precluded as a unique explanation by the simple fact of the relatively great amount of micrite and small amount of bioclastic debris within the mound core. Any explanation of the mud accumulation must also take into account the surrounding halo of coarse, winnowed, crinoidal, lime sand.

The fourth possibility is a combination of processes 1 and 3. Perhaps thickets of stalked crinoids, which supported masses of fenestrate bryozoans growing intertwined with them, offered effective baffles to gentle currents which commenced to heap up fine mud in lee areas in roughly conical piles. Once started, the perimeter of such piles was progressively colonized by crinoids and fenestrates which thrived as relative sea level rose. Their colonies gradually formed a wall-like thicket which surrounded and grew atop an increasing mound of mud. The quiet water within the growing ring of organic thickets accumulated mud and the coarser bioclastic debris drifted down the flanks. Exposed to gentle currents on the slope and around the outside of the mounds, this flank material was winnowed clean of lime mud. A steadily rising sea level is envisaged, the top of the mound never reaching very close to wave base and even remaining below the photic zone (Fig. V-15).

Conclusions

In several ways the Waulsortian facies is unique among carbonate buildups. Buildups made chiefly of lime mud are reasonably common in the geologic record principally in pre-Cretaceous time. Most of these, however, possess at least one type of fairly large organism whose bulk dominates the faunal assemblage and could form a bafflestone texture in the buildup, e.g., sponges in the Cambro-Ordovician, Permo-Triassic and Jurassic, tabular stromatoporoids in the Late Devonian, phylloid algae in the Permo-Pennsylvanian and the organ-pipe coral, "Thecosmilia" in the early Mesozoic.

Waulsortian mounds, which are just as large and more steep-sided than most, possess no major large organisms, only tiny fragments of crinoids and bryozoans constituting hardly more than 20% of the bulk. Additionally, most Waulsortian mounds appear to lack the ecological zonations described in other mounds.

Waulsortian mounds commonly contain strikingly multiple, regular and parallel stromatactoid structures (sheet spar) compared with such features seen in other calcite mud mounds. This appears especially true in Mississippian to Early Pennsylvanian mounds, although such regular stromatactoids are known in Silurian and Devonian buildups. One may argue that during the Middle Paleozoic a particular soft-bodied organism, with stromatactoid shape, was able to encrust and stabilize a sloping mud bottom. On the contrary, perhaps the multiple and parallel stromatactoids result from regular slumping and transmission of sheer stress in a more or less homogeneous lime mud accumulated without any large amount of internal shelly debris to disrupt the fabric.

The Waulsortian facies are essentially characteristic of Lower Carboniferous strata. However, very similar micrite mounds with bryozoans are known in strata as young as Middle Pennsylvanian age in the Arctic of Canada (Davies and Nassichuk, 1973). It remains to be demonstrated how important a role fenestrate bryozoans play in construction of such mounds. Do they provide necessary baffles for construction or are they merely accessory organisms growing on crinoids? Could their mesh-work be instrumental in stabilizing the steep slopes?

A basic but unsolved problem is what localizes the individual mounds. This is not unique to Waulsortian accumulations but is made more difficult by lack of any obvious baffling or framebuilding organisms. Possibly the answer lies in the distribution of certain organisms, perhaps originally controlled by gentle currents. At present, however, this cannot be adequately demonstrated.

Waulsortian downslope and basinal micrite mounds form a recognizable type of carbonate buildup repeated in nature in various parts of the geological column. In many areas they form one example of Type I shelf margin profile. It is tempting to cite as a partial analog the echinoderm-covered lime mud and sand accumulations discovered at about 700 m depth in Florida Strait off Bimini by Neumann et al., (1972). These, however, are in the area of Gulf stream current and are oriented and shaped by water movement. No evidence of this has been recorded for Waulsortian buildups.

Chapter VI

Pennsylvanian-Lower Permian Shelf Margin Facies in Southwestern United States of America

Coal measures of the Appalachians and cyclothemic deposits of the midcontinent Pennsylvanian are well known, but equally characteristic are several types of carbonate buildups produced in the clear and shallow seas around the southern extension of the North American craton. The sections containing these buildups remain lithologically varied because this was a time of general tectonic instability and considerable terrigenous sedimentation. Many of the carbonate-producing organisms which formed these offshore banks, shelf-margin buildups and shelf-interior mounds are unique to the Late Paleozoic. They appear in the Pennsylvanian and evolve continuously in the early Permian and even to the Triassic. They principally include types of algae and foraminifera, with several algal-like forms and certain sponges and stromatoporoids. Very few large frame-building organisms occur.

Paleotectonic Setting, Geologic History, and Climate

Widespread Late Paleozoic tectonic activity was as characteristic of North America as of the rest of the world (Fig. VI-1). The huge triangular extension of the craton which extends southwest through the U.S.A. stretches from the present Great Lakes region to New Mexico, Arizona, and Sonora. This was bordered throughout Paleozoic time by geosynclinal troughs along which sporadic folding and uplift proceeded in various places. Foredeep basins formed persistently between the active orogenic belts and the craton, e.g., the central Appalachians, the Black Warrior, Arkoma, Fort Worth, Valverde-Marathon, and Chihuahua-Pedregosa basins. Many thousands of meters of terrigenous sediment poured into these basins and, at intervals, out of them across the shelves of the cratonic interior. Similarly, the mid-Paleozoic Antler orogenic belt of the northern U.S.A. Cordilleran geosyncline remained somewhat active well into Pennsylvanian time in Idaho and northwestern Utah. Apparently, however, orogeny was not so violent here as in the southern and eastern perimeters of the shield.

The North American craton itself records equally impressive orogenic activity through the Late Paleozoic, but of a very different style. After general marine regression in Middle Mississippian (Visean) time terrigenous clastics of the Meramecan and Chesteran Series began to spread across the craton from the south and east, coincident with orogenic activity in the southern geosyncline. Only in the deeper tectonic basins adjacent to the southern trough did continuous sedimentation occur. Cratonic uplift removed part of the thin argillaceous Late Mississippian strata of the continental interior and the rising of isolated blocks continued coincidentally with Pennsylvanian marine transgression. The tectonic pattern

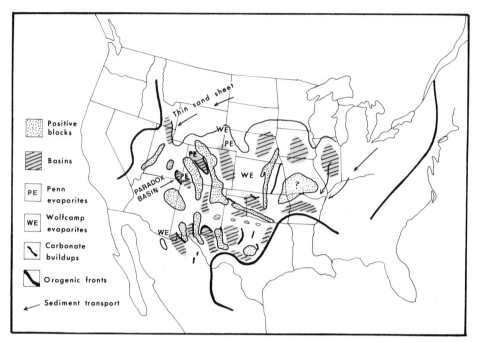

Fig. VI-1. Pennsylvanian-Wolfcampian paleotectonic map of United States with evaporite basins and carbonate shelf margin buildups indicated

of these long, narrow uplifts is crudely rectilinear. All have extremely steep sides and are clearly faulted blocks. They shed extensive feldspathic debris from the eroded Precambrian granitic and gneissic basement. Figure VI-1 shows them, their intervening basins, and areas of rimming carbonate margins.

The southwestern extension of the shield was the most affected by this fragmentation. The northern areas of the Midcontinent remained stable. Much sandy and muddy terrigenous sediment was shed to the southwest from the distant eastern Canadian area through the Ohio-Illinois basins, east of the Ozark dome, but the stable western side of the shield supported shelf deposits of clean orthoquartzite (Tensleep-Wells-Quadrant). In addition to the narrow, linear horsts like the Nemaha Ridge and Ancestral Rocky trends, three moderately uplifted equidimensional domes formed near the southern and eastern geosynclines: the Llano, Ozark, and Cincinnati-Nashville uplifts. These did not shed much sediment into adjacent basins but remained mildly positive from Pennsylvanian to the present time. They subtly affected cyclic sedimentation on the shelves separating them.

The unstable tectonic setting described above is clearly and extensively recorded by Late Paleozoic facies because the activity proceeded during major marine incursions beginning as early as Late Mississippian. These incursions reached a maximum in Des Moinesian-Missourian time and a gradual, though intermittent, tendency toward regression occurred thereafter. Clastics shed from uplifts in geosynclinal belts and from horst blocks within the craton, become increasingly prevalent in Virgilian and Wolfcampian strata. Halos of carbonate rim many of the uplifts lying seaward of the terrigenous sediments and form narrow carbonate shelf margins. Shelf areas are wider in the east, e.g., the Kansas shelf, the north central Texas outcrop region, and the eastern shelf of the Midland basin. These contain cyclic sequences of mixed clastics and carbonates and have been extensively studied by many geologists and are discussed in Chapter VII.

Climatic differences, as well as tectonics, left a strong imprint on Pennsylvanian sedimentation. Most of the seas were warm and tropical and of normal salinity as indicated by various types of calcareous algae and large foraminifera, as well as by the varied higher invertebrates which flourished in them. Pennsylvanian evaporites are known only in three western areas, along the axis of the Transcontinental arch which forms the crest of the southwesterly extended craton. Tidal flat conditions and stratigraphically extensive dolomites are unknown except in these evaporite troughs which are, from northeast to southwest: the northern Denver, central Colorado, and Paradox basins. The evaporites are generally of Desmoinesian age and coincide with the maximum limestone deposition of this same age farther southeast in Texas and New Mexico. Somewhat later, Wolfcampian evaporites, dolomite, and tidal flat strata formed in the more easterly basins. These lie in the subsurface of the northern midcontinent (Pine Salt of Montana and the Big Blue series of Kansas) subsurface and the Earp Formation of the shelf north of the Pedregosa basin. In most basins and shelves, however, open marine circulation prevailed despite intermittent ingress of large amounts of fine terrigenous material. Pennsylvanian strata are, in general, not very dolomitic in this whole area. A more humid climate is indicated farther east. Coal appears in Kansas and north central Texas and becomes important in the Appalachian foredeep. Sediments derived from the cratonic uplifts west from Kansas and Texas tend to be red. The easternmost traces of coal occur in these states and coal becomes more prevalent northeast to Pennsylvania. Paleomagnetic studies situate the Pennsylvanian equator diagonally across North America from West Texas to Newfoundland, placing the coals along this equatorial belt and the evaporites and red beds in the northern hemisphere desert. The extensive sand sheet of the Tensleep-Wells-Quadrant Formations also falls in the northern desert area and includes some sand dunes as well as marine sands.

Special Organic Communities Forming Carbonate Buildups in Late Paleozoic Strata

The special microfacies recognized within cyclic shelf sequences and described in Chapter VII are equally applicable to the strata surrounding Permo-Pennsylvanian carbonate buildups. In addition, at least a half dozen special assemblages of organisms exist which were capable of forming impressive and distinct masses of carbonate sediment during this time. These organisms belong mostly to quite different groups from those seen commonly in the geologic record before and after the Late Paleozoic. Some of them are as yet poorly known biologically and most have no Recent analogs in present carbonate areas. Whereas none of the forms seems capable of construction of an impressive organic framework, their prolific growth, encrusting habits, and potential for baffling and stabilizing sediment, must have been considerable. Several paleontologists in the Soviet Union, as well as in the U.S.A., are specialists in such organisms. The biological assemblages are described briefly below and illustrated in the Plates. Their preferred ecologic distributions can be ascertained from examining the diagrams of several buildups that have been well studied (Figs. VI-5, 7, 9, 10).

Algal Plates or Phylloid Algae (Plate XXII)

Study by Wray (1968) and Konishi and Wray (1961) and other specialists has enabled identification of abundant thin potato chip-like, irregularly crinkled plates, as probable codiacean algae with variously shaped cortical tubules and a type of red alga, *Archeolithophyllum.* Articulated growth forms with consistent patterns are unknown. The occasionally preserved cortical layers and orientation of the crystals of internal calcite mosaic indicates that only the external parts of the pads were calcified in the recognized codiacean(?) genera *(Anchicodium, Eugonophyllum, Ivanovia).* (See Horowitz and Potter, 1971, for taxonomic summary.) The plates themselves exist generally in a jumbled, brecciated mass in lime mud matrix and are never articulated—indeed, they may have grown as separate entities and as upright forms. They occur in horizontally bedded strata but commonly form micritic mounds and presumably were able to form sediment baffles and to trap lime mud. Since they generally become more abundant higher in the mounds, the mound itself may not have originated by such trapping, but once it was established, platy algae thrived on it and encouraged its growth. Their abundant growth possibly choked out other forms of life and they thus appear as a climax community. The delicate form of the plates, their common association with lime mud and the thickness of mounds affords evidence of approximate depth ranges for these algae. Codiaceans flourish today down to 20 m and may range to 70 m in the tropics. The fact that individual cores of mounds are about 25 m high at most and that these have in places an encrusting boundstone cap, indicative of wave base, gives an approximate minimum depth equal to about the mound height for initiation of platy algal growth. The extensive diagenesis through brecciation and solution alteration of the codiacean plates causes such rock to be excellent petroleum reservoir.

The algae may also occur in various states of fragmentation. This is commonly seen in flanking beds of bioherms or in piles of loose algal plates where mud matrix is essentially absent. Usually the crinkled form of algal plates is better preserved within bioherms or in beds with considerable micrite matrix. The *Archeolithophyllum* genus of red algae appears also in an encrusting habit and presumably grew as irregularly lamellar sheets on top of lime mud accumulations. It forms flatter lens-shaped structures in Kansas.

Platy or phylloid algae appear early in the Pennsylvanian, both as codiacean(?) forms and red algae and seem to flourish progressively through Late Pennsylvania. The Late Pennsylvanian beds seem dominated by the codiacean(?) genera *Eugonophyllum-Anchicodium,* with larger, more irregular pads. Earlier Pennsylvanian forms seem small, thinner, and appear to occur in somewhat straighter segments. The Pennsylvanian codiacean(?) genera still flourished in the Wolfcampian but gradually disappeared and are rare or absent in Middle Permian strata. Boundstone laminated crusts formed presumably by blue-green stromatolitic algae, may also be part of the algal plate mound facies, most commonly occurring on the tops of mounds.

Opthalmidid-Calcitornellid Tubular Foraminifera (Plates IV B, XX, XIII B)

(Synonyms or closely related forms (Toomey, 1972) are *Apterinella* and *Cornuspira.*)

The foraminifera occur in two basic growth forms:

1. Small masses of such foraminifera coat bioclastic debris of all sorts both within mound sediment and in normal marine, bedded limestone.

2. They also form low-lying plumose, hemispherical masses interstratified with blue-green algal stromatolites or codiaceans. The algae are generally coarse, neomorphically crystallized, and most internal structure is lost. Such structure of "forams" resemble modern *Nubicularia* and also *Renalcis* from the older Paleozoic. Although the growth form is low and encrusting, as much as 8 m of such rock has been measured topping algal plate mud mounds. It forms a true boundstone. Dunham (1969b) has demonstrated that it and *Tubiphytes* are responsible for growth constructed cavities in the Townsend-Kemnitz reservoir in Lea County, New Mexico. When weathered, the coarser crystallized algae take on a clear, white, or yellow (iron-stained) color and the foraminifera are black. The foraminifera have microporous tests and, when in unweathered rock, are white. This characteristic rock caps bioherms on the seaward or west side of the Sacramento Virgilian complex. It also forms small downslope bioherms, "satellites," on slopes within flank beds of larger mounds. Recent opthalmids are known in Florida waters but are not abundant (Moore, 1957). They flourish only 7–10 m deep in the relatively quiet shelf water of the Florida Reef Tract, behind the reefs and down the foreslope of the reef to 40 m depth. They occur in water of normal marine salinity and circulation. Bioherms in the outcropping Paradox Formation of Utah are composed chiefly of such forms intergrown with plumose algae (codiaceans?) and contain abundant scattered *Chaetetes* and *Caninia*.

Tubiphytes Maslov, 1956 (Nigriporella of Rigby, 1958; and Hydrocorallines of Newell et al., 1953) (Plates IV A, XXIV B)

The form is defined as tiny laminated growths which are, like the tubular foraminifera, epiphytic, encrusting and of low relief. Their biological affinity is not known. They were originally described in Russia (*Shamovella*, Rauser-Chernoussova, 1951 *n.nudem* and as *Tubiphytes* Maslov, 1956) and later termed *Nigriporella* by Rigby who believed them to be perhaps calcified tunicates. A good review of the genus was published by Toomey and Croneis, 1965, who consider them algae. They commonly contain clear calcite areas about which faint laminae are concentric. These areas may be openings within the structure of the organism (conceptacles?) or merely holes, representing vanished elements of the encrusted biota. Several holes may be seen within a single laminar mass. Like the tubular foraminifera, *Tubiphytes* was microporous (porcellaneous). It is dead white in reflected light and dark in transmitted light. Silicified material from the Big Hatchet Mountains, New Mexico, indicates that the organism may occur in small irregular balls a centimeter or less in diameter and arranged in chains or baskets.

On a world-wide basis, *Tubiphytes* occurs first in the early Carboniferous and is common into Late Jurassic reefoid strata. The type is of Kungurian (Permian) age. In the southwestern U.S.A. it appears in the Late Pennsylvanian, but becomes an important rock builder only in the Wolfcampian where it may dominate the biota of bioherms, growing intertwined with tubular foraminifera and existing in flank beds whose sediments derived from organisms growing on top of the mounds. The exact ecological position of *Tubiphytes* within such reefs is not certain; it is believed to form a boundstone with tubular foraminifera on exposed crests and seaward flanks of many carbonate buildups.

Komia Korde, 1951, a Tiny Dendroid Stromatoporoid (Plate XXIII B)

Wilson et al., (1963) gives a description and proper taxonomy. *Komia* was assigned to the red algae by Johnson (1961). Toomey and Johnson erroneously assigned certain branched forms of *Komia* to *Ungdarella* (Wilson, 1969). It apparently has a branched and flabellate form. Mostly it exists as tiny broken twigs much like the larger Devonian *Stachyoides* and *Amphipora* which it greatly resembles in internal structure. *Komia* is generally present in bedded limestones in the Middle Pennsylvanian (Derryan and Des Moinesian) in the southwest. In

West Texas, at Middle Pennsylvanian shelf margins, it is known to form porous grainstones and packstones on tops and seaward sides of buildups. Surprisingly, it has not been reported from extensively exposed and studied Middle Pennsylvania strata from the Paradox basin.

Palaeoaplysina

The form constitutes a genus, also described in Russia, which has recently been identified in the Canadian Arctic (Davies, 1971; Davies and Nassichuk, 1973) forming important bioherms of early Permian age. This organism is tentatively assigned to the hydrozoans and bears some similarity to stromatoporoids. It is described as forming thin plates or tabular pieces a few cm long, with internal branching tubules, a cellular skeleton and external protuberances. The plates occur in micritic matrix and also in debris piles as grainstones with platy algae and *Tubiphytes* boundstone. Mounds exist up to a few tens of meters thick and a hundred meters across. They may be stacked or offlap each other. The *Palaeoaplysina* mounds are known in the western Urals (as oil reservoirs), Idaho, the Yukon, and on both sides of the Sverdrup basin in Ellesmere Island outcrops. Like *Tubiphytes* the organism is known in Late Pennsylvanian strata but is more common in Early Permian beds. It is a member of the shallow shelf community, best developed in mounds of basin margin carbonates and clastics. Associated biota (foraminifera and algae) is that normally found in Late Paleozoic carbonates. Abundant dasycladaceans and oolites in such buildups indicate shallow water. *Palaeoaplysina* mounds are also interbedded with clastics in more basinal situations. Its distribution in the northern areas of both hemispheres and its apparent absence in the well-known Late Paleozoic of the southwestern U.S.A. indicates. that it is a member of the boreal fauna of Permo-Pennsylvanian age.

Encrusting Organisms

A veneer of encrusting organisms exists on the tops of some micritic bioherms in the southwestern U.S.A. Some of these were mentioned by Parks (1962), but the assemblage has never been described. Such fossils are commonly silicified and poorly preserved. Arcuate stromatolitic forms, stromatoporoids?, and sponges are known. The coarse fabric of the sponge *Stereodictyon* is recognizable at the top of the mounds in both the Sacramento and Big Hatchet Mountains of New Mexico.

Examples of Permo-Pennsylvanian Carbonate Buildups

The above described biotas and eleven basic microfacies described in Chapter VII comprise several types of carbonate buildups which exist in various tectonic settings throughout the southwestern U.S.A. Many have been carefully studied from outcrops and, as well, in the subsurface where the buildups provide extensive petroleum reservoirs. Several of these are described and figured below.

Middle Pennsylvanian Beds of the Carbonate Shelf Facies, Paradox Basin

Narrow but well-defined biohermal trends were first described and mapped in the Paradox basin by Wengerd (1951, 1955, 1963) from outcrops in the Goosenecks of the San Juan River where it crosses the Monument Upwarp in Utah. Tectonically

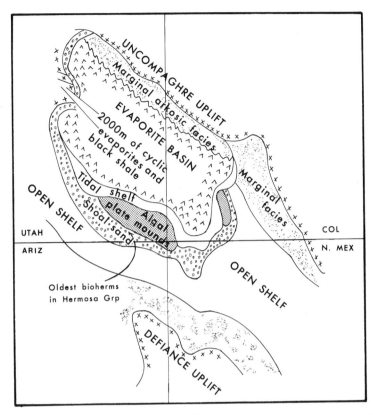

Fig. VI-2. Facies of Desmoinesian (Mid Pennsylvanian) strata in Paradox basin of Utah and Colorado. Open shelf, tidal shelf and shoal sands are largely composed of cyclic carbonate strata

these lie on a shelf southwest of the basin axis. Figure VI-2 shows the asymmetrical Paradox basin with its thick mid-Pennsylvanian central evaporite sequence (Paradox Formation) which grades northeastward to even thicker arkosic sandstones derived from the Uncompahgre uplift. These Desmoinesian beds are more than 2000 m thick in the basin and thin to 300–700 m of shelf carbonate strata within a relatively short distance (Hermosa Formation). The shelf strata consist of depositional cycles separated by dark shales and shaly dolomite which can be traced basinward into the evaporites (Fig. VI-3). The bioherms occur toward the top of the cycles in at least three levels in the formation. Earliest Desmoinesian bioherms lie farthest to the southwest on this shelf. As the basin filled progressively in later Desmoinesian time, younger bioherms developed farther northeast across the basin. The trend of the outcropping bioherms is northwest, parallel to the strike of the buildup trends on the southwest side of the basin. In general this applies also to the oil fields developed in younger buildups. Wengerd's (1963) map (Fig. VI-4) shows them to be long, straight, narrow, and somewhat stacked vertically. E.g., bioherms in the middle level are from 150 to 300 m long and 20 m thick.

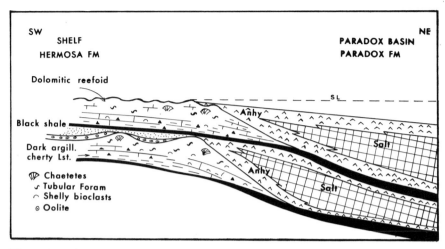

Fig. VI-3. Early mound development on southwest shelf of Paradox basin with cyclic and reciprocal sedimentation. Mounds may be capped by oolite and quartz sand or tops may be marked by an unconformity

Lithologically, the oldest buildups consist of cores of tubular foraminifera with plumose codiacean? algae forming a boundstone (Fig. VI-5). No algal plates are known. Scattered *Chaetetes* and *Caninia* are common in some of the masses. Red algae occur as accessory elements but *Komia* is not reported, although the buildups are within its stratigraphic range. Flank debris is developed impressively but disappears a short distance away; the bioclastic-lithoclastic beds grade rapidly to spiculite. The bioherms have a vertical relief of a few meters and steep slopes of several degrees (Fig. VI-5). Not enough sampling has been done to ascertain whether the basinward northeastern sides of the features have a biota that differs from the other parts.

Some interpretations can be made. The boundstone of tubular foraminifera and algae is indicative of steeper shelf slopes because this assemblage is similar to that capping some platy algal mounds seen in later Pennsylvanian strata. The rapid gradation to spiculite indicates the same steep slopes. Porosity must have been good originally; surface weathering shows much vuggy porosity despite chertification (particularly of *Chaetetes* heads).

Algal Plate Mounds at Shelf Margins

The most common Late Paleozoic shelf margin buildups are lime wackestone with abundant algal plates. These mounds are known from beds of middle Pennsylvanian to Wolfcampian age and occur throughout the southwestern U.S.A. They rim many of the subsurface oil basins in Texas, Oklahoma, and New Mexico and are known in outcrops at a basin margin in central Colorado (Minturn Formation), and in New Mexico (Sacramento, Hueco, and San Andres Mountains) rimming the Oro Grande basin. The facies provides excellent reservoir rock

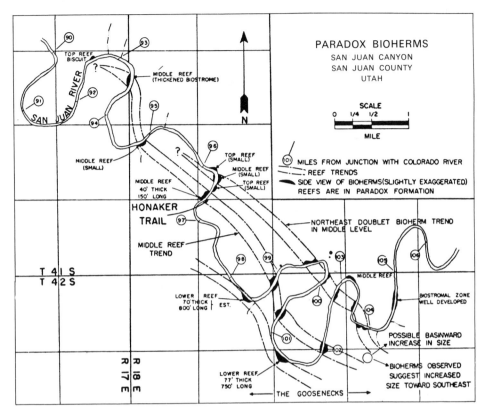

Fig. VI-4. Trends of early mounds in San Juan Canyon equivalent to Paradox basin evaporites. These trends consist of individual bioherms. The trends are about parallel to the strike of facies around the Paradox basin. From Wengerd (1963, Fig. 6)

also on the southwestern flank of the Paradox basin (Aneth, Desert Creek, Ismay fields) where mounds are mainly grainstone accumulations of platy algae (Fig. VI-2). Algal plate micrite mounds show relief of only 30 m or so but locally slopes of as much as 25° exist on edges of individual bioherms. Regional slopes into the basins may be much less, only 1 or 2 degrees at most. Beds on such gentle inclines are traceable for several km in both the Sacramento and Big Hatchet mountains of New Mexico. The shelf margin buildups may develop along strips only a few km broad and on shelves bordering channel-ways into the basins, e.g., Paradox and Oro Grande. The basins downslope of these shelves were filled reciprocally with fine clastics and evaporites at times when sea level had dropped exposing the shelf margin buildups to vadose diagenesis. Basinward sides of these micritic mounds along the shelves are somewhat steeper. Generally the mounds grew as chains, in bread-loaf shapes with long axes parallel to the depositional strike, although exceptions occur. The Ismay field in the Paradox basin shows a trend of buildups parallel to the shelf edge but with thin, lens-shaped individual mounds (12 m thick and up to 300 m long) oriented perpendicular to the basin margin, as

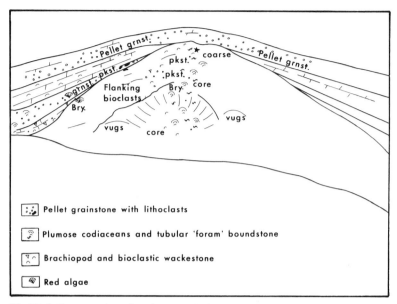

Fig. VI-5. Cross section of early mound of tubular foraminifera and algae in San Juan River across Monument Upwarp. (Lower bioherm mile 101.5, see Fig. VI-4 for location.) Mound rock (core) is of plumose codiacean algae and calcitornellid boundstone. Some mounds have *Chaetetes*. Red algae are prominent in flank beds. Lower flank beds are of dark spiculite

if they were tidal bars. Chains of the bioherms may be multiple, at least two or three in a row at the same stratigraphic level (Figs. VI-6, VI-11).

Bioherms are also affected by local structure along the shelves as well as by regional structure. Algal plate mounds tend to accumulate downslope from anticlinal axes, i.e., from tectonically induced shoals, in the Paradox shelf of Utah, the west side of the Florida high exposed in the Big Hatchets, and the area west of the Pedernal uplift in the Sacramentos. They are particularly well-developed on basinward sides where they tend to offlap each other downslope in a shingle effect (Fig. VI-7). Field observations show that mound development may have been affected by early compaction of underlying shales. Often an overlying mound drapes down the flank of an earlier-formed one, showing a tendency to accumulate in slightly deeper water just below wave base. Even in places where they are vertically stacked, and where flank beds are carbonate packstones instead of shale, this relationship is common. A drill hole placed through such a composite buildup would obviously penetrate both flank beds and mound core, as well as capping beds of individual bodies.

Algal plate (phylloid algae) mounds commonly present a characteristic growth history. Diagramatic cross sections are presented through algal plate mounds in the Sacramento Mountains on the eastern shelf of the Oro Grande basin, New Mexico (Figs. VI-7, 10, 11), and through the largest mound complex discovered in the Paradox Basin, the Aneth field in Utah (Figs. VI-8, 9). Variations in facies are obvious and many such mounds are lithologically complex although basically all are accumulations of platy algae in a brecciated micritic matrix.

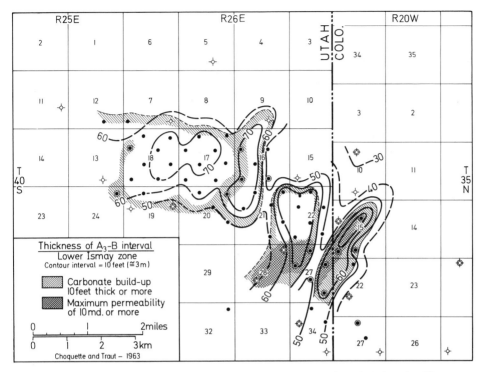

Fig. VI-6. Thickness of algal plate mounds at Ismay field, Utah and Colorado. The pronounced northeast trend of the buildups is at right angles to the overall productive trend. Darker crosshatch areas mark optimum permeability. From Choquette and Traut (1963, Fig. 4), courtesy of Four Corners Geological Society

In the Sacramento shelf cycles (Wilson, 1967a) the mounds have a preferred position in the vertical sequence of rock types. The cycles formed due to alternating transgression and regression across the shelf and the mounds began growth after the initial transgression, coincident with terrigenous influx, had concluded. Thus, the carbonate buildups occurred in a clay-free environment, at times when relative sea level was either stable or slowly dropping. They represent offshore sedimentation at the inundative, nonclastic phase of the cycle at the "turn-a-round" time between transgression and regression (Fig. VI-12). Generally, the algal plate facies developed after deposition of normal marine bioclastic lime wackestone or shale. What caused the accumulations to localize is not clear; perhaps hydrographic factors are responsible. A moundlike core (bioherm) developed in quiet water as an accumulation of lime mud with a variable content of algal plates with bryozoans and a scattered normal marine fauna. Algal plates appear to increase in abundance above the base constituting 20–40% of the volume of the rock. Later the sequence of additional facies developed which are summarized at the end of this chapter and diagrammed on Fig. VI-25.

In the Big Hatchet Mountain area, where steep depositional slopes prevailed toward the Pedregosa basin, slumps off bioherms resulted in rubble accumulations of coarse conglomerates and breccias on basinal sides of the mounds

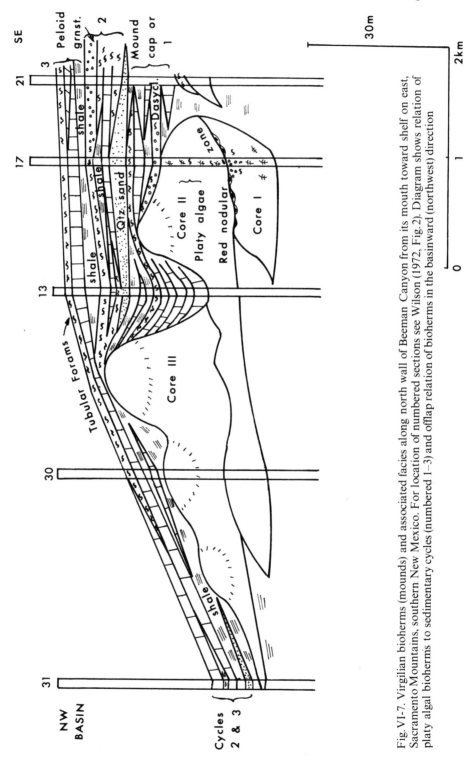

Fig. VI-7. Virgilian bioherms (mounds) and associated facies along north wall of Beeman Canyon from its mouth toward shelf on east, Sacramento Mountains, southern New Mexico. For location of numbered sections see Wilson (1972, Fig. 2). Diagram shows relation of platy algal bioherms to sedimentary cycles (numbered 1–3) and offlap relation of bioherms in the basinward (northwest) direction

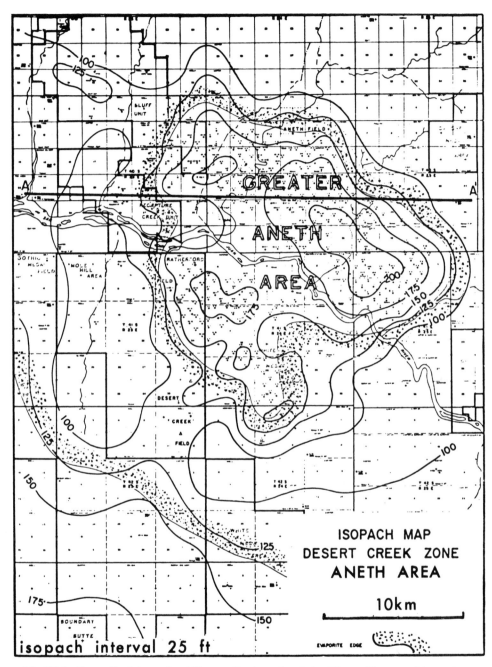

Fig. VI-8. Generalized isopach of Desert Creek cycle of the Upper Hermosa Formation. From Peterson and Ohlen (1963, Fig. 12) after Peterson (1959). Interval isopached (25 foot contours) is the carbonate-evaporite sequence between black shales. The map outlines the carbonate buildup forming the Aneth field. The trend is irregular but crudely aligned northwest-southeast parallel to the strike of facies around the Paradox basin. Illustration courtesy of Four Corners Geological Society

SW   NE

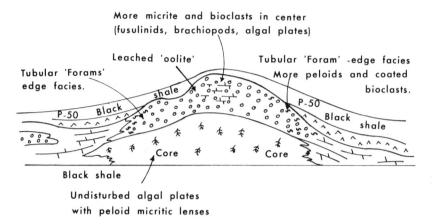

Fig. VI-9. Aneth buildup facies, Desert Creek zone. After Peterson and Ohlen (1963, Fig. 11). See isopach map, Fig. VI-8. The low mound is as much as 70 m thick and 25 km across. Vertical exaggeration is X 135. Steep side of mass facies into basin to northeast; slope is between $\frac{1}{2}-\frac{1}{3}$ degree. Illustration courtesy of Four Corners Geological Society

WEST   EAST

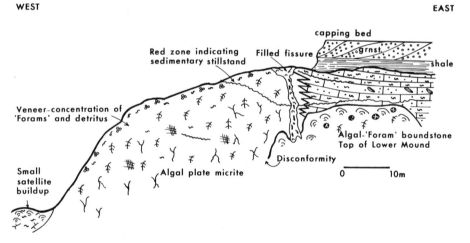

Fig. VI-10. Sequence of facies in Late Pennsylvanian mound. Interpretation of facies in Yucca mound, basal Holder Formation just north of Cloudcroft-Alamogordo highway, front of northern Sacramento Mountains, New Mexico. See map of area, Fig. VI-11. For legend see Fig. III-1. Note flank facies on lee side composed of peloid detritus with fusulinids and tubular foraminifera

(Fig. VI-13). Such beds are not seen in most shelf margin buildups; they must represent a development on unusually steep regional slopes.

Normally on shelf margins this sedimentation of cores and attendant flank beds, constituting the mound complex, was rapid enough to construct a platform or ramp by lateral accretion. In the Sacramentos, basinward of this shelf margin,

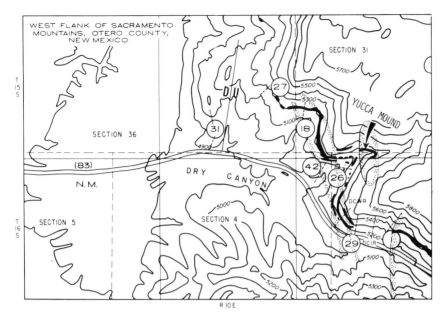

Fig. VI-11. Detailed map of Yucca mound area west flank of northern Sacramento Mountains, Otero County, New Mexico. Circled numbers are measured sections through lower Holder Formation of Late Pennsylvanian age. Wilson (1967a, 1972)

small biohermal cores are seen downslope (Fig. VI-11); these diminish in size basinward and are enveloped by grey-green silt and sand; forestet flank beds are missing. On the other hand, across the shelf, shoal-water conditions developed when the total accumulations had reached wave base. Horizontal capping beds of cross-bedded lime grainstones developed over the mound-flank bed complex. They contain battered debris of foraminifera, dasycladaceans, and gastropods, in places considerably altered by surface exposure. This final stage in development of such a mound complex is seen on Figs. VI-10, VI-25.

Porosity within algal plate-mound facies is complex and irregular. The sedimentary fabric is explained best by exposure to meteoric water before much lithification had occurred. Collapse and flowage of lime mud within a framework of delicate, but rigid, algal plates and skeletal fragments resulted in a jumbled brecciated mass of sediment with many syngenetic fractures. Stromatactoid cavities were formed and partly filled by geopetal mud. Leaching and solution acted on this heterogenous mass and attacked particularly the poorly calcified aragonitic algal plates; probably, at the same time, partial infill by internal sediment and drusy calcite was occurring. Multiple porosity levels in reservoirs show that this happened many times. On outcrops many biohermal cores are composite showing erosion before deposition of higher parts. This is probably due to sea level drop and exposure because zones of reddish conglomerate occur at such horizons. Thickness of such reservoirs in the subsurface is not great but locally good interconnection of pore space is present through the nontectonic and secondarily leached fractures.

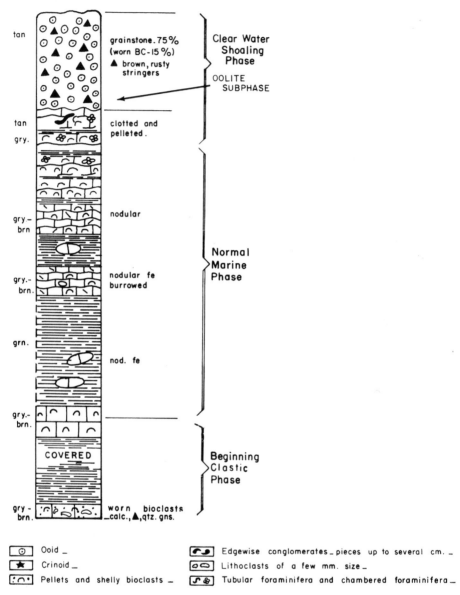

Fig. VI-12. Virgilian cycle 10 m thick, on south wall, near mouth of Beeman Canyon, Sacramento Mountains, New Mexico. Algal plate mounds typically develop in clotted-pelleted foraminiferal strata just beneath oolitic phase of cycle. From Wilson (1967a)

Little dolomitization occurs in such reservoirs except in the Aneth field of the Paradox Basin. Indeed, the cause of porosity is different in the Aneth field where a thick lime sand top is partly dolomitized and where considerable primary porosity is preserved within a framework of winnowed algal plates and pelleted mud lenses.

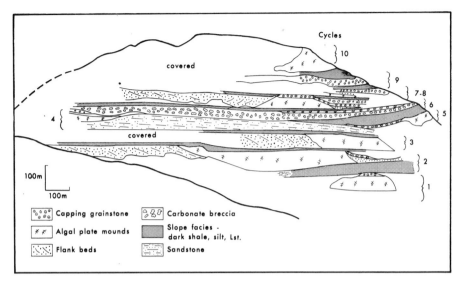

Fig. VI-13. Cyclic sedimentation of Virgilian strata at shelf margin in Horquilla Formation at Cement Tank Canyon, west side of Big Hatchet Mountains, southwestern New Mexico. Terrigenous slope sediments alternate with algal plate mounds and carbonate slope breccias and flank beds. Cycles are in brackets. Not much vertical exaggeration. From dissertation by Martin Schüpbach, Rice University, Houston, Texas

Platy Algal Mounds in Basins and on Shelves

Platy algal mounds are also part of large biohermal masses developed as major offshore banks in the subsurface Midland basin. These have steep sides (8–10 degrees) and many form very thick masses. For example, the Scurry County, Jameson, and Nena Lucia buildups and oil fields of northwest Texas may have had original depositional relief of a hundred meters (Fig. VI-14). Such large banks developed in basins owing to rapid carbonate sedimentation over the crest or on flanks of previously existing topography—either erosional or tectonic. In areas of rapid subsidence, only the highest parts of any area managed to remain in the photic zone long enough for algal and carbonate mud accumulation to be maintained. Large isolated banks thus developed in the Midland basin away from the major carbonate shelf margin. In a faunal study of strata comprising Nena Lucia bank, Toomey and Winland (1973) indicate that this Middle Pennsylvanian buildup had 100 m of relief on its open sea (northwestern) side. It is about 20 km long and 3–5 km across. The buildup consists principally of brecciated lime wackestone with abundant algal plates. The northwestern side has a concentration of porous, partly winnowed algal plates and peloidal foraminiferal wackestone to packstone. Paleontological changes across the bank were noted. Fusulinids, smaller mobile foraminifera, calcispheres, encrusting foraminifera (*Tetrataxis*), red algae, and encrusting bryozoans are concentrated with the phylloid algae on the open water side of the bank. After formation the buildup was buried by transgressive silty crinoidal grainstone and later by dark shales and limestone.

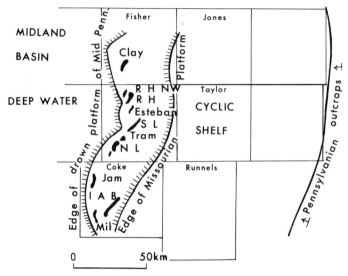

Fig. VI-14. Large carbonate banks of Late Pennsylvanian age in Midland basin off the west edge of cyclic Eastern shelf, subsurface of north central Texas. Counties are named. Offshore banks from north to south are termed: Claytonville, Rowan and Hope NW, Rowan and Hope, Esteban, Stone and Lake Trammel, Nena Lucia, Jameson, I.A.B., and Millican

The greatest such bank is the "Horseshoe Atoll", a complex of buildups rising from a Middle Pennsylvanian platform encompassing about eight counties in north central Texas and lying in the northern part of the Midland basin. Figures VI-15 and VI-16 are isopach and stratigraphic layer maps of this field. The initial large, equidimensional platform (40 km in diameter) accreted on opposite sides of a gentle arch trending north-south in the basin. About 100 m of marginal relief developed chiefly through accumulation of detrital shallow water carbonate building up above dark limestone and shale. During later Pennsylvanian time continued subsidence in the basin was matched by continued organic carbonate accumulation around all margins of the platform except the north. This arcuate buildup plus the added Wolfcampian accumulation on the southwestern side of the Horseshoe totals 900 m. The equivalent basinal strata total only 150 m. The 750 m on the western side and about half that in the Pennsylvanian on the eastern side may represent the approximate total relief of the buildups; part of this could be erosional since major disconformities are discerned at the base and on top of Cisco (Virgilian) strata. Numerous shale "breaks" can be traced across the buildup showing that the thick carbonates are basically detrital masses some of which had only limited relief during their formation.

Petrographic study across the eastern and Pennsylvanian prong of the Horseshoe found algal plate mounds and detrital wackestones with abundant fusulinids, oolitic grainstones, and encrinites forming flank beds. Sedimentary bodies within the major buildups have not been distinguished but configuration of isolated areas would indicate elongate mounds a few km long and 100 m high, more or less parallel to the shelf margin trend of the platform. Channels have been

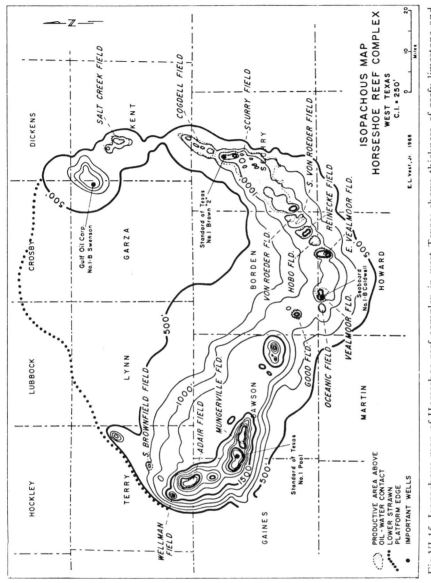

Fig. VI-15. Isopachous map of Horseshoe reef complex in west Texas; shows thickness of reefy limestone and location of significant fields producing along crest of atoll. Basinal limestone detritus of Wolfcampian age is not considered part of Horseshoe complex. Counties are about 50 km square. From Vest (1970, Fig. 8), with permission of American Association of Petroleum Geologists

noted through the barrier rim. These biostromal detrital buildups contain as many as 20 levels of porosity separated by shales. Presumably the levels represent former water tables and the porosity results from leaching and solution. This giant complex of multiple reservoirs had reserves of $2^1/_2$ billion barrels of oil, 90% of it in reefy buildups!

More common is the formation of platy-algal mud mounds across the shallow shelves. These have been extensively studied both in Kansas and in north central Texas where they are especially well developed in Missourian strata which repre-

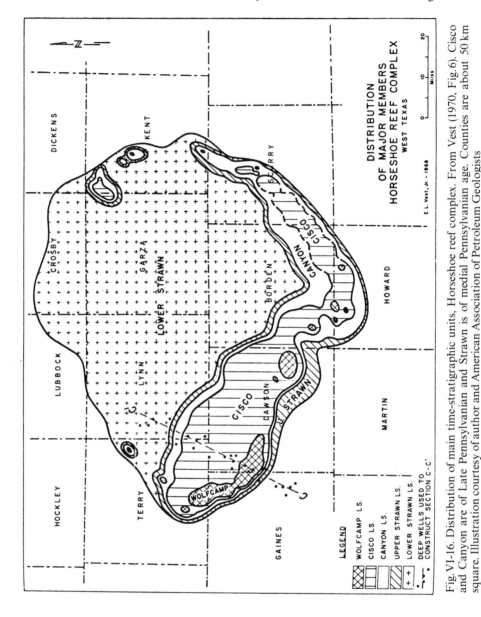

Fig. VI-16. Distribution of main time-stratigraphic units, Horseshoe reef complex. From Vest (1970, Fig. 6). Cisco and Canyon are of Late Pennsylvanian and Strawn is of medial Pennsylvanian age. Counties are about 50 km square. Illustration courtesy of author and American Association of Petroleum Geologists

sent in both areas maximum shelf inundation. Work by Heckel and Cocke (1969) and Harbaugh (1964) has described these as consisting of both *Archeolitho-phyllum* (encrusting platy red algae) and the familiar codiacean? platy genera. The accumulations in Kansas are flat lenses of phylloid algal wackestone with numerous spar-filled cavities. They are only 3–10 m thick, tens of km wide and separated by channels. Lateral slopes are gentle except at sides of channels where a maximum of 25 m relief appears. The channels cross the presumed strike of the mound-bearing strata. They may be 1–2 km wide and one has been traced as far

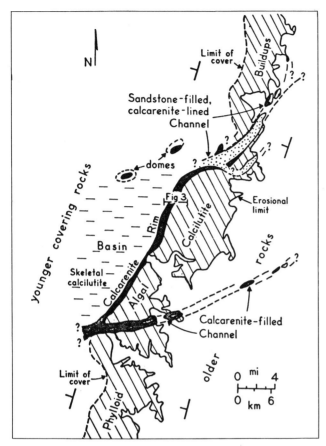

Fig. VI-17. General facies map of portion of Stanton Limestone outcrop in southwestern Kansas showing algal lime mud area and lime sand filled channels and western rim of low bank. From Heckel (1972c, Fig. 2)

as 35 km. They are characteristically filled with coated, skeletal grainstone (Fig. VI-17).

Regional relationships are indicated on Fig. VI-18. The mounds appear to be stacked, i.e., repeated vertically at the southern end of the outcrops along the Kansas-Oklahoma state line. Here the section changes mainly to shale southward into the clastic-filled Anadarko and Arkoma basins. To the east lay the low land of the present Ozark dome. The mound accumulations probably grew westward off the shoulder of this feature. The channels cut through the complex were also directed westward toward the open sea shelf. An apt analogy is made by Heckel (1972c) with modern geography along the Trucial Coast. Heckel also offers evidence that the banks represent a transgressive condition followed by a still-stand or slight regression. The mud mounds developed atop algal stromatolite heads surrounded by oolite. They seem to represent mud banks developed in very shallow protected marine water. Following development, a rim of bioclastic grainstone formed at their outer, shelfward edge. The grains are abraded, indicat-

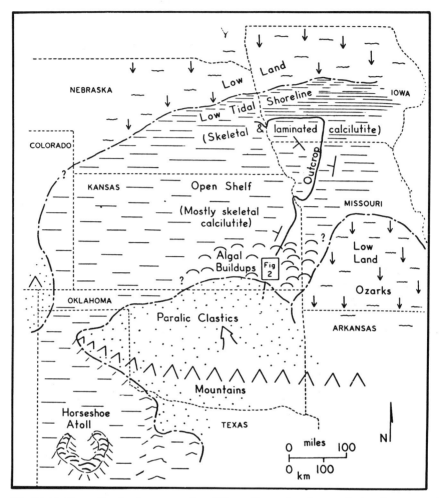

Fig. VI-18. Generalized paleogeography of Upper Pennsylvanian sea during limestone phase of cyclic deposition in midcontinent North America. From Heckel (1972c, Fig. 1)

ing wave action in shallow water. One mound flank dips westward 15° and at least 50 m of relief is indicated. The normal dip into the neritic shelf, however, was probably more on the order of 1°. The open carbonate shelf sediments northwest of the buildups are of burrowed argillaceous bioclastic wackestone with fenestrates, other bryozoans, brachiopods, sponges, and echinoderms.

The same types of flat lenses of micritic phylloid algal buildups are common across the Pennsylvanian outcrops of north central Texas. As in Kansas, these represent low carbonate banks occurring in the shelf lagoon between sandstone channels and clay mud ponds. It is significant that Chico Ridge near Bridgeport, Texas appears to represent oolitic lime sand accretion into a channel lying south of a platy algal mound oriented north-south; this situation is probably a duplication of the Kansas mound-channel relations described by Heckel.

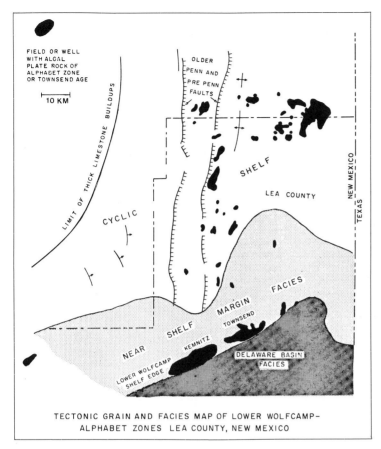

FIELD OR WELL
WITH ALGAL
PLATE ROCK OF
ALPHABET ZONE
OR TOWNSEND AGE

10 KM

OLDER
PENN AND
PRE PENN
FAULTS

LIMIT OF THICK LIMESTONE BUILDUPS

SHELF

LEA COUNTY

NEW MEXICO
TEXAS

CYCLIC

FACIES

MARGIN

TOWNSEND

SHELF

KEMNITZ

NEAR

DELAWARE BASIN
FACIES

LOWER WOLFCAMP
SHELF EDGE

TECTONIC GRAIN AND FACIES MAP OF LOWER WOLFCAMP-
ALPHABET ZONES LEA COUNTY, NEW MEXICO

Fig. VI-19. Paleotectonics and facies of Lower Wolfcamp Alphabet zones, Lea County, New Mexico on shelf north of Delaware basin

Understanding of shelf mounds both from Kansas and northern Texas (e.g., Possum Kingdom Dam area) suffers from a lack of published information about the western downdip (subsurface) relations. On the other hand, extensive three-dimensional subsurface study of Wolfcampian algal plate lenses has been made across the shelf north of the Delaware basin. In New Mexico the Baugh field "Alphabet" zones A through D (Fig. VI-19) are developed in thin cyclic lime-stones, the middle and upper portion of each being brecciated algal plate bio-stromes. Reservoir development is best where this facies is developed; it preferen-tially occurs on downthrown sides of growth faults and slightly downslope from structural crests. The upper portion of each cycle is a reddish, brecciated "soil zone" developed at marine regressive stages.

Thus, while platy algal mounds are best developed on shelf margins, the plants actually inhabited a variety of paleogeographic settings and their distribution and size of accumulation varies in part according to structural movement and in part according to vagaries of climate and terrigenous influx.

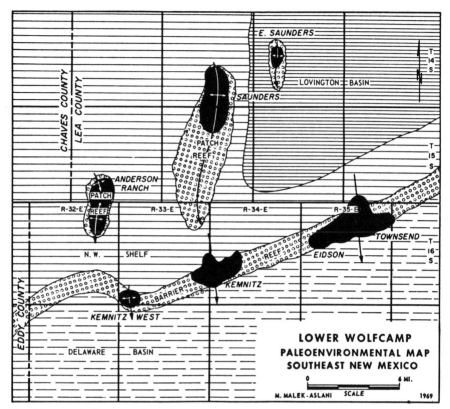

Fig. VI-20. Early Wolfcampian paleoenvironments, southern New Mexico showing details of Fig. VI-19. From Malek-Aslani (1970, Fig. 8), with permission of American Association of Petroleum Geologists

Townsend-Kemnitz Field—A *Tubiphytes* and Tubular
Foraminiferal Reef (Dunham, 1969b, Malek-Aslani, 1970)

Another type of organic buildup occurs in lower Permian strata in the southwestern U.S.A.—a linear mud bank with a boundstone cap. A well-studied example occurs in the subsurface around the northern end of the Delaware Basin in Lea County, New Mexico. Earliest Wolfcampian beds here form an east-west curving trend of carbonate buildups which cross several structurally high areas trending north-south and formed by early Pennsylvanian block faulting (Figs. VI-19, VI-20). Continued movement along these faults caused gentle scarps and shallow water and resulted in carbonate buildups along them throughout the Pennsylvanian. Where early Wolfcampian shelf margins cross the fault blocks, a structural culmination causes the Townsend-Kemnitz field. The trend has been extensively studied using abundant core material available from numerous wells. Regionally, the lowest Wolfcampian beds constitute a wedge thinning northward from the thickest and most porous development in the shelf margin trend. Basinal beds to the south are thin, dark, spiculitic limestone. The porous equivalent of

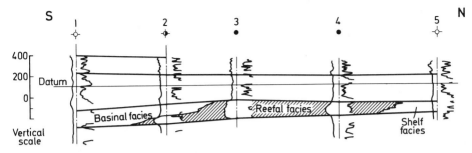

Fig. VI-21. Typical facies changes from electrical logs from north to south several kilometers across Kemnitz Field. For detailed petrography of darkened area see Fig. VI-22. From Ma-lek-Aslani (1970, Fig. 4), with permission of American Association of Petroleum Geologists

these along the shelf margin reaches more than 30 m and thins within 2 km north of the margin to approximately 25 m. Farther northward, it continues thinning for many km and disappears. These relationships can be easily ascertained from electrical logs (Fig. VI-21). In detail the porous buildup is formed by two half cycles separated from each other by shale, the upper, thinner unit being transgressive over the lower. The basinward slope is from 2 to 4 degrees. The total length of the recognized carbonate shelf margin is more than 50 km (Fig. VI-20).

Malek-Aslani (1970) described the chief facies developed in the buildups. The reef front and top is formed by a narrow belt (about 2 km wide) composed of rock with boundstone lattice made of 30–40% *Tubiphytes* and tubular foraminifera. *Tubiphytes* is also particularly abundant in flanking bed packstones. Its growth habit was that of an encruster in agitated water; it apparently lived best on the exposed flanks and tops of algal plate mounds.

The rest of the sediment in the shelf margin belt is bioclastic detritus. There is some fine lime sediment constituting matrix. The very top beds of the shelf margin may be composed of grainstones with rounded fragments (Fig. VI-22). Constructed cavities within organic framework form shelters for the fine matrix, affording abundant geopetal fabrics. There is little spar cement and considerable primary porosity, according to Malek-Aslani.

Shelfward of this belt lie facies considered backreef and protected from wave action by the extended reef flat formed along the very gentle slope. These sediments consist of "backreef" grainstones with much *Tubiphytes* and packstones with some matrix and less *Tubiphytes*. More micritic sediments contain dasycladaceans and may represent shoals in the backreef lagoon. Essentially, however, these backreef sediments represent deposition in normal marine, openly circulating waters showing that the shelf margin was at times completely inundated and did not restrict circulation behind it. Platy algae also constitute an important part of these sediments, appearing in more protected places behind the reef as well as down its foreslope. They are also more abundant in lower and middle parts of the sequence.

The foreslope facies contains an apron of packstone and wackestone. Some of these beds are depositionally crossbedded with as much as 45° dips. The sediments are rather coarse, bimodal, bioclastic debris mixed with black, spiculitic

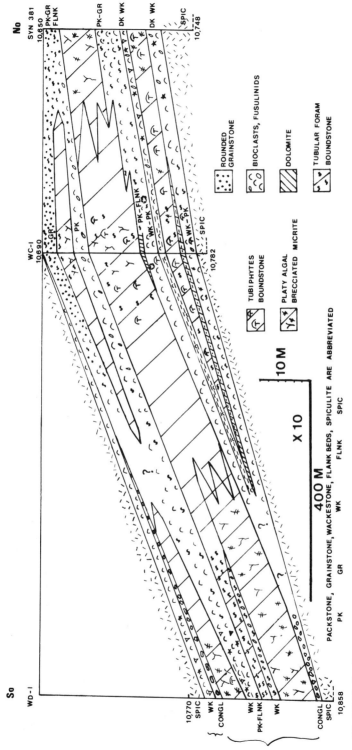

Fig. VI-22. Section across Kemnitz shelf margin Lea County, New Mexico in Lower Wolfcampian interval indicated on section in Fig. VI-21. Line of section indicated on Fig. VI-23. Petrographic study indicates algal plate development down slope and in immediate backreef with *Tubiphytes* boundstone and rounded bioclastic grainstone at shelf margin. Conglomerates are interbedded in slope deposits and mark two breaks in deposition within and at top of buildup

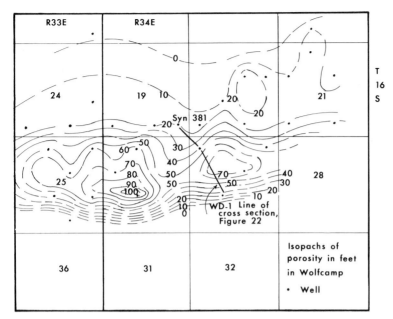

Fig. VI-23. Porosity isopachous map of Kemnitz field, New Mexico. Isopach in ten foot intervals of beds with between 8 and 18% porosity. Line of cross section on Fig. VI-22 indicated. Porosity clearly outlines fairway and reef at shelf margin. Numbered squares in grid are square miles. (Sections within Townships)

micrite. Fusulinid and crinoid coquinas are common in places within this facies. Very coarse talus also exists in the form of rudites of poorly sorted, varied lithoclasts and bioclasts. The lithoclasts are from both shelf and shelf margin sediments and the bioclasts are large, broken, abraded, and somewhat blackened fossils of brachiopods, horn corals, bryozoans, crinoids, and fusulinids. The Kemnitz north-south cross section (Fig. VI-22) shows a greater development of algal plates downslope, presumably in slightly deeper water. Basinal facies consist of argillaceous, dark, spiculitic micrite with lenses of purer carbonate micrite and much ball and flow lamination.

According to Malek-Aslani porosity is best developed within the *Tubiphytes* boundstone fabric. It reaches a maximum of 18%, but averages about 8%. There is fair interconnection within the boundstone so that permeability may be good although variable. By far the best production from Townsend-Kemnitz was from the very narrow reef front belt or "fairway". Malek-Aslani considers that most of this is primary porosity in a reef fabric. Some reservoir rock was formed by interconnected algal plate porosity where the large amount of original pore space was not filled by much micrite matrix and where brecciation, plate fracture and mud flowage occurred. See Fig. VI-23 for isopach of porosity.

Dunham (1969b) presented compelling evidence of both leaching and infilling of the fairway sediment in the vadose zone. This interpretation must be related to an understanding of the typically complex geological history of Pennsylvanian buildups which commonly record several sea level fluctuations and periods of subaerial exposure. The cross section of Townsend field (Fig. VI-22) demonstrates this geologic history. Along the axis of the Townsend-Kemnitz trend at the beginning of Permian time, sedimentation of black, cherty spiculitic mud gave way to normal marine, bioclastic lime wackestone. These sediments accumulated along with some of the algal plate mud-mound wackestone and built up a few tens of meters of section. Apparently, no sedimentation downslope was taking place; here a dark, thin, lag conglomerate formed, bearing eroded carbonate blocks derived from upslope shallower water sediments which were already cemented. Argillaceous and dolomitic beds in the lower part of the buildup represent breaks in reef sedimentation during the times that the conglomerate was forming downslope and perhaps an early period of subaerial exposure. During an ensuing transgression both the axis of the trend and its basinal slope developed the characteristic micritic facies with algal plates and *Tubiphytes*, the latter best developed with cornuspirid foraminifera over the crest. The more northerly areas behind the crest received the bioclastic debris. Finally, as sea level lowered, or as the mound accumulated into wave base, a cap of grainstone and tubular foraminiferal boundstone developed. The cornuspirid boundstone, with *Tubiphytes*, is particularly thick and well-developed on the foreslope area. The relief of the Townsend-Kemnitz trend at this time was at least 30 m because about 15 m of compacted shale filled in the basin in front of it and lapped unconformably against the foreslope facies. This unconformity, also developed over the top of the mound, records another period of subaerial exposure and diagenesis. A second conglomerate is developed basinward at this stratigraphic interval. A higher capping unit of pure carbonate debris and boundstone occurs above the main Townsend-Kemnitz reef and is separated from it by a "shale break" in immediate backreef areas. This upper unit represents a transgression which continued into Wolfcampian time as evidenced by the overlying blanket of euxinic and open circulation lime sediment which filled in the basin south of the reef trend and evenly overlaps the area above the reef crest.

According to Dunham (1969b), leaching in the vadose zone is the major diagenetic factor responsible for the porosity, although early collapse brecciation which aided the leaching and dolomitization is also important. Dolomite is found mainly in the lower part of the mound and is very fine-grained and pervasive through both grains and matrix. Some very coarse-grained void-filling dolomite also occurs. Sucrose texture and dolomitization preferentially in the original lime mud matrix does not occur. In general, dolomitization is not important in reservoir development. The vadose processes are related chiefly to the lowering of relative sea level, exposure of the mound, and development of a fresh water lens at the end of reef growth. The presence of the "fairway" with its thick section of more or less continuous porosity probably results from the leaching of the main mass of poorly bedded organic carbonate which was the topographically highest part of the mound and which, because of its high content of boundstone, developed the best pore-connected network following diagenesis (Fig. VI-23). Thin layers of porosity are developed in the northern backreef area and result from the bedded character of the strata deposited behind the reef front. About 80% of the estimated 14.5 million barrels of ultimate recoverable oil came from the fairway demonstrating excellent reservoir connection and drainage in this narrow belt.

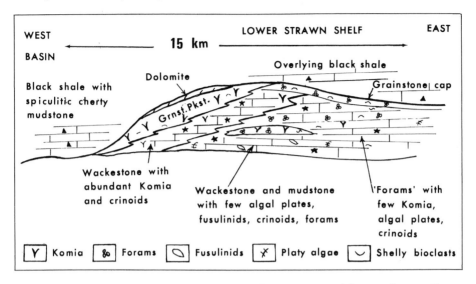

Fig. VI-24. Section across Middle Pennsylvanian field, west-central Sutton County, Texas. Section is 15 km across field. Interval diagrammed is a few tens of meters thick—a minor shelf margin buildup capped with *Komia* grainstone. It is of lower Desmoinesian age, zone of *Wedekindellina* and *Fusulina*

Description of an Early Pennsylvanian Field in Sutton County, West Texas

The small Denison field in West Texas is known to produce oil from a reservoir consisting of detritus of the tiny dendroid stromatoporoid, *Komia* (Fig. VI-24). This is one of the many fields of Desmoinesian or Strawn age which developed in shelf margin carbonate buildups along the shifting western margins of the eastern shelf of the Midland basin. The textural and biological makeup of these fields has not yet been well described in the literature. (See, however, Toomey and Winland, 1973.)

Komia was probably not capable of frame construction despite its known branching growth form when seen intact (*Ungdarella* of Toomey and Johnson, 1968, see Wilson, 1969, for correction). It is a delicate organism and is commonly found in micritic bedded limestones. In the Denison field it forms crinoid-rich flank debris on the seaward, exposed side of a lime mud mound containing a few algal plates and many foraminifera. The reservoir is composed of grainstone sediment with original porosity preserved and with some diagenetic alteration of *Komia* bioclasts. A good analogy may be made with the similarly shaped modern red alga *Goniolithon* whose debris forms a seaward fringe on Rodriguez Bank in the Florida Straits (Baars, 1963) and with the rapidly eroding red algae fringing reefs around the northern end of the Qatar Peninsula in the Persian Gulf. Both this field and the Nena Lucia buildup described by Toomey and Winland (1973) show that the porous reservoir facies may be preferentially developed on the open sea, winnowed side of these low-lying buildups.

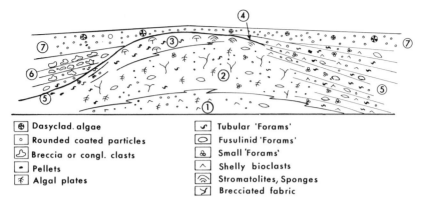

Dasyclad. algae	Tubular 'Forams'
Rounded coated particles	Fusulinid 'Forams'
Breccia or congl. clasts	Small 'Forams'
Pellets	Shelly bioclasts
Algal plates	Stromatolites, Sponges
	Brecciated fabric

Fig. VI-25. Idealized Late Paleozoic buildup showing distribution of seven commonly associated microfacies: 1. Basal bioclastic micrite pile; 2. Algal plate micrite core; 3. Crestal
boundstone of foraminifera or encrusting algae, *Tubiphytes*; 4. Organic veneer; 5. Flank beds
of tubular foraminiferal debris; 6. Marine talus breccia (rare); 7. Capping bed of shoal grainstone with dasycladaceans and gastropod shells. See also Figs. XII-4 and XII-5

Conclusions

Late Paleozoic carbonate buildups display a relationship of facies to growth
history and to a limited extent their tectonic settings. Useful analogies may be
made with some Holocene carbonate mud accumulations. The genesis and particular characteristics of porosity and permeability in these Late Paleozoic facies
render them particularly important oil reservoirs.

Growth History

Seven stratigraphically significant microfacies in the buildups result when their
normal development is uninterrupted. These are generalized for all shelf buildups
in Chapter XII.

1. The initial accumulation in many places was a *basal bioclastic micrite pile*
of debris.

2. The *algal plate micritic core* facies forms as a moundlike accumulation
(bafflestone of Klovan) formed below wave base in quiet water as much as 25 m
deep where the plants trapped lime mud (as determined by the average maximum
height of individual cores).

3. Some mounds of platy algae grew into active wave base where they acquired a *crestal boundstone* facies of encrusting tubular foraminifera and/or *Tubiphytes*.

4. Mounds in lagoonal or other geographically protected areas developed
only a *veneer* facies of certain sponges, stromatoporoids and stromatolitic algae
(Parks, 1962).

5. When the top of a mound remained at, or close to, wave base for a considerable time, extensive *flank beds* formed owing to abundant production of bioclasts from organisms on the mound top. These commonly include small mobile foraminifera, fusulinids, tubular encrusting foraminifera, *Komia*, etc. Volumetrically, these steeply dipping foreset beds may constitute more than half the bulk of the massive "reef" limestone.

6. In some mounds on steep slopes, a *marine talus breccia or conglomerate* is present down the seaward flank of the mounds and contains lithoclasts and bioclasts derived locally from the mound.

7. Commonly, after construction of a ramp or platform by the combination of core and flanking beds, any period of stabilized sea level resulted in the production of a horizontal *capping bed of shoal grainstone*, often cross-bedded and oolitic. This bed commonly overrides the top of a mound-flank bed complex.

Figure VI-25 illustrates an idealized mound with the seven component facies. Such a sequence of related facies is seen in many other mounds in other parts of the geologic column and are summarized in Chapter XII.

Orientation of Mounds Relative to Shelf Margins and Their Origin

Orientation and mound origin are interrelated and are important in both exploration and oil field development. Many individual mounds in the geologic record (e.g., around the Oro Grande basin, New Mexico) are oriented with long axes parallel to the trend of depositional strike and shelf margins. Also, there are some at right angles to the general facies trend, e.g., Ismay field in the Paradox basin of Utah, and perhaps in the outcrops of some mounds in the Big Hatchet Mountains in New Mexico on the eastern flank of the Pedregosa basin. The elongate mound shapes, their consistent orientation in any one area, and the detrital sediments in their basal beds would indicate an origin as bars or sedimentary piles formed by currents or wave action and later colonized by the platy algae and/or tubular foraminifera. Tidal bars in passes, terrigenous bars in off-delta areas, piles of molluscan or crinoidal debris all could have afforded "starter areas" for organically induced lime mud accumulations.

Analogy with the modern Florida Reef Tract accumulations is significant here. Tidal pass bars between the middle and lower Florida Keys and in the "Safety Valve" of Biscayne Bay, are oriented normal to the coast line, whereas Tavernier and Rodriguez bioclastic mud banks on the seaward side of the Keys have an orientation at a low angle to that of the coast. These two banks may have been formed at slight slope breaks and shaped by longshore currents. Tavernier and Rodriguez Banks make a good analogy with many Late Paleozoic mounds. They were formed as bioclastic lime mud accumulations in 3–5 m of water, following the post-Wisconsin marine transgression, grew to wave base, and are capped by a bioclastic debris formed from sessile organisms growing abundantly on the windward (seaward) side. These include corals, green algae, and a red alga, *Goniolithon*, with a growth form much like that of *Komia* (Baars, 1963). Their present shape and dimensions, except for thickness, are about those of many platy algal mounds. But analogies with other modern shallow marine lime mud banks, show how complex the origins of individual mounds may be. Not only is current orientation probably a controlling factor, but also the absence of currents allow some modern mounds to form. Where gentle currents meet and are damped out, mud mounds form, often opposite seaward openings of major lagoons. This is seen in slack water areas as at Bulkhead Shoal described by Pusey (1964) in the northern lagoon of British Honduras and Cayo Sucio from the Isla Blanca lagoon along the northeastern Yucatan coast (Brady, 1971). Furthermore, wind-produced gyres may also produce mud accumulations in completely closed lagoons.

Variations in Tectonic Settings for Pennsylvanian-Wolfcampian Buildups

The characteristic biota of the buildups described in this Chapter had the capacity for development in all the major tectonic settings. Composite mounds have developed large offshore banks in the Midland basin on which perhaps a few hundred meters of relief developed owing to rapid subsidence and the ability of carbonate sedimentation to keep pace with it, e.g., Horseshoe Atoll of northwest Texas. These and shelf marginal buildups appear generally to consist of distinct loaf-shaped mounds, formed preferentially downslope from paleotopographic crests (Type I buildup as outlined in Chapter XII, and Wilson, 1974). They tend to offlap each other downslope if subsidence was not great. No great linearly persistent barrier reef trends are known along the margins but only irregular chains of mounds commonly with abundant algal plates and caps of tubular foraminifera and *Tubiphytes*. Up on the shelves extensive flank beds of limestone developed surrounding mound cores. In contrast, mounds down the depositional slope have few carbonate flank beds and intermound areas are filled with shale or siltstone.

In addition, far across the shelves and removed from the basins, lens-like beds of the same organic facies as mounds occur but are merely a few feet thick. These formed in shelf areas distant from influence of terrigenous clastics, usually in the middle phases of the typical Pennsylvanian cyclothems.

Types of Porosity and Permeability in Pennsylvanian-Wolfcampian Buildups

1. Grainstone accumulations of algal plates, as in the Middle Pensylvanian of the Paradox basin, afford primary porosity.

2. Most algal plates accumulated in a lime mud matrix, but this also may develop good porosity through brecciation of the plates and of the partly lithified matrix. The mud collapsed, fractures formed through partly lithified portions, and rigid shells and algal plates themselves rotated and fractured.

3. Early solution porosity through internal alteration and decay of the algal plates offered channelways for subsequent fluid migration. Only the cortex of the plates were originally calcified, increasing susceptibility to break down. Leaching of additional void space followed through the action of meteoric water. Infilling of some voids with internal sediment formed from mud swept through the fabric as it collapsed. Part of this infilling is the vadose silt of Dunham (1969b).

4. *Komia*, *Tubiphytes*, and tubular calcitornellid foraminifera are composed of finely porous tests. *Komia* was probably composed of metastable aragonite on the assumption that it was a stromatoporoid. Such microporous forms were susceptible to diagenesis. In addition, these supine, encrusting forms constructed cavities during growth which offered good channelways for fluids and hence a permeable reservoir. Thus boundstone tops of mounds and some flank beds may have good porosity and permeability.

5. Subaerial exposure and the influence of meteoric water is important. The considerable sea level fluctuation through the Late Paleozoic gave good opportunity for leaching at vadose zones above old water tables. The reservoirs often contain multiple oil-water contacts in relatively thin zones. There is also much lithological evidence of sea level fluctuation. Even within the cores of many bioherms are reddish, oxidized zones containing conglomerates. Shelf cycles may contain further evidence of sea-level fluctuation.

6. Porosity developed through dolomitization is rare in the Pennsylvanian-Wolfcampian beds of the southwestern U.S.A. except where evaporite basins occur. Little stratigraphically controlled dolomite is present and tidal flat deposits are rare except in the Earp Formation, Lower Permian of Arizona and New Mexico, and in the Minnelusa Formation, Pennsylvanian of the northern High Plains.

Chapter VII

Late Paleozoic Terrigenous-Carbonate Shelf Cycles

Introduction

Chapter II emphasized the importance of the study of cyclic sedimentation and discussed the basic principles involved in its formation. Most characteristically cyclic sediments are found in shelf areas although regular changes in sedimentary conditions may be reflected in many types of basinal sediments (evaporites, flysch, carbonate-shale rhythms). Cyclic shelf deposits are particularly common in the Late Paleozoic of the northern hemisphere. The persistence of platforms situated just at sea level, the tectonic activity which provided periodic incursions of terrigenous clastics, and the enhanced possibilities of eustatic sea-level fluctuations (either glacially or tectonically induced) all coincided to make this time in earth history favorable for cyclic sedimentation.

This chapter compares two types of shelf cycles in which terrigenous clastic influx from distant deltaic regions is interlayered with carbonate strata: (1) the Yoredale Lower Carboniferous of the British Isles, and (2) the Pennsylvanian cyclothems of the midcontinent area of North America. Previous chapters have discussed the thick pure carbonate deposits formed on the shelf margins and basinal slopes at the same time as these cyclic sequences. The description here of the shelf deposits is brief as an extensive literature on both sequences already exists.

Both the Yoredale and the North American Pennsylvanian cyclothems must have been deposited across very extensive flat shelves in tropical seas whose carbonate deposition was periodically interrupted by clastic influx from a distant terrigenous source. Differences in stratigraphic sequence might have resulted (1) from differences in timing of clastic influx relative to sea-level rise and fall, or (2) to the extent to which deltaic progradation was able to overcome subsidence.

Thus, in the Yoredale case this influx of clay and sand either forced a gradual marine regression or coincided with an independently caused sea-level drop. Transgression is represented by a single limestone sheet. In contrast, in many midcontinent Pennsylvanian-Wolfcampian cycles and in related cycles in New Mexico, marine submergence commonly took place during limestone deposition interspersed with incursions of clay and sand. The traditional interpretation is that terrigenous deposition mainly coincided with the marine invasion and was, in each cycle, eventually overcome by the advancing marine conditions. This results in an interbedded sequence of limestone and terrigenous clastics preceding the record of maximum transgression and obviously in a much more complicated cycle. This explanation is founded on the interpretation of the channels below the limestone portion of the cycles.

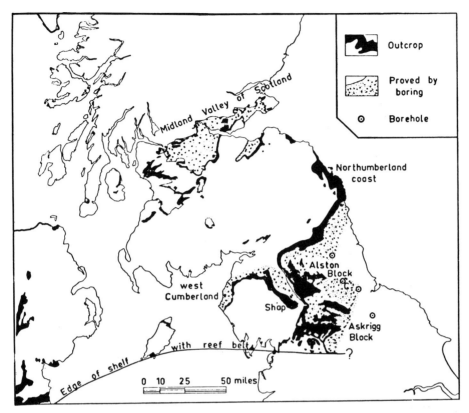

Fig. VII-1. Distribution of Yoredale Series and equivalents in northern England and Scotland from Moore (1959, Fig. 3)

Yoredale Cycles

The first cycles described in the geologic literature are those of the Yoredale-Wensleydale area in the Pennines of the English Midlands in beds of Visean age. About eight cycles, each as much as 30 m thick, occur as widespread sheets across the North Pennine block north of a basin in the Hercynian belt (Fig. VII-1). The cyclic sediments of the platform are spread northward through Northumberland in beds exposed along the east coast, north of Newcastle-On-Tyne and into the Scottish Lowlands near Edinburgh. Along the southern edge of the North Pennine block (Engleboro-Settle area) part of the Yoredale cyclic beds grade into encrinite of the upper Great Scar Limestone (Fig. V-10).

The typical cycle consists of a basal limestone sheet overlain by an upward shoaling and coarsening sequence of argillaceous and sandy sediments. The limestone, described below, grades into a black shale or siltstone which is transitionally overlain by grey siltstone, in turn by cross-bedded sandstone capped by a "seat earth" with rootlets and an overlying coal.

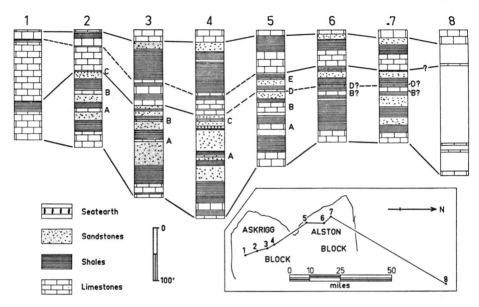

Fig. VII-2. Large scale variations in two Yoredale cycles (IV and II) south to north across northern England from Moore (1959, Fig. 13). The letters A through E mark traceable limestone beds. Boundary of the two cycles is marked by top of terrigenous clastics below C or D. For location of Askrigg and Alston blocks see Fig. VII-1

The clastic deposits clearly were formed as a typical outbuilding deltaic sequence as outlined by Moore (1959) and reviewed succinctly by Selley (1970) (Fig. VII-2). The next overlying limestone rests with sharp contact on the coal (Fig. VII-3). This is why the cycle is considered conventionally to begin at the limestone base. The limestones are dark, open marine wackestones-packstones with varied bioclasts, corals, brachiopods, bryozoans, crinoids, algal onkoids, and even some oolite beds. Fine argillaceous seams and stylolites are present. Cross-bedded oolite grainstones, encrinites, and small patchy "reefs" consisting of various types of organisms, are known. Chert nodules and siliceous limestone occur in the upper parts of limestone beds. The limestone grades upward to laminated black, pyritic, micaceous shale and siltstone with siderite concretions. Productid brachiopods are its only common fossils.

The Yoredale cycles represent a succession of marine incursions across the unstable shelf blocks north of the Hercynian trough. Each transgression occurred without any terrigenous incursion. The limestone represents water of open circulation and hence moderate depth over considerable shelf areas. Probably the deepest inundation occurred during deposition of the prodeltaic black shale-siltstone unit which heralds the clastic influx resulting from deltaic advance from the north. The fine, rippled, cross-laminated sands and the cross-bedded, coarser sands represent distributary and alluvial channel deposits. The coal and rootlet beds represent delta plain swamps, capping the sequence.

The part of the sequence consisting of coal, widespread thin limestone, and black shale-siltstone, much resembles the deposits beginning the inundative phase

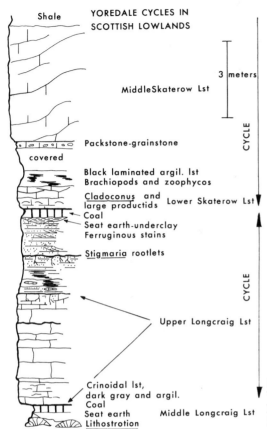

Shale

YOREDALE CYCLES IN
SCOTTISH LOWLANDS

3 meters

MiddleSkaterow Lst

CYCLE

Packstone-grainstone

covered

Black laminated argil. lst
Brachiopods and zoophycos

Cladoconus and
large productids   Lower Skaterow Lst

Coal
Seat earth-underclay
Ferruginous stains

Stigmaria rootlets

CYCLE

Upper Longcraig Lst

Crinoidal lst,
dark gray and argil.
Coal
Seat earth   Middle Longcraig Lst
Lithostrotion

Fig. VII-3. Lithologic details of two Yoredale cycles from the Scottish Lowlands, Firth coast east of Edinburg, near Dunbar

of the midcontinent Pennsylvanian cyclothem and can be similarly interpreted. The transgression represented by the limestone preserves the underlying rotting vegetation as coal.

Several theories have been proposed to account for the regularity of the Yoredale cyclothems. These include (1) a tectonic hypothesis of shelf-tilting (Bott and Johnson, 1967), (2) an idea of periodic compaction of argillaceous and organic matter under a requisite load of sediment (Westoll, 1962), and (3) what is perhaps a simpler sedimentologic explanation, the steady subsidence of the shelf with periodic ingress of terrigenous material caused by shifting of distant deltaic distributaries (Moore, 1959) (Fig. VII-4).

Other cyclic patterns resembling those of the Yoredales are known. In all of these, marine transgression occurs under nonterrigenous conditions and a distinct, well-defined carbonate marker bed with a sharp lower contact is formed. A very similar model to the Yoredale cycles is described by Van Siclen (1972) and Galloway and Brown (1973) from north central Texas Pennsylvanian strata.

Furthermore, a similar cyclic pattern abounds in essentially terrigenous sediments. In such cases the upper clastic unit is much thicker than the transgressive

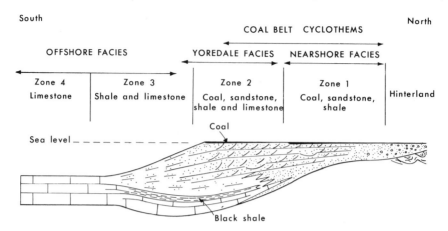

Fig. VII-4. Stratigraphic model showing variation in facies from basin to shore (south to north) in a single Yoredale cyclothem. Vertical scale and dip on deltaic front very exaggerated. Illustration after G.A.L. Johnson (1961)

unit and tends to be deltaic; an upward shoaling clastic sequence is typical. Examples include the Eocene cycles of the Texas Gulf Coastal plain (Bornhauser, 1947; Fisher, 1964) which contain a thin, shelly greensand widely transgressive over depositional surfaces and a regressive phase consisting of thick, prograding deltaic sand wedges. In the Lower Triassic Moenkopi Group (Virgin Formation) of southwestern Utah (Poborski, 1954) several marine transgressions are recorded during periods of no detrital influx. They are represented by thin, open marine limestones, overlain by thicker gypsiferous shales and siltstones. The classic Middle Devonian facies pattern of the Appalachians consists of thin limestones succeeded regularly by regressive deltaic facies and so follows the Yoredale pattern (Cooper, 1957). The Rocky Mountain Cretaceous contains many regressive sandstone wedges progradational from the rising cordillera eastward into the Pierre-Mancos shale basin of the Midcontinent. The transgressive phases of these cycles are relatively thin westward-projecting shale tongues with a few calcareous concretions (Young, 1955).

Pennsylvanian and Wolfcampian Shelf Cyclothems of the Midcontinent and Southern Rocky Mountains

Well known cyclothems of mixed clastic and carbonate sediments are also general throughout North America during the Late Paleozoic although the sequence of events recorded by them differs slightly from that of the Yoredales. They have been described by many writers, beginning with Charles Udden (1912), from outcrops more or less parallel to strike from the midcontinent states of Nebraska, Iowa, Kansas, Oklahoma, and north central Texas (Fig. VII-5). The algal plate mud mounds of the Pennsylvanian shelves, described in the preceeding chapter,

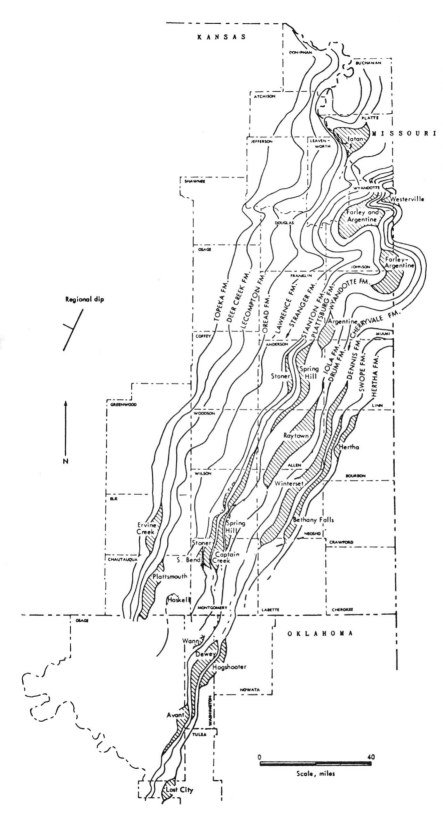

Fig. VII-5. Outcrop belt of Late Pennsylvanian strata in midcontinent area of North America showing areas of algal plate lime buildups. This long line of outcrops trends roughly parallel (only slightly oblique) to the strike of facies bordering the Kansas shelf lagoon to the west. This accounts for the impressively regular sequence of thin lithologic units along the strike of these strata. From Heckel and Cocke (1969, Fig. 2), with permission of American Association of Petroleum Geologists

usually lie in the upper half of these cycles. The cycles were formed across what Krumbein and Sloss have termed "unstable shelves" and are remarkable for widespread and uniform distribution of thin beds of varying rock types. Much shale, as well as limestone and sandstone, occurs in these strata—almost as much as in the basinal strata off the shelves. This imparts a characteristic erosional pattern and the rapid vertical alternation of thin beds of sandstone, shale, and limestone is a hallmark of Pennsylvanian sedimentation in North America (Mathews, 1974, Fig. 16-7) and also see Figs. VII-6, VII-7 for typical cyclic patterns of the Midcontinent. These diagrams have great vertical exaggeration and may not adequately convey the impressive repetitive alternations of the thin, and very different, strata.

Petrography of Cyclothem Beds

Two types of sandstones are commonly seen: (1) On the shelves thin deltaic sequences occur with channel conglomerates, shallow marine bars and beaches, etc. These may occupy channels of distributaries or occur as thick, localized pods of clastics in areas of great sediment influx. Petrographically the sands are variable, mostly slightly feldspathic, lithic arenites. Detailed studies on lithologic features of these have been published by Wanless (1964, 1972); Wanless et al., (1963); Brown (1972) and Galloway and Brown (1973). (2) Fine-grained, dark, and more uniform sands are known in the subsurface basins in contrast to such variable shelf sandstones. These sands represent sediment by-passed from the shelves during low sea-level stands (cyclic and reciprocal sedimentation; see Galloway and Brown, 1973).

Thin black shales: Studies of the very thin (1 m) and widely distributed Upper Pennsylvanian black shales of Kansas, such as the Heebner (Evans, 1966) and the Excello (James, 1970) have added much to a logical understanding of the cyclothem sequences. Such fissile, platy, organic-rich shales contain mainly a pelagic nektonic fauna of conodonts, fish teeth, spores, and limited benthos, including even a few brachiopods and mollusks. The shales contain phosphatic concretions and significant trace amounts of uranium. Their environment is clearly euxinic marine. They occur most commonly in the middle of the megacyclothem sequences of the Kansas Upper Pennsylvanian, lying east of the dominant limestone buildup in the subsurface. Similar black shale formed in coastal marine ponds and swamps in Illinois, has been extensively described by Zangrl and Richardson (1963).

Carbonate microfacies types: Special organic constituents and textures in Pennsylvanian-Early Permian shelf and shelf-margin limestones have already been described in Chapter VI. Environmental determination of the limestone members of shelf cyclothems is also of critical importance to interpretation of their geologic history and much attention has been given to this in the last 15 years. The organic composition of these Late Paleozoic limestones is particularly varied and despite much study, not all of the fossil organisms have been properly identified biologically. The abundance of organic debris creates packstone texture which is commonly associated with a microspar matrix of problematic origin. The following is an attempt to define and interpret 11 microfacies types common in Pennsylvanian-Wolfcampian strata. It is based principally on petrographic studies in the midcontinent region of the U.S.A. (e.g., Toomey, 1969a and b, 1972; Troell, 1965) and on the author's experience in West Texas, New Mexico, and Utah. Excellent illustrations of the microfossils and textures are found in papers by Toomey. Despite obvious limitations of study confined to certain provinces, these rock types are believed basic and useful for a general environmental interpretation of Pennsylvanian shelf sediments. They are widely repeated in many sections throughout the western and midcontinent United States. The rock types are numbered P1–P11 to distinguish them from the standard microfacies types.

P1. Molluscan argillaceous, ferruginous lime mudstone-*Myalina* beds. Outcrop expression: thin, yellow, soft-weathering limestone, a shell hash of large *Myalina* shells.
Environment: shallow, muddy water; significance of iron concentration is not known.

P2. Very fossiliferous, marly shale. Grey nodular-weathering shale with well-preserved fossils and a varied marine fauna. Chonetids, productids, *Neospirifer, Composita, Derbya, Enteletes* are all common brachiopods; horn corals; cyclostomate bryozoans forming large lumps; disarticulated pieces of crinoids and echinoids are common; some mollusks, fusulinids, and smaller foraminifera. Much of the fauna is vagrant benthos. These shales provide most of the well-known Pennsylvanian-Lower Permian fauna from the Midcontinent. Environment: soft substrate, but at times moderately clear water, open circulation and normal marine salinity but very little current or wave action; moderate depth—a few meters to tens of meters.

P3. Normal marine wackestone with 20–25% bioclasts.
Same fauna as contained in the above shale, plus algal plates, a few dasycladacean algae and many foraminiferal-onkoidal structures. Generally contains a highly varied biota. *Syringopora* occurs. Echinoids are more abundant than crinoids. Matrix is micrite or microspar, generally the former. On outcrop, bedding is wavy-nodular to planar with shale partings. Texture homogenized by extensive burrowing. Organic debris commonly somewhat macerated but with well-preserved internal structure.
Environment: Clear, calm water of open circulation and of moderate depth (a few meters to tens of meters determined from facies gradations downslope from bioherms); substrate soft. Standard microfacies type 9 (Plate VIB).

P4. Foraminiferal limestones. Poorly sorted debris with many sorts of foraminifera and of packstone or wackestone texture. Calcitornellids most commonly occur as flat-sided coiled tubules, which encrusted soft-bodied forms, and also are found as tiny masses or irregular balls. Endothyrid concentrations with packstone texture are known in earlier Pennsylvanian beds. Many mobile foraminifera are present, the same genera occurring with normal marine limestones: *Bradyina, Hemigordius, Tuberatina, Climacammina, Paleotextularia, Globivalvulina*. Outcrops of such limestone are wavy bedded and are indistinguishable from the above normal marine limestone.
Environment: Believed to be almost the same as for the above normal marine limestone; perhaps the foraminifera are concentrated by gentle winnowing of currents or waves; they appear to be more or less ubiquitous. Perhaps slight restriction or higher water temperature inhibits other biota. A variety of standard microfacies type 9.

P5. Fusulinid coquina. Fusulinids concentrated into coquinas but also commonly found in limestone with other members of the varied normal marine biota. They occur in well-bedded strata. Such coquinas are generally packstones with some intergranular solution and microstylolitization, particularly when the matrix is slightly argillaceous. Environment: Apparently normal marine, clear water, on shelves which ranged from depths of a few meters to a few tens of meters. Fusulinids must have been able to live high on biohermal slopes and hence in quite shallow water, for major detrital accumulations of them occur down slopes between bioherms standing as much as 25 m above the sea floor (Plate XXIIIA).
Fusulinids formed resistant particles, which were easily moved, and are found in many types of sediment but most commonly in carbonates. Standard microfacies type 12.

P 6. Platy algal mud facies. Lime wackestones (also described in Chapter VI), containing 10–50% platy algae. These leaf-like (phylloid) forms represent both red and codiacean(?) algae and occur in well-bedded limestones as well as in mound-shaped buildups. The plates are delicate and crinkled, in many places broken and somewhat abraded, but are also found whole. They were calcified only along the cortex; the wall structure is rarely preserved and mainly as micrite fill of the tubules. Other normal marine organisms may occur with the plates but the latter dominate; apparently their rapid growth discouraged many other organisms. The plates seldom occur together in any regular arrangement and one may conclude that they are partly calcified, light-weight pads, loosely articulated, and, upon death and decay, easily moved, heaped up, and deposited with fine sediment. The bedded limestones containing abundant plates are traceable to massive lens-shaped bodies. Associated with the algal plates, usually in mounds, are certain encrusting foraminifera such as *Tetrataxis*, *Tuberatina* and cornuspirids Plate XXII).
Environment: Quiet, clear, photic zone water as much as a few tens of meters deep but possibly much shallower; perhaps at times restricted circulation eliminated many invertebrates.

P 7. Flanking beds. Bioclastic, lithoclastic packstones and wackestones with organic debris derived from growth on tops of slopes and mounds and accumulated in wavy, foreset beds on depositional slopes.
Depositional dips from a few degrees to as much as 30 degrees, about the angle of reponse for such debris. Much of the bioclastic debris is worn, broken, and coated. Previously consolidated or agglutinated sediment is intermixed with finer sediment which infiltered the coarser framework. Geopetals are common. The biota is believed to be essentially that of the reef or mound tops. Foraminifera are abundant, particularly cornuspirid tubular forms, along with broken algal plates and much other bioclastic matter. Certain indigenous, special communities formed meadows or filled pockets on the flanks, e.g., crinoids or brachiopods. Environment: moderate wave energy-winnowed finer lime-mud sediment depositing it downflank where it was organically pelleted, or occurs as micrite matrix in interbioherm wackestones. Normal marine circulation and salinity is indicated by the varied biota. In part, standard microfacies type 5. This constitutes the example chosen for description and interpretation of carbonate microfacies (Chapter III, Plates I, IV A).

P 8. Capping beds of mounds and calcarenite shoals. Grainstones, with coated particles, ooids, peloids, and rounded and worn, rotted bioclasts. Robust, thick-shelled forms, such as euomphalid snails and dasycladacean particles, are almost the only recognizable bioclasts. Often the matrix is calcite mosaic, obscuring or obliterating any original isopachous and probable marine cement. Extreme compaction caused by early dissolution of grains may occur in some beds. These beds are rare in the Pennsylvanian-Wolfcampian but significant; they occur at tops of biohermal strata. They are generally cross-bedded sands, forming well-defined beds only a few feet thick.
Environment: Very shallow, shoal water with perhaps restricted marine fauna, strong wave and current action. These beds were often subjected to nonmarine diagenesis after deposition, having been exposed above sea level during or shortly after deposition. Standard microfacies type 11.

P 9. *Caninia-Chaetetes* beds. Massive micritic and pure limestone beds common in older Pennsylvanian strata with many corals. The organisms are not necessarily in growth position nor are they attached to each other to form boundstone. Probably they grew in soft, calcareous substrate where they were later buried. Both the colonial *Chaetetes* and large solitary *Caninia* are known in argillaceous sediments as well as in carbonates. If especially associated organisms exist, these are not known. Environment: Represent coral rich mud banks or shoals of very shallow water in clear open marine conditions without clay contamination; relatively quiet water. Standard microfacies type 7 and 8 (Bafflestone).

P.10. Onkoidal micrite—algal ball beds, "*Osagia-Ottonosia*". Thin (1 m) beds of lime mudstone or wackestone with rounded nodules, a few cm in diameter. These are masses of crinkly laminated, presumed stromatolitic algae, intergrown with dark plumose encrusting calcitornellid foraminifera. Some of these masses encrust and are interwoven with spirorbid tubes (Toomey, 1962). *Girvanella* tubules usually abound in the laminae. Environment: Matrix may be of coarser texture but is commonly micrite, showing quiet-water deposition. Probably accumulated in areas of slow deposition as loose balls on the sea floor and later were deposited with lime mud or silt. The specific gravity of these balls may have been low because of the large amount of plant matter. Such beds occur commonly in cyclic shelf deposits of all ages, are known also at disconformities, in tidal channels, and in low places behind bioherms. Modern onkoids, formed as lime-mud accretions by blue-green algae, are known in Florida and the Bahamas, scattered in tidal channels and shallow bays or straits, but true beds of the structures such as are seen in the geologic record have not yet been observed in the Holocene. These constitute standard microfacies type 22.

P 11. Basinal, somewhat argillaceous lime mudstones-wackestones. Dark, well-bedded strata with dominantly normal marine fauna. Indigenous fusulinids are very rare; apparently fusulinids did not normally drift so far downslope. Brachiopods are represented mainly by productids and their spines, and spicules. Trilobite and ostracod tests and mollusk shells are common. Echinoids are more common than crinoids. Detrital algal plates, when present, are well-preserved and may show cortical structure. Such shallow basinal limestones, when farther away from the shelf, may consist mainly of spicules, presumably from sponges.
Environment: Quiet, normal marine water, below shallow wave base and at or below common O_2 level. Turbid water induced slightly reducing conditions by inhibiting the normal in-fauna. Some burrowers were present. A somewhat euxinic variety of standard microfacies 9 (Plate II).

Interpretation of Midcontinent Pennsylvanian Cyclothems

Numerous authors in the Kansas Geological Survey Bulletin (1969) and the West Texas Geological Society volume on cyclic sedimentation in the Permian Basin (1969, 1972) described these Late Paleozoic shelf cycles and discussed their possible causes. These papers derive in large part from life-long investigation by R.C.Moore, J.M.Weller, H.R.Wanless as well as more recent studies of D.C.Van Siclen, J.Imbrie, J.W.Harbaugh, L.F.Laporte, E.G.Purdy, D.F.Toomey, A.R.Troell, L.F.Brown, and H.C.Wagner. Figure VII-6 is based mainly on work of Wanless (1972) and describes facies within an ideal shelf cycle across Illinois and Kansas. Figure VII-7 similarly describes a cycle from north central Texas (Van Siclen, 1972). Here the shelf lay closer to the source of clastics and had a steeper gradient.

Most of the above studies have been made following the long outcrop of these cyclothems which stretches from Iowa into north Texas. Any reasonable explanation of the strikingly repetitive vertical sequence must in addition consider the total east-west facies sequence, much of which is downdip in the subsurface of western Kansas and central Texas. Only in the last few years have explanatory models for cyclothems adequately considered downdip relations. See particularly the papers of H.R.Wanless and L.F.Brown. The above figures have attempted to show the complete facies pattern.

Modern studies of midcontinent cyclothems have accepted the following keys to interpretations:

1. Careful tracing of eastward-derived terrigenous beds shows that these thin deltaic units may be environmentally subdivisible into channels, distributaries,

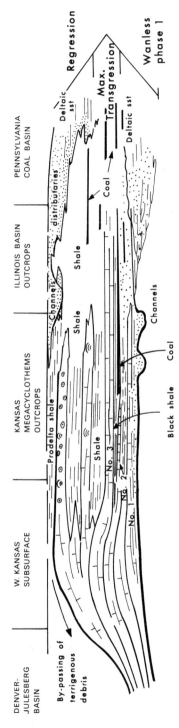

Fig. VII-6. Stratigraphic facies model of a single midcontinent cyclothem from Appalachians to Kansas. Note the extreme vertical exaggeration (up to 50000). The cycles are a few tens of meters thick and the distance is more than 1500 km. Note the complexity of the cycle compared to that of the Yoredale strata. After Wanless (1964, Fig.10)

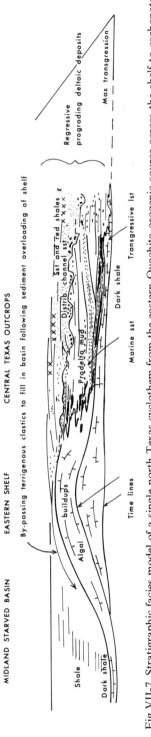

Fig.VII-7. Stratigraphic facies model of a single north Texas cyclothem from the eastern Ouachita orogenic source across the shelf to carbonate buildups at the edge of the Midland basin, a distance of about 200 km. From a model by Van Siclen (1958) modified by Brown (1972). The deltaic complex to the east caused a progradation of environments which coupled with continuous subsidence, differences in sedimentation, and differential compaction, resulted in vagaries of sedimentary fill-in across the shelf. Despite these irregularities the basic pattern much resembles that of the Yoredale cycles with a discrete marine limestone bed at the base overlain by marine shale, progressing upward through a shoaling terrigenous sequence to a channel sandstone and reddish shale

sand flats, and beach-bar units. The traditional lower clastic unit of the Kansas megacyclothem is the most variable and correlation errors are best eliminated by careful delineation of its units. Channels downcutting more than 30 m are known in such strata. Multiple sources of terrigenous influx are recognized. In an eastward direction less regular sequences occur (Wanless, 1972).

2. The black shale, its underlying thin (number 2) limestone, and its eastward equivalent coal facies represent the maximum transgression across a remarkably flat terrain during a time of tectonic stability when there was no ingress of clastics from the east. These beds then offer a key for tracing the cyclic sequences up or down the paleoslope. This triplet changes facies less rapidly than other members. It constitutes the most reliable time stratigraphic marker. Sediments of the other parts of the cycles were formed during a substantial migration of depositional environments during deltaic progradation and retreat. Each of these rock types has facies equivalents in time in both shelf and basinward directions.

3. The downdip limestone buildups (largely buried in the subsurface of Kansas and north central Texas) represent low-lying but extensive organic banks which must have strongly restricted circulation in the Kansas and north Texas eastern shelf by sheltering the lagoons from the open sea to the west. This offers an explanation for the thin black shale sheets.

4. Petrographic distinction of the various limestones in the megacycles of Kansas is important to their interpretation; they differ greatly in depositional environment. The lower limestones, where argillaceous, contain a marine molluscan fauna; the middle limestones (numbers 2 and 3) record clear, warm water with open circulation; and the higher limestones are indicative of shallower, locally more restricted seas, being commonly oolitic, onkoidal, or algal.

In the midcontinent United States there exist between 50 and 60 records of what must have been extensive transgressions and regressions across the central part of the North American craton during the 25–30 million years of the later Pennsylvanian and Wolfcampian. Many explanations for the cyclicity have been advanced (Chapter II and references above). Basic agreement exists that, whatever the causes, considerable sea level fluctuations occurred over the midcontinent shelf during this time, that the cyclicity is unusually apparent because of terrigenous influx, and that probably both eustatic sea level change and tectonic activity contributed to the repetitive pattern. The sequence of a half dozen or so complicated megacyclothems in Kansas (Fig. VII-6) indicates a surprising order to this complex and argues for a simple and systematic mechanism. To date, however, no general theory as to the basic cause for the megacyclothem patterns is completely satisfactory.

One may compare the midcontinent cyclothems with those of the same age described by Van Siclen (1972) and Galloway and Brown (1973) in north central Texas (Fig. VII-7). Where terrigenous influx is important, purely sedimentologic processes may be responsible for cyclicity and are superimposed on eustatic or other extrabasinal causes; in some places the sedimentological controls are dominant. Shifting of deltaic distributaries on a steadily subsiding shelf, restriction of sedimentation by plant growth, and differential compaction of coals, shales and limestones are all mechanisms which control changes in rates of sedimentation relative to sea level and which may be cyclically repeated.

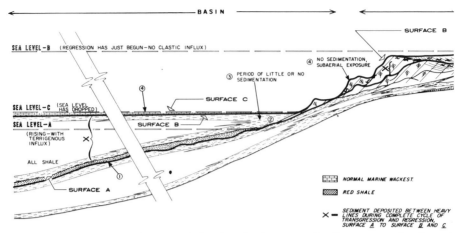

Fig. VII-8. Stratigraphic facies model showing development of sedimentary cycles through reciprocal sedimentation on the Sacramento shelf and eastern Oro Grande basin, southern New Mexico. The shelf is only a few km wide and each cycle only a few tens of meters thick. Sedimentation occurred at surfaces A–B–C at times when sea levels A–B–C existed. Basinal clastics, shelf carbonates, and basinal carbonates were deposited in turn. From Wilson (1967 and 1972, Fig. 4)

Galloway and Brown (1973) pointed out several facts indicating local (intrabasinal) controls on cyclicity during continuous subsidence of the north Texas area, rather than eustatic sea-level changes of great magnitude.

1. No old soil zones or widely traceable disconformities are observed on the shelf.

2. The general stratigraphy indicates almost continuous fluctuations of environments instead of widespread traceable individual cyclic units. There is much variability. Even the limestones on this shelf are not continuous beds but can only be correlated as general zones. Much marine reworking of terrigenous clastics occurred and a continuous interplay of terrestrial and marine sediments existed.

3. Channels do not get deeper toward the shelf edge, as would be expected in cases of drastic sea-level drops. There is no general dissection of the shelf margins.

4. The sandstone patterns are clearly of distributaries crossing the shelf giving the possibility of prevalent lateral shifting of clastic sources.

5. Despite the tendency toward basinward migration, the shelf margins are localized and vary somewhat in stratigraphic position. Clastic wedges on basinward slopes of the shelf margins are also localized and correlate with five deltaic lobes within the Virgilian strata alone.

Galloway and Brown (1973, p. 1212) conclude that "depositional or erosion events occurring simultaneously over the entire eastern shelf are not indicated ———. Although extrabasinal controls probably were operative to some extent, the depositional fabric was determined by interplay of locally active deltaic lobes".

This explanation for cyclicity is essentially that of Moore (1959) for the Yoredales whose sedimentary pattern is very similar to that of the North Texas Penn-

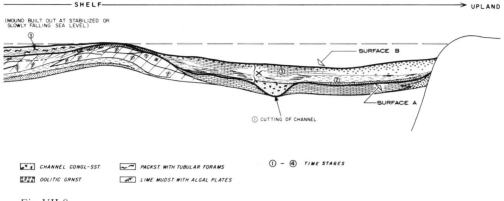

SHELF ───→ UPLAND

(MOUND BUILT OUT AT STABILIZED OR
SLOWLY FALLING SEA LEVEL)

SURFACE B

SURFACE A

① CUTTING OF CHANNEL

☐ CHANNEL CONGL-SST.   ☐ PACKST. WITH TUBULAR FORAMS   ① - ④ TIME STAGES

☐ OOLITIC GRNST   ☐ LIME MUDST. WITH ALGAL PLATES

Fig. VII-8

sylvanian except that perhaps the single basal limestones are more persistent. In contrast, the traditional explanation of midcontinent cyclothems has emphasized extensive east-west regression and transgression caused by extrabasinal sea-level fluctuation or by tectonic "tilting" or "bobbing" of the shelf. The reason for this is the much wider cyclic shelf in the Midcontinent, the extensive downcutting of certain channels and the complexity of the section interpreted to represent initiation of transgression.

In the Midcontinent, conventionally the base of the cycle is marked at the most striking lithologic boundary; this is commonly the base of a channel conglomerate. This disconformity has been interpreted to represent a significant and sudden sea-level lowering. A few channels in the Midcontinent are large and widespread, filled with coarse conglomerate, and cut down through tens of meters of section. The beds above the channel are a mixed clastic and carbonate section interpreted to have formed when the following marine transgression was barely successful in overcoming clastic influx. Many channels, however, may represent merely meandering distributaries during periods in which regressive and progradational terrigenous sedimentation occurred. Mathews (1974, Figs. 16.1 and 16.7) compares the two interpretations.

The north-central Texas cycles and those of the Yoredales are conventionally portrayed to begin with a widespread marine limestone whereas those of the midcontinent United States are conceived to commence with a variable section of limestones and terrigenous clastics.

If the lower part of the Kansas megacyclothems actually represents a part of a prograding regressive sequence, the Midcontinent, Yoredale, and north central Texas patterns are very similar. In any event, the triplet of coal, thin limestone, and black shale appearing in the basal Yoredale and in the middle or basal Kansas megacyclothem represents the most inundative and stable phase of sedimentation and hence the best time-stratigraphic marker in these complex sections.

Other Pennsylvanian cyclothems occur farther west and south in the Rocky Mountain areas. They continue far out into the desert southwest and up the

Cordilleran geosyncline. They have been intensively studied in New Mexico in the Sacramento Mountains by L.V.Cline (in Pray, 1959), Pray (1961), and Wilson (1967a, 1972). The author has outlined an example of cyclic and reciprocal sedimentation here between the sediments of the Oro Grande basin and the narrow carbonate shelf surrounding the eastern side of the Pedernal uplift (Fig.VII-8). The widespread cyclic character of these strata which extend throughout the area surrounding the cratonic blocks of the ancestral Rocky Mountains, induces, almost automatically, a belief in a widespread eustatic mechanism for the cyclicity. (Even the evaporitic Paradox basin contains sedimentary cycles.) On the Sacramento Mountain Pedernal shelf, however, close scrutiny of the section and its paleotectonic history indicate that a reasonable explanation may lie in the periodic uptilting and subsequent downfaulting of the tectonically unstable edges of the Pedernal block. In this area practically all cycles show that terrigenous influx occurred as sea level tended to rise. As in the traditional interpretation of the Kansas megacyclothems, the cycles begin with transgressive terrigenous clastics. Soon after clastic influx ceased, sea level reached its maximum. When the seas cleared, carbonate accretion, perhaps with continued rise of the shelf, caused a retreat of the sea off the shelf into the basin. Here again the timing of clastic influx relative to sea-level fluctuation constitutes the basic control over the sedimentary pattern.

The widespread cyclicity so evident in Late Paleozoic strata may be a consequence of some eustatic sea-level changes whose effects are enhanced by local tectonics which caused unstable shelves and periodic influx of terrigenous clastics which shut down the carbonate production and rapidly altered the depositional environments.

Chapter VIII

Permo-Triassic Buildups
and Late Triassic Ecologic Reefs

The Permian Reef Complex, surrounding the Delaware basin of West Texas and New Mexico, and the Middle Triassic (Ladinian) of the Dolomites of the South Alps are two of the worlds greatest and best known ancient carbonate buildups. They are closely related in age and organic content. Surprisingly enough, they still present unsolved problems of origin and subsequent diagenetic history. This Chapter briefly reviews and compares these famous models of carbonate facies focusing on both the deposition and diagenesis of the sediments.

Permian Reef Complex

The almost featureless plains of West Texas and southeastern New Mexico conceal rocks of Late Paleozoic age with a complex structural framework. The Central Basin Platform, a north-south horst, separates two structural depressions, the Midland basin on the East and the Delaware basin on the West. During Pennsylvanian time these basins, now covered by younger strata, were deep, euxinic and starved. During Permian time the Midland basin subsided slowly and was rapidly filled; in contrast, the Delaware basin sank more quickly, was filled partly with fine sand and silt, but remained relatively deep until the end of the Permian. A carbonate rim almost completely encircled this basin, separating it from the surrounding shelves on which accumulated a shallow water lagoonal and evaporitic tidal flat facies. The general facies and structural relations have been known since the 1930's and are reviewed by King (1942, 1948). An interesting chapter on the development of ideas on facies changes and "reef" recognition in the complex Permian strata of the southwestern U.S.A. is found in King's textbook, *The Evolution of North America* (1959). Lloyd (1929) is generally given credit for recognizing that the Guadalupe Mountains on the western side of the basin contained exposures of a kind of fossil "reef" which rimmed the entire subsurface Delaware basin. Lloyd's idea was an important one, for the northern and eastern sides of the basin lie below the almost featureless surface of the plains and furnish no hint of the underlying structure. The eastern border of the Delaware basin was discovered to form the western edge of the Central Basin Platform, one of the world's great oil producing structures. Permian strata produce most of this oil (Galley, 1958; Hills, 1972) from beds just behind the carbonate shelf margin. It is a rare opportunity when strata of the subsurface can be studied in nearby outcrops, in the same facies, and on the opposite side of the same basin.

Rising west of the Pecos River in New Mexico, the Guadalupe Mountains, which contain the Permian reef outcrops, form the front of the Rockies. They are a great triangular block, tilted gently northward and bounded on the west by a large

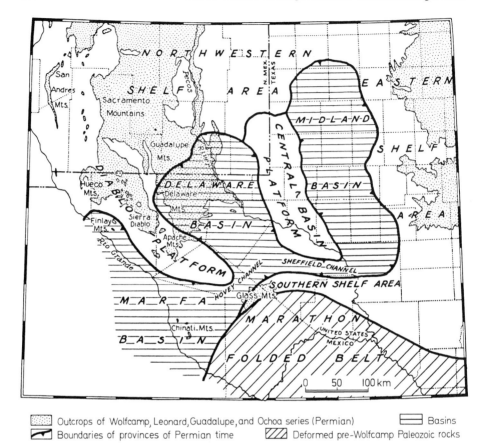

Outcrops of Wolfcamp, Leonard, Guadalupe, and Ochoa series (Permian)   Basins
Boundaries of provinces of Permian time   Deformed pre-Wolfcamp Paleozoic rocks

Fig. VIII-1. Permian structural provinces in western Texas and adjoining part of New Mexico. From King (1934 and 1942), courtesy of Geological Society of America and American Association of Petroleum Geologists

normal fault. The eastern side of the fault is marked by spectacular cliffs, hundreds of meters high which expose a slightly oblique section across the area of facies change. Relief on the eastern side of the Guadalupe Mountain block mirrors the original depositional surface which is beautifully exposed particularly to the south. Here one may walk the 25° slope of the old sea floor from basin to shelf, trudging upward, following the beds for a vertical distance of almost 700 m. The desert weathering conditions improve greatly the geologist's ability to trace beds from aerial views and to understand the major stratigraphic relationships.

Outcrop work in the Guadalupe Mountains by P.B.King, N.D.Newell and associates, P.Hayes, and R.J.Dunham was stimulated by knowledge of the relationships of these beds to subsurface strata and offered an opportunity to project facies models to the subsurface. No better example exists of cooperation between government and industry in resource development than in the exploration of the Delaware basin for oil and gas, salt, potash, and sulphur.

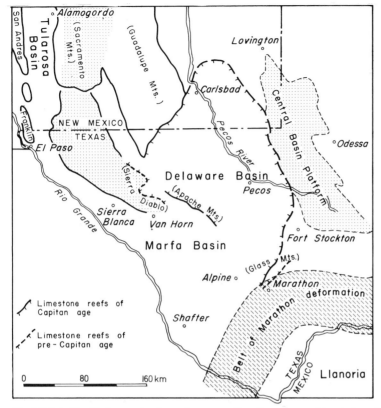

Fig. VIII-2. The Delaware basin and surrounding Permian reef tracts forming shelf margins off flanks of Pennsylvanian uplifts

Tectonic Controls on Sedimentation around the Delaware Basin

The general structural framework was first mapped by King (1934) and his illustrations have been republished many times (Figs. VIII-1, VIII-2). The shape of the Delaware basin is controlled by the trends of the Diablo Plateau and Central Basin Platform. The Pedernal uplift, the northern extension of the Diablo Plateau, extends eastward, widening into a shelf along whose flank Pennsylvanian and Early Permian shelf margins were built progressively into the Delaware basin. Both the east and west sides of the basin are more sharply defined than its threshold onto the northern shelf. Here the regressive outbuilding of Permo-Pennsylvanian shelf margins continued but with wider geographic spread. A map of the configuration of the Late Permian reef surface (Fig. VIII-3) indicates the possibility of open circulation of waters across a more gradually shoaling area to the north than was present on the east and west sides of the basin. Not such complete restriction occurred on the northern shelf and mounds of lime mud and bioclastic particles which may have been deposited as tidal bars are oriented north-south along the threshold. This gentler northern edge, and somewhat differ-

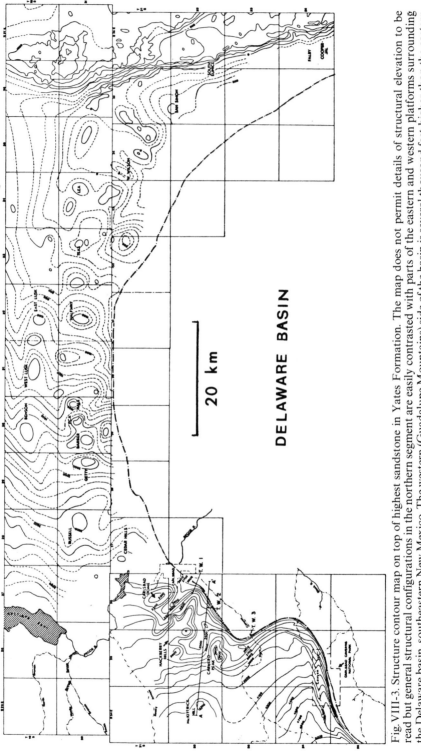

Fig. VIII-3. Structure contour map on top of highest sandstone in Yates Formation. The map does not permit details of structural elevation to be read but general structural configurations in the northern segment are easily contrasted with parts of the eastern and western platforms surrounding the Delaware basin, southeastern New Mexico. The western (Guadalupe Mountains) side of the basin is several thousand feet higher than the eastern side (Central Basin Platform) because of Tertiary uplift. Contours on the northern shelf outline numerous elongate domes—depositional mounds (?). Grid is of 6 mile (10 km) square Townships. Contour interval is 100 feet. From Motts (1972, Fig. 4), courtesy of Geological Society of America

ent facies, must be a response to a different pre-Permian tectonic framework; no basement faults are known to border the northern edge of the basin.

In the Guadalupe Mountains the western rim of the Permian Delaware basin is a fairly narrow accumulation of massive carbonate fronted by thick foreset beds which dip 30 degrees into the basin. Vertical depositional relief, that can be seen by tracing these beds down the front into the basin, increases stratigraphically upward to almost 700 m in the latest Permian. In the great canyons of the Guadalupes, the shelf margin can also be seen to have been built out basinward several kilometers as it grew upward. The same stratigraphic pattern is visible around the Delaware basin in the subsurface (Fig. VIII-4). Subsidence in the basin nearly managed to keep pace with carbonate accretion on the shelves throughout Permian time. The rate of sinking increased slowly until the carbonate shelf margin was almost vertical. This is a common occurrence in carbonate bank or platform history. Behind the rim, well-bedded strata of an evaporitic shelf environment developed. Regional facies patterns (Fig. VIII-5) are presented by Galley (1958), Hills (1972), and Meissner (1972).

Facies Sequence in the Permian Reef Complex

Several major works have described in detail the successive facies belts developed along the relatively narrow shelf margin (King, 1948; Newell et al., 1953; Dunham, 1972). Tyrell (1969) also described in detail the facies along the uppermost stratigraphic unit which can be traced from the shelf across the margin and into the basin. Earlier, Hickox (in Newell et al., 1953) traced and described a backreef bed, Yates A Dolomite. Further details were added by Achauer (1969). The stratigraphic succession and nomenclature of the shelf and basin deposits is outlined in Fig. VIII-6. The interpreted facies belts are reproduced as Fig. VIII-7. The facies belts have been more or less standardized into generally recognized units and the outcrops of the Permian Reef Complex in the Guadalupe Mountains have been developed into a classic model for carbonate shelf margin facies. Various unsolved problems exist, however, as to exactly how some of the facies fit in the original depositional profile. This is principally because of the great relief, inacessibility of many of the canyons and massive character of the "reef" itself. These circumstances have to date almost prohibited the detailed section measuring and sampling necessary to careful tracing of rock types and the clarification of facies transitions despite the spectacular mountain outcrops. It is a tribute to those persons who have accomplished part of this arduous work that a more or less coherent pictures has emerged despite the field difficulties. The facies belts are generalized in Table VIII-1.

The Permian strata also display a variety of sedimentary structures created during both deposition and diagenesis. (See comparison with the Triassic in this Chapter.) These strata illustrate beautifully the depositional facies across a typical narrow shelf margin (less than 10 km wide) with a steep (25–30°) edge and great vertical relief (700 m). These marginal belts separate a wide carbonate platform or ramp from a deeply subsiding basin. Application of the model to other carbonate facies complexes should be restricted by the consideration that: (1) the sediments formed in a strongly seasonal and evaporitic climate and endured strong contemporary diagenesis and (2) major fluctuation of sea level exposed the platform, its edge, and large parts of the slope periodically during Permian time.

The following key microfacies types are recognized and illustrated by Dunham (1972).

Pm 1. Spiculitic, radiolarian laminated calcisilt-packstone in beds of dark, platy limestone. Basinal, Facies belt 1. Standard microfacies type 1; Dunham, 1972, Figs. III-4–7.

Pm 2. Brown finely bioclastic packstone; from lower slope facies, deep shelf margin, Facies belt 3; standard microfacies type 2; Dunham, 1972, Fig. III-3.

Pm 3. Lithoclastic bioclastic packstone beds within coarser marine talus breccia; very fossiliferous with open marine biota, on foreslope of carbonate platform, Facies belt 4; standard microfacies type 4; Dunham, 1972, Fig. III-8–13.

Pm 4. Typical micrite of organic buildup with sponges, Facies belt 5. Microfacies type 7—bafflestone; Dunham, 1972, Fig. II-55.

Pm 5. Typical micrite of organic buildup with stromatolite-lined cavities, and encrusting organisms (*Tubiphytes*, spongiomorphs). Facies belt 5; standard microfacies type 7—bindstone; Dunham, 1972, Fig. II-59 (Plate XXV B).

Pm 6. Typical micritic bioclastic wackestone associated with vuggy boundstone of organic buildups, Facies belt 5; standard microfacies type 9; Dunham, 1972, Figs. II-53-54-60.

Pm 7. Lime grainstone of oolite and coated grains, Facies belt 6; standard microfacies type 11; Dunham, 1972, Figs. II-1–4.

Pm 8. Lime grainstone with dasycladacean and large gastropod particles, Facies belt 6; standard microfacies type 18; Dunham, 1972. Fig. II-48, Figs. I-2–5.

Pm 9. Lime grainstone, fusulinid coquina, Facies belt 6; special variety of standard microfacies type 12; Dunham, 1972, Fig. II-49, Fig. I-6.

Pm 10. Lime grainstone-pisolites, showing evidence for *in situ* growth; considered a diagenetic facies; Dunham, 1972, Fig. II-33–35 (Plate XV B).

Pm 11. Lime mudstone-wackestone with peloids and ostracods, mollusks, and calcispheres, Facies belt 8; standard microfacies type 19; Dunham, 1972, Figs. I-43, 55.

Pm 12. Lime mudstone-wackestone with well-developed fenestral fabric, Facies belt 8; standard microfacies type 19; Dunham, 1972, Fig. I-36 (Plate XIII A).

Table VIII-1. Facies Belts Permian Reef Complex

General facies name (Meissner, 1972)	Wilson's facies belt number and name	Tyrell (1969), Dunham (1972) microfacies designation
Bernal		Red shale and siltstone
Chalk Bluff	9. Platform evaporite	Evaporites; gypsum (parallel bedded, laminated, interbedded with fine sandstone) and dolomite wackestone
Carlsbad	8. Facies of restricted circulation on marine platform	Mixed carbonate environments; dolomite wackestone with calcispheres and ostracods, algal stromatolitic mudstone, unwinnowed shelf sand, pelletoidal dolomite grainstone. (Many marine to meteoric vadose sedimentary-diagenetic structures e.g., fenestral fabric, pisolites, tepees.)

Table VIII-1 (continued)

General facies name (Meissner, 1972)	Wilson's facies belt number and name	Tyrell (1969), Dunham (1972) microfacies designation
Carlsbad	6. Winnowed platform edge sands	Winnowed shelf sand; skeletal lithoclastic dolomite grainstone, with coated grains, foraminifera, dasycladacean algae, ooids, thick-shelled gastropods, cross-bedded lime sand. (Many marine to meteoric vadose sedimentary-diagenetic features.)
Capitan	5. Organic reef of platform margin	Shelf edge carbonate; lithoclastic skeletal lime grainstone and wackestone; bioclastic debris mixed with lime mud or silt and boundstone of abundant sessile benthos and encrusting biota. Bryozoans, red algae, *Tubiphytes*, foraminifera, sponges, brachiopods, crinoids. (Many marine to meteoric vadose sedimentary-diagenetic features, e.g., large druse-lined cavities, collapsed beds, large veins filled with coarse calcite druse.)
Capitan	4. Foreslope facies of carbonate platform (marine talus)	Basin slope carbonates; breccia or "reef talus;" vague and inclined beds with boulders and redeposited fossils. Finer grained beds are lithoclastic-skeletal wackestones, partly dolomitized
Delaware facies. Thick tongues of limestone intercalated with fine sandstones (some mega-boulder beds)	3. Deep shelf margin or basin margin "toe of slope"	Basin margin carbonates; dark well-bedded lithoclastic skeletal wackestone-packstone, some very fine-grained. Finer bioclastic debris includes foraminifera and abundant pieces of organisms living up the slope; almost all are redeposited. Brachiopods, bryozoans and crinoids prominent. Many sedimentary structures including inclined foresets, slumps, boulder beds, channels, load casts, carbonate mounds
Delaware facies. Thin carbonate units intercalated with fine sandstone	1. Basin	Basinal carbonate; black well-bedded, platy, laminated micrograded, fine-grained wackestone-packstone with sponge spicules, globular foraminifera and radiolarians, ammonoids
Delaware facies, black shale within fine sandstone	1. Basin	Starved basin, dark radioactive silty shale with carbonate beds a few cm. thick

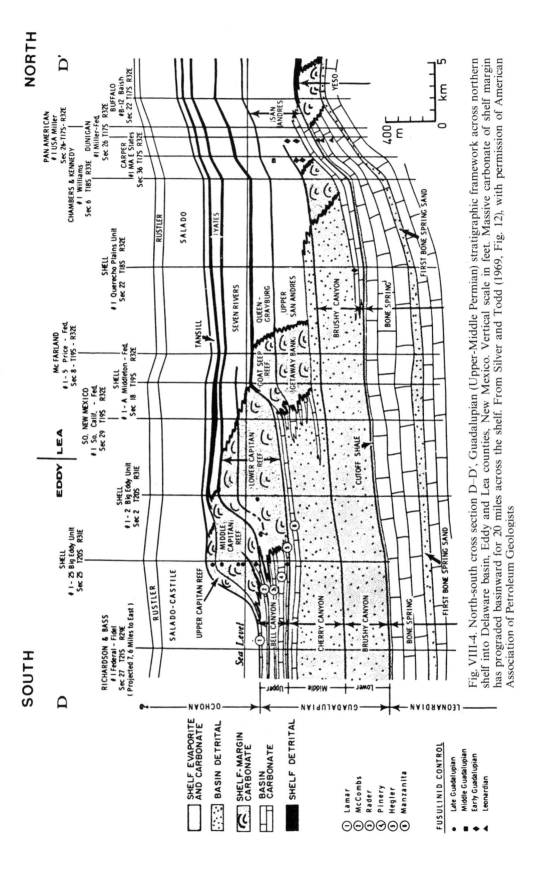

Fig. VIII-4. North-south cross section D–D', Guadalupian (Upper-Middle Permian) stratigraphic framework across northern shelf into Delaware basin, Eddy and Lea counties, New Mexico. Vertical scale in feet. Massive carbonate of shelf margin has prograded basinward for 20 miles across the shelf. From Silver and Todd (1969, Fig. 12), with permission of American Association of Petroleum Geologists

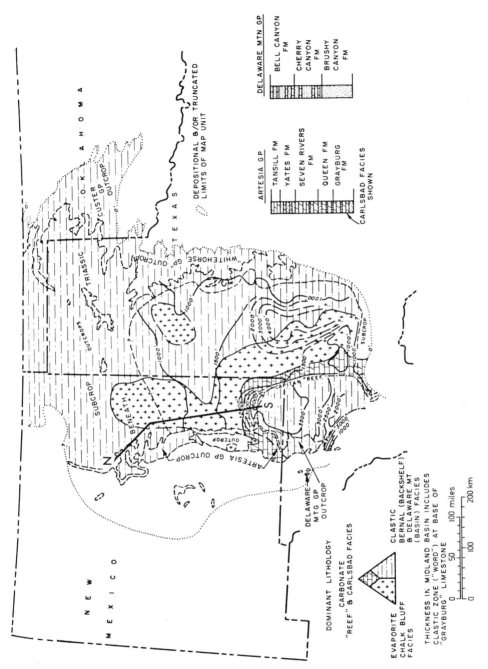

Fig. VIII-5. Lithofacies and isopach map of the Artesia and Delaware Mountain Groups and their equivalents in the southwestern United States. From Meissner (1972, Fig. 3), courtesy of West Texas Geological Society

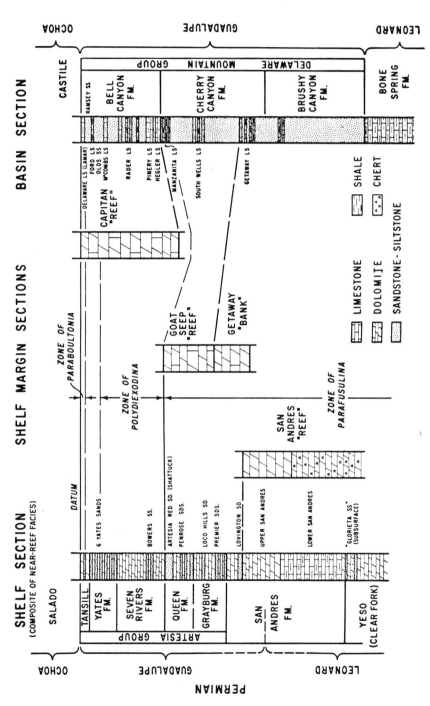

Fig. VIII-6. Typical Upper-Middle Permian stratigraphic sections in vicinity of the shelf margin of the reef complex around the Delaware basin, New Mexico. Shows correlations based on lithologic interfingering and on key fusulinid zones. Section is hung on a time datum presumed horizontal and thus depositional topography is not shown. Compare with Fig. VIII-7 which is a stratigraphic model based on this cross section. Line of cross section indicated on Fig. VIII-5. From Meissner (1972, Fig. 1), courtesy of West Texas Geological Society

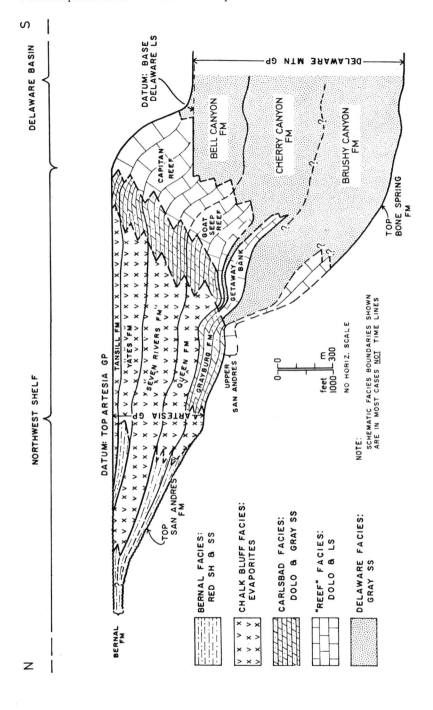

Fig. VIII-7. North-south section Sangre de Christo Mountains, New Mexico to Delaware basin of Upper-Middle Permian Artesia and Delaware Mountain Groups. Original deposition topography restored. Line of section shown on Fig. VIII-5 and basic stratigraphy on Fig. VIII-6. All from Meissner (1972, Figs. 1–3), courtesy of West Texas Geological Society

Problems of the Permian Reef Complex

The profile across the shelf: What was the original topographic profile across the shelf and its margin? Did the ramp-like shelf slope steadily seaward owing its restricted circulation to sluggish currents and lack of wave action across its vast expanse? Or did a barrier island complex lie at the outer margin of a platform—a barrier which was exposed at low sea level stands, subjected to diagenesis by meteoric water, and which offered a protection to a broad shelf lagoon? Both interpretations of the shelf environments are possible and both conditions may have prevailed at different times. Detailed mapping is needed to ascertain the relative amounts and distribution of lagoonal, tidal flat, and sabkha environments in the Chalk Bluff facies. More careful and stratigraphically controlled mapping in the Guadalupe Mountains could outline the configurations of old barrier islands by using signs of vadose diagenesis. Quite possibly, positions of mean sea level can also be ascertained from mapping distribution of algal stromatolites. Dunham's marginal mound hypothesis (Fig. VIII-9) suggests that a belt of these stromatolites existed on the back reef, lagoonal side, and marked the inner shore of a barrier island.

A further profile problem concerns the position of the massive limestone interpreted as "organic reef" by Newell et al. (1953) who conceived the shelf margin as forming a barrier reef in the breaker zone analogous to a modern coral rim. Further petrographic studies (Achauer, 1969; Dunham, 1972) have demonstrated a large amount of lime mudstone, fine broken debris, low-lying encrusting organisms, loose sediment dwellers, and in many places, almost complete absence of relatively large framebuilding biota (Fig. VIII-8). Sedimentologic interpretation would favor a downslope, quiet-water environment for such an accumulation. The position of the micrite on the original depositional profile is made unclear by the difficulty of tracing bedding planes and individual bodies of sediment through the facies and the gigantic size of the outcrops involved. A start on this was made by Smith (1973) who traced an organic framework of bryozoans on some canyon bed exposures and interpreted the shelf margin as an ecologic reef. Three interpretations of profile as epitomized by Dunham are given in Fig. VIII-9. The hypotheses are termed Barrier Reef, Uninterrupted Slope, and Marginal Mound. The author believes (following Dunham) that at some places careful observation of slopes and bedding planes indicates that the original high point on the profile is in a lime sand—barrier island facies and that the "reefy" wackestone, rich in sponges, *Tubiphytes,* and lined cavities ("organic reef" of Newell et al.) is a downslope deposit. The Permian Reef Complex is thus considered to represent a Type I shelf margin—a downslope mud accumulation rich in organisms (see Chapter XII and Wilson, 1974). Dunham's interpretation, that the mass developed rigidity more because of early diagenetic cementation than through organic binding, is discussed below.

Problems of sea-level fluctuation: Silver and Todd (1969) and Meissner (1972) demonstrated how cyclic and reciprocal sedimentation may be responsible for the strikingly abrupt Permian facies changes from carbonate-evaporite of shelf and shelf margin, to predominantly sandstone-siltstone strata in the basin. Dunham's work (1972) helps to document key facies relations in the Guadalupe Mountains. That the thin dark limestones of the basin may be traced up the slope of the shelf

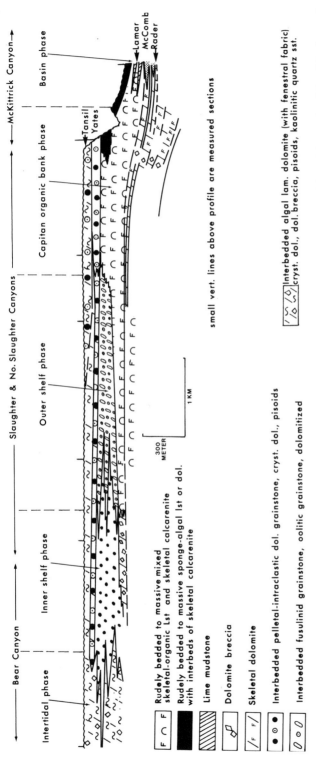

Fig. VIII-8. Composite section west to east across strike of Capitan Formation and into Delaware basin, Guadalupe Mountains, New Mexico and western Texas. Vertical exaggeration only X3. After Achauer (1969, Fig. 3), with permission of American Association of Petroleum Geologists.

margin through massive boulder talus to fossiliferous wackestone, offers compelling proof of the discrete periods of carbonate production, independent of terrigenous accumulation.

The fine sandstones constituting the bulk of the basinal sediments generally do not intergrade with the carbonates but lap out against them on the slope. Petrographically the sands show derivation from the granitic terrain of the Pedernal land mass to the north. They relate mineralogically to the widespread sheets of shelf sands such as Yates and Queen which intercalate with the lagoonal-tidal flat sediments of the platform. They must in some way have crossed the massive carbonate fringe at the shelf margin, now devoid of sandstone strata.

Figures II-20 and II-21 illustrate the process of deposition of these sands. They were formed by disintegration of the northern granitic terrain, spread as sheets across the shelf under marine-littoral conditions during marine regression, possibly windblown and washed across the shelf edge into the basin during low sea-level stands. Those that passed through the island barrier partly filled the cavities in it. Much sand may have come into the basin from the wider threshold at its northern end (Motts, 1972). Once in the basin the sands were spread by channelled distributaries downslope in relatively deep water.

Dunham (1972) advanced arguments for large vertical drops of sea level and considerable duration of times of exposure on the shelf and slope:

1. Extensive sedimentary structures indicative of vadose diagenesis (pisolite, tepee, giant polygons structure, druse veins, fenestral fabric, caliche-like cavity linings) pervade the sands of the crest of the shelf and the "organic reef"—bioclastic limemud-wackestone accumulation many tens of meters below the interpreted crest. Today these structures are commonly found in littoral environments and alternate wetting and drying and carbonate precipitation under marine vadose as well as meteoric influence is probably responsible for them.

2. Kaolinitized weathered feldspar from the quartz sands occurs in shelf margin sands, in silt fillings of cavities in the "organic reef", and in breccia and talus beds along the front of the buildup; unweathered feldspars occur in the sands of the shelf and basin. This distribution of kaolinite indicates that extensive subaerial exposure occurred from time to time at the margin of the shelf and down its foreslope.

3. Boulder beds in the conglomerates down the slope and into the basin include abundant cemented rocks eroded from the lower part of the slope. Some of these dark pebbles were soft when emplaced but many were lithified. The presumption that subaerial exposure was necessary to harden this fine sediment, and that wave action or cliff collapse was most probably the means by which these lithified blocks were initiated, argues for extensive sea-level drops (of at least 300 m) down the front. Other writers considering these deposits to be spectacular mud-boulder flows from channels in the front of the shelf, have believed that the blocks initiated under purely submarine conditions (Newell et al., 1953; Jacka et al., 1972). Secondary dolomitization is extensive in some parts of the slope rocks, particularly in brecciated talus beds. No dolomite pebbles are seen in the boulder beds of the Rader debris flow and other slides which ranged far into the basin. Dolomitization must therefore have followed the time in which the boulders were emplaced.

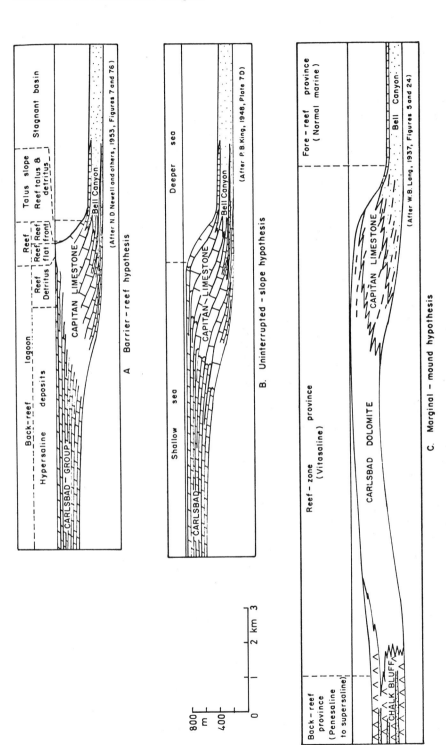

Fig. VIII-9. Alternative hypotheses of reconstructed profiles along shelf margin of Permian Reef Complex. From Dunham (1972, Fig.6)

The distinction between submarine cementation and meteoric cementation in the strata of the Permian Reef Complex is a problem currently being resolved in the light of studies of Holocene subtidal cementation of reefs and other littoral deposits. One must be cautious about interpreting the downward extent and influence of meteoric water within the shelf-margin strata until clearer relationships are established between the type and sequence of cementing carbonate crystals and the diagenetic environments (Schmidt and Klement, 1971). Coarse bladed or drusy void-lining cement is spectacularly developed in the Permian sponge-algal micritic facies. It is commonly an expanded and displacive fabric. In places it is dark and cloudy owing to inclusions. Remnants of tiny marine bioclasts may be included in it. It is generally later than the fine wavy lamellar rinds which line the cavities. The coarse fibrous-bladed druse could be calcite pseudomorph after aragonite. Masses of the latter are reported by Ginsburg and James (1974) from slopes 80 m below the surface of modern barrier reefs off British Honduras. The micritic lamellar wavy rinds could be the Mg calcite marine cements reported by Land (1971) and Ginsburg et al. (1971) from Jamaica and Bermuda. But both types of cement in the Permian Reef Complex, as well as in the Dolomites, are similar to certain cave deposits as pointed out by Dunham (1972). Perhaps the inference is that the diagenesis affecting these sediments is littoral. Wave splash zone, marine water saturation from high tides, interspersed with meteoric ground water and rainfall in a strongly seasonal climate may be the explanation. Caliche structures as well as thick rinds of cement may result from precipitation from marine, as well as from fresh water. Purser's and Loreau's discovery (1973) of extensive splash zone aragonite cementation along the Trucial coast may be important in evaluating the types of cement seen in the Permian Reef facies.

Problem of biologic composition of "organic reef": The book of Newell et al. (1953) is invaluable for description of the organic composition of the Permian Reef Complex. The uppermost lime sands of the latest Permian, interpreted to represent the crest of the shelf margin, are made up almost exclusively of piles of dasycladacean algae and oriented fusulinids with abundant scattered mollusks, principally large thick-shelled gastropods. Shelfward of the lime sand the biota is confined to a few mollusks, ostracods, calcispheres, and stromatolitic algae. This biological composition indicates conditions of abnormal and/or fluctuating salinity and water temperature. With the exception of the fusulinids, which probably lived slightly downslope from the crest and were washed up into sandbars, the fauna is euryhaline. The abundant dasycladaceans probably indicate extremely shallow water.

In many places, the organic reef is also dominated by a specialized sessile benthonic biota of primitive forms which could withstand great variation of marine conditions. These include abundant *Solenopora*, green algae, hydrozoans?, *Tubiphytes*, and sponges. Primitive encrusting forms of alga and/or foraminifera make up patches of boundstone and help to line cavities later filled with laminated crusts and coarse, drusy spar. The role of blue-green algal stromatolites in the "organic reef" is hard to evaluate because many structures in the past assigned to algae are probably inorganic encrustations resulting from vadose processes or even marine micritic calcite cement.

The sponges (Sycons, Sphinctozoans or beaded forms) dominate and are the only organisms larger than a few cm. Corals are practically absent in the latest Permian "organic reef". Fragments and whole shells of normal marine (stenohaline) forms such as brachiopods, bryozoans, crinoids, and cryptostomate bryozoan are abundant in places and have, in general, delicate forms adapted best to life in water at or below wave base. Fractures or large cavities filled with thick-shelled nautiloids and echinoids occur. Much lime mud and calcisilt (up to 50% of volume) occurs in the "organic reef". The micrite matrix, the low-lying encrusting habit and rather delicate form of the sponges and bryozoans, and the variety of normal marine biota in the upper part of the massive limestone, indicate that very special conditions prevailed to form what Newell et al. (1953) considered to be typical "reef" rock. Much more detailed study of vertical zonation, taxonomy, and growth forms of these organisms is necessary before the full story on environment of deposition can be adduced. The limited variety and primitive character of the sessile encrusting fauna in much of the organic reef may result from a spillover of saline water through the barrier bar sands. The rapid increase downslope of abundant, normal marine fossils (bryozoans and brachiopods) indicates water of open marine salinity. Practically all the well-known Permian faunas including the sessile "reef-dwelling" brachiopods, *Leptodus*, and the richtofenids, come from these slope deposits.

The evidence seems compelling that waves crashing on a massive organic barrier would not have resulted in the type of sedimentary and organic record now seen in the Capitan "reef-rock". More probably the sponge-algal micrite represents a quiet-water accumulation. Was this across a shallow flat extending out to a slope, dropping off into a sea with low wave action? Or did the slope decline rapidly into water a few tens of meters deep seaward of and below an exposed ridge of sand bars—a slope where sponges and other encrusting organisms struggled for survival in a lime mud substrate? Profile studies would indicate the latter is the more probable but more detailed field study is necessary.

The Middle Triassic of the Dolomites

The spectacular scenery of the Dolomites in the southern Alps (South Tyrol of Italy) has long inspired and fascinated mountaineers, skiers, and geologists. Huge vertical masses of blanched limestone rise like spires and immense castles from a dissected plateau of deep green meadows and dark soil. The steep vertical relief of these great masses is as great as 1000 m; much of it is due to original deposition, for these giant crags are mostly structurally intact remains of carbonate buildups of Middle Triassic age. The Dolomites have been studied as models of fossil "reefs" for almost 100 years. Mojsisovics' work, "Die Dolomit-Riffe von Südtirol und Venedig" (1879) still stands as a classic in the history of geological investigation of such phenomena.

The Dolomites are located in northern Italy, east of the Adige Valley, just southeast of the Alpine Brenner Pass separating Italy from Austria (Figs. VIII-10 and VIII-11). Structurally, they belong to the little deformed South Alps and are separated from the main Alpine deformation belt by the Giudicaria lineament, an

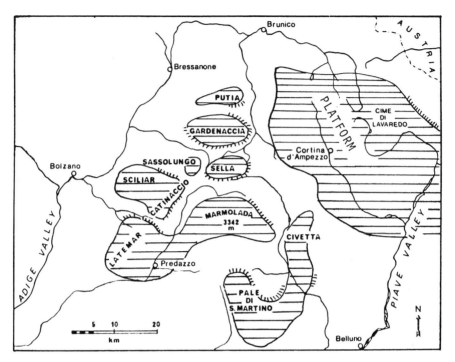

Fig. VIII-10. Location map of the Dolomites of the southern Tyrol, Italy showing the location of the major Triassic carbonate banks. After an index map of Bosellini and Rossi (1974, Fig. 1). Hachured lines indicate platform edges marked by inclined foreslope beds and talus. Location of banks from Leonardi (1967)

important fault zone. The great thickness of the carbonate buildups and the effusive basic volcanic sediments which filled in around them indicates geosynclinal subsidence on the outer continental margin (Bernoulli, 1972). This subsiding area block-faulted during Triassic time and carbonate banks developed extensively along it.

The area of the western Dolomites has been studied for more than 25 years by the Geological Institute of the University of Ferrara. The culmination of this effort is a two volume work by Pierro Leonardi in 1967, *Le Dolomiti: Geologia dei monti tra Isarco e Piave* and the following paragraphs review part of it. See also Bosellini and Rossi (1974) for a review in English.

The great development of major carbonate banks was in Middle Triassic time but a repetition of the shallow water restricted marine bank facies occurred also in the Late Triassic. Stratigraphically the South Alpine Triassic is identical to beds of this age in the complexly overthrust Northern Limestone Alps of Bavaria and Austria. These areas are separated from each other by the central and metamorphosed axis of the Alps (Hohe Tauern Window) which crosses Austria in an east-west direction (Fig. VIII-13). The precise area of derivation of the Northern Calcareous Alps is still one of the mysteries of the Alpine geology. Their root zone may be either north or south of the crystalline axis.

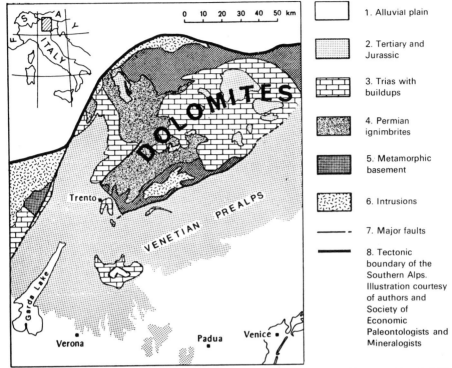

Fig. VIII-11. Regional geological setting of the Dolomites from Bosellini and Rossi (1974, Fig. 2).

The stratigraphic section for the Triassic of these regions is given in Fig. VIII-12. The total Triassic in the western Dolomites is more than 2000 m thick—and in the eastern Dolomites much greater.

Tectonic Control of Regional Facies and Thickness in the Dolomites

Three natural regions may be distinguished in the Dolomites: (1) an area of thick-bedded limestone and dolomite in the area of the Adige River Valley, west of the Dolomites proper, (2) the area of the western Dolomites, east of the Isarco River and the city of Bolzano (Bozen) where great isolated banks rise from a plateau, and (3) a formerly unified platform of thick-bedded massive Triassic limestone and dolomite also deeply dissected by erosion and lying still farther east around Cortina d'Ampezzo, the eastern Dolomites.

Apparently movement on or around a basement fault block controlled the location of the series of great banks comprising the western Dolomites but the cause of location of individual banks is not very clear. According to Bosselini (in Leonardi, 1967), the oldest Triassic Werfen beds, which constitute a variable sequence of terrigenous clastics and limestone, measure only a few meters in the Gardena Pass area of the western Dolomites but thicken eastward to about 500 m in the eastern Dolomites. A pronounced area of very thin, earliest Triassic beds is shown on Fig. VIII-14 and interpreted to represent a north-south positive area.

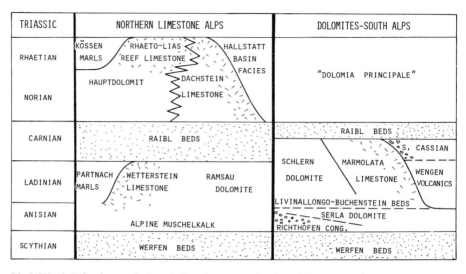

Fig. VIII-12. Triassic correlations in Southern and Northern Limestone Alps

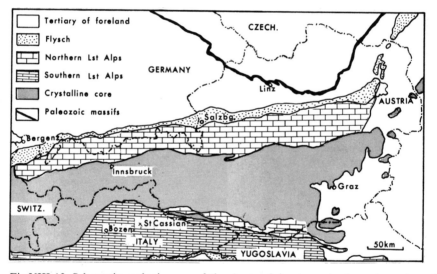

Fig. VIII-13. Schematic geologic map of the Austroalpine tectonic elements, eastern Alps. After Zankl (unpub.)

This is oriented directly across an east-northeast-trending thick area of earlier Permian sedimentation. This Triassic uplift, termed by Leonardi (1967) the Atesino structural high, also caused the development of the Richthofen conglomerate at the base of the Anisian beds above the Werfen. These coarse sediments are considered to represent a pulse of uplift and erosion of the high. The lower Anisian (Serla or Sarl Dolomite and its limestone facies, the Contrin) forms a

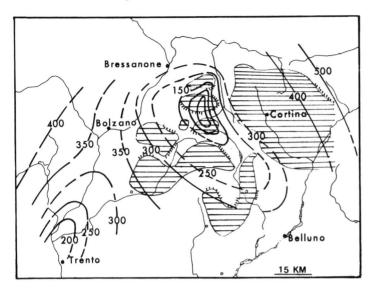

Fig. VIII-14. Isopachous map in meters of the Werfen beds showing earlier Permo-Triassic positive element in relationship to Dolomite banks. Larger banks seem to have formed down flanks of positive area. After Leonardi (1967, Fig. 218 by A. Bosellini)

carbonate platform across the total Dolomite area but thins from 30–100 m around the Atesino high. In the trough to the east the Serla is from 500–1000 m thick. From the Serla Dolomite platform rise the great isolated Ladinian banks, shaping the fantastic landscape of the western Dolomites. To the east, the Serla is separated from the Ladinian Sciliar (Schlern) dolomite by a relatively thin sequence of marine argillaceous strata with ammonites and brachiopods, the *Trinodosus* beds. In the western Dolomites these beds are of lagoonal facies. Thus Ladinian carbonate deposition was initiated in very shallow water and kept pace in many places with a pronounced subsidence. An arc of great carbonate banks formed in the western Dolomites. The northern banks, the Gardinaccia and Sella Groups, lie over, and just east of the crest of the Atesino high (Figs. VIII-14 and II-7). The southern members loop around the south flank of the positive area (Marmolada-Catinaccio-Sciliar-Latimar-Agnello Banks). West of the site of the great isolated banks, thick, more or less continuous Middle Triassic strata exist in the Adige River area. Here a wide platform existed whose limestones are characterized especially by diplopore dasycladacean algae, indicating banks with restricted circulation. Likewise, east of the western Dolomite banks a thick single platform of Ladinian dolomite formed, filling in the trough of the eastern Dolomites. These wider platforms east and west of the Atesino high seem to have grown around it as a halo; the individual, smaller banks developed on its top.

The Ladinian banks are now separated by a thick sedimentary sequence, mostly volcanic, which filled in deep areas between them. Under the volcanics in the intervening basins lies a dark basinal facies, termed the Livinallongo or Buchenstein strata. These rocks are mainly nodular and bituminous limestone with

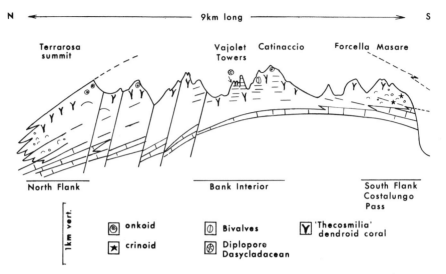

Fig. VIII-15. Fauna and flora across the Catinnaccio-Sciliar bank from north to south. Note presence of dendroid "*Thecosmilia*" and onkoids all along bank and dasycladacean and large bivalves in interior lagoon at Vajolet Towers. Vertical exaggeration only X 1.5. From Leonardi (1967, Fig. 128—a traverse by D. Rossi)

greenstones (altered volcanic tuffs). The limestone is highly siliceous and cherty and contains radiolarians; it has graded beds, lamination, microbreccias and slump structure. It contains abundant remains of *Daonella*, a bivalve of probable nektonic habit. These strata clearly represent deeper water euxinic deposits formed between the buildups as they grew. Beneath the banks the Livinallongo is thin as if the banks grew on areas of slight buildups of the underlying Serla Dolomite. Lateral to the banks, the Livinallongo strata thicken basinward up to 200 m, replacing the dipping forereef transition beds and forming the starved basin facies during the time of maximum subsidence (Fig. VIII-15) and bank growth (Bosellini and Rossi, 1974).

The great banks of the western Dolomites grew as much as 700–1500 m high. Some are roughly circular with a diameter of several kms. Owing to the gentle folding and faulting of the Teritary orogeny in the South Alps, they are still more or less in place. Foreslope talus dips off them from 10 to 30 degrees. Present erosion has preserved many of the original depositional slopes because the inter-bank, infilling strata are easily eroded compared to the massive dolomite and limestone. One can trace these transition beds, *(Übergußschichten of Mojsisovics)*, upward from the basin into completely unbedded dolomite along the edge of several banks. The profile of the bank edges range from almost vertical to inward-retreating or outward-growing, depending on the ability of organic growth to compete with subsidence and ingress of volcanic-clastic material (Fig. II-17). These flanking beds are composed of coarse nodular dolomite with abundant cavities and much coarse fibrous calcite infill (stromatactoid-structure or *Evinospongia* of earlier authors).

Organic Composition of the Dolomite Banks

Because of extensive and rather coarse dolomitization, the organic makeup and petrographic character of these masses is only just now being understood. A better understanding of the rock is actually obtained from studies in the Northern Limestone Alps where more limestone is present. However, not all the Dolomites have suffered this common disease of carbonates. Notably the Marmolada and Latimar are still limestone.

(As an aside, it is interesting that the mineral *Dolomite* was named for its discoverer, Count Dolomieu and that the mineral gave its name to these impressive mountains. Nature lovers, other than mineralogists, may be surprised that the circumstances should have not been the other way around.)

Although in the past the Dolomites frequently have been considered to be composed of coral reefs, estimates found in Leonardi (1967) are that about 50% of the bulk may be algae, 18% mollusks, 17% corals, and 6% foraminifera. Even these figures compiled by experienced modern researchers are hard to evaluate because dolomitization has obscured so much. Furthermore, the coarse, fibrous, veinous calcite *(Evinospongia)* appearing in the "reef" mass has been accepted in past years as an organism and probably also vadose pisolites have been confused with the alga *Sphaerocodium*. Corals that are present vary from bank to bank in importance with the algae. Those described and figured by Leonardi (1967), *Thecosmilia* and *Koilocenia* are branching, ramose, and tubular, some with tubules only a few mm in diameter. These occur commonly in micrite. The colonies are on the order of only a meter or so in size. Commonly they appear bedded as biostromes. Traverses made mainly by D. Rossi (Leonardi, 1967) across the Cattinaccio and Sasso Lungo show corals all the way across the bank (Fig. VIII-15). In some other areas, such as the Sella and Sciliar, corals seem more prevalent at the bank edges and on the upper foreslopes. Much coral material is seen in some bank-derived blocks in the higher basinal sediments. The peloidal micrite matrix, the scarcity and dendroid growth habit of the coral, and the recognition of the widely adaptable *Thecosmilia* (found in both backreef and forereef of the Northern Limestone Alps) are facts arguing strongly against any type of wave resistent coral reef rim for the Dolomite banks despite the steep slopes into the basin. Algal content is hard to estimate but much sediment-binding algae do exist. The micritic limestone Cipit blocks of the basin-fill, which were derived from bank edges, contain abundant evidence of crustose irregularly laminate algae(?), foraminifera, and the organism *Tubiphytes* (Plates XXIV B, XXV A). Sparse and isolated mollusks, including some large gastropods occur at bank margins. Brachiopods (commonly *Waldheimia*) occur, but bryozoans are not reported, and foraminifera and radiolarians are rare. Bulbous forms of hydrozoans or stromatoporoids are reported but are not commonly seen, perhaps owing to poor preservation in dolomite.

Crinoids are found on outer foreslope deposits in the Sciliar and Sasso Lungo Banks. Small mounds of crinoid particles occur in places. Crinoid remains are common in the Cipit blocks and cidarid echinoids also are present, indicating abundant echinoderms on the open sea sides of the banks.

Published traverses across the banks, by Rossi and Leonardi, show ubiquitous stromatolites (Leonardi, 1967). The alleged "Cyanophyte algal balls" *(Sphaero-*

codium) range up to 3–4 cm across and are stated by Leonardi to be common at the tops of the banks and in the overlying Raibler beds. The Sciliar and Cattanaccio traverses show these coated particles at the crests of the bank-interiors and preserved outer slopes. Some of these particles are essentially identical to the vadose pisolites of the Permian Reef Complex.

Several of the great banks of the Western Dolomites (Latimar and Sella) contain interior thin-bedded tidal flat and lagoonal strata showing that the organic rims of the banks surrounded large lagoons and have atoll or faro forms. Abundant mollusk shells and diplopore dasycladacean algae are present there. These strata preserve well-defined cycles of the typical fill-in carbonate-evaporite type and indicate successive periods of subaerial exposure. Tepee structures occur in such beds along with pisoids and other structures resembling the backreef of the Permian Reef Complex (see Chapter X and Bosellini and Rossi, 1974).

Organic Composition of Ladinian Wetterstein Limestone
in the Northern Limestone Alps

Much important information on faunal composition of Middle Triassic banks is available from studies in Bavaria and Austria and this knowledge can be most helpful in understanding the biological character of the Dolomite masses.

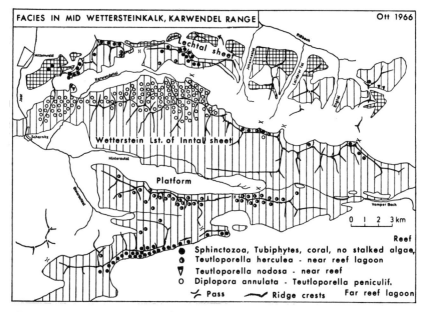

Fig. VIII-16. Biofacies subdivision of the Wetterstein limestone in the Karwendel Range. Lechtal and Inntal sheets are nappes, the great carbonate bank of the Inntal now overlying the Lechtal nappe. The facies are based principally on the distribution of dasycladacean algae which outline an asymmetrical bank with a wider "near back reef" facies on the south side. From Ott (1967, Fig. 7)

The Wetterstein limestone forms major offshore banks like those of the Dolomites but is much less dolomitized and has been carefully studied petrographically in recent years (Ott, 1967; Sarntheim, 1967; Toschek, 1968). The limestone is hundreds of meters thick (up to 1500 m) and shows generally regular upward shoaling major facies which also may be traced laterally in belts surrounding a series of east-west elongate carbonate banks (Fig. VIII-16). The exact distribution of the several known banks is debated because of tectonic displacements so common in this Alpine structure belt. The Karwendel bank, forming the high cliffs north of Innsbruck, is best known. Its boundary is a bioclastic girdle, wider on the south and perhaps indicating stronger wind waves and currents from that direction. Its narrow northern boundary is also clearly defined. Other banks are recognized south of Innsbruck.

All of these banks rise from a platform limestone of Anisian age—termed Alpine Muschelkalk. Their equivalent basinal facies (Partnach beds) is only 100–250 m thick and is dark, thin-bedded, slightly bioclastic-spiculitic limestone with chert nodules and several shale beds. This grades upward into the bank proper through some thick-bedded, dark gray, bioclastic wackestone with peloids and abundant crinoid stems, mollusks, and dasycladaceans derived from upslope. No real forereef breccia is present (unlike the beds in the Dolomites). The Wetterstein transition beds instead are a type of basin marginal deeper water limestone (Facies belt 3 of the author). Texturally, the bank edges consist of poorly sorted coarse bioclastic debris generally in a sparry matrix but with some bioclastic micritic infill. Much coarse spar coating is seen. Texture is essentially rudstone to floatstone. The biota of this bank edge facies includes large crinoids, echinoids, *Solenopora*, *Daonella* bivalves, the brachiopod *Rhynchonella*, gastropod *Euomphalus*, orthoconic nautiloids, dasycladacean *Teutloporella nodosa*, some few corals, many segmented or beaded sponges, and encrusting *Tubiphytes* (Plates XX, IV B, XXV A, XXVI A).

Ott's careful petrographic study (1967) shows that *Tubiphytes* is the most important reef builder. He lists five different finely textured, encrusting growths with delicate, tiny (2 mm) bushy networks in the sedimentary fabric. Microscopic study also discerns minute, 1 mm thick, stalks of crystalline calcite and includes tiny porous pipes in fasciculate or isolated form up to 3 mm in diameter. Those with nonporous walls are termed *Holocoelia*. Very delicate fine-branching septate corals, *Calamophyllia*, are up to 3 mm in diameter and encrust the epithecae of other forms. Larger forms are also present. Irregular meshworks and thick stalks of *Holocoelia* are present as are fasciculate forms such as "*Thecosmilia*", *Margarosmilia*, and *Pinacophyllum*. The relative abundance of such corals is hard to estimate. The megafauna is dominated by diverse groups of calcareous sponges. Inozoa or pharetrones are common but eight genera of columnar, segmented or beaded forms of Sphinctozoans have been described (Ott, 1967). The sponges seem to have grown in groups comprised of only a limited number of genera at one place and to have occupied different ecologic niches from the corals. Hydrozoans, both as tiny stalks and as encrusting concentric balls, are present but not common. Some few encrusting calcareous foraminifera and bryozoans appear to have been sponge parasites. Calcareous algae are not particularly important; some blue-green stromatolitic forms may occur as reef binders. The codiacean

Ortonella is known. Flat, encrusting red algae are as usual on the outer side of the reef and in detritus where they constitute 5–18% of the poorly-sorted bioclasts.

Ott pointed out that the sponge morphology of segmented, chambered, and vesicular structure is probably a reef growth adaptation. The segmented character results from rhythms of growth but also permits flexibility for waving in moderate currents. The fine and delicate growth forms, beaded and pipe-like form of larger organisms, and the lack of knobby, onkoidal or encrusting growth of sponges, hydrozoans or corals indicates carbonate buildup in water of some depth and moderate current. Optimum depth for the particular sponges is given as 4–18 m and for the corals as deep as 20 m. Such depths may be at or below wave base in inland seas. The width of the reef rim is not known but its outer fauna is clearly normal marine and its interior biota shows considerable salinity variation. A few zones of oolite may indicate very shoal water or even islands at the margins.

Two areas of dasycladaceans mark facies within the banks themselves. The *Teutloporella herculea* zone is positioned in the near backreef on both sides of the bank but may range 10–15 km into the bank. The algae stalks may be oriented. Codiacean and red algae also occur in this facies. The sediment represents muddy lime sand (packstone) at the crest of the profile or in a more protected environment just behind. The interior of the bank formed a considerable lagoon and tidal flat complex. It consists of 5–10 m thick cycles with a thick subtidal zone grading upward to mm algal laminites with reddish geopetals and fenestral fabric. These beds are dolomitic. Sarntheim (1967) listed grain aggregate lumps, fecal pellet mud, pure lime mud, and bioclastic muds with abundant broken pieces of *Diplopora annulata* as characteristic of bank interior facies, particularly in the subtidal part of the cycle. Some of these beds are shoal grainstones.

In summary, the Wetterstein limestone may be interpreted as consisting of major carbonate banks with margins mostly of bioclastic accumulations which contain downslope scattered encrusting and low-growing forms such as sponges and organpipe (dendroid) corals. Growth forms and general biological composition indicate accumulation just at or below wave base in only mildly agitated water (20 m). The author classes such profiles as Type I, downslope mud accumulations. The maximum height of the Wetterstein banks above the Partnach sediments in the basin is believed by most German and Austrian researchers to be only 100–150 m which is the extra thickness of the Raibl where it overlies the off-bank Partnach facies. One has, however, only to look at the undisturbed stratigraphic relations in the Dolomites to speculate on the possibility of much greater relief than this. On the other hand, the lack of talus and typical toe-of-slope lime mudstone facies (belt 3) might indicate more subdued relief of the Wettersteinkalk banks.

Later Triassic Deposits in the Dolomites

After the growth of the great banks in the South Alps was essentially complete, the intervening basins were filled with brown pyroclastics resulting from intrusion and eruption of basic volcanic material (plagioclase-pyroxine porphyry). Volcanic ash and pillow lavas are mixed with abundant terrigenous clastics and plant fragments. These beds constitute the La Valle or Wengen Formation. The literature describes them as "pseudoflysch" because of their characteristic monotonous, even, thin bedding. Certainly the strata were deposited below wave base. A few ammonites and *Daonella* are present. Exotic blocks derived from the banks rarely occur in them. The volcanic activity clearly postdates the maximum reef growth. There is no interfingering of the two facies although in places the volcanics cover the banks. Dikes and some major volcanic vents (e.g., in the Fassa Valley) cut through the buildups in places as well as fill in the interbank areas. The explosion craters contain blocks of Ladinian limestone showing that their age is late in the time of carbonate development.

The basin fill between the banks is formed by the St. Cassian beds. The strata have the same flysch-like bedding as the Wengen-La Valle formation, but are more calcareous, less tuffaceous and grade upward into varicolored marls and limestone with local conglomerates of volcanic pebbles. Some reworked tuff is also present. The strata are most noted for the presence of numerous blocks of bank-derived limestone, the Cipit boulders. These exotic, rounded boulders, up to a meter or so across, are rarely dolomitized and offer the best view of the original composition of the bank margins. Most are micritic and contain abundant crinkly-laminate algae(?), *Tubiphytes*, and other encrusters, small dendroid corals, and crinoid debris. The blocks contain many vugs lined with drusy cement and filled with internal oxidized silty ferruginous sediment.

The banks and interbank sediments of the western Dolomites are covered by a widely distributed sheet of argillaceous, dolomitic, and evaporitic strata, the Raibl, varying in thickness from about ten to several hundred meters. This upper Carnian marker is found also in the Northern Limestone Alps. It carries a molluscan fauna and much *Sphaerocodium*. It represents a time of clastic influx, general marine regression and widespread development of shallow lagoons. In the eastern Dolomites the Raibl is thick and bears evaporites. This is significant because evaporite ($CaSO_4$) deposition during its formation may well have caused the high Mg content of refluxing fluids which dolomitized the underlying Ladinian banks. This dolomitization is clearly secondary, and to some extent controlled by bedding and original sediment type. Finer-grained sediments are preferentially dolomitized. Corals are dolomitized but crinoids resisted the replacement process. Lithification of the strata had not been completed when dolomitization occurred because obvious permeability control on the process was exerted at several levels of magnitude. Peripheral areas of the banks are less dolomitized than the interiors. Certain levels in the foreslope beds are more dolomitized indicating preferential fluid migration. The Cipit blocks in the basin-fill sediment are rarely dolomitized showing that the process was not contemporaneous with bank formation but somewhat later. Apparently impermeable basin sediments protected the exotic blocks from the dolomitizing fluids. Large banks whose tops were covered by penecontemporaneous volcanics or by particularly thick and argillaceous Raibl Formation are now still limestone. The dolomitizing fluids passed not only through the Ladinian but also into the underlying Anisian Serla Formation.

Significantly, the strata next overlying the Raibl constitute an equally widespread sheet of tidal flat and lagoonal dolomites. The Norian Dolomia Principale or Hauptdolomit is about 250 m thick in the western Dolomites but thickens to

1000 m in the eastern Dolomite trough. These strata contain sparsely fossiliferous peloidal micrites with sedimentary structures indicative of intertidal and supratidal environments. The unit is thoroughly dolomitized and indicates an environmental complex which could easily have contained saline ponds in which evaporites were deposited. Evaporites are not seen but may have been removed by later weathering. Both the Dolomia Principale and the Raibl might have provided environments for the generation of dolomitizing fluids (see Chapter X).

Upper Triassic Reef-Lined Banks and Basinal Mounds of Austria and Bavaria

Introduction and Microfacies

The Upper Triassic of the Northern Limestone Alps may be instructively compared with the older Permo-Triassic carbonate buildups. There is a significant although gradual development of coral framework construction in the later Triassic reefs. Much the same type of evolution is seen in the progressive development in size, complexity, and abundance of corals and stromatoporoids from Ordovician to Devonian time.

Recently, regional interpretative facies studies of Late Triassic strata in the Northern Limestone Alps have been published both in German and English by H. Zankl, A. G. Fischer, E. Flügel, and F. Fabricius. As in the Middle Triassic, the Hauptdolomit, its limestone equivalent, Dachsteinkalk, and the overlying Rhaetic limestone were developed as a series of major carbonate banks. These likewise have been disarranged by northward thrusting. This has been known for many years for Mojsisovics also studied these strata. General agreement now exists that thrusting has resulted in a foreshortening of two to three times in the Northern Limestone Alps, compressing and disarranging the reef rimmed banks. Figure VIII-12 gives the stratigraphic nomenclature of these beds.

The rock is mostly carbonate. Flügel (1972) distinguished 12 basic microfacies types found in Late Triassic strata of the Northern Limestone Alps. The principles employed by Flügel in this work serve as the basis for delineating the standard microfacies types described in Chapter III of this book. Following Flügel's letter code, these Late Triassic microfacies are described below using as he did the Folk classification. Equivalent standard microfacies types are designated SMF.

MF 1. Biomicrite with sessil benthos, generally in life position. Subtypes are based on (a) preponderance of a certain fossil group, e.g., *Thecosmilia* or calcareous sponge micrite (bafflestone of Klovan), (b) amount of biogenic particles, (c) whether the micritic matrix is bioturbated or pelleted or has biogenic crusts such as algal filaments. SMF-7c and SMF-8. (Plates XXIV B, XXVI B).

MF 2. Biomicrite with sessile benthos generally in life position but bound by encrusting organisms. Subtypes are designated according to the type of encrusting organisms. Sessile foraminifera, alga-foraminifera consortium, calcareous sponges, SMF-7b, bindstone of Embry and Klovan (1971). (Plates XXIV A, XXVI A).

MF 3. Biomicrite to biosparite with associated bioclasts of sand to gravel size. Subdivisible on basis of homogeneity of the bioclastic fraction and on major faunal groups. SMF-5, bioclastic wackestone-packstone grainstone-floatstone.

MF 4.   Biosparite with bioclasts encrusted on all sides. Onkoidal grainstone. Subdivision into two types possible based on homogeneity of the bioclasts. SMF-13.

MF 5.   Fine debris biomicrite, detritus variably sorted, principally molluscan. Two subtypes based on whether fabric is homogeneous or bioturbated. SMF-9, microbioclastic variety of bioclastic wackestone.

MF 6.   Micrite or biomicrite with monotonous microfauna of foraminifera or ostracods, subdivisions based on biological grouping. SMF-23, a variety of lagoonal wackestone and mudstone with restricted marine fauna.

MF 7.   Unfossiliferous micrite. Subdivisions based on presence of grains consisting of micritic lithoclasts, biogenic micrite crusts, or intraformation pebble conglomerate and pellets. SMF-24, SMF-16, peloidal wackestone.

MF 8.   Algal mats with fenestral fabric. SMF-19 and SMF-20 (Plates XII, XIV).

MF 9.   Onkoidal micrite to onkoidal sparite with large alga-foraminiferal balls and lithoclasts. SMF-22 and SMF-13 (Plate IX).

MF 10. Pelsparite and biopelsparite with large peloids and nonencrusting green algae and foraminiferal particles. Two subtypes distinguished on presence or absence of frame-building organisms. SMF-18, foraminifera-alga grainstone.

MF 11. Intrasparite with preponderance of bahamith particles, grapestone lumps. SMF-17.

MF 12. Oosparite, subdivided on basis of associated bioclastic grains. SMF-15.

Examples of Banks and Reefs

From east of Innsbruck to west of Salzburg, the Norian-Dachstein forms a great reef complex hundreds of meters thick. According to Fischer (1964) there existed a barrier reef along the southern margin of the Northern Limestone Alps, parts of which are now located far north of the original southern margin in tectonically transported thrust sheets. Zankl (Fig. VIII-17) reconstructed the paleogeography as a series of basins surrounded by a number of thick limestone banks, rimmed on the south and southwest by narrow reef belts. The foreslope talus of these bank-rimming reefs disappears regionally to the south and on all sides away from individual banks. The off-bank facies consists of thin, reddish ammonite-bearing Hallstatt facies with *Halobia* coquinas interpreted as a deep-water starved basin sediment. Other basinal facies are known which include breccia beds, dark cherty limestones with turbidites, and shallower more argillaceous limestone with benthonic faunas.

The highest Triassic in the Northern Limestone Alps also displays significant facies changes. The Rhaetic parts of the Dachstein pass northward into the marly bituminous Kössen beds, evidently a sediment deposited in several slightly deeper basins rimmed in late Rhaetic time by small barrier and patch reefs. The height of individual bioherms within such basins and the relief at the front of the Steinplatte reef which borders a platform north of Lofer and Waidring in Austria, indicates a depth of at least 100 m for some of the basins.

Three forms of organic buildups may be delineated in these Late Triassic strata in which corals and spongiomorph hydrozoans appear in abundance. Two types of linear shelf margin buildups occur: Type II margins are composed of small reef knolls of patchy frame-building corals and spongiomorphs on gentle slopes described by Zankl (1969) on the Hohe Göll above Berchtesgaden in Bavaria, and Type III, true sharp reef rims at the Steinplatte described by Ohlen (1959) and by Fabricius (1966) in Rhaeto-Lias beds farther north in Bavaria. In

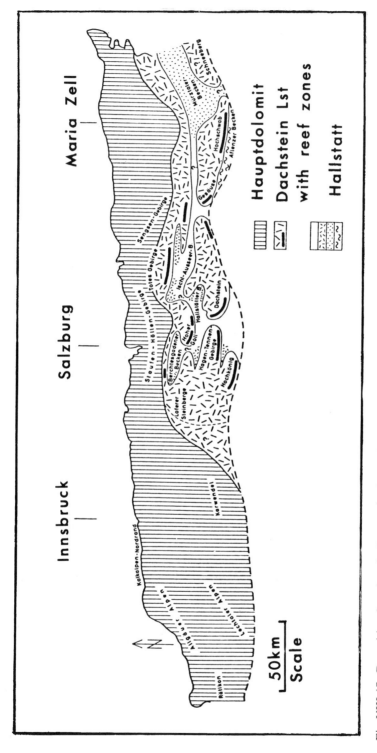

Fig. VIII-17. Depositional realms in Late Triassic Norian times in the Northern Limestone Alps before tectonic shortening (Zankl 1967, Fig. 1). Note interpretation of a series of Dachstein banks with reefy shelf margins on southern sides and intervening basinal facies

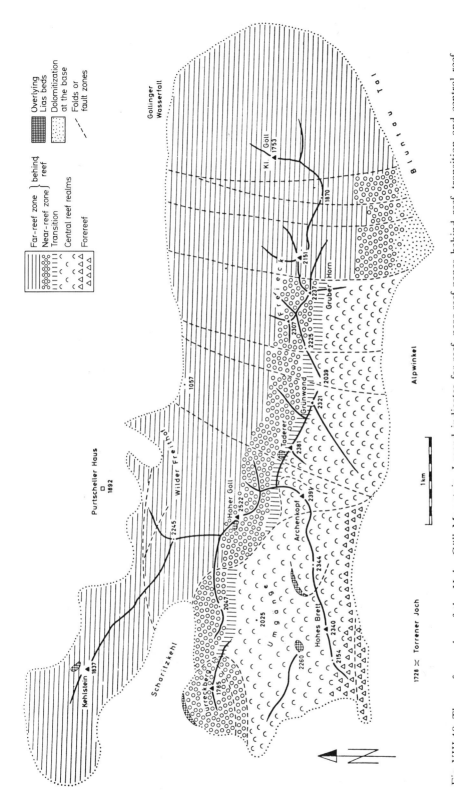

Fig. VIII-18. The reef complex of the Hohe Göll Mountain. Legend indicates far-reef, near-reef zones behind reef, transition and central reef realms and forereef. It also indicates overlying Lias beds, dolomitization at the base and folds or fault zones (Störungen). From Zankl (1969, Fig. 2)

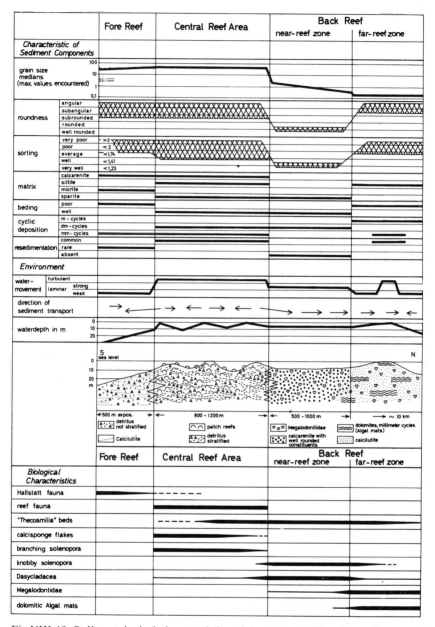

Fig. VIII-19. Sedimentological characteristics of the Hohe Göll reef complex according to Zankl (1969). From Zankl (1971, Fig. 5)

addition, in protected Kössen shale basins of Rhaetic age, large bioherms exist which are capped by reefoid organisms and surrounded by shale. These developed considerable sea-floor relief late in their growth. The Rötelwand and Adnet carbonate buildups may have been more than a 100 m high.

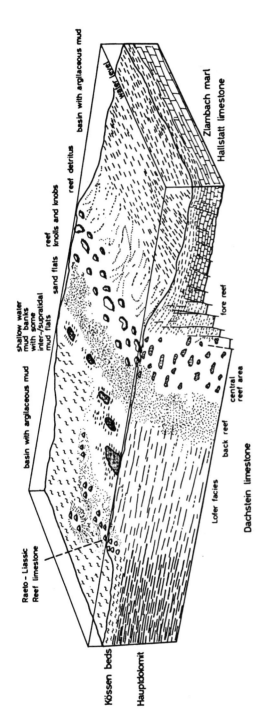

Fig. VIII-20. Paleogeographical interpretation of depositional environments during the Late Rhaetian. Note that all reefy shelf margins and bioherms are composed of superimposed reef knolls and knobs. From Zankl (1971, Fig. 3)

1. The Hohe Göll, a Reef Knoll Ramp—Type II

The Hohe Göll carbonate facies complex lies in a thrust sheet north of the Torrener Joch, high above the town of Berchtesgaden (Fig. VIII-18). Its southern boundary and original depositional slope is obscured by faulting so that no direct observation of original relief is possible. Zankl (1967) offered faunal evidence for a moderate depth (200 m) of the intervening basinal Hallstatt facies. He believes that only a gentle slope was present, 1 or 2 degrees into the basin. Part of this can be observed on the Hohes Brett just south of the Göll. The detrital slope facies has pockets of Hallstatt red limestone.

Much is known about the biological composition of the Hohe Göll margin, (Zankl, 1969, 1971). See Fig. VIII-19 and VIII-20. The ramp marking the southern and open-sea side of the great buildup is made up of a series of reef knolls, each characterized by one or two framebuilding species and with varying groups of organisms as subsidiaries. Biocoenoses cannot be well delineated; clearly their sequences in time and space are rather irregular. It is known that: (1) sponges along with the red alga *Solenopora* generally form lamellar crusts on reef sides; (2) "*Thecosmilia*", a large organpipe dendroid form, exists on more protected back-reef sides but also in quieter water basinal beds; (3) reef knolls are dominated (75% of total framework) by the corals *Astraeomorpha*, "*Thecosmilia*" and calcareous sponges. The coral colonies may be up to 10 m high. Other coral genera are the large solitary *Montlivaltia*, colonial *Thamnastrea*, and *Palaestrea*. Encrusting spongiostromes (blue-green algal crusts), stromatoporoids (spongiomorph hydrozoans), bryozoans, encrusting foraminifera, and *Cheiloporites* together make up the other 25% of volume of boundstone. Zankl has demonstrated how multiple generations of these encrusters grew intermittently with marine cementation. The interior of the knolls contains mainly calcarenitic-calcisiltite debris and not much micrite. Constructed voids filled with cement are present toward the tops of the knolls. This may or may not be marine or splash-zone cement. A vadose origin is possible.

The bulk of the outer platform strata are not reef knolls; 90% is estimated to be composed of bioclastic detritus. Most of the knoll-flanking material is bioclastic grainstone but resedimented particles, lithoclasts, are also abundant. The fore-reef detritus is mostly subangular and poorly sorted; more resedimented particles are present down the slope, and the matrix becomes more micritic (up to 30%). The biota is highly diverse, particularly with corals and encrusting organisms derived from the tremendous organic growth at the knoll-reef centers farther up the gentle slope. The immediate back-reef and reef crestal sediments in places contain very coarse rubble. Back-reef cycles of the Dachsteinkalk, which are present just shelfward of the bank margin, are described in Chapter X.

2. Rhaetic Organic Reef Rims—Type III

During very latest Triassic times a large scale separation of Rhaetic shale basins and intervening reef-covered swells occurred in the northern area where earlier the Hauptdolomit platform had existed. Fabricius (1966) gave a paleogeographical analysis and map of one of these basins and its reef margins. The argillaceous Kössen facies spread into the troughs and contained a benthonic fauna with characteristic mollusks. About half of the Dachstein reef species are

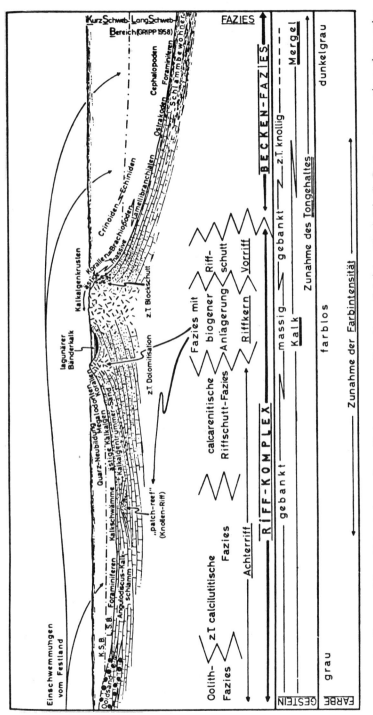

Fig. VIII-21. Facies diagram of Late Triassic in northern Tethyan geosyncline. From Fabricius (1966, Fig. 4). Bottom two sections show color intensity increasing in both directions from reef front and changes in gross rock type and bedding off shelf from well-bedded to massive to bedded and in part nodular in the direction of the basin (*Becken-Fazies*). Associated with this basinward change in bedding is an increase in clay content and a change from limestone to marl. The organic and petrographic categories in the German labels of the upper parts of the facies profile are easily read in English except for the following: *Kalkschwämme* are calcareous sponges, *ästige Kalkalgen* are branching calcareous algae, *lagunärer Bänderkalk* are lagoonal laminated limestones, *Kalkschlamm* is lime mud and *Kurz und Lang Schweb. Bereich* indicates the realms of short and long periods of sediment suspension and a reflexion of wave base. Illustration with permission of author and Brill and Company, Leiden, The Netherlands

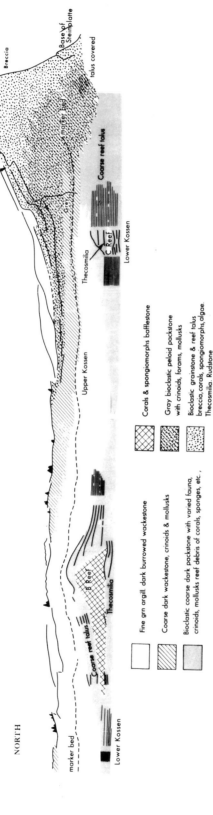

Fig. VIII-22. West side of Steinplatte, Tyrol in Austria. Forereef to basin facies according to Ohlen (1959). The reef front trends northeast at an acute angle to the plane of the cross section. Immediately east of the south end of the section lie extensive backreef strata. To the north lies the Unken Valley, a local Kössen basin filled with deep water Jurassic strata. The Steinplatte reef front indicates a depositional relief of 160 m and assuming surf zone near the crest. this figure indicates accurately the depth of deposition for the basinal Kössen marls into which the reef talus grades along the slope of the profile. Small initial reefs or mounds are present in strata just older than the main reef mass and consist of dendroid corals and spongiomorph hydrozoans. The cross section is about 1 km long. There is no vertical exaggeration

also found in the Rhaetic reefs. Figure VIII-21, from Fabricius, provides a useful synthesis of the whole gamut of facies.

Steep shelf margins with reefs were also studied in detail by Ohlen (1959) at the well-known Steinplatte-Sonnenwände complex above Waidring near Lofer, Austria (Zankl, 1971). The Steinplatte reef rises about 100 m above a Kössen dark shale basin and forms the northwestern border of an extensive carbonate bank whose preserved exposures trend to the east behind it along the Sonnenwände (Sunshine Walls). The growth history and biological evolution of the Steinplatte has been rather clearly worked out (Fig. VIII-22). Apparently reef knolls developed first in the argillaceous Kössen beds on piles of bivalves such as *Gervillia*, *Oxytoma*, *Avicula*, *Ostraea*, terebratuloids, and crinoids. The knolls grew by development of heads of *Thecosmilia* and the stromatoporoid-like hydrozoan *Stromatomorpha rhaetica* colonies in lime mud (Plate XXVI B). These early mounds formed in water of a depth of 30 m or more if one assumes a stable sea level. The adaptable *Thecosmilia* commonly exists in micrite matrix. It is found both in Kössen basins and in backreef areas. A middle phase of Steinplatte reef growth occurred when *Thecosmilia* proliferated and spread shelfward with the continued Rhaetic transgression. A diverse algal assemblage occurs with this large organpipe coral and includes *Solenopora* and *Sphaerocodium*. Sponges and spongiomorph hydrozoans are also abundant. Many varied forms existed on the forereef slopes at this later stage: the foraminifera, *Cheilosporites*, *Labyrinthia*, *Tetrataxis*, crinoids, certain sponges, *Solenopora*, and the corals *Thamnasteria*, and three other genera (Plate XXIV A). The last stage of reef growth reached 100 m above the sea floor and depositional slopes of almost 30 degrees can be traced from it down into the Kössen beds. The front of this last stage is mostly brecciated rubble and the reef mass is exclusively of *Thecosmilia clathrata*, sponges, and spongiomorph hydrozoans. Sediment equivalent to this last stage is largely calcarenite, typically in backreef "Lofer cycles" (Fischer, 1964), spread over a wide area along the Sonnenwände. Four or five species of dasycladacean algae occur, as well as the foraminifera *Triasina*, *Angulodiscus*, *Tetrataxis* (closer to the reef), rotalids, and *Labyrinthia*. The lime grainstone facies contains large megalodont bivalves and a variety of gastropods. Fenestral laminites are also common.

3. Rhaetic Bioherms

In addition to the bank-edge framebuilt reef rims described above, protected Rhaetian shale basins of the Kössen Formation contained large bioherms such as those found in the Salzburg area at Adnet, Rötelwand, and Feichtenstein. Several of these have been studied in detail. They developed much like the small reef knolls formed early in downslope position at the Steinplatte, probably originating on piles of bivalves which accumulated on the argillaceous lime mud bottom. This type of basal deposit may be seen at the Rötelwand bioherm. Initial relief above the sea floor was not great for these mounds but they built up in deep water by means of a framework of the ubiquitous dendroid *Thecosmilia*. Sieber (1937) described the Rötelwand as being a composite of small micritic reef knolls which accumulated to wave base but not in the zone of agitation. The top of the Rötelwand, however, contains abundant void-filling calcite, some with stromatactoid shapes and many encrusting organisms forming boundstone as well as large terebratuloid brachiopods, sponges, corals, and hydrozoans. Most of these Rhaetic bioherms are unbedded and surrounded by shale. Many of them possess fissures filled with reddish Liassic sediment. Perhaps these represent an old karst topography. It is easy to imagine that the bioherms grew to wave base and later came out of the water. Against this hypothesis is the observation that in places the reddish Liassic sediment contains fossils of deep water or open marine environments (brachiopods, crinoids, ammonoids). If subaerial exposure did occur, very rapid and extreme subsidence must have followed. The fissures in the bioherms may be tectonic and not solution in origin.

Similarities and Differences between Permian and Triassic Reef Complexes

A comparison between Late Permian and Triassic carbonate shelf margins permits an evaluation of the effect of a rigorous evaporative climate on sediments of great offshore banks. Our only well-studied Recent model for such a bank, the Bahamas, exists in a tropical climate. Also of interest is the chance to observe the slow evolution of reef-building, particularly of frame-constructing organisms during this time and the steepening effect this gradual change in organisms had on depositional profiles.

The Permian of the southwestern U.S.A. and the European Triassic are strikingly similar. Even the regional settings resemble each other. North from the Triassic Tethys, throughout western Europe, existed restricted marine shelf and shallow basin deposits of the Germanic Trias (Bundsandstein, Muschelkalk, and Keuper). These redbeds, evaporites, and extensive restricted marine limestones much resemble the north central Texas and Oklahoma back reef shelf deposits of Middle and Late Permian age. This is considered to result from similar climates. In each case the clastic-evaporite deposition on the shelves was interrupted by a major marine transgression (the San Andres-Grayburg of West Texas-New Mexico and the Muschelkalk of Europe).

Tectonic history is similar in the basins south of the evaporitic shelves. Here in both instances great carbonate banks developed. Both the basin areas were subjected to tensional or perhaps directional stresses leading to normal or strike-slip faulting which resulted in horst and graben structure. This seems to have localized bank development during a period of extreme subsidence which followed soon after tectonic fragmentation. The Permian banks developed over a fragmented foreland lying north of an active geosyncline (Ouachita-Marathon) whose orogenic development climaxed in the Carboniferous. The Austroalpine Triassic banks developed over or around uplifted fragmented continental blocks, probably over roots of a Hercynian (Carboniferous) orogenic belt, not over its foreland.

Subsidence in both cases was great. The Permian bank thickness totals close to 1000 m and the Triassic several times that. Later orogenic history was rather different in the two examples. Both bank complexes were caught in Tertiary fold belts but the compressive Alpine deformation in the northern part of the Tethyan trough was much more extreme than the block-faulting and open warping of the southern Rockies. The configurations of the great banks of the Permian of the southwestern U.S.A. and the Dolomites are well-preserved compared to the structurally distorted geography of carbonate buildups in the Northern Limestone Alps.

Individual major banks developed over earlier platform carbonates, the Leonardian (Middle Permian) paleotopography being somewhat analogous to the Lower Triassic Anisian. This common phenomenon is described in Chapter XII. Basins in between the banks were starved of sediment, were somewhat euxinic and probably had between 700 and 1000 m of water. In both areas basins were later filled—in the Triassic with volcanics and in the Permian with sands and minor volcanic ash contemporaneous with deposition and, in later Permian time, with 700 m of gypsum. In both cases exotic blocks tumbled off the 25–35 degree slopes and rest in fine clastics or volcanics well out into the basin.

The edges of the banks contain carbonate accumulations with specialized organisms which are very similar when Late Permian and Middle Triassic are compared and which contrast with those of the Late Triassic. Most of the faunal changes recognized at the end of the Paleozoic were in environments other than those of the reefy carbonate shelf margins. Similarities between organisms in these sediments are partly obscured by different identifications by workers in various parts of the world who publish in different languages.

For example, microscopic work shows that the bell-shaped chambered foraminifer *Tetrataxis* and encrusting tubular foraminifera are common to all the reef masses. Permian Reef Complex organic rich ("reefy") micrites have not been studied paleontologically in thin sections to the same degree as the Triassic in Bavaria and Austria. When this is done probably other similarities will result, as well as a few differences. Many microtubular organisms are known from the Triassic which have not yet been seen in the Permian. Additional faunal similarities between the buildups such as red algae, crinoids, and brachiopods on the foreslopes, ammonoids in basinal sediments, euomphalid gastropods and dasycladaceans in shelf marginal positions are common to many other carbonate banks. The faunal differences between Permian and Triassic assemblages are mainly in the lack of reefoid bryozoans and brachiopods and fusulinids in the Triassic and the significant addition of two groups of organisms—the mollusks and corals (Table VIII-2). Mollusca are fairly common in the Permian, but are important in Triassic reefs both as a foundation for boundstone and as contributors to framework. Specialized pelagic bivalves occurring in the Mesozoic basinal sediments in great abundance are absent in the Late Paleozoic. In addition, unique heavy-shelled megalodont bivalves form a common Mesozoic backreef biofacies not seen in the Permian. Corals are represented only by small solitary forms in the Permian forereef sediments; neither are they very common in early and middle Triassic strata in the Dolomites. They become generally more diverse, colonial, and very abundant in the Late Triassic. Late Permian and Mid-Triassic bank margins are held up principally by encrusting forms which trapped and protected vast quantities of fine-grain sediment mainly in down slope areas (Type I). The advent of abundant large coral colonies, plus dominance of hydrozoans in Norian and Rhaetic beds, resulted in development of reef knolls at the frontal margins (Type II) or in true reef rims (Type III) in places.

Table VIII-2. Comparison of biotas of Permo-Triassic buildups

Permian Reef Complex	Dolomites and Wetterstein Limestone	Norian-Rhaetic
Syconid sponges are major large organisms	Syconid (Sphinctozoan) are major large organisms; more than 50% of genera are the same as in the Permian. Clear phylogenetic relations are seen with Permian	Only Inozoan sponges *Peronidella*
"Encrusting Hydrozoan" of Newell *et al.*, is very important sediment-binding organism (*Tubiphytes*)	*Tubiphytes* is most important binding organism	*Tubiphytes* is rare but reappears in Late Jurassic reefs
Lamellar stromatolitic crusts lining cavities and knobby stromatolites. Coarse druse in cavities	Only wavy lamellar crusts lining cavities. Coarse druse in cavities	Lamellar stromatolitic crusts
No colonial corals	Colonial corals not common but range in size from small to about two meters. Forms are of dendroid construction, "*Thecosmilia*" types in micritic bedded limestone	Colonial corals are the most important large organisms. Mainly organpipe forms, "*Thecosmilia*", up to 10 m high
Hydrozoans (stromatoporoids) present	Hydrozoans-spongiomorphs (stromatoporoids) present	Hydrozoan spongiomorphs may be dominant with corals. Large masses
Unusually large and bead-like dasycladacean algae: *Mizzia* and *Macroporella*, forming grainstones at top of buildup	*Teutloporella*, *Poikiloporella* and *Diploporella*. Common in lime sands at bank margin and in bank interior	Dasycladaceans also important as sands in immediate backreef

It is notable that the Type I bank margins in both the Permian and in the Dolomites were commonly eroded and afforded a large supply of exotic blocks down the slopes. If most of the organisms and their matrix of fine sediment indicate growth in quiet water at or below wave base, what is the origin of the blocks? Perhaps periodic drops in sea level exposed and partly cemented the tops of the banks so that hard chunks could be eroded off by wave action during periods of shallow water. Climatic similarity probably has much to do with the generally similar sediments in the Permian and Triassic. The redbeds, dune sands, dolomites, evaporites, and important thick carbonate banks, large sponges, dasycladacean algae and corals, and interior tidal flat-lagoon sediments with spectacular vadose diagenetic features all indicate strongly seasonal climate with extended warm and arid seasons—probably interspersed with wet periods. This resulted in unusually rapid cementation of the shelf margin and downslope sediments. The striking diagenetic effects imposed on the cyclic sediments within the lagoons of the banks are discussed in Chapter X. The tidal flat and restricted marine strata of the Permian and Triassic lagoons resemble each other just as much as do those of the shelf margins.

Chapter IX

Reef Trends and Basin Deposits in Late Jurassic Facies of Europe and the Middle East

The Late Jurassic (Malm) of Europe and the Middle East together display some of the most interesting and varied regional facies known to geologists. Some of the significant and distinctive rock types in these strata include: coral-sponge reefs, Purbeckian fresh-water beds, Solnhofen platy limestone and the geosynclinal red nodular Ammonitico Rosso and pelagic Aptychus limestones.

Interpretations of some of these facies have puzzled sedimentologists for more than 100 years. Fortunately, no stratigraphic sections in the world are better known and no more detailed paleontologic zonation exists than in the Jurassic of Europe—thanks mainly to study of the prolific ammonite faunas. For this reason, facies analysis of these beds can be done with particular accuracy and within a reliable time-stratigraphic framework. Arkell (1956) and Holder (1964) presented compilations of stratigraphic data for these beds. The Late Jurassic stages (Table IX-1) include more than 20 widely correlatable ammonite zones as conceived by Arkell and later paleontologists. The regional review which follows describes the general depositional patterns across western Europe, the Mediterranean, and the Middle East, delineating shelf-margin reef growth, back-reef oolite and evaporites, and mentioning some paleoclimatic implications. An extensive European Jurassic literature exists. For this reason the review is brief and discusses in depth only a few of the most interesting rock types.

Table IX-1. Correlation table of Late Jurassic in Europe

Series	Arkell 1956 Stages	Tethyan Stages	Southern Germany Malm units	
Late Jurassic Malm	Purbeckian	Tithonian		Top reefs in Bavaria
	Portlandian		Zeta 6	Top reefs in Schwabia-Franconia
	Upper Kimmeridgian		Zeta 1–2	
	Lower Kimmeridgian	Lower Kimmeridgian	Epsilon	Solnhofen
			Delta	Top reefs in Swiss Jura
			Gamma	
	Oxfordian	Oxfordian	Beta	
			Alpha	

Regional Settings

Several major paleogeographic provinces are recognized: (1) the European shelf divided into several islands and basins, (2) a now missing North Atlantic source area of terrigenous clastics, (3) a shelf margin of carbonate buildups stretching west to east across southern Europe just north of (4) the Tethyan geosyncline whose Alpine nappes contain deep sediments of the continental margin, (5) the North African-Arabian Shield, (6) the thick miogeosynclinal platform areas of southeastern Europe and a southern carbonate shelf in Algeria, Tunisia, Iraq and Arabia, and (7) the starved and evaporite basins bordering the Zagros geosyncline to the east of the Arabian Shield.

The generalized maps of Figs. IX-1 and IX-2 include all the Late Jurassic. The Late Jurassic began in Europe with the Callovian-Oxfordian transgression which coincided with an influx of fine terrigenous material. Starved basins with thin, nodular, ammonite-bearing argillaceous limestone developed in the Mediterranean Tethyan trough, while dark ammonite-bearing marls formed widely in the earliest part of the Late Jurassic throughout central and western Europe. Oxfordian basins in the Middle East were also starved of sediment at this time. Neritic, oolitic and coralline facies are present in the Lower Oxfordian principally in the northwestern European areas. Reefs began developing in the Jura Mountains area late in Oxfordian time. Somewhat later (during Kimmeridgian) the amount of clay and silt across the European shelf diminished and in central Europe and the Middle East, much pure carbonate sediment formed.

The thickest carbonate strata formed around the northern and eastern perimeters of the Tethyan trough and to some degree along the northern and eastern edges of the North Africa-Arabian shield on the opposite side of the opening geosyncline. This neritic carbonate facies continued its development to the end of Jurassic time and the Tithonian reefy beds are widely regressive into the geosyncline, particularly to the east. At the same time, in northwestern Europe salt basins and fresh-water lakes with clay and silt deposits were forming. It is as if the whole of western Europe had tilted slowly and steadily up and gently warped in the process. The eastern areas sustained marine conditions to the end of the Jurassic. The interior of the Tethyan trough is now preserved in fragments in the Mediterranean borderlands and islands. These consist of carbonate platforms and seamount facies as well as the persistent starved basin radiolarite sediments. Major subsidence in many areas continued throughout the Jurassic with block-faulting as the Tethyan trough opened (Bernoulli, 1972).

Some of the areas of special sedimentation are mentioned below.

Evaporites

Late Jurassic time is one of the major evaporite periods in earth history. Throughout this time basins of the European warped shelf and those marginal to the North African-Arabian Shield, became periodically isolated and evaporite deposition occurred. Some of the basins were shallow, ephemeral lakes and lagoons;

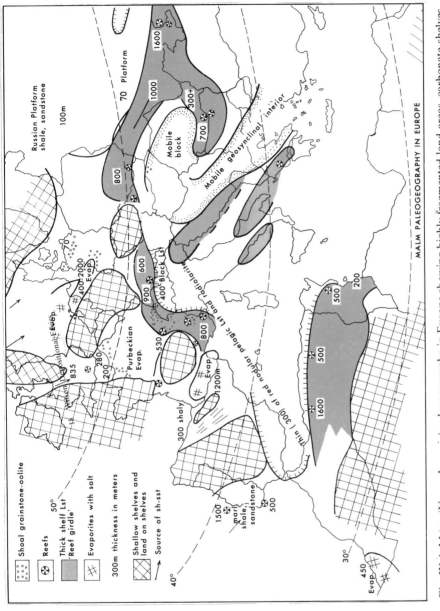

Fig. IX-1. Malm (Upper Jurassic) paleogeography in Europe showing highly fragmented land areas, carbonate shelves, salt basins, coral and oolite deposits, positive areas in geosynclines and general thicknesses. Carbonate banks in Mediterranean areas disarranged by Tertiary orogeny. Data from Arkell (1956) and Holder (1964). Hachured line separating leptogeosynclinal radiolarite facies from shelves marks present front of Alpine thrusting

others probably deep. A remarkable number of these basins exist through more than 35 degrees of present latitude—from Aden to the North Sea. Their evaporites are all essentially Late Kimmeridgian-Tithonian in age. They are present in the North Sea, North German basin, Paris basin, Aquitaine, western Moroccan coastal basin, northern Caucasus foredeep, Kara Kum foredeep north of the Elburz, Basrah basin of Mesopotamia in eastern Iran, and in the Rub al Kali and Aden basins of Arabia.

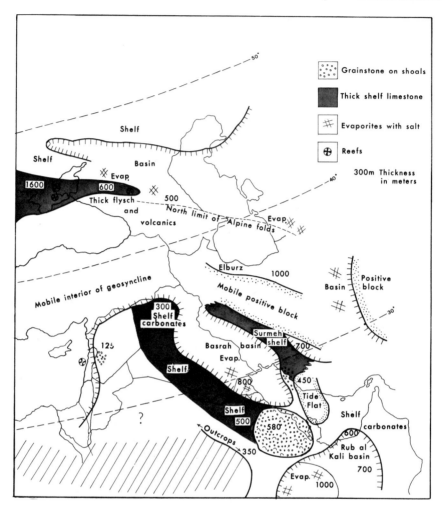

Fig. IX-2. Malm (Upper Jurassic) paleogeography in the Middle East showing basins with salt, geosynclinal areas, shelves with oolite and tidal flat deposits, and general thicknesses

Terrigenous Source Area

Throughout the Late Jurassic a source of clastics existed west and north of the European mainland. This source supplied more sediment toward the end of Jurassic time than earlier. The western basins of Europe were filled by fresh water and lagoonal evaporite facies during Purbeckian time. The North Sea and English basin contain thin lake deposits and the Spanish-Portugese coastal basins hold exceedingly thick sections of terrigenous sediment including coals. The fresh-water terrigenous clastic facies migrated across Europe to the Swiss-French Jura area which contains the southernmost exposures of silty ostracod and Chara-bearing Purbeckian facies at the top of the Jurassic.

Oolite

Figures. IX-1 and IX-2 indicate important areas of oolite accumulation in north-western Europe which are interpreted as patchy areas of irregular migrating shoals on shelves surrounding the shallow basins. In northwestern Europe these oolitic shelf deposits, like the earliest Malm patch reefs, are chiefly of Oxfordian age. As the southeastward regression of pure carbonate sedimentation developed during Late Jurassic time, oolite beds become younger in the section until they occur in Kimmeridgian strata in the Jura Mountains, partly behind the Tithonian reef tract. In the Middle East the famous oolite and shoal water grainstone of the Arabian Shield is essentially all Kimmeridgian and lies on a shelf east of the Arabian Shield.

Reefs

Lying in patches around shallow basins (Paris, English and North German basins) are coral reefs, mostly of Oxfordian age. At the edge of the European shelf-island-basin area, facing south into the Tethys, almost the total Malm (Upper Oxfordian-Tithonian) developed massive sponge and coralline limestone. This is particularly widespread around the northern edge of the Tethyan geosyncline in Latest Jurassic (Tithonian) time. The Jurassic section is capped by Tithonian reef facies overlying deep water sediments of early Malm age, especially in the eastern and south-central Mediterranean as far west as Sardinia and Sicily. Reefs developed on platforms within the geosyncline. None are known in the Jurassic sections of the Middle East.

Deep Lagoonal Carbonate

Immediately behind the reef rim in Franconia and Schwabia (southern Germany) and in Spain (Montsech) lay basins filled with quiet-water sediments, now preserved as interbeds of platy limestone and marl and including thin turbidites. These small, reef-surrounded basins have preserved remarkable faunas of nektonic marine, flying, and land animals in an environment completely inhospitable to normal benthonic creatures. These beds constitute a most remarkable facies of the Jurassic, the Solnhofen.

Miogeosynclinal Basins

An area of thick carbonate sedimentation (miogeosyncline) in southeastern Europe separated the interior (Tethyan) geosyncline of the Mediterranean area from the Russian platform and Scandinavian shield. Massive shallow water limestones, mainly of Tithonian age, reach almost 1000 m in this trough whose sediments are exposed in the Caucasus Mountains and the Elburz range of Iran. Similar shallow shelf deposits up to 600 m thick are represented in the Late Jurassic section in the

North African ranges of Algeria and Tunisia. At this time the North African-Arabian shield was mainly land. A prong of the continent extends north through Israel, Jordan, and Syria to the Mosul block in North Iraq. The Mesopotamian and northern Persian Gulf area is underlain by a Late Jurassic euxinic basin with salt, whose southern end is bordered by another wide carbonate shelf connecting the Arabian shield to an Iranian positive block, the Surmeh shelf. Within the Zagros geosyncline, 400–600 m of shoal-water cyclic carbonate sand and muds are present across this mildly positive area. The Rub Al Kali basin of southern Saudi Arabia contains euxinic, dark argillaceous limestone and salt. Other similar basins occur in Iran and Saudi Arabia. The Zagros geosyncline facies thus differs from the Mediterranean-Alpine Ammonitico Rosso and radiolarite facies.

Northern Tethyan and Continental Margin Facies

Bernoulli (1972) has subdivided the Late Jurassic Alpine and Appenine facies into several provinces. The interior or oceanic part consists of ophiolites, radiolarites, and pelagic shales with some volcanic sandstones and graywackes. The continental margin to the north is subdivided into two parts: the Helvetic black Quintner-kalk belt (Standard Facies belt 3) which grade into a deep trough and swell facies of the southern continental margin and slope. The latter facies occur in the external zones of the Appenines and Dinarids, and in the Southern Alps, and Austroalpine nappes. The paleotopography was complex with carbonate platforms, submerged sea mounts with abbreviated sections and adjacent deep troughs with continuous pelagic sediments. On the swells, thin pelagic limestones contain condensed faunal successions, early lithified red nodular limestone, hard grounds, and filled tectonic fissures. On the slopes of the swells peloidal limestone and encrinite pass to bioclastic debris limestone with pelagic bivalves and coccolith lime mudstones. The basins contain some clay but are mainly of pelagic limestone with radiolarite and redeposited and slumped sediment derived from up the slopes.

Basic Microfacies

Extensive petrographic work has been accomplished on Jurassic microfacies. Best references include the volumes of the International Petrographical Series edited by Cuvillier and Schürmann (1951–1969) and Carozzi et al., (1972) for the Aquitaine basin, France. Some of the most important of the microfacies are described and illustrated here, numbered J 1 to J 17 to distinguish them from the standard ones (SMF-).

Shelf and Shelf Margin in Europe

J1.   Light-colored lime mudstone with fenestral-peloid fabric, commonly laminated; pelletoids partly deformed–squashed together but retaining identifiable form. Fenestrules formed originally as holes caused by entrapped air or by gases from organic decay in the fabric. These became filled with sparry calcite. In some areas fecal pelleted mud became rapidly hardened and formed peloid sands. On areas subjected to periodic flooding and rapid drying (e.g., natural levees) lamination is graded with pellet sand overlain by finely laminated fenestral muds, in layers about a centimeter thick. Standard microfacies 19.

J2. Large foraminiferal grainstones and packstones, commonly with grains encrusted by algae. Foraminifera occur with rounded and worn particles in clean, washed, lime sand as well as in muddy sediment. *Parurgonia, Nautiloculina,* opthalmids, *Pseudocylammina Anchispirocyclina, Trocholina* are common genera; pure carbonate sediment with these large foraminifera probably represents deposition in ponds, lagoons, and associated shoals in areas of warm water with restricted circulation. Other forms such as *Alveosepta* occur in terrigenous, marly limestone probably representing brackish to fresh water deposits. *Trocholina* forms a particularly resistant particle, being of robust, chambered, and thick-walled construction and is common in grainstone and oolitic sparites. Standard microfacies 18.

J3. Onkoidal-pisolitic (Mumienkalk) limestone; light-colored, commonly of variable texture, grainstone to wackestone. These algal biscuits are unusually large (several centimeters) and their lime-mud coatings may bear multitudes of algal filaments and *Girvanella* tubules. They are closely related in microstructure to *Sphaerocodium*. Such sediment occurred in restricted marine shoals and lagoons where currents were strong enough to overturn lightweight fragments. Modern channels with moderate water movement in intertidal and shelf-margin shoals contain living blue-green algal onkoids which resemble the Mumienkalk particles. The Jurassic onkoids, however, were generally hard when emplaced, and in this more resemble fresh-water algal biscuits. A good marine analog to such hard algal particles has not been found except in the red algal nodules. Standard microfacies 22 (Plate XV A).

J4. Oolite grainstone with multiple coated ooids; nucleii of shell bioclasts and pelletoids. The excellent sorting of absolutely mud-free sediment indicates accumulation on bars and shoals with high energy-tidal currents which move sediments to and fro almost continuosly. Such well-formed ooids more commonly characterize shelf-margin positions than restricted marine shoals or belts of longshore sand sedimentation. Standard microfacies 15.

J5. Bioclastic grainstone-packstone. Brachiopod, bryozoan, echinoderm, coral, and molluscan debris occurring in grainy sediment associated with patch reefs and shoals are scattered across shelves. Most of the particles are worn and coated and in places ooids and lithoclasts are present with them. The fauna indicates normal marine salinity and open circulation. Standard microfacies 11.

J6. Coralline facies-boundstone. Both fasciculate and platy encrusting forms of corals make up patches of solid reef framework, along with the heavy-shelled pelecypod *Diceras*. The sediment between the coral heads generally consists of worn, coated, and rounded bioclasts of varied types indicating normal marine, open-water salinity and circulation. Standard microfacies 7 (framestone).

J7. Reefy limestones constructed mainly of spongiomorph hydrozoans. These organisms much resemble the Paleozoic stromatoporoids. Presumably such organisms are more tolerant of warm water and greater salinity than the corals. Standard microfacies 7 (framestone).

J8. Sponge facies with algal and foraminiferal (tuberoid) encrustations (Hiller, 1964). These are well known from the Middle Malm of Schwabia and Lower Kimmeridgian Treuchtlinger Limestone of Franconia. Large mounds with micritic matrix were formed in quiet and deeper water by sponges in the Late Jurassic. Standard microfacies 7 (bafflestone).

J9. Solnhofen Plattenkalk. This is extremely fine-grained, lithographic lime mudstone with fine calcite mosaic micrite. Grains of 3–4 microns are welded together. Many coccolith remains can be seen with the Scanning Electron Microscope. Such beds of limestone, *Flinze*, a few cm thick, are separated by thinner marls, *Fäule*. The bedding is almost perfectly planar and regular with the exception of local slumps. Preservation of whole fossils and lack of benthos or burrow structures indicates deposition below wave base in water deeper. more saline, and more anaerobic than normal neritic. These basins lay directly behind reefs and organic buildups which faced the Alpine trough and stretch completely across Europe. A variety of Standard microfacies 3 (Plate XXVIII).

Middle Eastern Shelf Facies

J10. Algal-bioclastic coarse grainstone. This sediment is particularly common in the Middle East Arab zone. The worn and rounded particles are up to several mm in size. Some are coated. There is a large amount of micritization, coarse crystallization and solution of particles. Cement consists of both even, drusy, grain-lining types and of blocky mosaic calcite. Organisms consist of badly altered pieces of red? algae, foraminifera, gastropods, and occasional bivalves or brachiopods. The sediment represents accumulation in current-washed shoals or bars, probably in a somewhat hypersaline environment. Little oolite is present. Variety of Standard microfacies 11 (Plate VIIA).

J11. Dasycladacean grainstone-packstones. The Late Jurassic dasycladacean *Clypeina jurassic* is well known. It occurs in diversely textured sediments and in various stages of disarticulation. More robust forms such as *Actinoporella* and *Salpingoporella* are also common. These are especially typical of the Middle East grainstones (Arab Formation). Dasycladacean algae at the present time thrive in extremely shallow water (only a few meters deep) but will tolerate considerable salinity variation–from brackish water to salinities of more than 60 per mil. Standard microfacies 18 (Plate VIA).

Alpine Basin-Slope Microfacies

J12. Argillaceous and silty dark lime mudstones and wackestones with scattered benthonic as well as pelagic bioclasts. Sponge spicules are common. Mollusks, echinoids, brachiopod and crinoid fragments occur. Arm plates of pelagic crinoids such as *Saccocoma* and ophiuroids are abundant in some layers. Bedding is thin and planar with rhythmic marl intercalations. Such slope limestones are transitional from shelf margins into basins. For example, Effinger marls and Birmenstorfer beds of the Central Swiss Jura. Standard microfacies 1 or 3

J13. Dark micropelletoid micritic limestone. Composed of calcitic silt and fine sand-sized calcareous particles with lime mud or spar matrix. Thin-graded laminations of a few cm represent distal parts of turbidites. The sand-size layers occur within a dominantly micrite matrix. Such limestones are thin, planar and regularly bedded. The Quintnerkalk of the Helvetic Alpine nappes is an example. In the Middle East, such black limestones are interbedded with evaporites and deposited in basinal environment. Standard microfacies 1.

J14. Light-colored pelagic (cherty) pure lime mudstone. This sediment is common in the Austroalpine nappes and Southern Alps, variously Aptychuskalk, Oberalm, Maiolica, Biancone. It is thin and rhythmic-bedded micritic limestone with very thin marl partings. Some beds may be spiculitic but basically the limestone is composed of calcareous nannoplankton rich in calpionellids and coccoliths. The fauna is pelagic: ammonite aptychi, *Nannoconus*, *Globochaete*, tintinnids and radiolarians. The key-hole brachiopod, *Pygope*, occurs in some of these beds. In some sections thicker beds of fine-grained allodapic (turbidites) limestones and even locally coarse slump breccias are interbedded, such as the Barmstein Limestone of the Salzburg area. Very early diagenetic or depositional solution of aragonitic material at its compensation depth is indicated for some beds and the possibility exists for their formation in troughs of several thousand meters (Garrison and Fischer, 1969). Standard microfacies 3 with some beds of Standard microfacies 4 (Plate III).

J 15. Limestones with very thin tests or shells of bivalves and brachiopods; also termed "filamentous micrites." *Halobia, Halollela, Posidonia (Bisotra)* are pelagic bivalves with extremely thin tests. Such organisms accumulated in great abundance forming shell coquinas (lumachelles) in micrite matrix on the bottoms of Mesozoic basins. They are an especially distinctive deposit in Triassic and Jurassic geosynclinal strata. A pelagic variety of Standard microfacies 8. (Plate XXIX C).

J 16. Radiolarites. These irregularly thin-bedded (ropy or nodular) red, purple and green strata are presumed to be derived from the very slow deposition, at great depth, of siliceous and calcareous tests. Much early mutual replacement of SiO_2 and $CaCO_3$ took place. The opaline silica is altered to cristobalite, chalcedony, and crystalline quartz. Matrix is generally clay-silica; any calcareous matrix was silicified in diagenesis. The radiolarians which formed the bulk of the original sediment are usually poorly preserved. A microscopic species of unknown biological affinity, *Globochaete alpina*, is commonly recognized in these deposits as well. The colors derive from trace amounts of Fe and Mn undiluted by the normally high carbonate or clay content. The radiolarite is generally only a few tens of meters thick, the product of basin starvation, below compensation depth for most carbonate. In places it must have been deposited in troughs with steep sides, for breccias and clastic turbidites with silt, shale and lithoclastic limestone microbreccia may be interbedded with it. Flute casts and graded bedding may occur in such strata, formed as the extraneous clastic material moved into the troughs from swells in the geosyncline. Massive mud-flow breccias some meters thick may also be present, disturbing the even, thin bedding below them by plowing, loadcasting, and channeling (Plate XXIX A).

J 17. Red nodular to conglomeratic limestone, rote Knollenkalk, or Ammonitico Rosso. Red conglomerate with nodular clasts usually in a micritic matrix (occasionally with spar cement between them). Bedding is thin and wavy to nodular. Microstylolites and ferromanganese crusts commonly occur at clast boundaries but some boundaries between clasts and matrix are vague. Many clasts appear to be only slightly displaced, others are more clearly transported. Early diagenesis and sediment flow is indicated. Redeposited sediment in slumps and turbidites are recognized (Plate XXIX B).

Laminated crusts attributed by some authors to stromatolitic algae(?) coat some pebbles. Burrowing structures of *Chondrites* and *Zoophycos* types are reported. The fauna is dominantly pelagic, although allochthonous brachiopods, crinoids, and mollusk remains occur. Calcareous microplankton such as *Saccocoma, Globochaete* and *Stomiosphaera* may be present. In some places almost all calcareous microplankton is absent within the nodules, as if dissolved at an early stage—even during deposition. Lime mudsteinkerns of ammonites are prevalent. Aragonitic shells are not preserved even as calcitic replacements, and careful petrographic study indicates that this type of shell material was dissolved away before burial. Molds of ammonites are corroded at former sea water-substrate surfaces, now forming irregular bedding surfaces resembling hard grounds. Commonly, crusts of ferro-manganese carbonate occur at these levels. Such evidence indicates very slow deposition in relatively deep water. Slow deposition is confirmed by the thinness of such strata, generally only 25 m or so; stratigraphic condensation is indicated by the numerous ammonite zones such strata bear.

Several possibilities exist for the cause of the nodular fabric and its depth of formation. Suggestions listed below are not necessarily mutually exclusive.

a) Solution of crusts to form nodules on very deep ocean floors—below present compensation depth for aragonites which is about 4000 m. Hollmann (1962) and Garrison and Fischer (1969) termed this process "subsolution".

No one can be certain that the Jurassic compensation depth for carbonates was the same as that in the present oceans.

b) Concretionary growth of nodules of high Mg calcite concomitant with aragonite solution in shallow buried sediment—hence segregation of lime in a somewhat marly substrate. Water depth within the photic zone, not more than a few hundred meters (H.C.Jenkyns, personal communication).

c) Nodules formed by segregation of lime early in diagenetic history by solution—compaction and pressure solution under some overburden and under water of variable depth, less than 1000 m (H. Zankl, personal communication).

d) Nodular fabric resulting from mass movement of partly lithified sediment off high areas of sea floor, nodules being in part true clastic particles.

Such limestones are widespread in the Mediterranean from Morocco to Greece, being characteristic of strata from Lias to Malm age. They have been found also in the western Atlantic basin just east of the Bahamas. Most geologists consider these peculiar sediments as having formed slowly at depths below wave base but in an oxidizing environment. Bernoulli (1972) concludes that these deposits formed on tops and slopes of submarine swells at variable depths on the southern part of the northern continental margin of the Tethys. They were underlain probably by continental crust and hence were deposited in depths shallower than a few thousand meters.

The Reef Girdle of Central Europe

An area of thick and massive carbonates encompassing the whole of the Upper Jurassic (600–900 m) trends east to west from the Caucasus, southern Poland, southern Germany, and curves south through the Jura Mountains of Switzerland and France (Fig. IX-1). In southern Germany the old names Malm and Weiss Jura are used for this section. The trend is a wide reef belt but does not include all the coral reefs in the European Late Jurassic strata. Thousands of coral biostromes and patch reefs exist behind (north and west) of this fringe, mainly in Oxfordian strata. They are well known in Britain and have also been described in the Yonne Valley of central France (Rutten, 1956). In the Swiss Jura Mountains, a classical area for detailed study of Jurassic reefs, the old stage named Rauracian designates the reefy rock units, and the Argovian the off-reef basinal facies.

The reef facies extends from the Jura across the essentially unfolded strata of the Schwabian and Franconian Alb or plateau north of the Alpine belt. A regional paleotectonic cause for this reef girdle is partly clear. The reefy carbonates are apparently shelf-margin deposits facing into the pre-orogenic Tethyan trough. In Switzerland, transitional facies can be seen in outcrops where the east-west reef trend changes into deeper water carbonates and shales to the south. In Switzerland the reef belt may have been localized by the Late Paleozoic (Hercynian) ridges and massifs, some of which stood as islands on the Late Jurassic shelf. Reefs were probably developed along subsiding margins of these structural trends.

The outcrops of the reef belt are wide—as much as 100 km, chiefly because reef development is regressive, moving from northwest to south and east; in the latter directions, reefs rise in the section from upper Oxfordian through Tithonian at the top of the Jurrasic. Jura reefs of Portlandian-Kimmeridgian age may be seen near Belley and Yenne in France, 80 km east of Lyon, and are known in the Basel region of northeast Switzerland in Oxfordian beds through work of Ziegler (1962), and Bolliger and Burri (1970). In the Schwabian plateau, sponge reefs of upper Oxfordian to mid-Kimmeridgian (Malm Beta to Epsilon) are capped by coral reefs of late Kimmeridgian (Malm Zeta) age. In Franconia, farther east along the plateau near Neuburg an der Donau, Eichstätt, Kelheim, and Regensburg coral reefs of very youngest Jurassic (Tithonian) are developed atop sponge-bearing beds of early Kimmeridgian age.

Swiss Jura Mountains

Figure IX-3 is a northwest-southeast trending section across the Swiss Jura some kilometers west of Basel. It is located along the northeast corner of the map of Fig. IX-4, roughly through Tiergarten and Moutier. A question exists about the correlation of the Effinger marl—whether it is a fill-in of a starved basin and just younger than the lower reef-lined shelf margin (Bolliger and Burri, 1970) or whether it is a facies equivalent of the reef as interpreted by Ziegler (1962). This is not of consequence to the facies interpretation.

Three main facies provinces are seen, regressive to the southeast.

1. To the northwest, a shallow shelf with tidal flat and fresh-water pond sedimentation.

2. A northeast-southwest trending belt with corals and bioclastic reef debris, including abundant oolite deposits.

3. To the southeast slope limestones and marly sediment with open marine faunas including ammonites. Equivalent beds in the frontal Alps—both in the Helvetic nappes and the autochthonous section over the Aar massif are found in the Quintnerkalk, a black somewhat argillaceous (?) limestone with a pelagic fauna.

Owing to the above stratigraphic studies, and those of Gygi (1969), one may precisely correlate upper Oxfordian strata across these facies belts in the Jura Mountains from south to north. These strata, diagrammed on Fig. IX-3, are discussed below in detail because they offer an excellent example of the range of facies across a shelf margin of Mesozoic age and because microfacies and paleontology are particularly well-known.

The basinal reef-equivalent strata (Birmenstorf beds) are brown, spiculitic, pelleted mudstones and wackestone with chert concretions formed from decay of hexactinellid sponges. Typical fossils, in addition, are rhynchonellid and terebratulid brachiopods, large bivalves such as *Lima*, echinoderms, gastropods, belemnites, and perisphinctid ammonites. The beds become more argillaceous farther from the reef area. Northward, toward the reef edge, the argillaceous off-reef strata pass into microbioclastic wackestones, in places containing some minor coarser reef debris. No real reef talus occurs. The fauna is much like that of the basinal slope beds; cidarid echinoids and shallow water bivalves occur in addition to the cephalopod and sponge biota.

The 25 m of basinal-slope beds thicken northward to a bank of coral-bearing nodular limestone termed Liesburg beds. These strata are transitional from underlying sandy argillaceous beds and are extremely fossiliferous with many plate-shaped coral colonies of *Thamnasteria* (as much as half the rock), crinoids, cidarid echinoid, pectinid bivalves, terebratulid brachiopods, and lithistid sponges. The fossils are commonly silicified. Overlying the Liesburg unit is massive coral-bearing limestone. The lower corals of this bank are plate-shaped *Thamnasteria* in a micrite matrix. Higher up this genus is replaced by roundish coral heads and large branching types of corals such as *Latomaeandra;* only a few recognizable echinoderms, brachiopods and bivalves are associated. Although no lithoclastic reef talus is seen, much of the upper reef limestone is formed of coarse bioclastic packstone. The coral buildups appear to consist of a series of individual patches

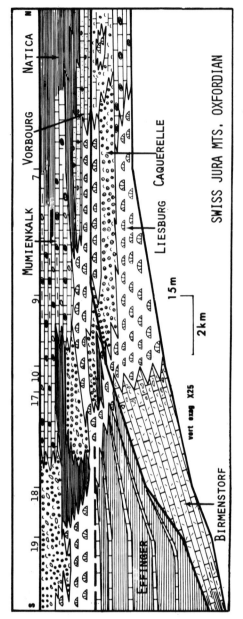

Fig. IX-3. Cross section of Oxfordian strata in Swiss Jura Mountains. For legend see Fig. III-1 and for line of section Fig. IX-4. Shows silty and onkoidal lime mudstone facies of lagoons passing through oolitic grainstone with patch reefs of coral across a coral bank to thin-bedded brown bioclastic wackestone and argillaceous beds of the basin. Progradation of facies indicated over 15 km across Jura Mountains near Basel to Gorges de Pichoux, Moutier. After Bolliger and Burri (1970)

or knolls forming a gently sloping ramp inclined at only 1 or 2 degrees. The ramp was covered by mildly turbulent water, a few meters or tens of meters deep. No crashing surf occurred to distribute much coarse debris downslope. Moderate wave action winnowed fine material away from around the reef knolls surrounding them with halos of bioclastic sand. Some lime mud was trapped within the coral framework.

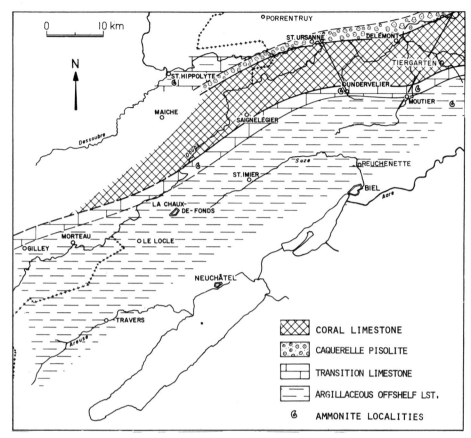

Fig. IX-4. Facies of "Middle Rauracian" (Lower Kimmeridgian) showing 10 km width of coral ramp near edge of carbonate platform and narrow belt of Caquerelly pisolite probably at crest of shelf margin. Another narrow belt of darker thin-bedded (transition limestone) lies basinward of the coral reef. Line of crossection of Fig. IX-3 is at extreme northeast corner of map through Moutier. From Ziegler (1962)

Another facies, oolitic onkolitic bioclastic lime packstone-grainstone, is developed slightly shelfward along the coral knoll ramp (Caquerelle pisolite). Whereas bioclastic debris may be coated or uncoated, well-formed ooids are seen and onkoids are common (algal encrustation on gastropods, bivalve, bryozoan, and ossicles of crinoids). Serpulids occur at the base and coral debris is common. Some vague cross-bedding is seen and this sediment probably formed in a turbulent environment as rapidly accumulated sand and gravel bars. This facies is interpreted as forming the crest of the reef profile, sediment piled up to sea level shelfward from a platform of coral heads and bioclastic debris. The low profile, coral growth forms, and internal reef construction much resembles that seen in Devonian strata and form a Type II shelf margin of the author's classification—a knoll reef ramp.

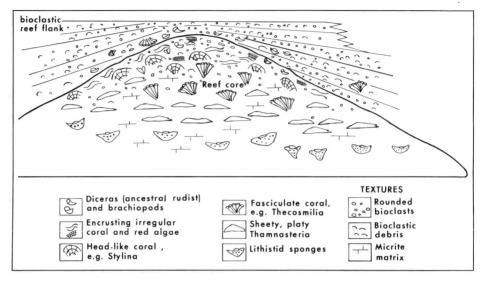

Fig.IX-5. Idealized Upper Jurassic patch reef or reef knoll showing upwards changes in biological sequence and growth forms of corals. Note change from sheety to dendroid to massive encrusting, and from sponge to corals to red algae and *Diceras*. Matrix changes from micrite at base to rounded grainstone at top

Several backreef facies are equally well-defined and, owing to regression, are widely exposed both behind the upper Oxfordian reef ramp and stratigraphically higher above and in front of it. Behind the platform chalky limestone (St. Ursanne unit) overlies and grades into the pisolitic sediments. Patch reefs occur in these fine-grained chalky beds. Their corals show an interesting ecologic sequence discerned by Pümpin (1965). Lower beds contain compact corals such as *Thamnasteria* followed by fasciculate forms such as *Calamophyllia* and *Cladocora*. Between the corals red algae such as *Ptychochaetetes* and large *Solenopora* occur. Higher in the reef are found large branching corals such as *Latomaeandra* and *Thamnasteria gracilis*. Reef growth ends with large head-like corals of *Stylina* and the red alga *Solenopora*. Some irregular encrusting corals may also cap the sequence. Many thick-walled mollusks, such as *Diceras, Cardium, Nerinea*, lived in the reef. The same sequence of coral growth may be discerned in Oxfordian patch reefs in the Yonne Valley in the southeastern Paris basin at Merry-sur-Yonne (Fig.IX-5).

Strata known as Vorbourg beds, form another backreef sequence of light gray, dense, homogeneous mudstone and wackestone, containing scattered onkoids. The upper Vorbourg beds are laminated algal mat tidal flat beds with windblown quartz silt and highly resistant trace minerals such as zircon, rutile, and tourmaline. Marl intercalations contain freshwater ostracods and charophyte oogonia and indicate the presence of old lakes, ponds, and salinas. Black pebble breccias and hard grounds mark sites of exposure or very slight deposition. More argillaceous backreef strata (*Natica* marls) overlie the Vorbourg unit. They contain some ferruginous oolite and calcarenite with local algal mats. Some normal marine, open circulation organisms are found. These include snails: *Natica, Nerinea;*

bivalves: *Pecten, Trigonia, Mytilus;* brachiopods: *Rhynchonella, Terebratula;* and crinoids and echinoids: *Cidaris* and *Stomechinus.*

Another characteristic backreef microfacies is the well-bedded, light-colored onkoidal micrite ("mummy beds"). In this sediment algal laminae encrust fragments of large fragmented fossils such as corals, crinoids, gastropods, brachiopods and mollusks. Characteristic large foraminifera occur in the micrite in between the nodules: *Pseudocyclammina, Nautiloculina, Cristellaria,* miliolids, and textularids.

The algae preferentially are backreef, not reefoid forms. The calcareous algae include Codiaceans in the form of small tufts (up to 1 cm across): *Cayeuxia, Marinella* and *Lithocodium.* The blue-green encrusting forms are bean and biscuit shaped, with irregular protuberances or concentric arcuate crusts impregnated with *Girvanella* tubules. The algal-foraminifera association is interpreted to indicate very shallow, warm, clear water of variable salinity. Circulation must have been capable of moving the onkoids, but insufficient to winnow the mud. The onkoids were formed with durable, calcified crusts. The main algal bed (Hauptmumienkalk) is extensive over the whole Jura Mountains.

The Schwabian Alb

Beyond the folded Swiss Jura, across the Rhine and north of the Danube, Jurassic rocks are exposed in a broad limestone plateau trending eastward across southern Germany, and constituting the Schwabian and Franconian Alb. The Upper Jurassic (Malm) outcrop here is 40 km broad and extends for several hundred km, continuing the broad reef girdle exposed in the Jura Mountains of France and Switzerland. These outcrops constitute the classic area for ammonite zonation worked out by Oppel in the early 19th century. Recently, facies study has been carried out by Gwinner (1971) and Aldinger (1968) in Schwabia, and Barthel (1969) in Franconia. The reefs exposed here are interesting because of the facies progression they display. Older buildups are sponge-algal micrite mounds deposited in downslope positions and constituting a Type I shelf margin. During the course of Late Jurassic time, corals succeeded the sponges, reefs built up into wave base, and shallower water facies appeared (Type II shelf margin). These facies are diachronic and migrate eastward as well as southward.

Generally the "normal" section of the Malm in Schwabia is composed of alternating marls and evenly and rhythmically bedded limestones. The sedimentary texture is wackestone with ill-sorted bioclasts of mollusks, brachiopods, serpulids, sponges, and a few onkoids. The megafauna is principally ammonites and thin-shelled mollusks; foraminifera, spicules and many coccoliths also occur in the micrite matrix. Beginning in the higher Oxfordian layers, a wide belt of micritic mounds occur. These mounds are rich in siliceous sponges with much encrusting algae and tubular foraminifera. Such beds are time equivalents of the coral reefs of the Jura but represent somewhat deeper marine waters, a sag in the reef girdle. The mounds, termed the Massenkalk facies, developed as irregular masses scattered across the belt, and later (in Malm Delta and Epsilon time) as a wide sheet of massive limestone. The belt was possibly 100 km wide by late

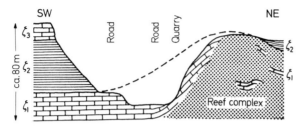

Fig. IX-6. Sponge mounds in Kimmeridgian beds of Schwabia from near Münsingen. From Gwinner (1962). Mound is micritic with scattered lithistid sponges. Bedded limestones onlap flank. The height of the mound at end of growth was at least 50 m

Kimmeridgian; it now extends under the Molasse plains to the south. The individual mounds show some intergrading and interbedding of massive and normally bedded limestone and marls. In places, however, they accumulated as very large masses of rapidly accreted and uncompacted carbonate. In these mounds a sea floor relief of as much as 100 m may be demonstrated by observing stratigraphic onlap against the sides of the buildups. The slopes of individual mounds average 10–15 degrees but may be much steeper (Fig. IX-6).

The constituent sponges are flat or cup-shaped siliceous forms now largely replaced by crystalline calcite; they occur in a micrite matrix. The total megafauna, considered in variety of species, consists of 8–10% sponges, 10% bryozoans, 20% brachiopods, 35% serpulids and other encrusting forms, and 13% anmonites, various bivalves, belemnites, etc. Volumetrically, the bulk is sponges with abundant encrusting algae and tubular foraminifera, the invertebrate shelly fauna consisting of small forms inhabiting the underside of sponges. Abundant calcareous crusts coating the sponges are presumed to be symbioses of alga-nubicularid foraminifera; they preferentially grew upward (like stalagmites) to form knobs and pellets (Hiller, 1964). Additionally, algal? stromatolites in the higher mounds encrust in place of the alga-tubular foraminifera association.

These sponge-algal mounds are interpreted as developing in quiet and somewhat deeper water than one normally associates with carbonate buildups for the following reasons:

1. Relief of as much as 100 m can be stratigraphically demonstrated by interreef fill-in against sides of mounds; depths of interreef areas of at least this much must have existed at the end of mound accumulation and possibly at its beginning.

2. Equivalent off-mound beds contain even alterations of marls and limestones and a glauconite bed that can be widely traced across the Schwabian Alb— essentially indicating sedimentation below wave base.

3. Micritic matrix of mounds indicates quiet-water deposition.

4. Siliceous sponges commonly grow at considerable depths.

5. There is little or no evidence of extensive boring organisms.

6. There is no talus or widespread bioclastic debris surrounding the mounds.

7. The tops of these mounds were colonized by algae but not by corals. They were, therefore, in the photic zone but below wave base. As much as 50 m of water may have existed above the mounds.

Higher in the section (Malm Zeta, upper Kimmeridgian or Tithonian) the sponge-algal mounds are overlain and colonized by reef corals like those of the Swiss Jura Mountains. A backreef facies of cross-bedded bioclastic oolitic sediment developed to the east and north, at this later time (Gwinner, 1962). The particles are of echinoid, bivalve, bryozoan, coral and calcareous sponge fragments and lithoclasts. Micritized rinds occur on the particles. The matrix may be micritic or sparry. The texture is generally packstone but some grainstone is common. A fauna of thick-shelled mollusks is present (*Nerinea*, and *Diceras*), along with algae, associated with the corals. Also, as mounds increased in relief, the reefs grew into wave base and the basins between became filled with "cement marls" and in places by thin platy beds with a pelagic and nektonic fauna, but with no benthos. This Solnhofen type deposit is known at Nusplingen in Schwabia. The considerable submarine relief which developed at this time is indicated by widespread, though relatively uncommon, fine-graded beds, graded breccias and glide and slump masses. These masses are mixtures of reef debris and broken pieces of lime mud from flank beds.

Franconian Reefs

The plateau continues north of the Danube, farther east in the German land of Franconia. The lower Malm consists there of sponge mounds and bedded limestones; higher in the Malm occur abundant, very fossiliferous coral reefs in scattered knolls having individual relief of a few tens of meters. The reef patches vary from 25 to 70 square km. Locally, 1 or 2 km of flank beds are seen around an individual core. They are well exposed in the valleys of the Altmühl and Danube Rivers (Fig. IX-7) and have been paleontologically studied in various places (Bausch, 1963; Bausch and Zeiss, 1966; Barthel et al., 1971). The reefs extend through the highest Jurassic (upper Tithonian) and are regressive southward toward the Tethyan trough (Barthel, 1969). The highest Jurassic reef facies is interbedded with black lime mudstone resembling the Helvetic Quintnerkalk, in wells which penetrate the molasse near Munich. Farther south, age-equivalent beds in the Austroalpine nappes of the Northern Calcareous Alps are in an entirely different geosynclinal facies.

The composition of the reef knolls varies somewhat according to stratigraphic position within the latest Jurassic. The older sponge-algal knolls at Kelheim, near Regensburg on the Danube, are of lower Kimmeridgian age, and according to Bausch and Zeiss (1966) are capped by the encrusting coral *Microsolena* and by spongiomorph hydrozoan framework (18% of volume) with minor (less than 2%) megafauna consisting of crinoids, snails, brachiopods, solitary corals, and sponges, also occurring in the central core. Reef detritus constitutes 80% of the bulk volume of the accumulation and consists of crinoids, snails, solitary corals, and large bivalves, including *Diceras*. About half of the debris is within the coral-hydrozoan framework and half is reef-flank debris dipping steeply off the core. Other beds at the southwest edge of Franconian outcrops near Neuburg a.d. Donau, display somewhat different reef facies. Here the core of the Laisacker reef (of lower Tithonian age) rarely contains spongiomorph hydrozoans but much encrusting coral *(Microsolena)* which is easily mistaken for them; more than 20 coral genera are recognized here. The core contains large coral stalks up to $1^1/_2$ m in diameter with many *Lithophaga* borings. Some of the corals are badly broken, the delicate form *Enallhelia* making up much detritus. More than 100 species of mollusks have been noted from these beds. Especially large forms occur such as *Trichites*, limnids, pectinids, *Praeconia*, and pleurotomarids. Much interstitial peloidal micrite is present in the core facies as well as coarser bioclastic debris, and encrusting blue-green algae appear to be important sediment binders. The flanking debris contains bioclasts and lithoclasts in grainstone and packstone texture and many large displaced colonies of *Thecosmilia*-like fasciculate coral. Some of these beds are very coarse-grained. Much solution and dolomitization of these beds is evident, apparently owing to their permeability. Unlike the older reefs at Kelheim, few ooids or onkoids are seen in the interreef beds and the thick-shelled *Diceras* is lacking. Dasycladacean algae are absent as are coralline red algae. From both core and flank facies, *Cliona* and *Entobia* sponge borings are ubiquitous, infesting both sessile organisms and in-drifted shells of ammonites. These differences in biological character may be caused by the younger reefs' location which was somewhat removed from the open sea.

Summary of Malm Reef Belt

1. Along the whole belt, reef cores are in patches surrounded by wide halos of bioclastic reef flank deposits.

2. The reef belt is very wide (tens of kilometers) and its seaward slope very gentle (1–2 degrees).

3. There is a progressive regression toward the Alpine-Tethyan trough upward in the section—a buildout toward the south and east from the Vosges block and Vindelician ridge, roots of the Hercynian orogenic belt.

4. Early development was of Type I downslope sponge reefs with encrusting alga-foraminifera, probably in water tens of meters deep. These existed below wave base rising as much as 100 m above the basin floor. Such buildups were followed by a period of shallowing and by coral-hydrozoan colonization and organic encrustation. Large mollusks *(Diceras)* were common as the buildups grew; in general the buildups grew into wave base and extensive debris derived from biota accumulated on top. Such construction developed a Type II shelf margin.

5. In several places along the reef belt the organic accumulations surrounded, protected, and helped restrict basins deep enough for deposition of unusually fine carbonate mud which preserved a fauna derived from both the land and open sea.

The Solnhofen Facies

A series of world-famous, fossil-bearing deposits of platy to thin-bedded, very fine-grained Late Jurassic limestones occur in southern Germany in Franconia and Schwabia. These were formed in deep lagoonal basins lying behind the outer

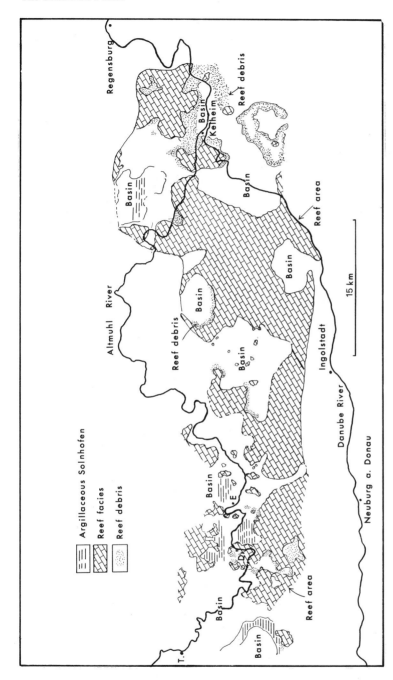

Fig. IX-7. Outcrop map of Malm Zeta-2b (Kimmeridgian) in Altmühl and Danube Valleys of Franconia, Germany showing Solnhofen basins surrounded by coral and sponge reef facies. After Freyburg (1968)

reef barrier, and separated from each other by reef knolls. These Solnhofen deposits were quarried in earlier times for lithographic stone and more recently for flagstones. The depositional basins of these strata actually stretch all along the reef girdle, lying behind its front and being generally of Kimmeridgian-Tithonian Jurassic age. They are known from Montsech in the Pyrenees, Cerin in the French Jura, Nusplingen in Schwabia, and most importantly, in the valley of the Altmühl River, in the Franconian village of Solnhofen and from Eichstätt. Here the somewhat irregular distribution of the basins has been illustrated by Freyberg in 1968 (Fig. IX-7). The swales lie roughly between north-south trending reef knolls and platforms. The platy basinal limestones are not very fossiliferous but what fauna is present is strikingly preserved with many soft parts moulded as delicate impressions, for example insect wings and bird feathers. The extensive collections in museums in Eichstätt, Munich, Frankfurt, Stuttgart, and Berlin result from 200 years of painstaking hand quarry operations.

Many persons have studied the fossil collections and recently sedimentological studies have added greatly to our understanding of these fascinating strata. Earlier some paleontologists considered the Solnhofen limestones to be tidal flat deposits but now the consensus favors an origin in quiet, somewhat deep local basins. Controversy also continues over the origin of the extraordinarily fine micrite sediment. The description of the deposits and their fossils given below follows modern sedimentological interpretations in reviews of the problems by Barthel (1970), Van Straaten (1971), and De Buisonje (1972).

Lithology

1. The limestone is very fine-grained micrite, with calcite mosaic of 3–4 microns. Electron scanning microscope studies have recently shown that the micrite contains abundant coccolith plates (Flügel and Franz, 1967).

2. It occurs in thin planar beds from a few mm to 30 cm thick.

3. The layers of pure micritic limestone (Flinze) are locally interbedded with marls (Fäule).

4. The limestone are about 97–98% $CaCO_3$, the 2% residue, being Fe and silica. The interbedded marls are actually 85–91% $CaCO_3$ and contain clay minerals.

5. The rhythmic bedding resembles that of certain deep-water and slope limestones found in basins and geosynclines. Perhaps the limestone beds represent kinds of turbidity or density current deposits, rapidly deposited following periodic storms (Plate XXVII).

6. Marl layers seem more common over older high areas; limestone beds thicken into the intervening basins.

7. Individual limestone beds are traceable from quarry to quarry over as much as 8 km.

8. The strata lack the sedimentary structures found in recent tidal flats; no algal stromatolites, mud cracks, tidal channel or washout deposits, fenestral fabric, or flat pebble conglomerates occur. Ripple mark is extremely uncommon.

9. Rare, deformed, slumped and brecciated layers (Krummelagen) are present. Two of these are widespread over the Franconian area. Plate XXVII shows one of these at Nusplingen in Schwabia. Very small-scale graded bedding is recognized throughout. Rare groove casts occur. These sedimentary structures indicate deposition in considerable depth with some slope to the sides of the basins.

Diagenesis of Strata

More compaction appears in the Solnhofen than is usual for many ancient limestones.

1. Cohesiveness of the bottom sediment was sufficient for organisms such as fish and ammonites, to leave marks as they settled upright on the bottom and later fell over. A few arthropods left sharply preserved claw and telson tracks in the mud.

2. Syneresis of the lime mud occurred at an early time, presumably by dewatering in a subaquatic environment. This causes a rough top surface of the beds and a smoother bottom surface. The roughened top surface was erased by movement of a decaying fish carcass, indicating early formation of the rough surface (De Buisonje, 1972).

3. Compaction occurred in the limestones away from fossils. Such beds traced away from the fossil remains became $^1/_4$ to $^1/_6$ as thick compared to where present with the fossil. The noncompacted fossil is often forced into the overlying limestone Flinze bed and rests on a socle (pediment) of uncompacted, early cemented limestone. On the other hand, flattening of ammonite and other shells is observed in tops of marly layers just under overlying Flinze beds.

4. Microscopic study of the micrite reveals many planar and sutured grain contacts, apparently the result of pressure solution.

Biological Content

1. The strata are actually poor in fossils but the biologic composition is varied and unusual. Only the swimming crinoid *Saccocoma* is common.

2. Fossils generally occur along bedding planes in the Flinze and in the marly (Fäule) layers.

3. The soft parts are excellently preserved in some specimens showing very slow decay, in an environment lacking currents and scavenging benthos. Periods of very rapid burial may be responsible for the excellent preservation—the extent to which this is true is not ascertainable.

4. In all, 750 species of animals and plants have been described. The original list of Johannes Walther (1904) has been updated and revised—of this 99% of the organisms are of marine nekton or pelagic environment. In order of diminishing frequency these are *Saccocoma* (free swimming crinoid), ammonites, fish, crusta-

cea, squids, belemnites, reptiles and jelly fish. Some of the fish are very large (1–2 m) and rays and sharks and coral-browsing higher developed fish are represented. Ropy fecal material (*Lumbricaria*) exuded by pelagic feeders is fairly common. Most of the material is comminuted remains of *Saccocoma*. Even the micritic matrix is largely of pelagic organisms—fragments of coccoliths are estimated to be as abundant as 0.5 million per cubic mm (Flügel and Franz, 1967)!

5. Rare benthonic animals also occur and tell much about bottom conditions within the deep lagoonal basins. Many of these are arthropods of which 70 species and 25 genera are known. Even arthropods, such as the horseshoe crab *Limulus* and the spiny lobster *Mechochirus* which are very tolerant of salinity variations, could not live long on the bottom. Some tracks and trails of *Limulus* are irregular, inferring disorientation of the creature before death. The remains of the horseshoe crab in places are found at the end of the tracks. Bivalves are also found at the end of tracks, having died soon after emplacement in the environment. Actually, tracks and trails are very rare on bedding planes. This, plus total lack of burrowers, indicates a lethal bottom environment. Other benthonic organisms are such as could have floated into the lagoons attached to other organisms: seaweed, pectinids, oysters, stalked crinoids. Many of the preserved benthos are of smaller size, juveniles, which drifted in. In only one quarry (at Zandt) do true benthonic organisms occur, apparently buried in life position on the bottom. Here arthropod moulted skeletons are present and a group of ophiuroids occur together. Foraminifera of bottom-dwelling assemblages occur in several places. As many as 60 species are known with some Radiolaria, sponge spicules, and tiny juvenile forms of the megafauna. Many of these derive from the outer reefs and were washed in as tiny detrital particles. Others are considered indigenous by specialists. Granting a certain tolerance of foraminifera to higher salinity, temperature, and stagnation, their presence may indicate some limit to the lethal conditions prevailing on the bottom—conditions which effectively killed off megafauna and permitted extraordinary preservation of those animals and plants which sank to the bottom.

Bottom conditions must have been quiet. In many cases an ammonite or fish touched and grooved the bottom in a head-on, tail-on, or upright position and later toppled over to lie flat. Certain fish are found commonly with the caudal fin broken off, this having stuck in the cohesive mud before the body fell into a prone position. Animals were buried in various stages of decay but many are well preserved, from which it has been inferred that the lime-mud influx was periodic but rapid.

6. An account of the Solnhofen facies would be incomplete without mention of the striking finds of terrestrial and flying animals. Four skeletons and their remnants of the earliest known bird *Archaeopterex* are known and 200 specimens of *Pterodactyls* and other flying reptiles. Most of these carcasses bear signs of deformation by drying before final deposition and may have been washed into the deep lagoons from the nearby shore to the north. Land reptiles are rare but present as are land plants such as conifers and cycads. At least 180 species of fossil insects are recorded. This biological assemblage is fully as foreign to the environment of deposition as is the number of open-marine animals derived from the outer reef track and the open sea coccolith spores making up much of the micrite.

Interpretations

Explanations of the Solnhofen environment have varied. Prior theories include derivation of the mud from the northern coastal regions as windblown dust or from restricted marine origin in backreef shallow platforms and lagoons. Most evidence indicates a southern source and an origin as fine detritus from the reef barrier mixed with pelagic coccolith ooze. Buisonje (1972) proposed that sudden death of coccoliths due to "red tides" caused great floods of fine carbonate mud. Other theories have postulated deposition on tidal flats with mud made sticky by periodic drying under subaerial exposure. Lack of the characteristic and ubiquitous sedimentary structures of the tidal flat preclude this interpretation. The following theory of origin is derived from Barthel's (1970) discussion.

1. Deposition occurred in small partly isolated basins which originated by infill of a sea floor relief of 30–60 m between sponge-algal reefs; the top of these bioherms may well have been below wave base and thermocline, an estimated minimum of 20 m water depth.

2. The land mass to the north uplifted, and as water shallowed, rapid growth of coral reefs occurred, colonizing some of the sponge-algal mounds, accentuating the basin-swell topography and most importantly forming a southern barrier cutting off the basins.

3. Carbonate sediment was derived principally from plankton in the open sea to the south and as fine detritus from the reef barrier.

4. Accumulation took place in a hot, seasonally evaporative climate as indicated by contemporaneous anhydrite and salt in northern Germany, Paris, and Moroccan coastal basins. Flora and fauna also indicate a warm, subtropical climate; there is an abundance of cycads, reptiles, insects.

5. The basins became stagnant with hypersaline brines, derived from evaporation of inflowing sea water mixed to some extent with fresh water from land to the north. The brine flowed out continuously through the reef barrier and never reached the necessary salinity to precipitate sulfates (100–150 ppm). It was, however, sufficiently saline to kill and preserve (pickle) organisms in a largely anaerobic environment.

Several interesting questions appear unanswered in regard to the Solnhofen facies. Why was there insufficient organic matter to color the sediment black, particularly in view of the known subtropical climate on the land to the north? How did the bottom mud become cohesive enough to retain prints?

Modern lagoons with photic zone bottoms contain a soupy layer of blue-green filaments of organic slime and aragonite-sediment insufficiently coherent to be grooved. The Solnhofen lagoons indicate very early consolidation of bottom mud, a process which has not taken place in modern shallow water muds since their formation during the several thousand years since post-glacial sea-level rise. Compaction of the Solnhofen lime mudstone is greater than one would expect by comparison with modern shallow-water lime muds. The estimated 4 to 6 times volume compaction of laminae away from early calcitized fossils and the flattening of many organisms seems too high for pure calcilutites.

This Chapter has concentrated on European shelf marginal carbonate build-ups and associated lagoonal facies. In addition, Jurassic shelf strata, the world over, are also noted for oolite development. This facies is discussed in the following Chapter. Of equal interest is the wide latitudinal spread of Jurassic evaporites—from more than 50° North to less than 20° North. This is true in both European and the Middle East and in North America. The European reef belt falls within this 30 degree wide evaporite area. Coals are known in Portugal and in the Caucasus indicating more humid climate north and west of the major evaporite and limestone belt. In addition, in the European and Middle East Tethys, as well as in the basinal Smackover facies of the Gulf of Mexico coast, some of the most distinctive types of deep water carbonates are present. The microfacies part of this Chapter lists six varieties including the radiolarites. Other generalizations about basinal carbonates are found in Chapter XII.

Chapter X

Shoaling upward Shelf Cycles and Shelf Dolomitization

Three types of sedimentary cycles are characteristically developed during periods of pure carbonate sedimentation and are remarkably widespread features of broad platforms and interiors of major offshore banks. Most shelf or "backreef" strata contain such cycles even though they may not be obvious if terrigenous beds are rare and do not serve to accentuate the stratification. Cycles consisting of mixtures of terrigenous and limestone units are discussed in Chapter VII.

Pure carbonate shelf cycles almost invariably are asymmetric and consist of upward shoaling lithologic sequences, most of the deposition having occurred during periods of marine regression following through the environments represented by facies belts 6 through 9. It is as if a relatively rapid rise of sea level occurred repeatedly on a steadily subsiding shelf and was followed persistently by sedimentary progradation and a fill-in of the inundated area over some period of time. A rhythmic or hemicyclic pattern would logically stem from this process, but the same effect could result just as easily from episodic and abrupt shelf subsidence as from independent regional sea-level fluctuation.

Such cycles tend to multiply over tectonically "neutral" areas. In locations far within the shelf, or near its landward side where uplift is persistent, exposure surfaces are prominent and cycle members are incomplete through non-deposition. On the outer edges of shelves, subsidence is too continuous and water is too deep for the effect of sea level fluctuation to be reflected in the sedimentary record (Fig. X-1).

Carbonate shelf cycles usually belong to a general regressive earth history; they tend to multiply upward in a major stratigraphic sequence, becoming thinner, more restricted marine in character, and less regular.

Shelves with pure carbonate deposition are hardly affected by terrigenous influx and the variations in clastic sedimentation cannot be part of the cause for cyclicity in such strata. Nevertheless, several other interrelated controls operate to complicate the fundamental mechanism of relative sea level fluctuation and to

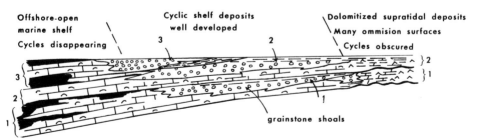

Fig.X-1. Three cyclic patterns from shelf to basin showing best developed cycles in the intermediate area of moderate subsidence

modify somewhat the basic cyclic patterns. In addition to local tectonism mentioned above, these include: changes in degree of restriction of water circulation over the shelf, tidal variations, over-all amount of shelf submergence, climatic changes, degree and frequency of periodic drops of sea level, and variation in bathymetric relief, i.e., height and steepness of carbonate platforms or shoals. Chapter II has outlined the causal mechanisms for cyclic sedimentation and refers to current works on cyclic sedimentation.

Observations of many such shelf carbonate cycles shows that three general but not completely distinctive types of upward shoal sedimentary cycles are present.

1. Oolite or grainstone cycles:

These contain a major grainstone body, commonly oolitic, at or near the top; in some examples lagoonal deposits occur above. Generally a hard ground surface occurs at the top.

2. Lime-mud cycles with essentially low-energy carbonate deposition throughout:

The top is commonly thin-bedded tidal flat carbonate or sabkha (salt-flat) evaporites.

Cycles 1 and 2 are formed on wide, shallow cratonic shelves and may grade into each other, the first type being more toward the open sea side of the shelf. Nevertheless, on a wide regional basis there is a tendency for the two distinct types to exist individually and examples of both are described below.

3. Platform interior cycles with intense early diagenesis:

Such cycles may resemble the other two types but commonly have coarse, poorly sorted Bahama-type peloidal, grapestone- and onkoid-bearing sediment and abundant fenestral fabric, druse veins, tepees, pisoids, and other strong imprints of vadose diagenesis. Large offshore banks or atolls in rigorous, seasonal, evaporitic

Table X-1. Outline of recurring environmental sequences in upward-shoaling carbonate cycles

Standard facies belts	Subenvironments	Phase of cycles
9	Marine hard ground or surface of subaerial exposure	Terminal phase
	Sabkha-pond evaporite	
8	Tide flat	Clear water carbonate-shoaling phase
	Lagoon and tidal channels and shelf mud mounds and low banks	
6–7	Immediate back reef shoals and tidal bars	
2 or 7	Widespread marine water of depth sufficient for circulation, at or approaching wave base	Normal marine, open circulation phase
	Below wave base, terrigenous influx < subsidence	Beginning terrigenous clastic phase
	Lagoonal – littoral, terrigenous influx > subsidence	

climates and with considerable relief and good underground drainage form special areas for development of this type of cycle. Both marine and meteoric vadose-phreatic alteration is possible.

Table X-1 outlines the idealized sequence of upward shoaling environments displayed by the above cycles. Few individual cycles follow this pattern completely but the schema is useful in predicting sequence once careful petrographic description has been made.

Oolite-Grainstone Cycles

Shoaling cycles with rounded, worn, coated, bioclastic grainstones and some well-formed oolite are especially common in Jurassic and Early Carboniferous strata of the northern hemisphere. Grainstone members of many other cycles lack well-developed oolite and have only coated particles, particularly in Devonian and Cretaceous strata.

Oolite Cycles on the Mississippian Shelf around the Williston Basin

Description of the cycles: The Williston basin and Montana area contains a 400–700 m thick Mississippian sequence consisting of the Lodgepole, Mission Canyon, and Charles formations of the Madison Group. These units together represent a major upward shoaling sequence in which the sea spread out of the shallow Williston basin over the Central Montana high and at first deposited dark argillaceous limestone alternating in several cyclic sequences with oolite. Above strata which indicate extensive inundation, the cycles become highly oolitic with a light-colored, peloidal unit at the top. In later Mississippian (Charles) time, the cyclic pattern continued to form as the sea retreated gradually from the shelves, confining its deposits to the center of the Williston basin. During this period, the upper part of each cycle became more evaporitic and finally salt was deposited in the basin center.

The Lodgepole-Mission Canyon cycles are best developed along the margins of the very large, shallow, marginal cratonic Williston basin in North Dakota, Saskatchewan, and its adjoining shelves on the west. A prominent east-west element on the Montana shelf west of the basin (Central Montana high) today coincides geographically with major Rocky Mountain uplifts exposing the shelf equivalents of the Williston basin strata. Very detailed stratigraphic work was done in these outcrops and in wells throughout the Williston basin in the two decades from 1950–1970 because of successful petroleum exploration. Both the regional stratigraphic framework and sedimentological details are well understood. The first good analysis of the facies pattern was by Edie (1958) and the most recent work is by Smith (1972).

More than 15 cycles can be traced in the Madison Group along the part of the Montana shelf exposed in the Big Snowy and the Little Belt Mountains uplifts of

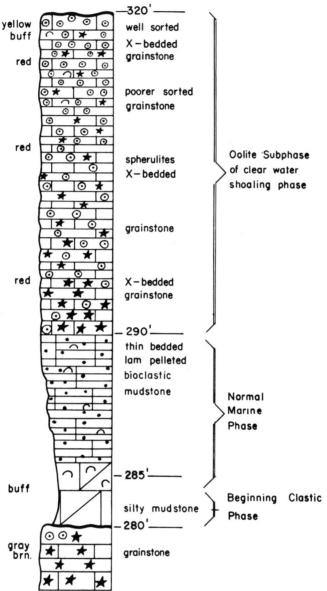

Fig. X-2. Upward shoaling cycle at Timber Creek, Big Snowy Mountains, Montana, in Mississippian Lodgepole Formation. Numbers indicate footage in measured section—about 13 m thick. Contacts between cycles are probably marine hard grounds because no evidence of subaerial cementation exists in the oolite

the Central Montana high and in the Bridger Mountains to the south. Figure X-2 illustrates a typical Lodgepole oolitic cycle. Smith's detailed petrographic work and tracing of the cycles across the Central Montana high from north to south adds regional knowledge to the facies pattern (Fig. X-3). He recognized five cycles in the formation, the lower of which (the Paine Member) in the Big Snowy and

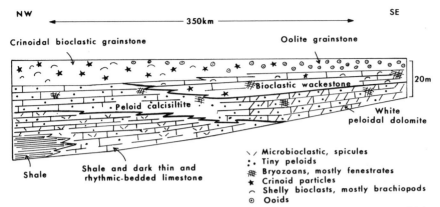

Fig. X-3. Idealized Lodgepole Formation cycle from Madison Group across central Montana from northwest near Great Falls to southeast. After Smith (1972)

Bridger Mountains contains Waulsortian bioherms surrounded by rhythmically bedded, dark, shaly "deeper water" limestone. The four typical, overlying cycles have been traced in wells about 350 km from north central Montana, near Great Falls, across the Central Montana high onto a more shoaling area in southern Montana (Fig. V-14). Each cycle has a lower member composed of fine-grained dolomitized unfossiliferous laminated mudstone and ripple cross-bedded calcisiltite, a peloid lime packstone, and a grainstone. These lower beds of the cycle are lighter in color to the south and more dolomitic in places. This grades upward through very fossiliferous wackestones often with bryozoans and brachiopod fragments to bioclastic crinoidal grainstone and/or cross-bedded oolite often with overpacked textures. The oolite-crinoidal grainstone beds cap the cycle. To the south, and across both the Bridger and Big Snowy Mountain outcrops, oolite is more common at the tops of the cycle, but to the north and northwest of the Central Montana high the cycles are capped more commonly by crinoidal bioclastic grainstones. Diagenetic studies show very early cementation but no evidence of vadose diagenesis or clearcut subaerial exposure (Jenks, 1972).

Interpretation of the cycles: Figure V-14 is a somewhat interpretative regional facies map of Lodgepole cyclic units around the western Williston basin and the Montana shelf. Repeated and apparently somewhat rapid marine invasions of the shelf area from northerly directions are responsible for the cycles. The lower member, of fine, rippled, pellet mudstone with few bioclastic fragments, is a shelfward equivalent of the basinal environment of black, argillaceous, bryozoan wackestone and siliceous mudstone. The cleanly washed oolite and crinoidal bioclastic grainstones capping each cycle must represent shoals and banks accreting during regressive phase off the shelves.

Middle Jurassic Cycles of the Paris Basin

In outcrops around the southeast Paris basin, Purser (1969, 1972) described very similar cycles to those of the Mississippian in Montana. These cycles correspond approximately to each of the three Jurassic stages of the Dogger: Bajocian, Bath-

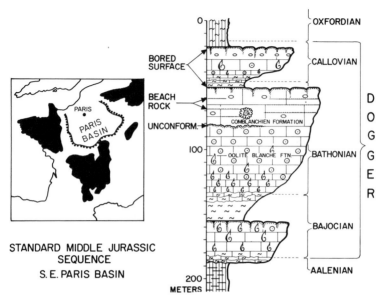

Fig. X-4. Standard Middle Jurassic sequence of the southeast Paris basin. Drafted at Laboratory of Comparative Sedimentology, Fisher Island, Miami, Florida—see also Figs. 5 through 8. Courtesy of B. Purser (1972)

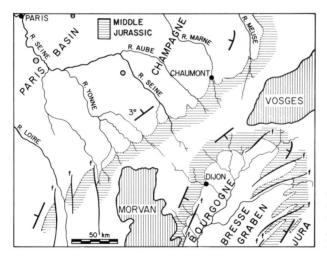

Fig. X-5. Index map of the southeast Paris basin. Circles in basin represent oil tests penetrating Middle Jurassic. Courtesy of B. Purser (1972)

onian, and Callovian (Fig. X-4). The cycles resemble those in certain Liassic beds of Lorraine and England (Klüpfel, 1917; Hallam, 1964). They are each a few tens of meters thick. A lower argillaceous, thin-bedded bioclastic unit with brachiopods and crinoids, and in places spiculitic and pelletal like the Lodgepole, is overlain by a thicker middle unit of strongly cross-bedded oolite consisting of well-formed, multiple-coated ooids. Foreset beds off sand bars up to 10 m high can be seen and festoon cross-bedding occurs with troughs some meters deep and tens of meters

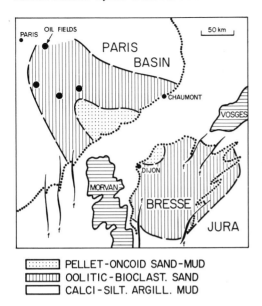

PELLET-ONCOID SAND-MUD
OOLITIC-BIOCLAST. SAND
CALCI-SILT. ARGILL. MUD

TOP OF CALLOVIAN CYCLE

Fig. X-6. General facies at upper surface of Callovian cycle outlining a low relief bank, southeast Paris basin. Courtesy of B. Purser (1972)

across. This impressive oolite is overlain by a thick-bedded, light-colored, peloidal grapestone unit, with mudstone to packstone texture, and with large algal onkoids. Hard grounds begin to appear in this upper unit and the top of each cycle consists of a widespread surface showing oxidized and blackened particles, large oysters plastered along it, and in places, evidence of early submarine cementation which has prohibited stylolitization and prevented later compaction. The oolitic and bioclastic grainstone of Callovian age may be traced around the outline of a large, low carbonate bank which existed over a weak structural high. The wide sheet-like grainstone distribution indicates outward diachronous migration of the shoal sands and construction of a wide flat bank with gently sloping ramps at the edges. It had no well-defined shelf margin and only 20 m of relief. It is about 200 km across. Figures. X-5 and X-6 are maps by B.H. Purser of the Jurassic outcrops and configuration and facies of this low buildup. Figures X-7 and X-8 diagram the facies through this buildup at a very high vertical exaggeration of 2500 X. Types of microfacies can be illustrated from traverses across this bank and in vertical sequences in any one of the cycles (Purser, 1972). They are:

1. Well-bedded, argillaceous, silty lime mudstone—offbank on the shelf.

2. Well-bedded bioclastic peloidal calcisiltite packstone-grainstone—offbank on the shelf. Standard microfacies types 2 and 9.

3. Bioclastic oolitic grainstone—just off the bar at the edge of the bank. Standard microfacies type 11.

4. Oolite grainstone—bars at bank edge. Standard microfacies type 15.

5. Bioclastic onkoidal coated particle packstone. Standard microfacies type 13.

6. Peloidal slightly bioclastic wackestone—bank interior. Standard microfacies 17 and 19.

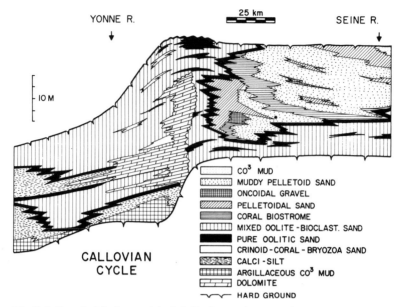

YONNE R.   25 km   SEINE R.

CALLOVIAN
CYCLE

CO³ MUD
MUDDY PELLETOID SAND
ONCOIDAL GRAVEL
PELLETOIDAL SAND
CORAL BIOSTROME
MIXED OOLITE-BIOCLAST. SAND
PURE OOLITIC SAND
CRINOID-CORAL-BRYOZOA SAND
CALCI-SILT
ARGILLACEOUS CO³ MUD
DOLOMITE
HARD GROUND

Fig. X-7. Detailed facies model of Callovian cycle across low relief carbonate bank. Section trends down outcrop belt on southeast side of Paris basin. The black zigzag belt merely outlines the vertical line pattern of oolitic-bioclastic sand. The pure oolite beds which occur as intervals within the former facies and are also marked in black. Fig. X-7 and X-8 show respectively the south and north sides of the bank and fit together, overlapping at the Seine River. The original relief was low (far less than $1/2$ degree slope) but sufficient to create a facies differentiation across the broad bank interior which progressively became more restricted during Callovian time. Vertical exaggeration 2500 X. Courtesy of B. Purser (1972)

Although the Callovian facies outlined by Purser circumscribe a somewhat local bank, the well-formed oolite of the cycles is much more regional and is widely exposed in the Dogger of the Jura Mountains of Switzerland and France as well as far north in England. No major reef belts are known to be equivalent to this extensive oolite, although coral patches and biostromes occur within it in places.

Late Jurassic Arab Zone Lime Sands of Arabia

Regional setting: Late Jurassic grainstone beds contain important oil reservoirs in the Middle East and currently produce annually at least one-quarter of the vast petroleum supply from that part of the world—principally from Saudi Arabia. The Arab zones (Kimmeridgian strata) form a remarkable cyclic series containing coarse lime sand and gravel spread across part of the Arabian shield.

This area lies across a tectonically neutral shelf between the eastern side of the Arabian Shield and a positive area, the Qatar-Surmeh high, which lay farther east, out in the Zagros miogeosyncline. Late Jurassic basins were situated north and south of this shelf. The Basrah basin in the northern Persian Gulf area was the site

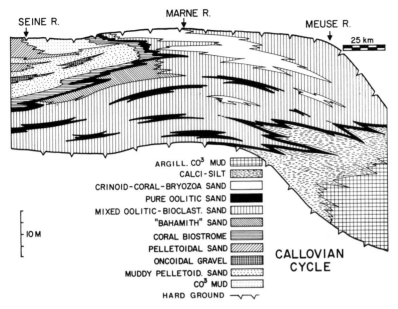

Fig. X-8. North side of low relief Callovian bank, southeast Paris basin. For additional explanation see Fig. X-7. Courtesy of B. Purser (1972)

of very considerable post-Jurassic subsidence. To the south lay the Rub al Kali basin, an immense but more shallow subsiding area which accumulated about 5000 m of post-Permian sediment (Fig. IX-2).

Late Jurassic beds were laid down during sporadic marine advances and retreats at the time of a major transgression over the periphery of the shield. Calcareous sandstone is seen in places along the edge of Late Jurassic outcrops in inland central Arabia, indicating proximity to shoreline west of the present Tuwaiq escarpment. Essentially, however, terrigenous sediments are absent across the Late Jurassic shelf and sediments are pure carbonate and evaporites, except in the two basins where dark, argillaceous limestone predominates. Tectonic movements intensified during the Late Jurassic and the basins became more strongly differentiated and probably starved of sediment in Oxfordian time. Basinal strata of this age are merely reduced sections of dark shale and limestone although shelf equivalents are transgressive, fossiliferous limestones. In latest Jurassic time (Kimmeridgian-Tithonian), perhaps owing to a change of climate, gradually increasing marine restriction resulted in the characteristic carbonate-evaporite Arab zone cycles over the shelf areas. Progressive evaporite conditions formed a terminal Jurassic anhydrite sheet (Hith Formation of Arabia) and final marine retreat into the basins formed salt deposits, a stratigraphic situation comparable to the Mississippian Charles Formation of the Williston basin.

Description of the cycles: There are about eight discernible cycles in the Late Jurassic of the Arabian Hasa coast, but four prominent ones, high in the sequence, have long been designated Arab zones A through D from top down. Each consists of a lower unit of bioclastic lime mudstone-wackestone with normal marine fossils (brachiopods, mollusks, echinoderms, foraminifers), a middle grainstone member of clean, washed calcarenite composed of rounded skeletal particles, in a few beds oolitic, and a thin, tidal flat, dolomitic mudstone grading within a few feet to a thick anhydrite bed. Figure X-9 is a lithologic cross-section of Arab D, the lowest and thickest of the cycles possessing a persistent and widespread anhydrite above it. Capping anhydrites of the several older cycles beneath Arab D are confined to the more basinal areas and cannot be recognized in the Jubaila Formation of the Saudi Arabia and Qatar shelf areas. A very detailed petrographic description of the Arab D cycle was published by Powers (1962) whose regional study shows it to consist in places of two cycles. The upper one consists basically of a lower, well-developed grainstone unit grading up to a tidal flat dolomite and thence to a thick anhydrite which forms the essential seal of the reservoir. The lower Arab D subcycle is more normal marine; its upper member is composed of dense dolomitic lime mudstone, no evaporite being present. This dense lime mudstone was used by Powers (1962) as a key marker. It is stratigraphically equivalent to upper Jubaila beds on outcrops. Below this dense micrite cap is a grainstone facies best developed along the northern Arabian Hasa Coast. The lowest beds of the lower subcycle are lime wackestones which merge imperceptibly into the open marine, dark gray, slightly bioclastic Jubaila or Darb Formation. Thus no clearly defined base exists for the Arab D cycle and it cannot be accurately isopached. It is generally less than 100 m thick.

Arab C cycle contains rock types similar to those of Arab D. It is also a double cycle south of the Qatar-Surmeh high and its lower half bears a thin capping anhydrite like that of its upper subcycle. Of particular interest in the cycle is the basal grainstone which is thin but persistent. The maximum development of this rock type in both subcycles coincides with the Hasa Coast, shifted slightly eastward from that of Arab D, and indicates a regression away from the Arabian shield out into the Rub al Kali basin and up the flanks of the Qatar-Surmeh high. The cycle is generally indicative of shallower, more restricted seas than is Arab D. More oolite is present. Little or no normal marine wackestone occurs. The trend toward more restricted and evaporitic conditions is continued in the overlying cycles B and A.

Geographic variations of the cycles: ARAMCO work (Steinecke et al., 1958; Powers, 1962) indicated that lime mudstone, dolomite, and anhydrite dominated Arab zone outcrops in the Arabian interior and graded downdip to lime sand in the subsurface along the Hasa Coast. The cross-section (Fig. X-9) trends down the coast from the Basrah basin of the northern Persian Gulf, crosses the major grainstone area of Saudi Arabia, and passes toward a tidal flat and dolomite facies underlying the Persian Gulf eastward from Qatar. Figure X-10 shows that the major grainstone development in the Arab D and C cycles occupies a passageway between two basins north and south and two positive areas east and west. It was a wide threshold area in which strong currents must have passed between the two basins and which lay between the shallower tidal flats on the positive elements. The more clearly defined shelf margin at the northern end of the grainstone area was apparently the site of the most active tidal exchange because only in wells located along this margin does any true oolite appear. The northern facies change from basin to oolitic shelf must occur within 35 km (Fig. X-9). Most of the shelf

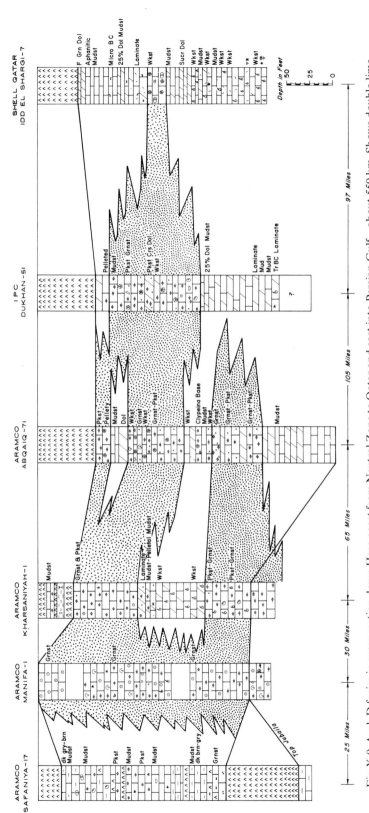

Fig. X-9. Arab D facies in cross section down Hasa coast from Neutral Zone to Qatar and east into Persian Gulf—about 550 km. Shows double lime-sand bodies in Arab D with northern sand buildup and its eastward disappearance into tidal flat dolomite at Id el Shargi

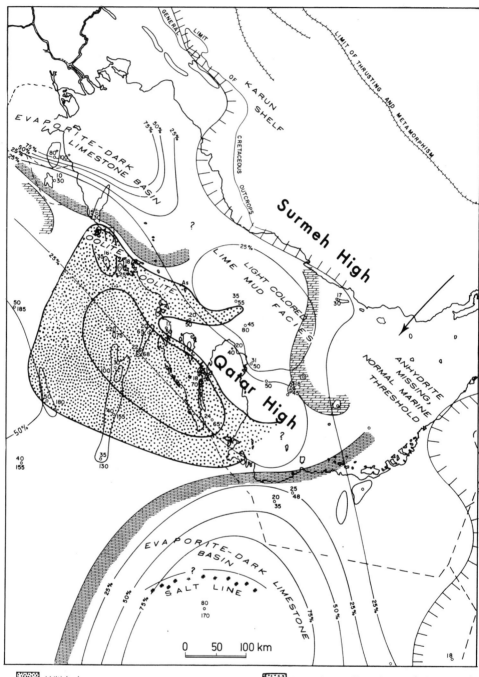

Within inner area, more than 50% carbonate is lime sand.

More than 25% carbonate is lime sand.

—25%— Evaporite percentage of total cycle

Well ⌐20⌐ Percentage anhydrite of total cycle
      └45┘ Thickness of anhydrite

Practically all carbonate is dolomite.

Edge line of dense brown lime mudstone facies

Thickness of cycle (including anhydrite) varies from 300–200 feet (91.5–61 m) thinning away from west central area.

grainstone consists of lightly cemented rounded and micritized skeletal fragments, many only identifiable now as peloids. Some of the sediment is very coarse. The biota is restricted to dasycladacean algae, rare foraminifers, algal onkoids, gastropods, and rare, worn fragments of normal marine organisms, such as brachiopods and crinoids. By analogy with Recent carbonate sands, such a grainstone sheet might be composed of a series of individual tidal bars but well spacing is not close enough to discern this.

Percentage lines on the facies map (Fig. X-10) indicate relative thickness of the capping anhydrite of Arab D Cycle. The evaporite cap is thinnest over the eastern edge of the lime sand buildup but may constitute about half the cycle westward over the Arabian shield. Eastward it thickens across the Qatar-Surmeh high and then thins, probably because of later dissolution. The thickening of the anhydrite toward the positive areas on each side of the grainstone sheet, and its development there in association with laminated tidal flat dolomite, affords evidence that it was developed as a sabkha deposit on the surrounding highs. In addition, the areas of thickest anhydrite along with interbedded salt (more than 75% of the total cycle) lie within the two basins. These beds also contain thin dark shale and carbonates. They change shelfward to lime grainstone through an intermediate facies of dark brown, more or less homogeneous lime mudstone. These strata have even planar laminations and contain scattered lenses of fine peloidal grainstone. Wells penetrate this facies in the Basrah basin. A few scattered wells in the Rub al Kali basin show dark lime mudstone of the Dijab-Darb formations, mainly along the Trucial Coast. Both east and west from the lime sand sheet the cycle changes to restricted marine carbonate facies: light-colored, dolomitic and chalky lime mudstone with sedimentary structures characteristic of tidal flats is found in wells in Qatar and the eastern Persian Gulf (Wood and Wolfe, 1969). As shorelines on old positive elements are approached, this section becomes thoroughly and finely dolomitized.

The Smackover Formation

The Arab zone facies have much in common with those of the Jurassic Smackover Formation, a major sedimentary cycle of Oxfordian age which surrounds the Gulf of Mexico and lies in the deep subsurface from Mississippi to Tamaulipas, Mexico. It crops out only in northern and east-central Mexico, where it is termed the Zuloaga Formation. It forms a 100 m thick belt of shelf oolite and grainstone about 30–80 km wide with a narrow slope facies of peloidal bioclastic wackestone. This belt grades gulfward into much thicker basinal facies (300–600 m). The distribution of facies and thickness indicates that an underlying tectonic hinge line, probably the edge of the eroded roots of the Ouachita Mountain chain, controlled this narrow grainstone strip.

◀ Fig. X-10. Facies map of Arab D cycle indicating evaporite and dark shaly limestone in northern Basrah basin and southern Rub al Kali basin. Oolite and grainstone buildup is in a shelf area between these basins and between the Arabic shield on west and Qatar—Surmeh high on east

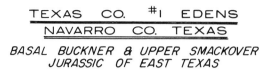

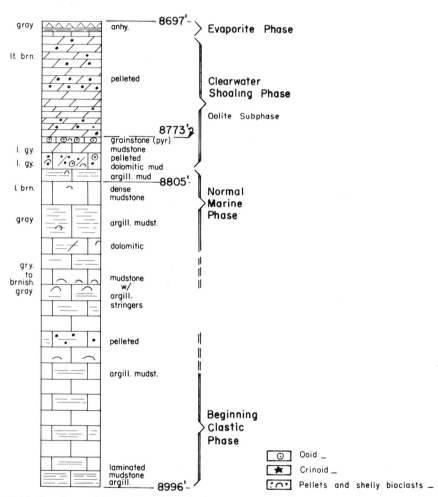

Fig. X-11. Smackover (Upper Jurassic) facies in east Texas well in a typical shoaling upward cycle with about 30 m (100 feet) of upper oolite and dolomitized peloidal grainy sediment and an evaporite cap. Lower 60 m is dark brown argillaceous bioclastic wackestone with brachiopods and crinoids. Depths on log are marked in feet

The very thick lower dark lime mudstone or wackestone of the gulfward Smackover much resembles the Diyab-Darb of Arabia. It contains the pelagic crinoid, *Saccocoma*, *Favreina* fecal pellets, and miscellaneous crinoid fragments. This dark limestone with the upper Smackover oolitic pellet grainstone and overlying tidal flat dolomitic mudstone and anhydrite (Buckner Formation)

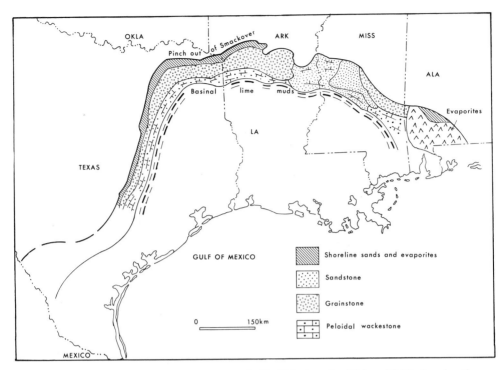

Fig. X-12. Upper Smackover facies, northern Gulf of Mexico after Bishop (1968), showing the peripheral grainstone buildup a few 100 m thick and about 75 km wide around the Gulf grading into very thick dark lime wackestone and basinal lime mudstone which reach more than 1000 m

forms a single sedimentary cycle which is identical to an Arab zone unit, but is much thicker (Fig. X-11). In Louisiana an overlying thinner cycle of shelf lime-stone is termed Haynesville A. The two cycles grade shoreward, as well as up-ward, into the tidal flat pellet mudstone and sabkha anhydrite. A shoreward sandy and redbed facies occurs against the edges of the old Paleozoic orogenic belt.

The upper Smackover limestones are more oolitic than those of the Arab zones. They were formed along a narrower belt and probably along a steeper slope. The sand and gravel grains are largely dasycladacean or other algal parti-cles, onkoids, and coated peloids. Shelly bioclasts are comparatively rare. A few wells in northern Louisiana have penetrated Smackover limestone with hydro-zoan and mollusk fragments but apparently backreef brines restricted normal marine faunas on the narrow shelf. As in the Arab Formation, no reefs are known from the Jurassic of the Gulf coast although the tectonic settings would seem ideal for their development. Figure X-12, based on Bishop's studies (1968, 1969), shows the facies variations along the Smackover belt. The facies outlined by Bishop in the upper Smackover of northern Louisiana are as follows:

Shoreward:
    Buckner shaly anhydrite, quartz, sand, peloidal oolitic packstone.

Crest of shelf margin:
> Oolitic coated pellet, lump-bearing well-sorted and partly cemented grainstone, the main reservoir facies. Pisolitic, onkoid, lump-bearing coarse-grained packstone with abundant codiacean and dasycladacean remains, few bioclasts (Plate X B).

Slope and basin:
> Peloid mudstone-wackestone with argillaceous partings grading down to brown lime-stone with few bioclasts, the typical lower Smackover lime mudstone and dark shale.

Extensive dolomitization has affected the shelf facies of the Smackover, partic-ularly in Texas, and is responsible for development of considerable porosity and permeability and good oil reservoir rock.

Very similar Jurassic cycles are also known from the Lias of Lorraine where they were first well described by Klüpfel in 1917. These have been discussed more recently by Hallam (Duff et al., 1967). The Helvetic nappes of central Switzerland contain thick Lower and Middle Cretaceous limestone whose sedimentary cycles of this type were described by Fichter in 1934. The upward shoaling pattern of one of these cycles was described in detail by Ziegler (1967).

Hard Grounds and Emersion Surfaces

Many oolitic grainstone cycles are capped by hard ground surfaces which are widely traceable and form an integral part of the cyclic history. These surfaces can be formed under both marine and subaerial conditions and represent secular stillstands in sedimentation or major regressions punctuating depositional his-tory. They are marked by numerous early diagenetic features which have been thoroughly described both in Europe and North America. Recognition of the significance of shallow marine hard ground surfaces stems from work by Shinn (1969) on Holocene beds in the Persian Gulf and also by application of this study to Jurassic strata in France (Purser, 1969, 1972).

Criteria indicative of lithification of hard ground under marine conditions include: isopachous druse or palisade intergranular cement, organic planation of carbonate surfaces by browsing invertebrates, surfaces pitted by echinoids, bor-ings in hardened sediment by pholad bivalves, sparry calcite geopetals in borings which have been scoured by later borers, oysters plastered on hardened rock, reworking of hardened pebbles from the surface, and common micritization of the upper centimeters of the surface by algal and bacterial action. In additon, below such marine hard grounds evidence of a pronounced slow-down of sedimentation may exist. More abundant burrows may appear or there may be more common vertical burrows which formed in the slowly hardening substrate. Concentration of glauconite, phosphatic nodules, and trace amounts of iron pyrites and man-ganese oxide are prevalent along such surfaces. Oxidation of the increased iron content on later weathering commonly results in a reddish zone marking the hard ground whether or not there was originally oxidation by subaerial exposure.

Emersion surfaces showing subaerial exposure occur at the tops of some shoaling cycles and are marked by certain petrographic characteristics controlled in part by climate. Sands may show evidence of pendent or meniscus cements if lithified in the vadose zone. On the other hand, intertidal splash zone areas form beach rock, aragonitic micrite coatings, and other features indicative of true marine diagenesis. In areas of intense periodic dryness carbonate sediment of all textures may develop fenestral fabric. Intertidal desiccation pans may develop evaporites, dolomite, and iron carbonate crusts and blackened breccia form around saline ponds.

In humid climates previously formed salina evaporites are commonly dissolved after subaerial exposure and form laterally extensive breccia. Intense rainfall results in microkarst and regoliths of red soil and collapse solution breccia in deep fissures. In semiarid climates with seasonal precipitation caliche surfaces may occur with breccia, pseudopeloids, vadose pisoids, reverse grading downward to micritic fabric, and downward convex crinkly laminae. Bacterial action may induce thorough micritization along these surfaces just as in marine hard grounds. Caliche micrite crusts may form on rootlets, root hairs, and algal and fungal tubules through continuous evaporation and transpiration.

Most of the above sedimentary features have been discussed in the glossary section of Chapter III.

Summary

All the above cycles and many similar examples have the following characteristics in common:
1. They generally occur on the parts of wide shelves marginal to basins. The prominent grainstone member must result from seaward progradation of the shelf margin environment (belt 6).
2. Well-developed oolite particles and coated, worn bioclastic grainstone dominate. The well-formed ooids are, by analogy with Holocene sediments, indicative of strong and regular tidal currents.
3. The cycle may grade upward and shelfward to restricted marine facies but in places a hard ground surface of non-deposition directly overlies the grainstone—i.e., a surface of lag deposition over a long time-period.
4. When restricted marine facies do occur at the top of the cycle they tend to be thick-bedded lagoonal muds with local hard grounds rather than thin-bedded intertidal and sabkha deposits.
5. No reefs are prominent on the seaward margin of the shelves. Such oolite cycles do not form behind a well-developed barrier reef which would restrict too much the flow of tidal currents.
6. Crinoidal limestone in places is important on the seaward side of the shelf. This is particularly true in Mississippian and Jurassic strata.
7. The cycles commonly constitute part of a megasequence which shoals upward; the cycles multiply and also become individually thinner and more restricted marine in character higher in the section.

Lime Mud-Sabkha Cycles

A second type of upward shoaling cycle is composed mostly of micritic sediment whose faunas and sedimentary structures show a progressive upward change through restricted marine carbonates to laminated evaporites of sabkha origin. In

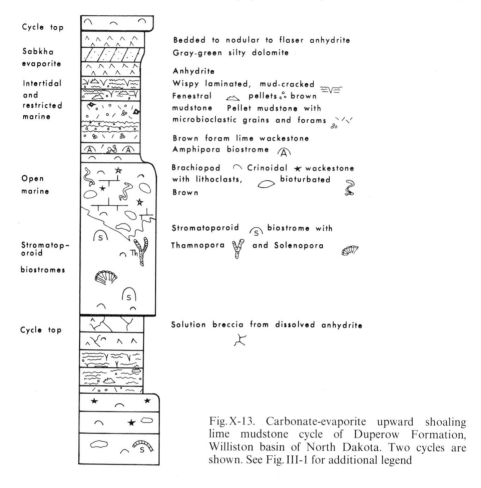

Fig. X-13. Carbonate-evaporite upward shoaling lime mudstone cycle of Duperow Formation, Williston basin of North Dakota. Two cycles are shown. See Fig. III-1 for additional legend

these cycles a sharp upper contact occurs with open marine beds at the base of the next cycle but evidence of erosion or prolonged subaerial exposure is rarely found. These cycles may be facies equivalent to oolite-grainstone cycles which lie on the seaward exteriors of wide shelves; the micritic facies (belts 8 and 9 of the ideal profile) usually lies more shoreward. The Arab zones of the eastern Persian Gulf illustrate this very well. Wood and Wolfe (1969) described in detail a sequence of tidal flat and sabkha evaporites from the Umm Shaif field, a sequence equivalent to the Arab zone grainstones of the Saudi Arabian mainland.

There are, however, extensive shelf areas covered by such cycles which almost completely lack grainstone of any type. Two of these are described below.

The Duperow (Devonian) of the Williston Basin

The author (Wilson, 1967b) studied the cyclic Duperow (Frasnian) sediments of the Williston basin in some detail. These strata formed in a vast backreef lagoon stretching southeast behind the Cooking Lake platform and Leduc reefs of Al-

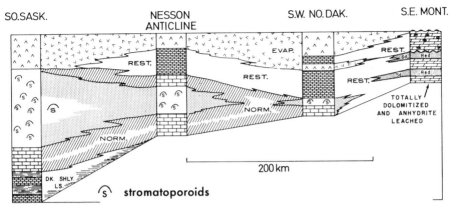

Fig.X-14. Interpreted Duperow facies in ideal cycle from north to south across Williston basin. Normal marine-burrowed bioclastic lime wackestone (oblique lines) and stromatoporoid patch reefs grade to restricted marine laminated lime mudstone with a microfauna (white area). This probably graded in southeastern Montana to sabkha evaporites which were leached from the most shelfward areas. From Wilson (1967b)

berta. The sea covering this shelf and shallow basin was more than 1000 km across in any direction. Sedimentological analysis of its sediments show it to have been generally very shallow. Boundaries of the Duperow cycles are marked by sabkha anhydrite and radioactive silty gray-green dolomite mudstone and about a dozen cycles may be traced from southern Alberta across the Williston basin of Saskatchewan and southern North Dakota using gamma ray-neutron logs. These beds are known in outcrops of the Jefferson Formation of the Montana shelf where the same cycles are identifiable.

Each cycle (Fig. X-13) consists of a lower member with two alternative phases; either dark brown, burrowed, lithoclastic-bioclastic brachiopod-crinoidal wackestone or a stromatoporoid biostrome with a few corals and red algae. In Saskatchewan a lower unit of dark shale occurs below these normal marine limestones. The middle part of each cycle is brown lime mudstone lacking megafauna and bearing a restricted marine or brackish microfauna of ostracods and calcispheres interbedded with laminated beds of peloidal or homogeneous lime mudstone. The cycle is capped by bedded irregular nodular and laminated anhydrite and gray-green silty, very fine-grained dolomite displaying intertidal and supratidal sedimentary structures. The only grainstone appears in a few thin beds above the stromatoporoid biostrome.

The Duperow cycles are exceedingly widespread and constituent beds only 3–5 m thick can be traced for several hundred km across the Williston basin (Fig. X-14, X-15). Deposition occurred in a vast back-reef lagoon south of the Alberta reef belt and stretched to a sandy shore in South Dakota and Wyoming. This very shallow basin was periodically and apparently rapidly flooded with marine water, permitting certain benthonic organisms to flourish and even patch reefs to grow at times. Gradual shallowing as sedimentation filled in the basin resulted in extensive tidal flats and evaporitic sabkhas; extensive dolomitization occurred on

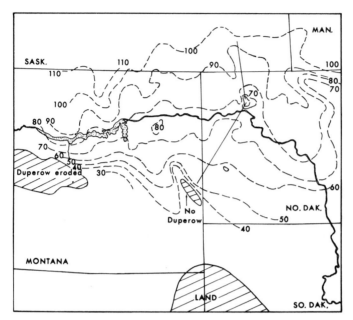

Fig. X-15. Isopachous map in feet of Duperow cycles IIIa and IIIb showing gradual thickening from Montana shelf into axis of Williston basin in North Dakota and Saskatchewan. Cross section of Fig. X-14 trends from southeastern Saskatchewan across Nesson anticline (closed 70 foot contour north of Missouri River) and through western North Dakota into southeastern Montana. After Wilson (1967b)

the peripheral shelves (Fig. X-16). The time of deposition of each cycle can be estimated from 500000 to a million or so years assuming a constant rate of sedimentation through the Late Devonian. Using the rate of rapid progradation of very similar Holocene sediment along the Trucial coast (20 km in 5000 years) and correcting for the thick Duperow cycles, it may be estimated that the 1000 km width of the Williston basin might have been filled in by lateral progradation during the half million to a million year span of a single cycle. That this repeatedly and regularly happened in a vast non-orogenic area and in sediments lacking evidence of erosion of previously deposited strata, suggests some climatic or world-wide eustatic mechanism as a cause.

Middle Permian of West Texas Shelves

Very similar cycles are known in restricted marine shelf and platform strata of the widespread Clear Fork and San Andres Formations of West Texas and New Mexico. These beds represent one of the major marine transgressions through southwestern North America, extending from central Texas far west to the Grand Canyon (Kaibab Limestone) and south of central New Mexico into Chihuahua (Concha Limestone). Extensive dolomitization has resulted in good porosity and permeability in the subsurface and more than half the oil in this vast province has

Fig. X-16. Generalized facies of a typical Duperow (Late Devonian) cycle in the Williston basin of North Dakota and southern Canada. Note stromatoporoid banks and brown normal marine wackestone facies to north and restricted marine lime mudstone and wackestone south of the heavy dashed line. Evaporite at top of cycle increases to the south in proportion to carbonate but suddenly disappears owing to subsequent solution. The exact configuration of the southern land area is unknown. Its presence is inferred from increased terrigenous content and complete dolomitization of the Duperow in the southern Montana-Wyoming area. After Wilson (1967 b)

come from these strata. Regional facies (Fig. X-17) are mapped by Meissner (1972) and sedimentary cycles were described by Meissner (1972), Chuber and Pusey (1972), and Lucia (1972). The latter paper on shoreline carbonate sedimentation offers useful descriptions and illustration of the Clear Fork cycles on the Central Basin platform. Table X-2 is from Lucia (1972, p. 169).

Summary

Numerous other examples of lime mud-sabkha cycles occur in shallow basins and adjoining shelves in North America: e.g., the Ordovician and Silurian Williston basin strata well described by Roehl (1967), the El Paso Ellenburger-Arbuckle groups of southwestern U.S.A. and the Glen Rose beds of the Texas craton. A

Table X-2. Sequence of sedimentary features in the upper Clearfork, Flanagan field, Texas (Lucia, 1972)

Interpreted sedimentary environment	Sedimentary structure	Fossils	Particle size
Supratidal (Facies Belt 8–9)	Irregular laminations, lithoclasts, desiccation features, quartz silt beds	*Rare* Thin-shelled small forams, ostracods, molluscs	Lithoclasts to lime mud
Intertidal (Belt 8)	Distinct burrows, churned-to-wispy, mottled structures. Quartz silt beds. Algal stromatolites. Discontinuous fractures	*Very few* Thin-shelled small forams, ostracods, molluscs. Filamentous algae	Fine sand-size pellets to lime mud
	Current-laminated rocks, cross-bedding	*Very few* Echinoids, small molluscs,	Fine sand-size pellets to mud with some lithoclasts
Infratidal (Belt 7)	Churned rocks, burrowed rocks	*Locally abundant* Echinoids, large fusulinids, molluscs, algal-forams (?) bryozoans	Coarse sand-size pellets to lime mud

sedimentologic study by Shinn et al. (1969) described the environments of such a cycle formed in the last 5000 years on the vast mud flats west of Andros Island, Bahamas. Fine sediment derived from the Bahama platform was brought eastward and trapped against the lee shore of Pleistocene eolianite on the island. The mud flat has prograded westward (toward the source of sediment) for 10–15 km at a thickness of 5 or 6 m in this brief period. The same process is clearly seen in the Persian Gulf lagoons around Qatar and the Trucial Coast where the prograding sequence is capped by sabkha anhydrite and gypsum.

Many great carbonate sequences of the geologic column consist of such cycles. They have the following special characteristics:

1. They tend to occur on wide shelves and across shallow intracratonic basins particularly where far removed from major shelf margins. In certain cases, however, major carbonate ramps extending into miogeosynclines may also accumulate such deposits.

2. Micritic textures are general throughout. The only common packstone or grainstones are found in thin channel and natural levee deposits.

3. The earliest beds in the cycles are generally open marine or partly restricted sediment. The upper part is always a tidal flat sequence whose sedimentary structures generally hold the key to environmental interpretation—specifically algal stromatolites marking the ancient level of high tide. Evaporitic climates commonly result in a cycle with a capping sabkha anhydrite sequence; tropical cli-

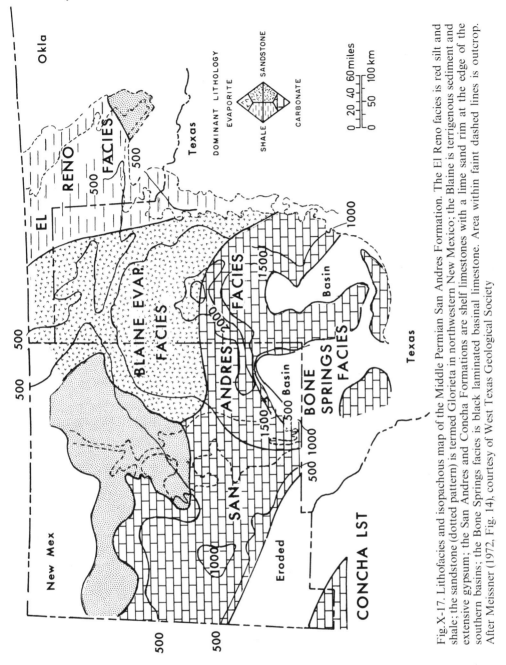

Fig. X-17. Lithofacies and isopachous map of the Middle Permian San Andres Formation. The El Reno facies is red silt and shale; the sandstone (dotted pattern) is termed Glorieta in northwestern New Mexico; the Blaine is terrigenous sediment and extensive gypsum; the San Andres and Concha Formations are shelf limestones with a lime sand rim at the edge of the southern basins; the Bone Springs facies is black laminated basinal limestone. Area within faint dashed lines is outcrop. After Meissner (1972, Fig. 14), courtesy of West Texas Geological Society

mates induce the formation of characteristic crusts and minor caliche surfaces at tops of the cycles.

4. Sharp lithologic contact occurs at the top but there is no evidence of downcutting channels or much prolonged or intense action of meteoric waters.

5. Dolomitization of such cyclic sequences is very common.

Platform Cycles with Intense Diagenesis

A third type of upward shoaling cycle is seen on major limestone banks or platforms where rapid carbonate buildup has created considerable bathymetric relief and where these buildups have been exposed to frequent sea-level fluctuation. Such cyclic beds were termed *Loferites* by Fischer (1964) who emphasized their fenestral fabrics. Some grainstone members of the cycles are very coarse and contain onkoids, coated particles and grapestone lumps. Such sediment may result from deposition and early diagenesis in warm hypersaline water of moderate circulation and from conditions of intermittent exposure. In general such sequences are distinguished from the wide shelf cycles not so much by dominance of certain textural types as by paleogeographic position and diagenesis. The impressive diagenetic effects could result from abrupt and sustained drops of sea level. At least in the Lofer example much of the sedimentary record is transgressive instead of regressive.

The opportunity for good drainage and splash zone cementation coupled with strongly seasonal periods of rainfall and aridity results in such spectacular alteration of these carbonates that certain sedimentary-diagenetic structures become hallmarks of this type of cycle. Two of these sequences are well-known backreef or bank interior strata of the Permo-Triassic banks of North America and the Alpine region discussed in Chapter VIII. Their great similarity substantiates the general analogy already made between these buildups.

Bank Interiors of the South Alpine Triassic

Assereto and Kendall (1971) illustrated spectacular exposures in the Val Seriana quarries in the Bergamasc Alps north of Bergamo, Italy. Here in cycles a few meters thick, burrowed bioclastic micrite with tiny tubules and gastropods constitutes a lower phase. A middle phase consists of gastropod, onkoid, tubules, and lithoclast-bearing wackestone, and an upper laminate fenestral dolomite mudstone exists. The lower unit is considered shallow lagoonal, the middle intertidal, and the upper supratidal, affected by rigorous early diagenesis. Lithoclasts, coated grains, pisoids, fenestrule (shrinkage pores), disarranged crusts, vadose sediment infills, and giant polygons ("tepees") in cross section which show mud cracking through many feet of sediment are characteristic. These features offer evidence that strong evaporative conditions intermittent with marine deposition caused frequent periods of drying out and abundant precipitation of calcium carbonate. Leaching as well as desiccation and precipitation of carbonates occurred under both marine and fresh-water vadose influence.

Bosellini and Rossi (1974) have also described similar cycles in the Dolomites, particularly in the atoll-like interior of the Latimar Mountain group. Here the lower subtidal phase consists of fossiliferous, burrowed, peloid and dasycladacean-rich strata; the middle intertidal and supratidal member contains fenestral stromatolitic structures with Foraminifera and ostracods, and an upper soil member includes red vadose pisoids, sheet cracks, and tepee structures.

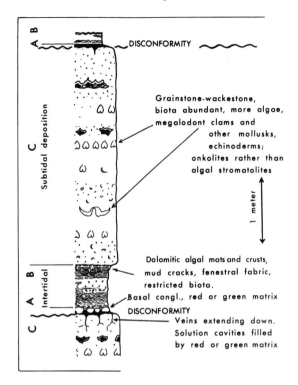

Fig. X-18. Diagrammatic representation of Lofer cyclothem, principally developed in near reef Dachstein Formation. *A*, basal, argillaceous member, representing reworked residue of weathered material (red or green), commonly confined to cavities in underlying limestone. *B*, intertidal member of "loferites" with algal mats and abundant desiccation features. *C*, subtidal "megalodont limestone" member, with cavities produced by desiccation and solution during succeeding drop in sea level. From Fischer (1964, Fig. 7)

The cyclic Lofer series of Sander (1936), Schwarzsacher (1948), and Fischer (1964) in the Dachstein limestone displays the same rock types in a sequence which varies somewhat depending on position in the regional facies. In the Northern Limestone Alps the Norian Dachstein facies with megalodonts and onkoids was deposited in large subsiding carbonate banks scattered within the Alpine trough. In central Austria northward and westward from Lofer towards Innsbruck, the bank facies is 1000–1500 m thick. In these directions the more bank-edge Dachstein facies passes stepwise into the more restricted Hauptdolomit which is distinguished from the former by the abundance of dolomitized intertidal and supratidal deposits and by less of the megalodont-bearing grainy limestone. The Hauptdolomit extended toward the Vindelician shoreline which lay north of the Alpine thrust belt and into the evaporites of the Keuper basin.

Fischer's important study (1964) of the Lofer facies resulted in the discernment of repetitive sequences such as that diagrammed in Fig. X-18. These are characteristic of immediate backreef or bank-edge deposits. The sequence consists of a conglomeratic boundary zone of red soil breccia marking a disconformity. Associated with this is fracture-filling marine sediment in places with nests of brachiopods and crinoid debris (A unit). A thin intertidal laminite (B unit) follows and grades in most places into an upper thick lagoonal grainstone-packstone member (C unit) with onkoids, coated particles of dasycladaceans, large megalodonts in place, gastropods, and variable amounts of micrite matrix (Plates IX B, XII B). The C unit has generally a sharp top showing erosion, minor karsting, fracturing,

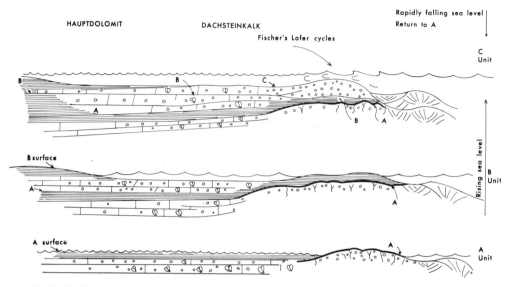

Fig.X-19. The formation of Hauptdolomit-Dachstein cycles, northern Limestone Alps of Austria. Possible explanation of Lofer cycles developed in Dachsteinkalk at bank edge position and their relation to bank interior Hauptdolomit cycles. The pattern is very idealized; the cycles may be very irregular because of numerous shoals and channels. Major deposition takes place near to reef during transgressive (high sea-level stand) phase during formation of unit C

brecciation, and soil formation. Many cycles cannot be traced more than a few hundred meters laterally indicating irregular topography on the surfaces of underlying C units. In some sequences a disconformity also occurs at the top of the laminated tidal flat B unit.

Sections farther away from the Dachstein shelf edge in the Hauptdolomit show persistent sedimentation without the erosional surfaces. The Hauptdolomit consists primarily of only two facies: (1) A light-colored thick-bedded, bioclastic, burrow-mottled, dolomitic wackestone with foraminifera such as *Angulodiscus* and *Permodiscus*, dasycladacean particles and scattered megalodonts in living positions. The unit may contain siliceous streaks and laminae. (2) These beds alternate with thinner units of tidal flat mm laminite, both planar and wavy (Plates XIA, XIIB). In places the laminae are torn into irregular crusts. The laminae are probably algal stromatolites but well-developed algal heads are rare. The unit is dark in places. Breccia beds are common, probably representing both residue of evaporite solution and some pebble conglomerates formed in shallow tidal channels. Fenestral fabric is not so well developed in the Hauptdolomit as in the Dachstein limestone which formed at the periphery of the bank. In some places pockets of stagnant marine water occurred on the platform and were filled with dark, highly bituminous laminated limestone with well-preserved fish remains, worms, and plant fragments (Müller-Jungbluth, 1968).

A plausible relationship between the Dachstein and Hauptdolomit alterations may be explained by a periodic rise and fall of sea level in the wide lagoon far behind the Dachstein reef rim (Fig. X-19). This interior area remained relatively undrained even at low stands of sea level when it became essentially a complex of tidal flats, sabkhas, and salinas. The best-developed diagenetic features, which mark the typical Lofer cycle, are seen in the better-drained, near backreef Dachsteinkalk, closer to the reef rim. This area was also more exposed to marine water of the splash zone.

Summary—Comparison with Permian Reef Complex

Combining Fischer's (1964) description of the *Loferites* with Dunham's observation in the Permian Reef Complex (1972) the following general diagenetic characteristics may be recognized. See also Smith (1974 b).

1. Common dolomitization (see following section).
2. Large scale development of aligned shrinkage pores, fenestral fabric or "birdseye" sparry blebs developed partly through desiccation in algal mat mudstones with or without *Girvanella* tubules, in peloidal or homogeneous lime mudstones.
3. Prism cracks—large scale desiccation polygons formed in muddy sediment or in hardened rock resulting in tepee structures formed through several meters of sediment. The process is clearly one of sediment expansion resulting in upward curling and breakage of edges of megapolygons. Vadose sediment may fill in the cracks.
4. Sheet cracks parallel to bedding, or oblique zebra cracks in host sediment. Such cracks, like the prism cracks, are generally filled with coarse calcite druse and/or internal sediment. Dunham believes this type of druse in the Permian Reef Complex may result from precipitation from meteoric water. Other researchers maintain a marine origin for this cement. Folk (1974) has pointed out that its coarse drusy crystal habit could indicate that its calcite mineralogy was pseudomorphed from aragonite. In the Dolomites many years ago bulbous and veinous forms of this cement were described as a fossil *Evinospongia*.
5. Neptunian dikes. Megafissures cutting down through many meters of strata and filled with sediment of either marine or meteoric water origin or an alternating combination of both. The origin of the dikes is complex. They may be in part tectonic as a result of a slump-fissuring, but may also represent karst solution along joints at edges of giant polygons (tepee structures). The oblique "zebra cracks" in sediment filling some of these may be caused by dewatering of fine lime-mud sediment, or possibly by slumping at the same time. Many of the dikes are filled with marine sediment bearing brachiopods in the Triassic and cephalopods and crinoids in both Triassic and Permian.
6. Red soil and caliche horizons are particularly common in the Dolomites and Permian Reef Complex. Breccias, peloidal fabric with reverse grading, and downward oriented crinkly lamination are common hallmarks of this structure.
7. Large coated particles have been interpreted as "vadose pisoids" in strata of the Permian Reef Complex. See Newell et al. (1953), Thomas (1965), Dunham (1969a), and Kendall (1969) for prevailing views. The downward elongation and fitting of pisoids, perched silt grains, drusy coating which envelop groups of individual balls, and smoothly curving laminae indicate that many of the particles are not original depositional entities but may be concretions. These are identical to some of the concentrically coated particles described as algal balls of *Sphaerocodium* in the Triassic of the Dolomites. Vadose zone caliche, marine-salt spray evaporation, concretions caused by fresh water seepage in coastal swamps, and cave-pearl formation have all been considered as alternative explanations for many concentrically coated lumps which have formerly been considered as onkoids.

True onkoids are also common features of these cyclic deposits. Kendall (1969) proposed a multistage origin for such particles in the Permian facies of onkoids, deposited in beds which are later additionally cemented by passage of vadose zone water which may be either of marine or meteoric origin. The process may have been reversed in places: concretions may have been eroded out, coated, and redeposited in a marine environment. To date no exact

TYPES OF CYCLES

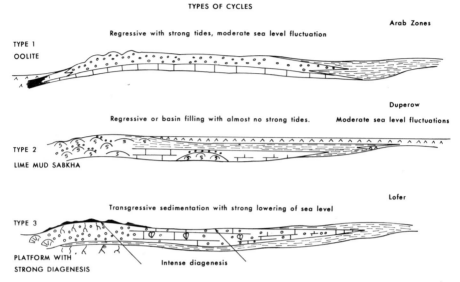

Fig. X-20. Hydrographic features controlling three types of shelf cycles Legend: brick pattern marks open marine limestone; open circles mark grainstone with indicated organisms; dashes indicate restricted marine limestone-dolomite

analog for these particles has been described in Holocene sediments. Whatever their precise origin, these large pisoids are characteristic of this environment of intense diagenesis.

Several factors could be responsible for the particular features listed above, all within a shallow marine and shoreline area in a strongly seasonal dry and humid climate with an overall record of more evaporation than precipitation:

1. Intense and sudden temperature changes and alternation of extreme dryness and torrential rains could cause expansion or contraction of the sediment.

2. Frequency, duration, and regularity of the processes could vary from several dry periods seasonally alternating with brief heavy rains to equal times of dryness and rainfall. The latter might result in solution, the former in more carbonate precipitation.

3. Substrate drainage enhances movements of fluids and karst development and is controlled by permeability of calcarenitic strata, elevation above the fresh-water table, amount of fractures within the substrate and frequency and extent of sea-level fluctuations.

4. The degree of access of marine saline water to the sediment while it is undergoing diagenesis by meteoric water may be important; amount of salt spray, intensity of storms, tidal fluctuations are considerations. Increased ionic activity of mixtures of salt and fresh water is probably important in speeding up diagenesis. Perhaps marine vadose and phreatic alteration is of equal importance to that by fresh water. Purser and Loreau's (1973) description of aragonite accretion and cementation in the splash zone of the Trucial coast is important in this context. Here periodic inundation by the sea and splash zones at the shoreline result in carbonate crystal growth within the sedimentary fabric.

Table X-3. Summary diagram of three cycle types

Cycle type	Lithic sequence	Environmental interpretations, climatic and hydrographic factors	Paleogeography
Shelf oolite	a. Lagoonal laminated mudstone or hardground over major oolite development b. Oolite bars, good cross-bedding c. Normal marine marl or wackestone at base with rare reefs	Tropical or arid climate, moderate circulation. High tidal range (at least 1 meter)	Wide shallow shelves around basin and wide, low, flat banks on such shelves
Shelf lime mud	a. Sabkha evaporites b. Laminated tidal-flat mudstones. Restricted marine lagoonal mudstones and wackestones c. Normal marine marl or wackestone	Evaporitic-arid or tropical climate, very poor marine circulation. Low tidal range	Wide shallow shelves around basins and terminal fillings of shallow basins
Platform with intense diagenesis, _Loferites_	a. Soil zones with caliche or sabkha-restricted marine-lagoonal to intertidal mudstones. b. Coarse lime sand, gravel, lumps, and onkoids. Little oolite or cross-bedded bioclastic sands c. Large heavy-shelled organisms in calcirudites d. Sand banks in wide marginal belts grading to bank interior muds	Strongly seasonable evaporitic climate, intermittent good circulation and very restricted circulation in lagoons; considerable sea-level fluctuation, abrupt subaerial exposure	Open-sea major banks or exposed platforms or ramps with considerable bathymetric relief. Good drainage during low sea-level stands and strongly fluctuating water tables

The environment reconstructed for _Loferite_ cycles is one of large, steep-sided well-drained porous carbonate platforms or major offshore banks standing high above sea level at repeated periods during their early history. Dunham's (1972) indication of large-scale sea-level fluctuations in the Permian Reef Complex becomes highly significant in terms of backreef diagenetic history.

The use of Fischer's term _Loferite_ for such characteristic strata is to expand its meaning beyond purely shrinkage pore or fenestral fabric, but this is a convenient and recommended term for such cycles. Similar strong diagenetic imprint on restricted marine sediment has been observed in Early Mississippian beds in the Williston basin and in some Cretaceous bank deposits in Mexico, but neither of these examples shows the degree of extreme diagenesis of the Permo-Triassic strata. Table X-3 and Fig. X-20 contrast certain lithologic characteristics of the platform _Loferite_ cycles with those composed largely of shelf oolite or tidal flat muds.

Dolomitization of Carbonate Banks and Interior Shelf Cycles

Origin of Dolomite

Dolomite rock is very common in the geologic record but American geologists have argued about its genesis, at least since 1911–1917, dates which marked the publication of Steidtmann's and Van Tuyl's papers on its origin.

Stratigraphic and petrographic evidence show that most dolomite (dolostone) in the geologic record results from a replacement process in which about half of the Ca in a $CaCO_3$ molecule is substituted by Mg, forming $CaMg (CO_3)_2$. The end result is a rock composed of a mosaic of rhombs varying from 5 to 250 microns in which original fabric is partly or completely obliterated. The mosaic may be loose-fitting or dense. Vuggy moulds of originally dense and resistant calcite particles may be present.

The kinetics of the process are slow, offering one explanation for the lack of dolomite on the sea floor and for its rarity in Holocene sediments as compared with its abundance in the geologic record. The conditions necessary for dolomitization are (1) a sufficiently porous and permeable calcareous sediment to act as host for the Mg replacement; (2) a fluid of the correct chemical composition to react, capable of dissolving $CaCO_3$ and releasing Mg; and (3) a long-enduring supply of Mg and (4) a hydrodynamic head to force great volumes of water through the sediment. This is important because all natural waters have a relatively low content of dissolved salts.

Several processes probably produce the dolomite observed in the geologic record. Two genetic types are readily distinguished in the field: (1) stratigraphically controlled, early diagenetic dolomite, and (2) coarse dolomite precipitated hydrothermally along faults and veins. A variety of dolomite-forming situations are known (Table X-4) and distinction between them may not be easy. Combinations of processes operating over long periods of time undoubtedly account for many complex relationships and inhibit clear-cut and consistent generalizations supported by field and petrographic observations.

Important advances in our understanding of dolomite genesis have been made from recent studies of Holocene dolomite (Table X-4). Three of the processes observed in the Holocene and Pleistocene are clearly associated with emergent platforms and their attendant shelf cycles and probably constitute the most important mechanisms for forming stratigraphically controlled dolomite. They are discussed briefly below, followed by discussion of the stratigraphic evidence bearing on environmental conditions and timing of the dolomitization process.

Adams and Rhodes (1960), Deffeyes et al. (1965) and Illing et al. (1965) provided a dolomitization theory based on the development of Mg-rich brines through evaporation. The first two groups of authors proposed that dense saline brines whose Mg/Ca ratios had been raised by loss of Ca through evaporative precipitation of gypsum and anhydrite in tidal flats, ponds, and supratidal areas (sabkhas) migrated regularly down through lime sediment and dolomitized it (evaporative reflux). Subsequent study of evaporative flats in other carbonate areas has indicated that Mg enrichment of brines might also occur when inflowing sea water repeatedly dissolves Mg calcite in earlier-formed carbonate sedi-

Table X-4. Genetic classification of dolomite formation (modified from Badiozamani, 1973)

Dolomite rock

Primary precipitation from standing body of water

Hypersaline lakes

Coroong, S. Australia (Alderman and Skinner, 1957)
Deep Springs Lake, California (Clayton et al., 1968)
Baffin Bay (Behrens and Land, 1972)

Early diagenetic: replacement before cementation of calcitic matrix

After or during emergence from the sea

Concentration of Mg by re-solution of Mg calcite in marine sediment by ground water and sea water mixing (Bathurst, 1971)

Migrating ground water mixing with marine water "Dorag" (Badiozamani, 1973)

Jamaica reefs (Land, 1972)

Recognized

Processes

Most common processes Mg derived from sea water

Evaporite precipitation in supratidal sediment

Dolomite crusts on Andros Isl. (Shinn et al., 1965).
Evaporative pumping (Hsu and Siegenthaler, 1968) and/or seepage refluxion (Illing et al., 1965) on Persian Gulf shores

Before emergence

Under a body of saline water

At sediment-water interface and in sediment

Within a rock column

Seepage reflux?

Adams and Rhodes (1960)

Coroong (v.d. Borsch, 1965)
Bonaire (Deffeyes et al., 1965)

Late diagenetic: replacement after cementation of calcitic matrix

Secular migration of connate waters?

Hydrothermal in veins, fractures, collapsed areas

Meta-morphism

ment and mixes with earlier-entrapped brines and salts saturating the bottom sediment (Bathurst, 1971). Under such conditions there would be a continual addition of the more soluble Mg salts to the brines and a continual loss of the Ca by organic or chemical precipitation of $CaCO_3$, steadily increasing the Mg/Ca ratio. The eventual seepage of such brine through sediment underlying supratidal areas or below the bottoms of shallow ephemeral lakes and ponds would induce dolomitization.

The necessary movement of large amounts of fluid for dolomitization in tidal flats poses a problem. The hydrodynamic head set up by the formation of dense evaporative surface brines over permeable sediment saturated with normal marine water has been considered sufficient to move the required volumes of water. But such water must move downward through relatively impermeable sabkha sediment in a topographically flat area where no relief is present to help form a hydrodynamic head. To date, field evidence has not demonstrated a great amount of downward movement of such water although the natural laboratory provides sediments deposited and altered by dolomitization only in the last few thousand years. Field evidence on Persian Gulf sabkhas does seem to indicate a mechanism for creating thin widespread stratigraphically controlled dolomite beds associated with intertidal sediments.

A good Holocene example of dolomitization, principally by refluxing evaporative brines through intertidal to subtidal sediments from supratidal areas, has yet to be described. However, the reflux mechanism can be inferred in several natural evaporating lakes and bays where gypsum is being abundantly deposited and where the more soluble and abundant NaCl and Mg salts must be escaping (Deffeyes et al. 1965).

A theory based on reverse water movement, upward pumping by evaporation of sea water on sabkhas has been proposed by Hsu and Siegenthaler (1969) to explain Holocene dolomite occurrences on evaporative flats in the Persian Gulf. Gypsum and dolomite replacements down to 1.5 m below sabkha surfaces have been observed in the Persian Gulf. Dolomite in Caribbean tropical regions also occurs on supratidal areas but is confined to thin hard surface or near-surface crusts, only a few cm thick. The dolomitization has taken place in the last 3000 years. Evaporative pumping offers a reasonable explanation for replacement dolomite associated solely with intertidal-supratidal sediments.

Another process of dolomitization related to persistently emergent areas has been proposed (Hanshaw et al. 1971; Badiozamani, 1973; Land, 1973a,b). It has been pointed out by Runnels (1969) and Matthews (1971) that fresh phreatic water with only a minor amount of Mg, when combined with marine water, forms a fluid which may be undersaturated with respect to calcite. The solubility of calcite is greatest with intermediate ionic strength.

Saturation with respect to dolomite increases continuously with increasing sea water added to phreatic water. Badiozamani (1973) calculated that in brackish water "in the range of 5–30% sea water, the solution is undersaturated with respect to calcite and many times supersaturated with respect to dolomite." The model proposed necessitates a continuous supply of Mg derived from sea water and mixing with meteoric water during constant fluctuations of sea level. During emergence, the interface where the phreatic lens of fresh water impinges on underlying marine or saline connate water would be a dolomitizing zone; this front could pass through a considerable thickness of sediment as sea level drops. The same conditions would occur during marine transgression.

Some additional Mg may become dissolved in ground water from Mg calcite in freshly deposited marine sediment on shelves which periodically emerged and were subjected to tropical rainfall. Wide, flat, low platforms such as Florida and Campeche might have undergone many episodes of drowning, deposition of a relatively thin sequence of marine sediment, emergence, and erosion without having left a trace in the geologic record. The thickness of preserved sediment even on great platforms of major subsidence is very limited when modern rates of accumulation of shallow-water calcium carbonate are considered. Goodell and Garman (1969) pointed out that, based on rates of modern Bahaman-type sedimentation, the Bahama Banks should have been the site of eight times as much shallow-water carbonate accumulation in the last 120 million years as they possess. This argues for persistently intermittent sedimentation, exposure, and perhaps cannibalization of carbonate sediments as they are deposited. During a general period of carbonate bank sedimentation there could have been many minor transgressions and regressions.

The theory of dolomitization through mixing of meteoric marine water does not necessitate a strong evaporative climate to concentrate Mg in interstitial brines by precipitation of $CaSO_4$. It rather utilizes a climate of tropical rainfall to provide a fluctuating lens of fresh water overlying saline brines as seen today in Yucatan and Florida.

Both evaporative reflux and the above model of a migrating front of brackish water (Dorag dolomitization of Badiozamani, 1973) require the flowage of ground water through supratidal coastal sediment and the dolomitizing of permeable sediment deposited in more normal marine conditions. Both models require a relatively positive area (mainland or island) to generate the particular type of water required and to furnish a head for the great volume of migrating fluid. In both situations extensive dolomitization through thick sections may be aided by the tendency toward prograding carbonate sedimentation—the formation of shoaling upward or fill-in sedimentation. Seaward building of land surfaces over poorly consolidated marine sediment also brings progradation of near-surface diagenetic environments to alter the underlying carbonate sediment.

Petrographic Evidence of Time of Dolomitization

A common problem in dolomitization is not only the origin of the fluid and its movement but its timing as well; some petrographic observations have important bearing on this. In some instances they indicate that dolomitization was not necessarily penecontemporaneous with sedimentation although its map patterns show it to be persistently associated with former shelves and positive areas, which could have supplied the proper water. Zenger (1972a,b) has pointed out that many stratigraphically controlled dolomites contain sedimentary and organic structures indicative of the supratidal-intertidal environments in which Holocene dolomite is now forming and in which the model of penecontemporaneous sabkha dolomitization may be reasonably applied. But he has also thoroughly dem-onstrated that there is fully as much dolomite, whose petrographic character indicates deposition in completely marine condition instead of in supratidal eva-poritic areas.

It is interesting (and somewhat puzzling) that petrographic observations show much regional dolomitization, even in originally open marine sediments, to have been relatively early in the diagenetic history of the sediment. Some control on amounts and location of dolomite is commonly evidenced by differences in per-meability of the original fabric. In many rocks showing partial replacement, micritic matrix, micritic peloids, aragonitic bioclasts, and dense calcitic shells are dolomitized, in that order. Presumably this sequence is based on ease of solubility and the accessibility to fluid of the replaced fabric. In sediments demonstrating such progressive dolomitization the process must have occurred before complete calcitic cementation. Its end product is commonly porous, fossil-moldic, sucrose-textured dolomite with all the calcite removed (Murray, 1960).

In other layers and in the same sections, however, may exist grainy beds of finely crystalline, dense dolomite, interbedded with partially dolomitized lime-stone and with beds of pure sucrose dolomite. In many such beds all types of fragments and matrix are replaced, preserving the most delicate internal shell structures as well as the micrite with tiny dolomite crystals. Such beds must have been more or less thoroughly cemented before dolomitization since fabric per-meability exercised no control on the degree of dolomitization. Chert nodules, which preserve delicate fossils incorporated in sucrose porous dolomite which has obliterated all unsilicified bioclasts, show that much dolomitization was late enough to have been preceded by the first stages of silicification.

In summary, petrographic evidence often indicates a complex history of multi-stage dolomitization in which the process may vary in time relative to other diagenetic events such as anhydrite replacement, silicification, and calcitic cementa-tion. Thus, in many instances petrographic evidence on timing of dolomitization is equivocal and contradictory, indicating replacement probably occurring over long time periods or during periods of rapidly changing condition.

The Stratigraphy of Dolomite

One way of ascertaining when dolomitization occurred, and what was its origin, is by its lateral and vertical stratigraphic patterns. Stratigraphic studies show gener-ally that shelves or positive areas are preferentially dolomitized. Many, but not all, had evaporites on their landward or interior sides. Sediments in all three types of shoaling cycles may be dolomitized. Observations also indicate that in some examples, despite its early occurrence, the process was not necessarily penecon-temporaneous with sedimentation.

Shelves are preferentially dolomitized compared to the basins which they bor-der: Examples are manifold in the geologic record: The Middle Permian of West Texas (Galley, 1958, p. 431, Fig. 31), the Madison Group of the Williston basin (Sloss et al., 1960, p. 28), the Red River Ordovician of the Williston basin (Fig. X-21), the Silurian of the shelves around the Michigan and Appalachian basins

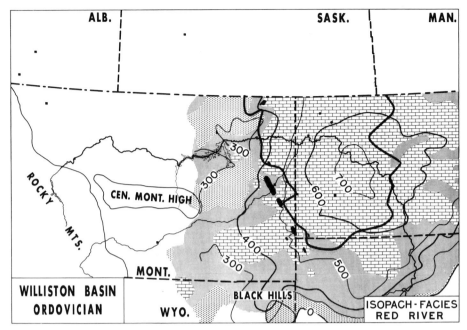

Fig. X-21. Dolomite and limestone facies of the Red River (Ordovician) Formation, Williston basin. Isopach lines in feet. Coarse dot pattern marks 100% dolomite; fine dots 50–100% dolomite, and brick pattern mostly limestone. Center of basin indicated by closed 600 and 700 foot contours and by heavy line which encloses discrete anhydrite beds in upper part of formation. The wide shelf peripheral to the basin is preferentially dolomitized; the formation is termed the Big Horn or Whitewood Dolomite south and west of the basin. Heavy black areas mark petroleum accumulations

(Alling and Briggs, 1961, Fig. 10), the Gatesburg-Conococheague facies from the Adirondack axis into the Appalachian miogeosyncline (Wilson, 1952, p. 319, Fig. 4), the Shuaiba limestone at the top of the Thamama Group around the fringes of its outcrop on the Arabian Shield, the Jurassic Arab Formation of the Persian Gulf (Chapter IX) and the Edwards Formation of the Comanche platform and San Marcos arch, central Texas (Fisher and Rodda, 1969, Fig. 7; Rose, 1972, p. 57, 58). Berry and Boucot (1971) demonstrated that regionally North American Silurian limestone surrounds the shelf area of the craton whose Silurian strata are almost totally dolomitized. The Late Jurassic *Gigas* beds of the north German subsurface were laid down along a narrow shelf south of the Pompeckj high and were thoroughly dolomitized shortly after deposition. They pass southwards into limestone of the Malm evaporite basin under the north German plain (Schmidt, 1965, p. 136, Fig. 12). Many other examples could be cited. In most of those noted above, strata representing evaporite or carbonate tidal flat lagoons with restricted circulation occur in the upper parts of the units.

Where cyclic sequences possess an anhydrite member, dolomite may be closely associated in vertical or lateral sequence: Accepting the idea of downward reflux of brines from evaporitic sabkha surfaces, one would expect that carbonates immediately below the evaporite members of cycles would be the more thoroughly dolomitized. This is commonly, but not always, so. Differences in original texture and in degree of cementation at the time of dolomitization in some instances caused differences in permeability which in turn controlled susceptibility to replacement. In the late Devonian Duperow Formation of the Williston basin, beds immediately *above* the anhydrite in the cycle are dolomitized for a meter or less. This perhaps indicates that the dolomitizing fluid from the evaporite itself was forced up into the overlying bed when the sulfate layer compacted. Examples cited in this chapter of a close vertical and lateral association of dolomite and anhydrite cycles include the Duperow formation, the Arab zones of the Persian Gulf, and the San Andres Middle Permian of West Texas, the Triassic Keuper and Hauptdolomite of central Europe.

A significant large-scale relationship between dolomite and evaporite deposition is seen in the regional distribution of the Middle Pennsylvanian in the western United States. The Desmoinesian is dolomitic only where it surrounds the three evaporitic basins of this age (northern Denver-Julesburg, central Colorado, and Paradox basins). Pennsylvanian strata surrounding other positive areas in the southern part of the North American craton are nondolomitic.

Some early diagenetic dolomitization is later than the depositional cycles in whose strata it is found: Conceivably, a period of tectonic subsidence of a basin resulting in relative uplift of its edges shortly after an episode of deposition would induce movement of fluids through the shelves and into the basin center. Such a history would permit dolomitization by the mixture of meteoric, connate, and marine water.

Examples of "somewhat later" dolomitization vary from broad regional patterns around shallow basins to areas of individual carbonate banks.

1. Wilson (1967b) demonstrated that the Devonian Duperow dolomitization pattern followed that of the Devono-Mississippian uplift and peripheral erosion around the Williston basin rather than precisely the depositional facies and thickness patterns of Duperow sedimentation.

2. Whereas dolomite is roughly peripheral to the Williston basin in the middle Madison Group and confined to the shelf facies belts, it is also highly developed in north-central and western Montana and cross-cuts isopach lines and facies belts in that area (Sun River Dolomite) (Fig. II-8). Here it must be post-depositional.

3. In the Dolomites of northern Italy, a few of the great Middle Triassic banks with central lagoonal and intertidal facies are not very dolomitized. The exceptions are those overlain by thicker sequences of argillaceous Raibl or buried by volcanic sediments instead of Raibl beds in the normal evaporitic facies. The inference is that the time of dolomitization followed the construction of the great Ladinian banks and was related to the subsequent evaporitic shelf deposits of the overlying Upper Triassic Raibl and Hauptdolomit (Leonardi, 1967).

4. Where basinal carbonate buildups are buried in later evaporites they are commonly dolomitized but the reason for this is not entirely clear. Examples include the pinnacle reefs of Michigan basin and the Zama-Rainbow area of northern Alberta, Canada (Chapter IV). Detailed subsurface correlations, using petrophysical logs, indicates that the evaporitic cycles between the buildups were formed during periods of somewhat lowered sea level and at times when at least the tops of the buildups were exposed subaerially. It remains an open question as to whether later fluid migration laterally from the overlying and surrounding evaporites, or ground-water movement at times when the pinnacles were exposed was responsible for dolomitization. Jodry (1969) argued that connate water derived from compaction of surrounding evaporites and limestones flowed upward through the Michigan basin pinnacle reefs and dolomitized them.

5. In several instances where carbonate buildups on shelf margins are well-dolomitized, there is little or no dolomitization of forereef talus blocks derived from the bordering reef or forereef. This clearly indicates dolomitization at a somewhat later period than reef growth. Examples are: the Canning basin foreslope exotic boulders of Devonian age, northwest Australia; the Miette-Ancient Wall debris flows (Late Devonian) in the Canadian Rockies (Chapter IV); the Triassic Cipit blocks in the Dolomites and the lower talus blocks of the Permian Reef Complex (Chapter VIII). Adams and Rhodes (1960) proposed a theory of penecontemporaneous reflux dolomitization for Guadalupian strata; it is also possible that dolomitization of the shelves surrounding the Delaware basin occurred toward the end of Permian time when the overlying Salado evaporites were extensively deposited beyond the confines of the basin. Dolomitization also affected the coarse and once permeable talus in the middle foreslope of the shelf margin rather than the micritic (less permeable?) sponge-algal facies of the upper slope.

A related example is found in the Ordovician exotic boulders in the Marathon fold belt of West Texas (Wilson, 1954; McBride, 1969; Young, 1970). The blocks derived from the Lower Ordovician Ellenburger group are practically all limestone whereas the environmentally equivalent shelf facies is highly dolomitized. The best interpretation is that Ellenburger dolomitization occurred after the boulders were emplaced (i.e., in late Canadian and Middle Ordovician time during the formation of the widespread North American pre-Simpson unconformity).

6. A consideration of dolomitization during geologic time also permits the inference that "somewhat later" diagenetic processes are at work to cause it. It is well known (Chilingar 1956) that Paleozoic and Precambrian carbonates are more dolomitized than those of the Mesozoic and later time and that the geologic record contains much more dolomite than can be accounted for when compared with what is forming today. Inasmuch as there is not much evidence that Paleozoic tidal flats were more extensive over the world than in the Mesozoic, one may assume that Paleozoic limestones have simply had more time for dolomitizing waters to pass through them—that much dolomitization is *not* penecontemporaneous but due to continuous processes operating over vast periods of time. These lines of argument obviously favor the hypothesis of ground-water or connate-water movement as the mechanism rather than refluxing or upward pumping of penecontemporaneously formed evaporitic brines.

Extensive shelves around the Alberta basin of Canada provided sites of dolomitizing fluids which migrated basinward: The Late Devonian banks of the Alberta basin are extensively dolomitized except where more or less isolated by shales and farthest removed from the platforms. Note the presence of limestone in Golden Spike, Redwater, and Swan Hills banks on Fig. IV-20. In the strata exposed in the mountains of Alberta there is not much evidence of penecontemporaneously deposited shelf evaporite whose formation may have produced the dolomitizing fluids. Slight evaporitic conditions occurred later at a time of regional shallowing and silt deposition (Alexo Formation). There is, however, abundant evaporite in the Alberta shelf south of the basin both contemporaneous with the Leduc reefs (Duperow Formation) and also later in the Devonian (Winterburn Group and Stettler beds).

There exist exceptions to all the above generalizations of stratigraphic relationships of dolomite and limestone. The chief reason for this may be the variations in timing between dolomitization and calcite cementation which inhibited permeability and early diagenesis. Often grainstones are preserved as limestone because early cementation rendered them less permeable than lime muds. Furthermore, subsequent diagenetic history may to an important extent overprint early, more regional patterns of dolomitization. Local leaching of calcite from partly dolomitized zones along faults, later episodes of secondary dolomitization, epigenetic dolomite formed along veins and fractures, preferential stylolitic solution of calcite, and dedolomitization by waters enriched in $CaSO_4$ all are late diagenetic processes modifying the results of earlier and more stratigraphically controlled dolomitization.

Conclusions

Stratigraphic evidence favors dolomitization associated with platforms and positive areas. These could have been preferred locations for at least two processes logically causing the replacement by migrating fluids: (1) density reflux of high Mg/Ca brines from sabkha evaporation which could have occurred penecontemporaneously, and (2) movement down a hydrologic gradient of a phreatic water lens mixing with connate water or purely marine water along old coast lines. This is a post-depositional process. The common cyclicity of upward shoaling or fill-in cycles of shelf strata offers evidence for repeated fluctuations of relative sea level to aid in either or both these processes. Study of distribution of dolomitized banks and reefs in basins shows that in some areas fluid migration from shelves was important and that dolomitization in other basins may have been associated with sea-level lowering. In several examples of regional dolomitization it appears to be postdepositional rather than penecontemporaneous and to have operated through fluid migration during tectonic adjustment.

Chapter XI

The Rise of Rudists; Middle Cretaceous Facies in Mexico and the Middle East

Geologists have long recognized the Cretaceous as a time of outstanding development of carbonate platforms and offshore banks. These are known in an equatorial belt from 40° N to 20° S along the Tethyan seaway through southern Europe, in the Middle East, across southern Asia and the sunken Mid-Pacific and around the Gulf of Mexico-Caribbean region. This Chapter focuses particularly on Middle Cretaceous strata in these areas and on the role played in carbonate accumulation by the rudist bivalves within them. Rudists added immensely to the volume of material in Cretaceous buildups, forming a framework both within shelf mounds and along shelf margins. They are significant constituents of reservoir rock in some of the world's greatest oil fields. In this Chapter a discussion of rudist morphology and paleoecology is followed by a description of some typical Middle Cretaceous facies patterns.

The Rudist Bivalves

In the Mesozoic, Pelecypods of the Pachydont group became well adapted to life in more or less restricted marine environments and to both sheltered and rough water. More than one hundred rudist genera evolved from the early rudist genus *Diceras*, a thick-shelled, coiled-beak clam often found capping Jurassic coral patch reefs and forming flank debris around such rises on the shallow sea bottoms.

In the Cretaceous rudists evolved into bizarre forms in which one valve rested on the substrate or became attached while the other served as a lid or cap; the valves usually articulated with massive teeth and sockets. When the rudists grew in crowded conditions the prevalent form was an erect, tall, twisted, slender shell, oriented during growth toward favorable food-bringing currents. Rougher water caused stubby, trunk-like forms to develop. In growth form, rudists resemble the smaller Paleozoic richtofenid brachiopods or even horn corals. The shoal water, tropical reefy environment is highly competitive and a premium exists for rapid growth. For rudists this resulted in many large species, as long as $1^1/_2$ m, having porous, almost vesicular shells. The shell structure varies from simple laminae, as in monopleurid and requienids, to thick, compact walls in some forms. The important rock-building caprinid group has a very thick wall containing open canals which occupied from 30 to 75% of shell volume. These porous walls, together with a large central cavity, gave the caprinid rudists great original pore space. The same high porosity was developed by radiolitid rudists whose thick wall was regularly cellular. Figure XI-1 illustrates some typical rudists; Fig. XI-2 shows their distribution in time.

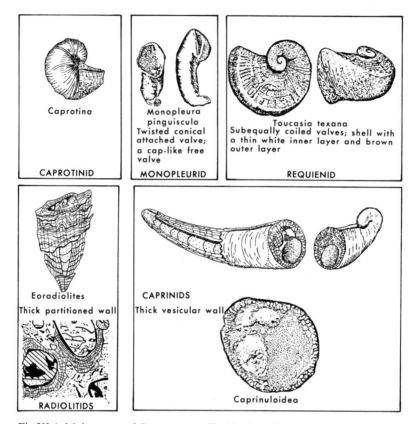

Fig. XI-1. Major types of Cretaceous rudist bivalves after drawings compiled by A.J. Coogan

Bob Perkins, Alan Coogan and others have studied the ecology of Middle Cretaceous rudists for many years. The observations below come mainly from their detailed analyses of backreef mounds in strata on the Texas shelf and from studies on shelf margins in central Mexico. Four ecologic assemblages are known. These are only partly related to major taxonomic groups.

1. The interior beds of carbonate banks and shelves contain bedded limestone formed in part by biostromes of *Requienia* and *Toucasia* whose left valves are large and attached and whose shells are thin and laminate. Dentition is normal, two teeth occurring in the smaller right valve. Both valves tend to coil. The requienid group is long-ranging and primitive in that it is directly related to the Jurassic ancestor, *Diceras*. It is found in a variety of environments and lived in association with other rudist types. Alan Coogan (personal communication) noted its occurrence with the radiolitid *Sauvegesia* in the El Abra of Mexico. But biostromes of requienids may also occur to the exclusion of other organisms except abundant miliolids and are interspersed with mm laminates, algal stromatolites, intraformational flat pebble conglomerate, dolomite crusts, and hardground surfaces. They thus form an important part of the restricted marine lagoonal to intertidal environment of the Cretaceous, resisting successfully the

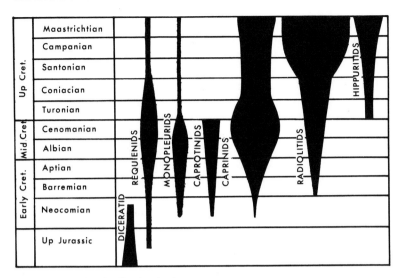

Fig. XI-2. Rudist distribution in time

frequent changes in salinity and temperature. Bob F. Perkins (personal communication) has also observed these same rudists in backreef bedded strata above caprinid mounds where they seemingly occupied tidal pools or formed patches of low relief on very shallow sea bottoms. Coogan cites other rudist groups which may form biostromes in the Middle Cretaceous strata including *Eoradiolites* and *Coalcomana* in the Texas Albian.

2. Monopleurid rudists are the most primitive of a second major group of rudists. In these the left valve is a cap and has the two teeth and single socket; the right valve is elongate, spiral, or conical and is attached. Monopleurid shells are thin and cellular. The genus seems to have been tolerant of clay mud for they are found in inverted conical clumps in marls generally to the exclusion of other rudists. They commonly are found in biostromes in beds as old as Barremian.

3. The caprinid rudists have long, slender, twisted, right valves with canal-bearing walls. They grew in abundance and formed moderately large mounds in shallow backreef positions. Many specimens appear in growth positions in these mounds, rooted in the ever-present mud matrix which accumulated in and around the rudist patches. Many other individuals are overturned and lie in irregular positions in these patches. The rudists were not colonial and did not form a framework by budding and branching. Their bulky, elongate, twisted, stem-like forms intertwined with and were attached to each other as well as rooted in the substrate. They must have formed a substantial baffle capable of growing in quieter water, but also of resisting surf action. Caprinid rudist patches, with some relief, are known to have also formed at shelf margins and in downslope positions as well as in backreef areas. The shelf margin patches generally have a micritic matrix and are surrounded by grainstone flank beds whose particles are exclusively rudist fragments derived from the mound inhabitants. Apparently the cap-

rinids were environmentally more tolerant than some other rudists and probably could withstand the quieter backreef realm of restricted circulation and variable salinity as well as inhabit waters of more open marine circulation at the shelf margin.

The caprotinids are a reef-forming group related to monopleurids. They are common in Europe as well as in the Gulf of Mexico province. They had a thick, two-layered wall and often a bulky overhanging free valve. The caprotinids in the Texas shelf built substantial portions of rudist mounds in the Edwards Formation, occupying the upper zone with the radiolitids. They look like caprinids in the outcrops. These forms have not been widely recognized in shelf margin positions in Mexico or in "Deep Edwards" on the Stuart City reef trend.

4. Another important group of rudists of Aptian and younger age is the radiolitids. They are the important reef-formers of the Late Cretaceous but also are common in Middle Cretaceous strata. These forms possess thick walls composed of closely spaced radial and concentric plates. Adaptation proceeded to the extent that the attached valve was more or less a heavy keg-like or conical form. The upper valve was a mere cap which could easily be clamped shut perhaps during subaerial exposure and permitted an intertidal existence for some genera. Ecologically the group was widely adapted. They are known in backreef biostromes, at tops of shelf mounds and, in the Middle Cretaceous, commonly at shelf margins and on foreslopes where they were associated with and gradually replaced by corals, spongiomorphic hydrozoans, red algae, and sponges, a biota inhabiting water of normal marine salinity. The robust construction of many members of the group as well as their abundance in the outer shelf marins would indicate a surf-resistant habit but other forms (*Durania* and *Sauvegesia*) existed in quieter water.

In a detailed study of a Cenomanian platform east of Rome (Polsak et al., 1970; Carbone et al., 1971) radiolitid dominance of the rudist faunas is reported to be internal to the shelf marginal caprinids although considerable faunal overlap occurs. It is possible that increased adaptation of the radiolitids during Cenomanian and later times permitted their existence in more restricted environments. The facies analysis by Carbone et al. (1971) indicated a subtidal environment for the radiolitids and caprinids existing on shoals at the platform margin studied at Roca di Cava. In this area, both seaward and bankward, more micritic sediments were deposited than at the shelf margin.

Accessory organisms in Cretaceous buildups include oysters and the oyster-like genus *Chondrodonta*. These occur in heaps of shell debris and are known in the upper parts of shelf mounds. Corals, mainly dendroid branching forms such as *Cladophyllia*, may form biostromes beneath the shelf mounds and occur on the foreslopes of rudist shelf margins. Knobby growth forms of coral also occur, like those found in Jurassic reefs, but the shallow-water coral forms seem to have been replaced at some quieter water shelf edges by the rudists. Hydrozoan spongiomorphs are known in foreslopes and outer shelf-margin positions as in the Jurassic. Cretaceous algae are abundant in the carbonate realm and the major algal groups generally occur with certain of the major rudist groups: radiolitids with the red algae in an outer shelf margin position, codiaceans with the tolerant caprinids over a wide environmental range, and algal stromatolites and requien-

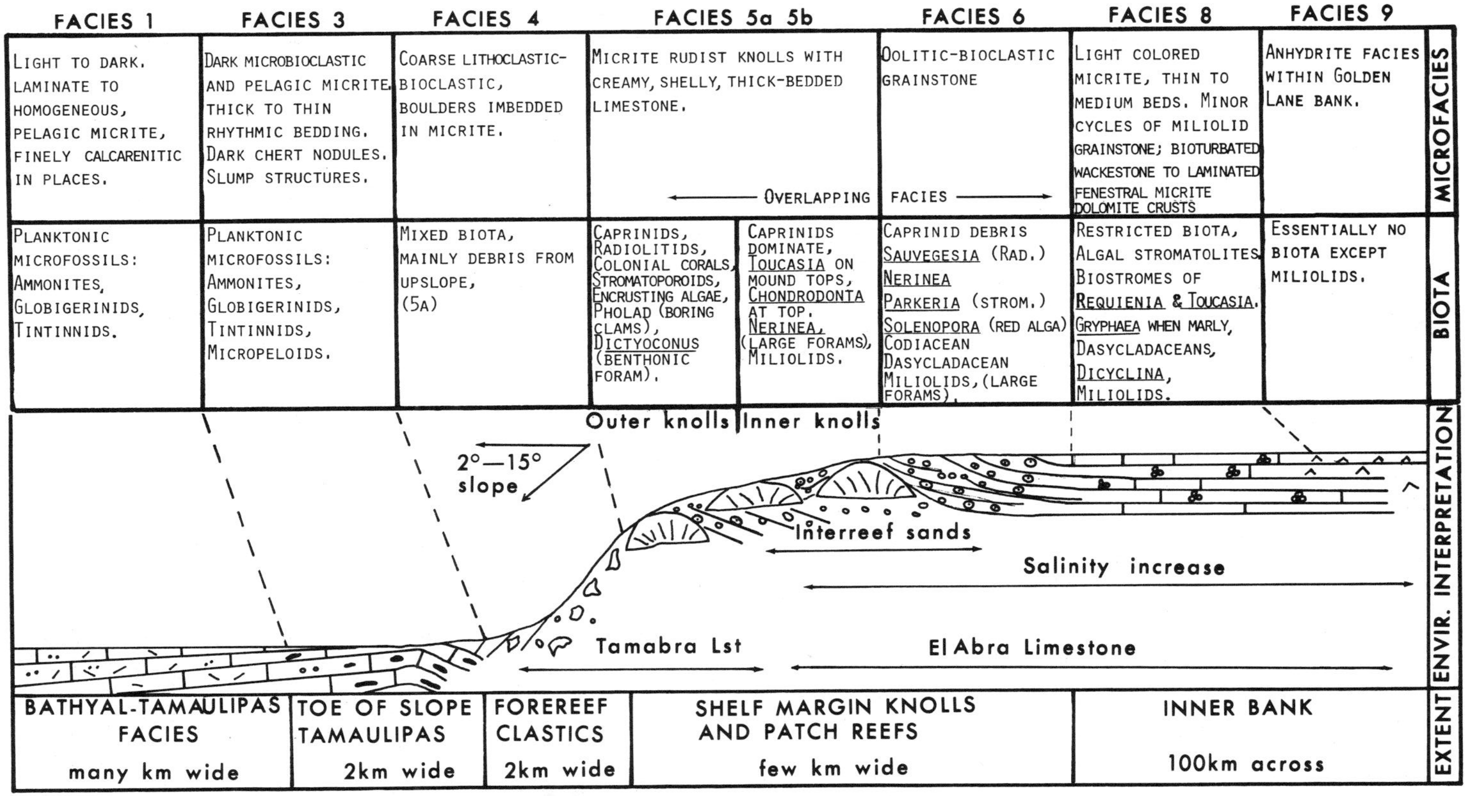

	FACIES 1	FACIES 3	FACIES 4	FACIES 5a 5b	FACIES 6	FACIES 8	FACIES 9
MICROFACIES	LIGHT TO DARK, LAMINATE TO HOMOGENEOUS, PELAGIC MICRITE, FINELY CALCARENITIC IN PLACES.	DARK MICROBIOCLASTIC AND PELAGIC MICRITE, THICK TO THIN RHYTHMIC BEDDING, DARK CHERT NODULES, SLUMP STRUCTURES.	COARSE LITHOCLASTIC-BIOCLASTIC, BOULDERS IMBEDDED IN MICRITE.	MICRITE RUDIST KNOLLS WITH CREAMY, SHELLY, THICK-BEDDED LIMESTONE.	OOLITIC-BIOCLASTIC GRAINSTONE	LIGHT COLORED MICRITE, THIN TO MEDIUM BEDS. MINOR CYCLES OF MILIOLID GRAINSTONE; BIOTURBATED WACKESTONE TO LAMINATED FENESTRAL MICRITE DOLOMITE CRUSTS.	ANHYDRITE FACIES WITHIN GOLDEN LANE BANK.
BIOTA	PLANKTONIC MICROFOSSILS: AMMONITES, GLOBIGERINIDS, TINTINNIDS.	PLANKTONIC MICROFOSSILS: AMMONITES, GLOBIGERINIDS, TINTINNIDS, MICROPELOIDS.	MIXED BIOTA, MAINLY DEBRIS FROM UPSLOPE, (5A)	CAPRINIDS, RADIOLITIDS, COLONIAL CORALS, STROMATOPOROIDS, ENCRUSTING ALGAE, PHOLAD (BORING CLAMS), DICTYOCONUS (BENTHONIC FORAM). / CAPRINIDS DOMINATE, TOUCASIA ON MOUND TOPS, CHONDRODONTA AT TOP. NERINEA (LARGE FORAMS), MILIOLIDS.	CAPRINID DEBRIS SAUVEGESIA (RAD.) NERINEA PARKERIA (STROM.) SOLENOPORA (RED ALGA) CODIACEAN DASYCLADACEAN MILIOLIDS, (LARGE FORAMS).	RESTRICTED BIOTA, ALGAL STROMATOLITES BIOSTROMES OF REQUIENIA & TOUCASIA. GRYPHAEA WHEN MARLY, DASYCLADACEANS, DICYCLINA, MILIOLIDS.	ESSENTIALLY NO BIOTA EXCEPT MILIOLIDS.
ENVIR. INTERPRETATION	BATHYAL-TAMAULIPAS FACIES	TOE OF SLOPE TAMAULIPAS	FOREREEF CLASTICS	SHELF MARGIN KNOLLS AND PATCH REEFS		INNER BANK	
EXTENT	MANY KM WIDE	2KM WIDE	2KM WIDE	FEW KM WIDE		100KM ACROSS	

OVERLAPPING FACIES — Outer knolls — Inner knolls — Salinity increase — interreef sands — 2°-15° slope — Tamaulipas — Tamabra Lst — El Abra Limestone

Fig. XI-3. Idealized Middle Cretaceous facies across large offshore banks in central Mexico. Biofacies from Bonet (1952), Griffith et al. (1969), and Coogan et al. (1972).

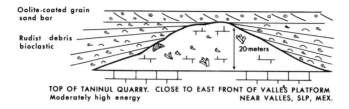

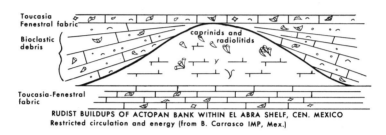

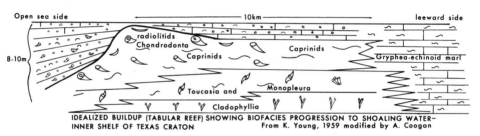

Fig. XI-4. Facies in three rudist mounds on shelves or along gently dipping shelf margins in Texas and Mexico

ids together in extreme backreef positions, e.g., in the El Abra of Mexico. Figure XI-3 shows a facies profile with the ecological zonation of the major rudist groups and associated biota. Figure XI-4 is an illustration of the composition of typical rudist mounds.

Rudists differ somewhat from other common reefoid organisms in the geologic record. This is true both in depositional environment and in habit of construction.

1. Generally in the Middle Cretaceous (except for monopleurids) they inhabited clear-water, pure carbonate environments without clay contaminants. In the Late Cretaceous some of the radiolitids and hippuritids were also tolerant of the clayey bottom.

2. They were somewhat tolerant of variable salinity and restricted circulation. The absence of brachiopods, ammonites, and echinoids from typical rudist facies indicates that the organism could withstand fluctuations in temperature, salinity, and oxygen beyond the limits of normal marine water. Discovery of numerous diastems and hard grounds within caprinid mounds in Texas, and the presence of coal directly above caprinid (?) biostromes in the Middle East, indicates that some rudists could survive intermittent exposure.

3. Some groups of rudists spanned the entire range from outer shelf oxygenated and agitated water to interior backreef protected environments (e.g., requienids and radiolitids).

4. In general, rudists could thrive better in lime-mud environments than filter feeders such as Coelenterates and sponges. Requienid biostromes are always in micritic sediments and caprinid mounds also contain much lime mudstone.

5. Many rudist fragments as well as the living shells were too large to be easily coated or encrusted and accumulations of them are not bound together organically as in coral reefs.

6. Although not truly colonial organisms as are commonly found in reefs, these peculiar mollusks, like corals, were capable of rapid growth into wave base and of large bulk. Their wave resistance is indicated by flanking halos of winnowed shell debris derived from their shells. The fauna is often monotonous where caprinids flourished as if all other organisms were choked out by thriving rudists.

7. In general, individual rudist mounds on more or less stable shelves and in many shelf margins are no higher than 10–15 m and many are much less, perhaps giving a maximum depth for inception of rudist growth.

8. Because of the above facts, rudist buildups in shelf margin positions may be expected to have more continuity than those formed chiefly by corals. They could more easily withstand seaward moving off-shelf water and may be expected to rim both lee and windward sides of offshore banks. These rims may be only a few 100 m wide and have variably gentle and steep slopes. Open sea sides of the banks can be ascertained from the diverse faunal content of accessory organisms such as corals, red algae, and hydrozoans.

9. The structure of rudists, and the fact that part of their shell was aragonite, resulted in both high initial porosity and later partial solution under conditions of meteoric water flow.

10. Rudists evolved rapidly from Late Jurassic to Late Cretaceous when they abruptly became extinct, leaving reef-building corals and calcareous algae as major framebuilders of the Tertiary. Sizeable Cretaceous rudist buildups are known as far back as Aptian or Barremian time.

Shelf Margin and Platform-Bank Interior Facies of Middle Cretaceous of Mexico and the Gulf Coast

Regional Facies around the Gulf of Mexico

In Middle Cretaceous time the Gulf of Mexico was almost completely bordered by large carbonate platforms and offshore banks. These formed as a result of extensive Albian and Cenomanian marine transgression which prevented influx of terrigenous clastics from western North America. The wide areas of clear, tropical water at the edges of the Gulf induced limestone accretion which built out ramps from several positive areas around its subsiding center. From these ramps, platforms soon developed (Fig. XI-5). Shallow shelf basins and intervening positive

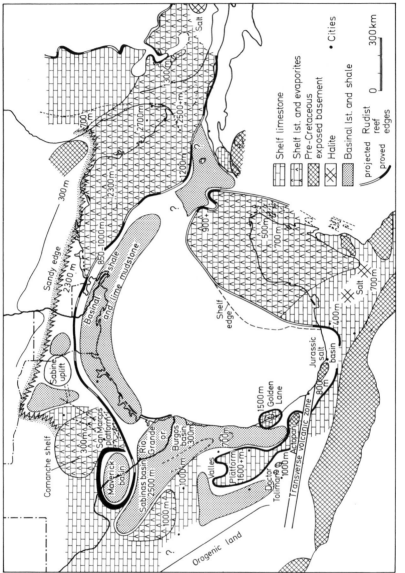

Fig. XI-5. Lower and Middle Cretaceous facies and thickness around Gulf of Mexico. Compiled from numerous sources, particularly Bryant et al. (1969), with permission of American Association of Petroleum Geologists. Recent data on facies in Southern Mexico courtesy of A.A. Meyerhoff.

areas on some of the platforms became the sites of evaporite deposition when periodically cut off from the open Gulf. These shelf areas were separated from the Gulf by narrow carbonate rims reaching nearly to sea level and composed of accumulations of rudists and biogenic lime sand. Slopes off these rims into the basin tend to be from 2 to 5 degrees around the Gulf and steeper along some of the offshore banks in Mexico. Bathymetric relief could have been as much as 1000 m. Thickness of shallow-water Cretaceous carbonate in these great platforms is even greater. In the Gulf and Atlantic, Lower and Middle Cretaceous fossils have been dredged from submarine outcrops far down the slopes below some of these rims. This would indicate continuous buildups of the platforms since Jurassic time or possibly widespread faulting along the escarpment bases. Arcer profiles across the platforms show that a pronounced ridge in Cretaceous strata generally trends parallel to and just within the platform edges, presumably evidence of marginal buildups (Bryant et al., 1969). The same geophysical technique outlines the outbuilding of Early and Middle Cretaceous strata along sloping ramps in the seaward direction and shows the final construction of the great platforms.

These Early-Middle Cretaceous bank and platform areas have remained as such until the present, and today's bathymetric configuration of the Gulf of Mexico clearly outlines the apparent Cretaceous shelf and basin areas. A narrow inlet must have entered the Gulf at the south, at about the position of the present island of Cuba. The Antilles orogenic belt was at that time a geosynclinal trough.

The western side of the Gulf was bordered by another area of great subsidence, the Mexican geosyncline which was filled by sediments. Thick carbonates, measured in thousands of meters, now observed in the eastern Sierra Madre, constitute evidence for a miogeosyncline in which basinal limestones and immensely thick carbonate banks formed. The western boundary of this area is buried beneath Tertiary volcanics along central and western Mexico. Presumably these strata mask a series of Mesozoic orogenic islands and a source area for both Early and Late Cretaceous sandstones.

Middle Cretaceous climate is comparatively well known. Cycad stumps and dinosaur tracks, occurring in littoral sediments of Albian age cropping out in the Edwards plateau country of central Texas, indicate a tropical climate north of 30 degrees latitude. Albian evaporites from the North Texas-Tyler basin and the Maverick basin in south Texas, as well as from Guatemala and Cay Sal off Cuba, indicate periodic dry periods as well as a hot climate. The great thicknesses of lime mud sediment also indicate tropical seas. Vast quantities of this sediment not only make up the shelf platforms but in addition accumulated in foreslope positions and in marginal offshore basins. Only in the northern coastal areas does much shale and sand occur. Sands appear to have been introduced from the Appalachian area in the east and from a western source in northern Mexico, north of Monterrey and Saltillo. The coastal stretch between these two areas is a narrow belt of sand, which crops out in southern Oklahoma and north central Texas and shows that minor river systems drained the midcontinent area much as today. Occasional influx of clay resulted in widespread dark shales which punctuate the dominantly carbonate stratigraphy in bands around the whole perimeter of the northern Gulf, e.g., Pine Island-La Pẽna-Otates of Aptian age, the Albian Kiamichi shale and the Cenomanian Grayson-Del Rio formation.

The Great Middle Cretaceous Carbonate Banks of Central Mexico

Figure XI-5 indicates the distribution of known banks of this age in central Mexico. They include the Toliman, El Doctor, Actopan, and Valles platforms, and the Faja de Oro (Golden Lane). The bank or platform facies is generally termed the

Table XI-1. Middle Cretaceous correlations in Mexico and Texas

Stage	Central Mexico	North Mexico	Texas	
Cenomanian	Upper Tamaulipas (Basin)	Cuesta del Cura	Buda lst.	
			Del Rio shale	
Albian	Tamabra forereef / El Abra (Shelf)	Upper Tamaulipas (Basin) / Aurora	Georgetown / Person / Edwards / Kainer / Edwards / Glen Rose	Stuart City reef / Georgetown Atascosa group
Aptian	Otates shale	La Peña shale	Pine Island	

El Abra Formation and the intervening basinal limestone, the Tamaulipas (Table XI-1). A northern shelf area marked by the outcropping Aurora Limestone exists in Coahuila and Chihuahua.

Inasmuch as one of the great older oil fields of the world exists along the edges of the Faja de Oro or Golden Lane platform, lying south of Tampico, the facies of these rocks have been much studied by both Mexican and U.S.A. geologists in the last 20 years, in hopes that the vast early production of the Mexican fields can be extended into offshore areas around the margins of the Gulf. The offshore extension of the Golden Lane by Petroleos Mexicanos and the gas fields of the Stuart City-Edwards Reef trend in south Texas have encouraged this study.

Facies have been described by Wilson et al. (1955) on the El Doctor bank; Barnetche and Illing (1956) on the Poza Rica field; Griffith et al. (1969) on the Cuesta El Abra and Deep Edwards reef trend; Bonet (1952) on the Golden Lane; Becerra (1970) on Poza Rica; Carrasco (1971) on the Actopan platform; Rose (1963), Keith (1963), and Bonet (1963) on the Golden Lane, Cuesta El Abra, and Edwards trends, and Coogan et al. (1972) on Golden Lane and Poza Rica. The following basic microfacies types are recognized in these strata. Despite some differences in terminology, agreement is general about the definition of the rock types and their environmental meaning. The outline below lists them in order of facies progression from basin to shelf.

Basinal and basin margin (Standard Facies belts 1 and 3) (Plate III).
Tamaulipas limestone of eastern Mexico; marginal slope facies of Carrasco, termed planktonic micrite by Griffith et al. (1969). In northern Mexico the thicker-bedded facies is included in the basinal Aurora and the thin-bedded, rhythmic Cenomanian facies of the Cuesta del Cura Formation.
Pelagic lime mudstone and wackestone. Dark to light gray, evenly bedded with stylolitic contacts along clay seams and organic films. Many dark chert layers are present. Thin (10–

30 cm) to medium-bedded, rhythmic bedded with slump structures (boudins) and surfaces of discontinuity in bedding caused by gliding or channeling. Micritic texture with scattered calcisilt particles such as micropeloids, tintinnids, calpionellids, and globigerinids. Laminae and micrograded beds, cut and fill structures, and occasional burrows. Micropeloid calcisilt particles are detrital and mainly of shallow water origin (SMF-3).

Marginal breccia and conglomerates (Standard Facies belt 4).

Tamabra Limestone is the stratigraphic term used in Poza Rica. This is the detrital facies of Coogan et al. (1972). Wilson et al. (1955) described it as coarse conglomerates marginal to the El Doctor bank. Carrasco (1971) described a massive body of this facies (30 m thick) at the base of the El Doctor or El Abra limestone.

Such conglomerates are both rudstones and floatstones. Clasts are from 25 cm to 1 m in diameter; they include fragments of previously lithified shallow-water limestone as well as micritic clasts with planktonic fossils. Clasts are rounded to subangular, and may be well graded. The boundaries with the micrite matrix are usually, not always, well-defined. Edges of the clasts commonly show truncation and erosion and may be somewhat weathered and oxidized. In some areas, strata downslope at basin margins are principally biogenic micrite with scattered large rounded blocks. Slumped beds and convolute strata occur, over- and underlain by horizontal planar strata and indicate penecontemporaneous deformation on slopes.

Shelf margin bioclastic facies (Standard Facies belts 4 and 6).

This term, used by Coogan et al. (1972) and Carrasco (1970) equates with the skeletal sand and silt of Griffith et al. (1969). The facies occurs in the Tamabra (foreslope debris) and in halos marginal to caprinid and radiolitid mounds on shelf margins where it is usually included in the El Abra Formation.

The strata may lack well-defined bedding. In other places thick, wedge-shaped foreset beds occur and cross-lamination is common. Interbedded are ill-sorted, coarse debris and much better sorted, finer, biogenic fragments. These sands may be mixed with biogenic calcisilt. The fauna in outer slope beds (e.g. Tamabra of Poza Rica and the narrow fringe of the Deep Edwards trend) contains fragments of caprinids and radiolitids (60% according to Griffith et al. (1969), red algae, spongiomorph hydrozoans, sponges, and colonies of dendroid corals, each group averaging around 5% each. Up to 5% lithoclasts may be present. The bioclasts are exclusively of rudists and gastropods when higher on the shelf, forming halos around caprinid mounds (SMF 4 and SMF 5).

Shelf margin rudist mounds (Standard facies belt 5) (Plate XXX E).

Caprinid micrite facies of Coogan et al.; organic reef of Griffith et al. and Carrasco. These are accumulations formed by individual mounds of caprinids with some radiolitids. Alan Coogan (personal communication) would distinguish the following biofacies subdivisions: coral-algal, coral-spongiomorph, *Monopleura*, and caprinid. The buildups are on the order of 3–15 m thick and a few 100 m across. They are superimposed on each other on open sea edges of the major banks. Examples are: northeast border of Actopan platform, the El Doctor Bank and the Cuesta El Abra on the east side of the Valles platform. Some of the rudists are in growth position but most are not. They are commonly disoriented in a micrite matrix. Accessory organisms include corals, boring pholad bivalves, snails, some requienid rudists, and various large foraminifera including *Orbitolina*, *Dictyoconus*, *Chofatella*, *Cuneolina*, and *Coskinolinoides*. At the end of the Cuesta El Abra east of Valles, San Luis Potosi, Griffith et al. (1969) estimated the following faunal distribution which indicates open circulation: caprinids 23%, radiolitids 15%, coralline algae, hydrozoans, and oolite 10% each, and dendroid coral 5% (SMF-7c bafflestone).

Shelf-margin and bank-interior facies of El Abra (Standard Facies belt 7).

This is a coated particle grainstone facies, following terminology of both Coogan et al. and Griffith et al. It is termed the oolitic facies by Carrasco. Beds are from a few cm to a meter thick and cross-bedded. The sediment is grainstone of sorted, rounded, superficially coated particles of lithoclasts, miliolids, and dasycladaceans. Alan Coogan (personal communication) reported the radiolitid *Sauvegesia* in this near backreef facies. Well-formed ooids are rare; grapestone lumps and peloids are present. In some places on the shelf margin, red algae such as *Solenopora* and *Jania* form the nuclei. These coated, grainstone beds are known to overlie and flank caprinid mounds on the Cuesta El Abra (SMF-11).

Bank interior facies of El Abra (Standard Facies belt 8).

These strata are light-colored and well-bedded, varying from 30 cm to 3 m thick. They are commonly cyclic; repetitions of strata progress upward in a shoaling sequence. Lowest beds are burrowed micrite with *Toucasia* or miliolid grainstones. These pass upward through laminated micrite to a cap of dolomite crust or intraformational pebble conglomerate. Eight common bank interior-restricted marine microfacies are recognized in the various studies noted above.

1. Miliolid grainstone: Thick-bedded (1 m) to massive strata, burrows in upper parts of bed, some peloid grains but mostly tests of miliolid foraminifera (SMF-18) (Plate XXX C).
2. Peloidal grainstone to wackestone with some grapestone: Sparry cement, generally 75% grains, less than 10% miliolids, 5% bioclasts. This sediment is derived from organic pelleting of lime muds in an environment where lime precipitation continuously hardens the grains and where gentle current partly winnows remaining lime mud (SMF-16).
3. Miliolid wackestone-packstone: Burrowed sediment which includes a few rudists (SMF-19).
4. Molluscan wackestone: *Toucasia* and monopleurid rudists, oysters, and scattered miliolids. In places the oysters or rudists form biostromes (Plate V B).
5. Dolomite-lime mudstone laminated crusts: Formed by evaporation causing mineral precipitation and replacement. By analogy with Holocene sediments, lamination is probably algal in origin. Disruption of the crust by drying and curling-up of the fragments results in the construction of intraformational breccia which fills channels and is a common lateral facies to the crusts (SMF-24).
6. Bulbous algal laminites with fenestral fabric: Mostly peloidal lime mudstone or homogeneous mudstone. Represent desiccated algal mats with scattered miliolids and ostracods (SMF-20).
7. Planar fine mm laminites, nonfenestral, often pure dolomite, containing scattered miliolids. Probably also induced by algal trapping of seasonal increments of lime mud.
8. Homogeneous lime mudstone or dolomite (SMF-23).

The outcrop areas of the platforms studied in Mexico all show strong facies differentiation displaying the above standard rock types and illustrating clearly their interpreted environments. The interiors of the platforms and large offshore banks are wide expanses of the well-bedded El Abra Formation. The margins contain knolls of caprinid-radiolitid rudist buildups and much biogenic lime sand. These facies are reasonably well understood. The foreslopes of the banks, on the other hand, provide problems in interpretation in almost all places where they have been studied. Is the common faulting at the edge of the massive carbonate of the banks contemporaneous with deposition, or later? Just how steep was the original slope? How much original relief was there on any one bank—tens of meters or hundreds of meters? To what extent can sedimentologic study of the basinal sediments provide some estimate of depth? Clear answers to these questions are important in developing exploration models and in comparing Cretaceous shelf margins with others in the geologic record.

One of the best areas for display of shelf margin facies lies along the Tampico highway about 10–12 km east of the city of Valles, San Luis Potosi (Figs. XI-3, XI-6). Here extensive quarries have been cut into the eastern edge of the Valles platform which has been uplifted along a faulted anticline in front of the Sierra Madre for a distance of more than 100 km. The frontal, east-facing scarp, the Cuesta El Abra, exposes knoll reefs close to the shelf margin at the railway loading station, Taninul. Large springs and caverns along the escarpment are present, making additional exposures possible. Detailed studies have been made

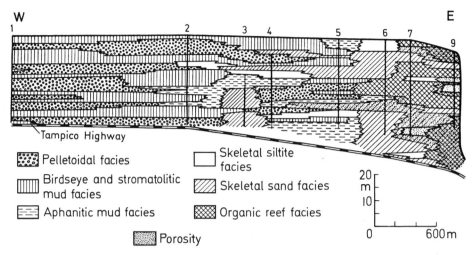

Fig. XI-6. Facies distribution of El Abra limestone as determined by factor analysis. From Griffith et al. (1969, Fig. 8) in a traverse across the Cuesta de El Abra, 15 km east of Valles, San Luis Potosi, Mexico. Compare with Fig. XI-10. Illustration courtesy of Society of Economic Paleontologists and Mineralogists

of these Cuesta El Abra exposures and are available from Bonet (1952), Griffith et al. (1969), and from the Corpus Christi, Texas, Geological Society Guidebook of 1963. Unfortunately, as is usual, the foreslope deposits are incompletely represented along this scarp, having been downfaulted into the coastal plain. A well 15 km east of the El Abra scarp (Coco no. 22) penetrated a basinal section of Tamaulipas limestone (Facies belt 3).

A particular problem in foreslope interpretation relates to the important reservoir trend of Poza Rica lying in the subsurface not far to the south of the Cuesta El Abra outcrops (Fig. XI-7). The Poza Rica trend parallels the western border of the Golden Lane, curving in an arc for 150 km and lies as much as 1000 m structurally lower than the latter. Is this great difference in relief structural or depositional? The interpretations are (1) that either the Poza Rica trend consists of downwarped (faulted) outer margin (reef knoll) sediments of the Golden Lane bank whose present hummocky configuration results merely from karst erosion and faulting or (2) that the Poza Rica trend is formed merely from great piles of forereef breccia in basinal lime mud. The problems connected with these interpretations and their significance are discussed by Coogan et al. (1972) who ascribed to the first view.

The following facts and ideas represent a general consensus despite some disagreement as to the Poza Rica trend:

1. The Golden Lane platform is an elongate bank of shallow water limestone (El Abra Formation) tilted gulfward. It is 145 by 65 km in size and its total post-Jurassic thickness may be about 1500 m. It is Albian and Cenomanian? in age. If true stratigraphic thickness is accepted for the Jardin 35 well in the Golden Lane, the El Abra Formation must be 3000 m thick; if not the thickness could be half that.

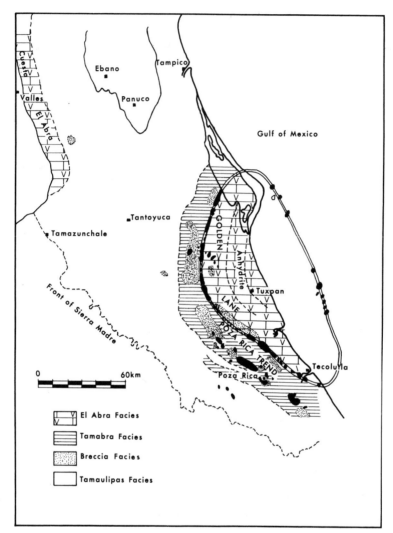

Fig. XI-7. Oil fields and facies of Middle Cretaceous Golden Lane and Poza Rica trends on Gulf coast of Mexico. Dotted pattern includes breccia and conglomerates of Cretaceous age along Poza Rica trend and of Eocene age above the El Abra on the Golden Lane bank. The Golden Lane lies in the subsurface. The eastern margin of the outcropping Valles Platform is shown at the Cuesta de El Abra in the northwest corner of the map. Dashed lines indicate major facies boundaries. After Lopez-Ramos (1955) and Coogan et al. (1972, Fig. 2)

2. The bank was formed offshore and isolated in the Tampico embayment of the Early Cretaceous Gulf of Mexico. Its most interior facies along the present Mexican shoreline includes several anhydrite beds. It is separated from the Valles platform in the miogeosyncline to the west by a trough only about 60 km wide (Chicontepec or southern Magiscatzin basin).

3. The Golden Lane bank developed on an igneous basement of sialic rock which is a southern continuation of the Tamaulipas structural arch and probably

lies over a structural culmination of the Mexican part of the Late Paleozoic circum-Gulf orogenic belt.

4. The edge of the platform consists of a string of individual paleomorphologic or structural hummocks in which oil accumulated under a seal of Tertiary-Late Cretaceous? argillaceous strata. These highs lie along the eroded top of the thick mass of El Abra Limestone. They are probably erosional and not purely depositional but their resistance to erosion may have resulted from original mound areas of thick-shelled rudists. Because of early drilling practices, the rock of the Golden Lane El Abra Formation is very poorly known.

5. The Middle Cretaceous limestone mass in central Mexico is regionally overlain by the pre-Turonian unconformity of the Gulf area, and much karstic collapse, cavern formation and porosity development in the El Abra occurred probably before Late Cretaceous deposition. Although it seems reasonable for the bank to have been exposed during formation of the regional pre-Turonian unconformity, this cannot be proved because of subsequent erosion of critical strata. The top of the Golden Lane bank was quite possibly intermittently emergent from Cenomanian to Oligocene time. A latest Cretaceous conglomerate derived from El Abra facies shows exposure and erosion in latest Cretaceous and before the Paleocene. Late Cretaceous through Eocene beds overlapping the great bank were removed by uplift and gulfward tilting of the mass in pre-Oligocene time.

6. The Poza Rica trend consists of a series of structural highs parallel to those of the Golden Lane but they are structurally much lower. They are covered by a complete section of Late Cretaceous and Tertiary beds. Any post-Cenomanian sea-level drops which exposed the El Abra banks could not have affected the Poza Rica trend.

7. The Poza Rica strata (the Tamabra) are, however, more dolomitic and fragmented than those in the Golden Lane. They have suffered extensive neomorphic alteration. The beds are thickest near the Golden Lane bank (500 m) but thin considerably within 20 km northwestward away from the bank.

8. The Tambara contains a mixed and interbedded biota of pelagic lime mudstone and shallow water benthos.

9. The thick Tamabra carbonate facies in Poza Rica is interrupted by numerous thin shaly units (perhaps bentonites) which can be traced over part of the field. Some of these have pelagic microfossils.

Barnetche and Illing (1956) and Coogan et al. (1972) considered the Tamabra facies as forereef shoals, *in situ* organic growth, pointing out that generally the fauna was more normal marine than that of the El Abra. It has radiolitid rudists, hydrozoan spongiomorphs, corals as well as caprinids, and almost no miliolids. It thus contrasts greatly with the Golden Lane El Abra which, although poorly known, is seen to be mostly restricted marine. Coogan et al. (1972) noted a sequence 90 m thick in one core which they interpreted as upward shoaling, thus *in situ*, accumulation. They believed that the chain of Poza Rica fields produced from rudist reef accumulations. These occupy a chain of positive uplifts now represented by thin areas of post Jurassic to Mid-Cretaceous strata over what appear to be fault blocks (Fig. XI-8). These are considered comparable to such offshore British Honduras banks as Turneffe, Glovers reef, and Chinchora.

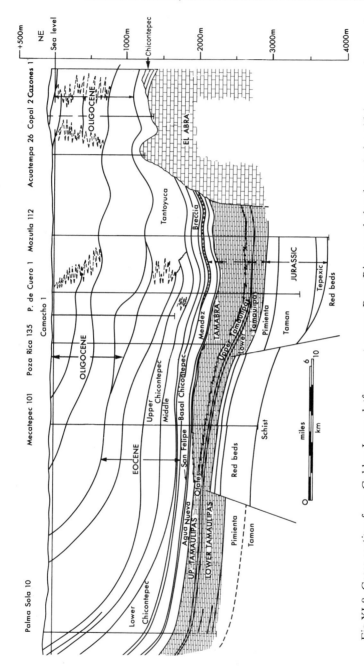

Fig. XI-8. Cross section from Golden Lane platform west across Poza Rica trend showing relative thickness and general stratigraphic relations of El Abra, Tamabra, and Tamaulipas limestones. Evidence for interfingering of Tamabra Limestone and Agua Nueva Formation as shown, is scant. After Coogan et al. (1972, Fig. 7) and modified from Pemex cross section by P. Carillo, R. Rocha, and A. Acuña. Illustration courtesy of American Association of Petroleum Geologists

On the other hand, arguments for an allochthonous origin of the Tamabra facies in Poza Rica (mainly from work by Becerra, 1970) are as follows:

1. Fragmentation of the rudists increases progressively westward away from the supposed source of the bioclasts on the Golden Lane.

2. Decrease in thickness of the Tamabra is progressive from east to west away from the supposed source.

3. General bad preservation of the rudists and their lack of orientation indicates that they are detrital particles and not of local derivation. Normal biological attrition of carbonate particles weaken this argument as well as that of item 1.

4. The amount of clay in the sediments shows that the rudists did not grow locally.

5. Interstratification of the Poza Rica Tamabra facies with the pelagic faunas of the typical Tamaulipas facies, the pelagic facies being in stratigraphic sequence from Albian through at least Cenomanian, shows that the shallow water carbonate facies could have been introduced gradually into the basinal environment. The presence of lime mudstone bearing Tamaulipas pelagic microfauna inside of rudist tests indicates a mixing of the two environments.

Evidence for the exposure and erosion of the top of the Golden Lane bank has been pointed out, whereas a complete section of Late Cretaceous-early Tertiary strata over the Tamabra belt is known. It has been considered that the Tamabra represents forereef shoal sediment deposited *in situ* on the outer rim of the Golden Lane bank. They are now 1000 m lower than the Golden Lane (Fig. XI-8). Certainly the rudists, corals, and hydrozoans found in the Tamabra did not exist in water of such depth. If the bank was downwarped structurally it must have been buried before Late Cretaceous time. Becerra's evidence for stratigraphic continuity within the Tamabra and a gradual buildup of the facies through Middle Cretaceous time is significant. A continuously penecontemporaneous origin for the Tamabra material as detritus from the lip of the Golden Lane bank seems to be a logical explanation.

Could the Middle Cretaceous relief have been the full 1000 m now structurally indicated? Surface work by B. W. Wilson, B. Carrasco, and P. Enos on the outcropping banks in the Sierra Madre have clearly indicated that great relief existed on the flanks of some of them. It is not clear just how much, despite the considerable difference in thickness between bank (1500–2000 m) and basin (300–400 m) facies. Field studies have indicated faulting at bank margins as a response to the Tertiary orogenic forces. This complicates discernment of stratigraphic onlap relations. From analogy with Cretaceous to Recent banks in the eastern Gulf of Mexico, it is tempting to equate the 5- to 6-fold thickness difference to an original relief constructed by differential rate of sedimentation from bank to basin. Facies show that the bank tops were barely awash, even intermittently exposed, and the fauna of the basin sediments is consistently pelagic, but we cannot ascertain the absolute depth difference. Perhaps there was a differential substrate subsidence between bank and basin but it is tectonically improbable that circular patches on the sea floor would subside five times more rapidly than surrounding areas. If the banks built up without a considerably greater subsidence than the intervening basins, they must have stood more than 1000 m in Cenomanian times. This is not an incredible figure when one observes the amount of original relief preserved in the Triassic Dolomites of the Tyrol.

Sedimentologic evidence of the peripheral facies indicates a considerable relief but is not definitive as to how much. Slump features exist in thin to medium-bedded limestones, coarse conglomerates contain clasts of previously lithified

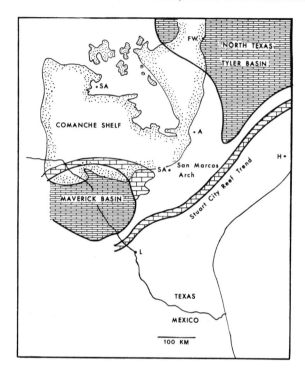

Fig. XI-9. Regional Middle Cretaceous (Albian) paleogeography of Texas during formation of Fredericksburg-Washita groups. Devils River and Stuart City trends consist of rudist buildups at shelf margins. For detailed facies on Comanche shelf see Fig. II-9. Outcrops are stippled. Names of cities are abbreviated: Houston, Laredo, San Antonio, Austin, Ft. Worth, San Angelo

limestones and show locally strong graded bedding. Some basinal facies consist solely of pelagic fauna in micrite with abundant black chert nodules, indicating sedimentation well below wave base.

Deep Edwards or Stuart City Reef Trend

A major carbonate platform, termed in Texas the Comanche Shelf and a narrow shelf margin (Stuart City reef—see Fig. II-9), closely related in facies to the Mexican offshore banks, exist completely along the northern border of the Gulf of Mexico. The platform has produced minor amounts of oil and considerable gas along its south Texas edge (Fig. XI-9) and carbonate facies have been closely studied here since the 1950s. The same microfacies types in the same lateral sequence existing in the Mexican banks, are recognized here by Keith (1963), Rose (1963) and Griffith et al. (1969). The latter study indicates the value of a comparative analysis of combined surface and subsurface strata. Very detailed study of microfacies of the carbonate shelf margin rocks of the Cuesta El Abra in Mexico was used, with the help of computer analysis, to determine the minimum significant parameters necessary to describe related facies in the Deep Edwards trend. Use was made of a modern analog offered by the Florida Reef tract to confirm conclusions about the environmental meaning of the facies sequence. Application to the Deep Edwards subsurface study, where data and time were limited, was made more accurate by such comparisons.

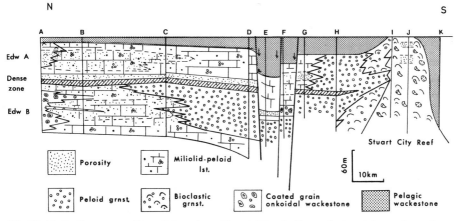

Fig. XI-10. Section across Deep Edwards or Stuart City shelf margin trend through southern Atascosa and northern Bee counties, Texas. After Griffith et al. (1969), courtesy of Society of Economic Paleontologists and Mineralogists

Figure XI-10 is a section across the shelf edge based on the above work (cf. Fig. XI-6). Essentially the facies progression is from peloidal, miliolid, and coated-grain micritic sediment, extensively dolomitized far behind the bank edge, to a marginal peloidal grainstone-wackestone facies in which grains with micritic rinds are common. This peloidal sand grades within a mile or so to shelf-margin facies, winnowed bioclastic sand banks interstratified with caprinid wackestones. In the outermost part of this 8 km wide marginal belt, the sandy debris contains a diverse fauna with button and clump-like spongiomorph hydrozoans (e.g., *Actinstromonia*), numerous species of corals, including the dendroid *Cladophyllia* and the compact *Diploastrea*. Some of the reefy organisms are also imbedded in lime mud. The marginal belt abruptly gives way to dark shaly pelagic lime mudstone younger than the shelf margin carbonates.

The following are pertinent facts of facies distribution and structural configuration:

1. The marginal belt is only 5–10 km wide and is very long and unusually regular. It stretches more than 1000 km along the coast from north of Laredo, Texas in Webb County, to north of New Orleans, Louisiana, whence it passes out into the Gulf.

2. It is indicated that the rim was a topographic high, rising above the shelf area, by the absence of Georgetown (Uppper Albian) strata over its top in some places and thinning and onlap of the Georgetown by the overlying Del Rio Clay. The "Georgetown" or Atacosa basinal mudstones are considered to have filled in on the gulfward side of the shelf margin and overlapped it. This interpretation would require considerable bathymetric relief to have existed, measured in hundreds of meters. Cross section reconstructions (Van Siclen, 1958) show the slope on the seaward side to have been about 2 degrees, somewhat less than on the great offshore banks of central Mexico.

3. The belt, where best delineated, shows faulting on its gulfward side just as is present at Cuesta El Abra on the east margin of the Valles platform and suspected

on the west side of the Golden Lane bank. As usual, it is difficult to distinguish faulting from original bathymetric relief on the front.

4. The facies are slightly regressive within the marginal belt. Wells drilled near its abrupt edge penetrate restricted marine carbonate of the shallow platform at the very top, followed downward by rudist wackestone and bioclastic debris, rudist and coral debris, and basinal lime mudstone. As much as 650 m of such sequences are known; the maximum of reefy rock known may be as much as 500 m.

5. In contrast, the backreef platform facies is extensive and well known. It includes both shallow basinal strata in the North Texas and Maverick basins and shelf sands and dolomitized intertidal muds on the central Texas, San Marcos platform. Serveral zones of faulting, marking grabens, trend northeast across the San Marcos platform. These began to accumulate sediment in Cretaceous time. Isopachs of the Comanchean strata show a persistent axis of thickening 10–20 miles shelfward of the narrow reef trend, marking the Karnes trough, which is one of these grabens. Numerous oil fields occur in backreef dolomites associated with the upthrown sides of these faults.

The geologic history of the platform begins with an Aptian transgression followed by major construction of the platform during most of Albian time. Upper Glen Rose and Fredericksburg strata show progressive restriction in the basins as well as over the intervening San Marcos platform. Minor sea-level fluctuations laid bare the San Marcos platform periodically. The Comanche shelf built up to sea level toward the end of Albian time but during the Albian-Cenomanian transgression was completely submerged by deeper water sediments, the dark pelagic "Georgetown" limestone. During the marine advance and shale influx which ended reef growth, the San Marcos platform shrank to half its original size and the pelagic and open marine benthonic facies of late Albian time shifted northward across the North Texas embayment and on to the Comanche shelf northeast of the Llano uplift. Presumably during this time the outer basin beyond the reef trend was starved of sediment until latest Albian-Cenomanian time (Van Siclen, 1958).

The large Cretaceous shelf of the northern Gulf of Mexico with its platforms and basins (Fig. XI-5) offers an instructive example of how structurally positive areas, such as the Ocala Uplift, San Marcos Arch, and Llano Uplift, served as the nuclei for the development of carbonate ramps and platforms. The presence of Jurassic and perhaps early Cretaceous salt in the intervening shelf basins shows that the tectonic framework was established early in the Mesozoic.

There exist some unanswered structural questions about the Deep Edwards-Stuart City trend. Why does it appear so regular? Was its formation induced by carbonate shoal facies developed over an old ridge in the Ouachita orogenic belt or over the fractured southern edge of a Gulf of Mexico plate which has migrated northward under the Texas-Louisiana coasts? Either explanation could account for the regular form and narrowness of the belt. Does the Albian-Cenomanian shelf margin trend coincide with that of the Lower Cretaceous Sligo-Cupido Formation? The prevalent regressive patterns of other shelf margins would indicate that the Middle Cretaceous margin should be somewhat gulfward of the Cupido-Sligo.

Many researchers have pointed to differences as well as similarities between the Mexican offshore banks and the Deep Edwards-Stuart City trend. No reef talus has been found on the foreslope of the Deep Edwards trend but actually very little subsurface data are available from this area. The measurable slopes along the trend are considerably less than on the sides of the Mexican banks (about 2 degrees compared with 5–30 degrees in Mexico). The backreef Albian and Cenomanian beds in Texas grade into rather argillaceous strata at the outcrops. However, the Mexican area was far beyond the northern source of this material and is essentially one of pure carbonate with any marl confined to very thin and stylolitic layers between basinal limestone strata.

Significant differences existed in the later geologic history of the two areas. Carbonate bank development lasted longer in Mexico and was completed only at the end of Cenomanian time rather than in Late Albian time as in Texas. Neither was the Deep Edwards-Stuart City reef trend exposed to as much weathering and karst-producing solution. No large cavernous oil pools are known in Texas as there are in Mexico's Golden Lane. Tertiary orogeny, which tilted and periodically exposed the Golden Lane bank and uplifted the western banks into the Sierra Madre, did not affect the Gulf coast. Only along the former topographic crest is secondary leaching porosity developed in the Stuart City trend. Dolomitization is the important factor in all the backreef Texas fields.

Middle and Lower Cretaceous Facies in the Middle East

At the present time about two-thirds of the world's known petroleum reserves outside of the USSR reside in the Middle East in areas surrounding the Persian Gulf. The reservoir strata include chalk, carbonate sands, dolomite, and quartz sand of Mesozoic and Tertiary age. An appreciable part of the productive section is of Cretaceous age. The Burgan Sandstone in Kuwait, which presently forms the world's largest oil field, is Albian in age. The equivalent Nahr Umr Shale in the Middle East is conjectured to be the source bed for much of the Cretaceous oil and possibly even for Tertiary production. Some of the most interesting reservoir facies in the geologically intriguing Persian Gulf area are found in Cretaceous rudist and lime sand shelf margin buildups. Such strata have been extensively studied petrographically, originally for paleontology and later for sedimentologic interpretation, some results of which are given below.

Stratigraphy and Tectonic Framework

Table XI-2 illustrates the major stratigraphic units for the Early and Middle Cretaceous. These are mapped on Fig. XI-11. The facies are largely controlled by clastic influx off the Arabian shield. The terrigenous facies gives way progressively east and south of the shield to massive, pure limestones deposited both north and south of the center of the Persian Gulf.

Table XI-2. Middle and Lower Cretaceous correlations in the Middle East

Cretaceous stage		Iraq	Arabia NO. Persian Gulf	Arabia Qatar	Iran
Lower Wasia	Albian	Upper Qamchuqa lst. / Jawan	Mauddud lst. / Burgan / Nahr / Umr	Mauddud / Nahr Umr shale	Bangestan / Kazhdumi
Thamama group	Aptian	Lower Qamchuqa lst.	Shuaiba lst.	Shuaiba lst. / Hawar shale	Dariyan / Gadvan
Thamama group	Barremian			Karaib lst.	Upper Khami group (Fahliyan)
Thamama group	Neocomian	Sarmord shale and lst.	Sulaiy mudstone	Yamaha grainstone / Sulaiy lime mudstone	

The major positive element is the Arabian shield which stretches north into Jordan and western Iraq. A northward projection, the Mosul block, remained high during the Cretaceous and was enough protected from terrigenous influx to act as a nucleus for a halo of carbonate platform sediments.

Two major troughs existed east of the Arabian shield. The northern one was a part of the Zagros geosyncline and stretched from Turkey to the west central Iranian coast. It is somewhat elongate. Its east side was sediment-starved, and deep, from Jurassic until well into Cretaceous as indicated by a thin section of dark, muddy, pelagic sediments, including radiolarites. The northern Arabian shield furnished a considerable amount of coarse clastics. These poured into the southwestern part of the geosynclinal trough which is generally termed the Basrah basin; as much as 1000 m of mixed terrigenous and limestone sediment accumulated here during Early and Middle Cretaceous. Quartz sands began to appear from the Arabian source area in Iraq during the earliest Cretaceous and the terrigenous influx progressed southward throughout the Cretaceous period. The Albian-Cenomanian sands of the Wasia Formation are best developed at the southern extremity of the Hasa coast at Bahrein Island. All of these sands, Garagu (Lower Cretaceous of Iraq), Zubair-Byadh (Aptian-Barremian of Saudi Arabia), and Wasia (Albian of Arabia) are alluvial to the west and have shoreline facies in the subsurface beneath the present shoreline of the Persian Gulf. The clays were washed eastward into the geosyncline where they form the dark shales and silts of the Sarmord, Ratawi, Nahr Umr, and Kazhdumi, the obvious source for much of the Middle East oil.

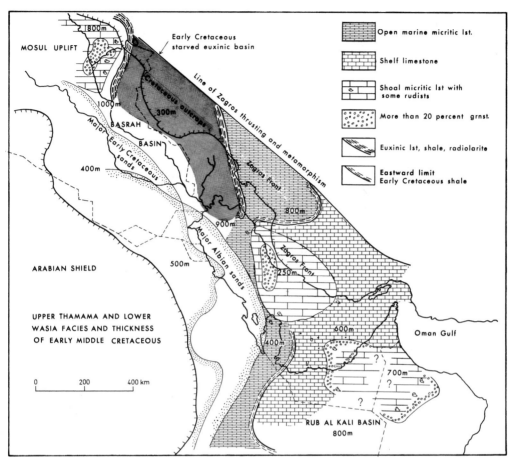

Fig. XI-11. Lower and Middle Cretaceous (Barremian to Albian) facies and thicknesses in the Persian Gulf area of the Middle East (Upper Thamama and Lower Wasia Formations). Based on data compiled by the author in 1964

The Basrah basin ends southward at an eastern projection of the Arabian shield marked by the Qatar peninsula. This and the southern Iranian coastal bulge are but surface reflections of a large uplift of Mesozoic age. This Qatar-Surmeh high acted as a nucleus for carbonate shoal sand and rudist development during Jurassic and Cretaceous time and great thicknesses of Mesozoic limestone in the Zagros geosyncline result from carbonate buildups on the flank of this subsiding but relatively positive area. The chalky, shallow water carbonate strata over the center of this high thin to a half or quarter of the basinal thickness.

The southeastern Arabian and eastern Persian Gulf area was the site of a second basin, a great shallow tectonic depression underlying the present Rub al Khali desert. Unstable troughs and borderlands of the Oman, Hadramaut, and Yemen areas lay to the south and east, but this immense equidimensional basin, 1000 km across, subsided comparatively quietly during most of Cretaceous times.

Toward the beginning of the Cretaceous, evaporite seas withdrew from the Arabian shield into the basin. Not much of the Early and Middle Cretaceous terrigenous clastic influx reached it and the section deposited during this time, is mostly limestone. In places farthest removed from the source of clastics in northwestern Arabia, shoal carbonates of Cretaceous age tended to develop on the eastern extremity of the Rub al Khali basin forming grainstone and rudist reservoirs for oil along the Trucial Coast and in the Persian Gulf itself. Some of the most detailed analyses of Middle East Cretaceous facies derive from studies here.

Facies Patterns

Mosul Block: The Aptian and Albian facies surrounding the Mosul Block have been described and mapped regionally by Dunnington (1958). Previously, extensive thin-section studies, which laid the groundwork for biostratigraphy and environmental analysis, were carried out by F.R.S. Henson of Iraq Petroleum Company. Detailed information on the Mesozoic geology of Iraq rests almost entirely on the published work of these two men, which also reflects a large amount of man-years by a talented staff of British and Swiss geologists.

The area over the subsurface Mosul block consists of about 200 m of partly eroded Lower and Middle Cretaceous, but almost 1000 m is present eastward along the edge of the shoal limestone facies. Still farther east the beds thin into the Zagros geosyncline passing into a dark argillaceous limestone facies. Along the Iranian-Iraqi border in the Pir-i-Mugrun ridge, Middle Cretaceous limestones are seen grading to globigerinid marls in a southerly direction (Henson, 1950, Fig. 14). The strike of the oil-producing structures of northern Iraq cuts directly across the belts of thick carbonate facies. Figure XI-12 is a greatly exaggerated cross section representing 700 m of limestone from the Mosul block southeast down the axis to Kirkuk anticline to the outcrops in the frontal Zagros Mountains of Kurdistan, a distance of 200 km. This section includes Albian, Aptian, and Barremian strata whose age was determined from extensive micropaleontology by the Iraq Petroleum Company and whose carbonate petrography was determined by the author from the same thin sections.

The Albian western facies (Jawan Formation) is essentially shelf evaporite. This gives way eastward to a dolomitized carbonate mudbank which farther northeast of Kirkuk grades to *Orbitolina* peloidal wackestone and in the Kurdistan outcrops is represented by thin ammonite-bearing argillaceous limestone. The Albian carbonate bank is 300 m thick and the outcropping dark limestone measures 80 m.

Aptian-Barremian strata east of Mosul show more complex facies. The most shelfward strata are not preserved, having been eroded in both pre-Albian and pre-Senonian time. These western beds are foraminifera-rich shoal mudstones and wackestones with an algal flora of *Lithocodium* (codiacean) and *Diplopora* (dasycladacean). The foraminiferal genus *Orbitolina* is also common along with other large forms such as *Choffatella* and *Pseudocyclammina*. Under the eastern end of the Kirkuk anticline lies the edge of the Aptian-Barremian bank whose full, uneroded thickness is 450 m. The shelf margin consists of grainstones and some

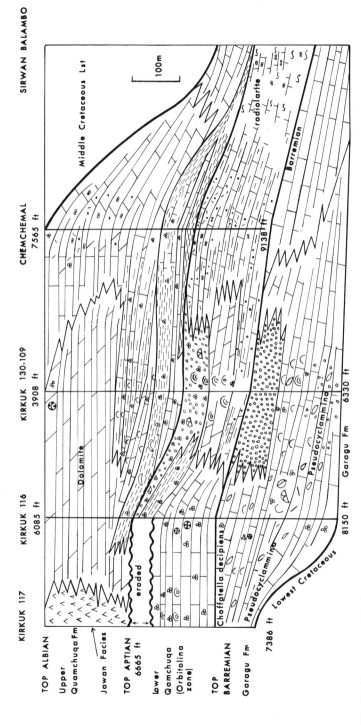

Fig. XI-12. Lower and Middle Cretaceous (Barremian to Albian) strata off the east flank of the Mosul block, Iraq in Iraq Petroleum Company wells along axis of Kirkuk anticline east to outcrops in Kurdistan, see Fig. XI-11. For legend see Fig. III-1. Depths in wells in feet. Distance along section about 200 km. Rudist reefs probably existed along the sand shoals indicated at shelf margins but the well data do not document this with certainty

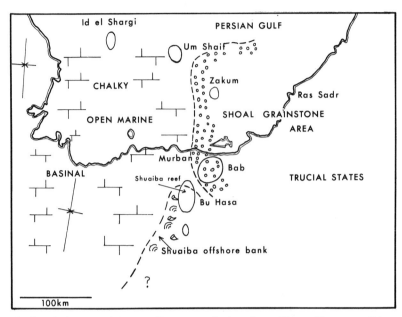

Fig. XI-13. Suggested Lower Cretaceous facies of the Upper Thamama Limestone of the Trucial Coast, Arabia. Shuaiba is uppermost unit of Thamama (see Fig. XI-14). The shoal grainstone area includes the whole Upper Thamama. It is not clear to what extent these shoal areas represent regional trends or to what extent they merely represent local shoal facies induced by structural growth over individual domes

reefoid strata with *Lithocodium* and rudists interstratified with shaly tongues of *Orbitolina* limestone and marl. Off the bank the facies changes rapidly to pelagic argillaceous limestone and in the outcrop to the east the equivalent is radiolarite and ammonite limestone totalling only 200 m. The facies pattern is typically regressive off the Mosul block, the Albian change of facies into thinner, true basinal pelagic strata occurring farther east than that of Aptian-Barremian beds.

Rudists appear to be present but not extremely common in either Albian or Aptian strata as interpreted from the well data, but subsurface studies may not give a correct impression. Henson observes major buildups and scattered patches of rudists forming what he termed "shoal banks or reefs." He interpreted the facies transition so remarkably displayed in the Pir-i-Mugrun, not far from the line of section on Fig. XI-12, as consisting of massive, much recrystallized rudist reef which grades through microdetrital forereef limestone with bioclastic rudist debris and *Orbitolina* to globigerinid lime mudstones.

Upper Thamama shoal area in the northern Rub al Khali basin: Over the central part of the Persian Gulf and into the basin, the upper or Aptian part of the Thamama, the Shuaiba Limestone, is separated from the main body of the Thamama (Karaib) by a widespread shale unit (Hawar). The Shuaiba was formed during an extensive inundation of the northern Arabian shield. During this event, extensive shoal deposits formed to the southeast in the Rub al Khali basin (Fig. XI-11). A series of thick grainstone-rudist buildups probably exists along this edge in the Shuaiba and Karaib limestones. However, it is not yet possible to evaluate the role of Cretaceous structural movement in localizing shoal facies on individual domes, major anticlinal trends, or shelf margins. It is known that local movements cause thinner deposition over structures but exact facies control is not easily demonstrated.

The usual, open marine facies of the Upper Thamama consists of chalky, peloidal, molluscan, and *Orbitolina*-rich wackestone and packstone with 20–40% grains. In places argillaceous streaks with attendant ball and flow compaction structures (causing nodular limestone) are present with hard-ground bored surfaces and represent still-stands of marine deposition. Burrowing is common in these micritic sediments. Porosity is uniformly above 20% in the micrite matrix but permeability is low. In the shoal limestone area a number of grainstone-packstone units occur in the upper 300 m of the Thamama. These average 30–40 m thick each and are separated by dense, stylolitic zones. The topmost of these units (A zone), the Shuaiba equivalent and the underlying Murban B zone produce considerable quantities of oil in the eastern Persian Gulf. Detailed petrographic study of the chalky but grainy reservoirs, has been made at Id El Shargi, Um Shaif, Zakum, and Murban Bab Dome. The main facies recognized are:

a) Rudist buildups or biostromes, probably mostly of monopleurids.
b) Molluscan-burrowed normal marine wackestone.
c) *Orbitolina* wackestones (Plate XXX B).
d) Rounded, coated, altered, chalky shell debris and algal lumps forming grainstones and packstones with high porosity and permeability. Dasycladacean remains occur but the most common alga is the irregular and encrusting codiacean, *Lithocodium*.

A particularly detailed petrographic study of the Shuaiba facies (Harris et al., 1968) has delineated a possible shelf margin trending across the northern edge of the major Bu Hasa structure south of Murban. Figure XI-13 is a sketch map of the area. The Bu Hasa dome is 35 km long and 18 km wide. It is gently arched with dips of about 2 degrees. The facies and related thickness outline an upper Shuaiba buildup which faces northward. The significance of such an obvious topographic relief at the very top of the Thamama is great. Porosity-permeability development is known regionally along the Albian-Aptian unconformity because of subaerial exposure and leaching of the chalky rock. The presence of depositional relief underlying the unconformity means an even better opportunity for reservoir development. Similar patchy rudist reef occurrences are known atop the Id El Shargi structure west of the major shoal limestone development.

In the Murban area, the Shuaiba reefy facies reaches 170 m and thins northeast off the flank of Bu Hasa dome to 40 m of dense basinal facies. The inclined slope of the original bank as interpreted from facies, is only $^1/_2$ degree. Equivalent strata in the large Bab dome, 20–30 km to the northeast, are wackestones indicative of offshore, open circulation. The Bu Hasa buildup may be an isolated low carbonate bank atop the structural culmination. Figure XI-14 is a cross section of the reefy margin taken from the above cited work by Abu Dhabi Petroleum Company geologists (Harris et al., 1968). The interpreted geologic history is as follows: during early Shuaiba time the whole area of Bu Hasa and Bab domes was the site of an algal biostrome of

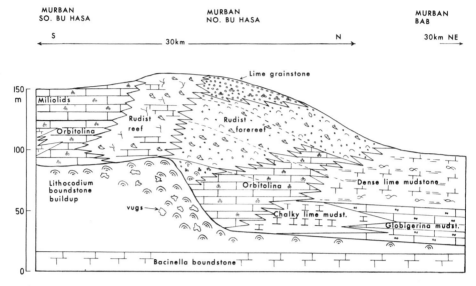

Fig. XI-14. South to north cross section of uppermost Shuaiba beds along Bu Hasa dome to Bab dome south of Murban on Trucial Coast of Persian Gulf. Vertical exaggeration X 200. For legend see Fig. III-1. For location see Fig. XI-13. After Harris et al. (1968)

Bacinella. This was followed by a deposit of chalky lime mudstone. Next a mat of *Lithocodium* developed across the whole area; at the northern end of Bu Hasa this grew rapidly enough to form a boundstone reef. This early reef stage shed some debris off to the deeper area to the north. This debris graded through *Orbitolina* wackestone to basinal lime mudstone. The *Lithocodium* reef was followed by a rudist reef, consisting of monopleurids and caprinids ? in a chalky micrite with much fracturing. Much rudist forereef debris accumulated on the open sea side. As reef growth stopped, presumably upon growth into wave base or to sea level, a detrital shoal facies developed which capped the forereef slope. This consists of bioclastic and lithoclastic grains. The southern end of Bu Hasa dome contains a facies interpreted as open backreef. It began with a lower *Orbitolina* wackestone and developed a miliolid wackestone at the top, probably indicating more restricted circulation. The facies described in these Aptian beds are almost identical to those of Mexico. The environmental significance of the double reef and why the higher rudist buildup replaces the *Lithocodium* reef is not clear.

A distinct facies sequence is also discernible in Cenomanian strata in both the Rub al Khali basin and in the northern Persian Gulf. These are associated with carbonate buildups also forming large banks such as are found in the Bangestan limestone of the Zagros in Iran and within major sedimentary cycles of the same age in Iraq. The typical Cenomanian bank facies consist of patches of whole rudists in a lime-mud matrix, or rudist-algal packstones (peri-reef deposits) or miliolid mudstones-wackestones of ponds or tidal flats. More open channels and lagoons on the banks are marked by wackestones with miliolids and the abun-

dant, large, elongate, foraminifera of genus *Praealveolina* (Plate XXX A). Offbank deposits consist of several facies; in sequence, away from the bank, these are:
1. Pure lime wackestone with mollusks, echinoids, and *Orbitolina* or marls with oysters, clams, and echinoids.
2. *"Oligostegina"* (calcisphere) limestone and marls.
3. Globigerinid marls.
4. Dark, bituminous, spiculitic radiolarian-bearing siliceous limestone and shale in the Zagros geosynclince basins.

The Cenomanian facies are thus like those of the Albian-Aptian with the notable addition of the alveolinid foraminifera in the bank deposits and the calcispheres in the immediate slope or bank margin marls. As in the Albian, identical microfacies are known in Mexico.

Chapter XII

Summary

This final chapter reviews the stratigraphic principles set down in Chapter II. It generalizes and describes in some detail nine standard facies belts developed along typical carbonate shelf margins; it characterizes several types of shelf margin profiles seen in the geologic record, and lists the common facies and trends displayed by individual buildups of both the basins and the shelves. At the end an attempt is made to interrelate the tectonic, hydrologic, climatic, substrate and organic parameters which control carbonate sedimentation. The prevalent tectonic settings of carbonate deposition are outlined, reviewing parts of Chapters II, IV, and VI.

Stratigraphic Principles

1. There is only a single common carbonate facies pattern. It is formed essentially from *in situ* accumulation of biochemically precipitated sediment. It consists of an inner, shallow, low-energy zone (undathem or shelf) and an outer, deeper, low-energy zone (fondothem or basin) which is separated by a shallow high-energy zone (clinothem or shelf margin).

2. This pattern is the natural development of maximum carbonate production and accumulation downslope from and peripheral to a positive element. Such a sedimentary pattern is typically initiated following a marine transgression and the inundation of a shelf. In initial stages a carbonate ramp may develop with a high energy zone close to shore but a platform with a seaward margin rapidly forms, dividing the facies into three parts.

3. The general arrangement of the above three belts may be subdivided environmentally into at least nine standard facies belts, described later in this Chapter.

4. The facies belts vary in width and uniformity, being narrower and more regular where the platform margin is steep. Where subsidence has been less and tectonic elements are not so well defined, the belts are wider and more diffuse.

5. Progradation: Thickness and stratigraphic patterns are the result of balance between subsidence and the rate of sedimentation. Calcium carbonate accumulation is very rapid when conditions are optimum. This results in prevalent regressive patterns even where considerable thicknesses of platforms and offshore banks attest to a great subsidence. Generally, the constructed depositional relief increases at the shelf margin as a basin and adjacent shelf subside.

6. Preserved sedimentary records formed during transgressions are known but are rare. Transgressive units generally are thin. Independent evidence exists that transgressions may be relatively rapid.

7. Carbonate basins may be filled by gradual deposition and regular sedimentary thickening from margin to center as continued subsidence takes place *or* by the progradation of shelf margins with either step-like or steady outbuilding of platforms.

8. A process of cyclic and reciprocal sedimentation fills basins which develop well-defined depositional topography at margins and which are affected by persistent sea-level fluctuations. The shelf grows up and out during high sea-level stands while the basin is sediment-starved; at lower sea-level stands the shelf remains above the equilibrium profile; sediment by-passes the shelf and fills in the basin.

9. On a simple isopach map of a carbonate time-stratigraphic unit, a positive area may be represented either by an abnormally thin or abnormally thick area, depending on the amount of carbonate buildup over buried structural highs, and on the depth of water over the "high" when carbonate sedimentation commenced.

10. In basins it is common for groups of "pinnacle reefs" to develop on the edges of submerged platforms parallel to major surrounding platforms. These are responses to later carbonate accretion either on structural or paleogeomorphic "highs."

11. In basins, positive linear trends (e.g., edges of basement fault blocks) may develop carbonate buildups. These linear buildups can be multiple and cut across basins or form their shelf margins.

12. Correlation along shelf margins may be difficult but electrical log tracing of clinoforms downslope, correlation of thin clastic zones of basin and shelf, careful use of detailed paleontology and seismic profiles aid in this procedure.

13. The principle of rapid prograding sedimentation is applicable also to the construction of the common shelf cycles characteristically seen on carbonate platforms. The calculated rates of shallow water carbonate accumulation derived from Holocene studies indicate that progradational accumulation is more rapid than paleontological zonation can resolve. Assuming that inundation is more rapid than gradual progradation, the best correlation lines are boundaries between cycles where a rapid change occurs upward to more marine conditions.

14. Shelf cyclicity is prevalent in many varied tectonic settings, in rock of many different types, and so widespread in the geologic record that one may admit it to be a normal geologic process.

15. Shelf correlations using sedimentary cycles may be checked by attention to key beds originating from presumed rapid inundation of planar surfaces (e.g., coals) or representing still-stands in sedimentation (concretion zones, glauconite beds, burrowed horizons), or representing geologically instantaneous events (zones of quartz silt formed by dust storms or bentonites formed from falls of volcanic ash).

16. Shelves also contain widespread, somewhat thin, carbonate rock units of fairly uniform facies and paleontology. These represent slow, extensive sedimentation beneath deep enough water so that open circulation prevailed and not much progradation of shallow water carbonate facies took place.

Nine Standard Facies Belts
in an Ideal Model of a Carbonate Complex

As indicated in Chapter II, recognition of a consistently recurring pattern of limestone facies in the ancient geologic record constituted an important advance in carbonate stratigraphy and sedimentology. The studies were further enhanced by using Holocene sediments to interpret these geologic examples. The evolution of the concept of carbonate facies belts has been in progress since about 1950 and has resulted in the development of essentially a single ideal model. This is applicable, with variations in thickness and regularity, to all the tectonic settings outlined in the latter part of this chapter. Figures II-5 and XII-1 depict this ideal sequence.

The reasons for development of the sequence have been given in Chapter II. The following section discusses the nine facies belts in more detail and should serve as a summary of the whole range of facies described in Chapter IV through XI. This model is patterned across an ideal shelf margin. Figure XII-2 illustrates profiles of some actual examples using this schema (see also Armstrong, 1974, Fig. 10). Types of basinal and deeper water carbonates are especially detailed in the following section. A latter part of this chapter describes the origin and facies sequence of carbonate mounds found principally along the open marine facies belts 2 and 7.

Belt 1 A. Turbidite and Leptogeosynclinal Deep Water Facies (Fondothem)

The deeper water facies belts described below are conveniently divided into those in: geosynclinal troughs (a) filled with continuously and rapidly deposited limestone turbidites, (b) quiet deep water, slow sedimentation with intermittent debris flows (leptogeosynclinal deposits of Trumpy, 1960).

a) Limestone turbidites: Geosynclinal troughs filled by allochthonous carbonates are not common in the geological record but they are impressive and several have been well described, e.g., Thomson and Thomasson (1969) on the limestone of the Marathon basin, West Texas; McBride (1970) on the Maravillas of the same area; The Flysch Calcaire and rhythmic limestones of the Alps (Lombard, 1956, Beaudoin, 1970), and the Apennines (Carozzi, 1955; Scholle, 1971) fall within this category. Meischner (1965) coined the useful term allodapic limestone for those basinal sequences of breccias, microbreccias, and lime sands derived from contemporaneously formed shelf and slope carbonate particles. These are commonly interbedded with calcareous pelagites and argillaceous strata. Thicknesses of such beds may vary but are often great. The sequences may contain exotic boulders and unusually coarse sediment. The allochthonous material may be emplaced by turbidity flows, by mass movement of debris, or even by volcanic eruption. Geosynclinal subsidence and sediment instability are apt to result in a thick and continuous record of deep-water environment with all the sedimentary structures and textures of terrigenous flysch. The troughs may be narrow and contain rapidly changing facies.

Facies Belt	Facies Profile	2nd Order Sedimentary Bodies	Standard Microfacies
1 BASIN (WIDE BELTS)			1 SPICULITE; 2 MICROBIOCLASTIC CALCISILT; 3 PELAGIC MICRITE; RADIOLARITE SHALE
2 OPEN SEA SHELF			2 MICROBIOCLASTIC CALCISILT; 8 WHOLE SHELLS IN MICRITE; 9 BIOCLASTIC WACKESTONE; 10 COATED GRAINS IN MICRITE
3 DEEP SHELF MARGIN		DEBRIS FLOWS AND TURBIDITES IN FINE LAMINATE STRATA. MOUNDS ON TOE OF SLOPE.	2 MICROBIOCLASTIC CALCISILT; 3 PELAGIC MICRITE; 4 BIOCLASTIC LITHOCLASTIC MICROBRECCIA
4 FORESLOPE (VERY NARROW BELTS)		GIANT TALUS BLOCKS. INFILLED LARGE CAVITIES. DOWNSLOPE MOUNDS.	4 BIOCLASTIC LITHOCLASTIC MICROBRECCIA. LITHOCLASTIC CONGLOMERATE. 5 BIOCLASTIC GRAINSTONE-PACKSTONE. FLOATSTONE. 6 REEF RUDSTONE.
5 ORGANIC BUILD UP		DOWNSLOPE MOUNDS. REEF KNOLLS. BOUNDSTONE PATCHES. FRINGING AND BARRIER FRAMEWORK REEF. SPUR AND GROOVE.	7 BOUNDSTONE. 11 COATED, WORN BIOCLASTIC GRAINSTONE. 12 COQUINA (SHELL HASH)
6 WINNOWED EDGE SANDS		ISLANDS, DUNES. BARRIER BARS AND PASSES AND CHANNELS.	11 COATED, WORN BIOCLASTIC GRAINSTONE. 12 COQUINA (SHELL HASH). 13 ONKOIDAL BIOCLASTIC GRAINSTONE. 14 LAG BRECCIA. 15 OOLITE
7 SHELF LAGOON OPEN CIRCULATION (WIDE BELTS)		TIDAL DELTAS. LAGOONAL PONDS. TYPICAL SHELF MOUNDS. COLUMNAR ALGAL MATS. CHANNELS AND TIDAL BARS OF LIME SAND.	8 WHOLE SHELLS IN MICRITE. 9 BIOCLASTIC WACKESTONE. 10 COATED GRAINS IN MICRITE. 16 PELSPARITE. 17 GRAPESTONE. 18 FORAM, DASYCLADACEAN GRAINSTONE.
8 RESTRICTED CIRCULATION SHELF AND TIDAL FLATS		TIDAL FLATS. CHANNELS. NATURAL LEVEES. PONDS. ALGAL MAT BELTS.	16, 17, 18. 19 FENESTRAL PELOIDAL LAMINATE MICRITE. 24 RUDSTONE IN CHANNELS. 21 SPONGIOSTROME MICRITE. 23 NON LAMINATE PURE MICRITE. 22 ONKOIDAL MICRITE.
9 EVAPORITES ON SABKHAS-SALINAS		ANHYDRITE DOMES. TEPEE STRUCTURES. LAMINATED CRUSTS OF GYPSUM. SALINAS (EVAPORATIVE PONDS). SABKHAS (EVAPORATIVE FLATS)	20 STROMATOLITIC MICRITE. 23 NON LAMINATE, PURE MICRITE. NODULAR-PEARL ENTEROLITHIC ANHYDRITE. SELENITE BLADES IN MICRITE.

Fig.XII-1. Synopsis of Standard Facies Belts reviewing second order bodies of sediment and standard microfacies associated with each belt. See Fig.II-4 for more complete description of the nine facies

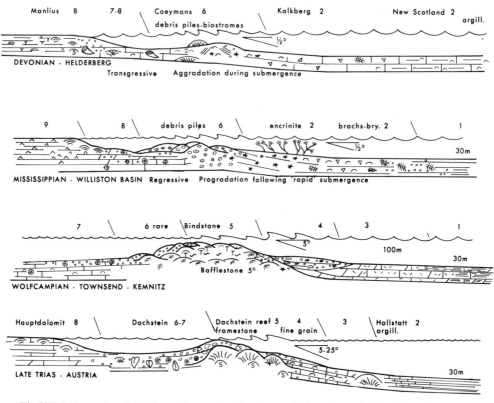

Fig. XII-2. Examples of stratigraphic profiles showing variations from the idealized and complete sequence of numbered facies belts. Note that steep profiles have narrower belts. The sections are from 160–300 km long. All the examples are discussed in the text except the Devonian Helderberg of New York (Laporte, 1969)

The Marathon-Ouachita geosyncline contains a 700 m thick record of such limestones and argillites of Ordovician age, capped by a dark cherty limestone (Maravillas Formation). The sections contain dark micritic limestones, microbreccias, shales, beds of conglomerates and exotic blocks, and some cherts. The whole Ordovician section has been described by McBride (1969, 1970) and Young (1970). Similar Ordovician strata are known in the interiors of both the Appalachian and Cordilleran geosynclines.

b) Leptogeosynclinal troughs: These do not contain extensive allodapic limestones, but remained deep during their history and were only intermittently sites of sediment from outside the basin. Their sediment is chiefly pelagic. Mostly siliceous sediment accumulated when protected from argillaceous influx and in water deep enough for extensive solution of calcium carbonate (modern compensation depth for calcium carbonate is 3000–4000 m). These sediments have much in common with those occurring in quiet cratonic basins (see below).

The Alpine-Mediterranean Tethyan troughs of Mesozoic age contain several special types of continental margin sediments which are discussed in Chapter IX and reviewed below. These thin, calcareous and siliceous sediments, rich in ferric iron, are considered to be products of basins removed from land and protected from much terrigenous influx. The soluble iron supposedly derives from a few rivers draining a tropical land. The hinterland contained extensive salt flats and probably not much run off occurred. Pelagic sedimentation existed widely and at varying depths. Several local Jurassic troughs within the Austroalpine nappes of Austria have been carefully studied (the Unkenthal by Garrison, 1967, and Garrison and Fischer, 1969; the Berchtesgaden area by Jurgan, 1969; the Glasenbach near Salzburg, Bernoulli and Jenkyns, 1970; and the Tauglboden by Schlager and Schlager, 1973). Bernoulli and Jenkyns (1974) reviewed these sediments thoroughly.

The following rock types are known, mostly in starved geosynclinal troughs and on the deeply submerged swells within them.

1. Radiolarites (J 12, Chapter IX): Siliceous, very fine-grained laminites with radiolaria and other tiny siliceous organic particles. Very fine, paper-thin layers of siliceous shale are interbedded with radiolarian layers.

2. Red biomicrite and red nodular limestone (J 17, Chapter IX): Beds of ferruginous micrite with less than 5% argillaceous material, with pelagic microfauna; largely dissolved former aragonitic bioclasts; some ferruginous manganese crusts. Peculiar conglomerate and rubble beds are caused in part by *in situ* solution on the sea floor, of calcitic sediment deposited at about the compensation depth for aragonite. Characteristically such sediment occurred on swells (high areas) within the geosyncline and may be traced in some areas as slumped and resedimented masses moved downslope to trough positions. A related microfacies is ferruginous pisolite in a red biomicrite matrix.

3. Light-colored pelagic lime mudstones with thin interbedded allodapic units (J 14, Chapter IX). Such strata are evenly and planar bedded. Much silt-size and finely calcarenitic material as well as micrite exists.

4. Dark basinal micrite and spiculitic limestone with some micropeloids (J 12, Chapter-IX). This limestone is similar to the light-colored pelagic facies but contains higher amounts of organic matter. It is commonly cherty and contains *Chondrites* traces (burrows). Probably such beds represent the distal ends of turbidites.

5. Pelagic micropeloid beds and tiny coated particles with ammonoids, Radiolaria, and calpionellids (J 13, Chapter IX). These beds may be allodapic and are distal turbidites.

6. Bioclastic grainstones and packstones composed of ammonites, small gastropods, pelagic bivalves, and echinoderms. Most of such sediment must be a product of current deposition, representing bars or megaripples formed at some depth on the sea floor. Two special types are listed below, numbers 7 and 8.

7. "Filamentous" microcoquinoids of very thin shells of pelagic bivalves such as *Halobia*, *Daonella*, *Posidonia* (J 15, Chapter IX).

8. Red to pink encrinites. This characteristic rock type represents accumulated echinodermal debris in a nonreducing environment. The significance of the red color of this and other rock types in deep water environments is discussed below.

9. Spiculite: accumulation of very light, mostly hexactinellid sponge spicules of opaline silica. Probably the spicules washed down into the basin centers from the slopes.

10. Sporadic debris flows of calcareous breccia, ranging from coarse-grained to micro-breccia. Fragments were mostly consolidated before deposition and were derived from nearby shelves and slopes.

11. Ferro-manganese crusts and nodules on limestone; these are products of very slow sedimentation and concretionary growth in deep water. Manganiferous nodules are present in the modern sea floors at great depth and are not uncommon in Austroalpine Tethyan Mesozoic strata.

The red-purple color common in some of these above sediment types is an indirect result of the slow sedimentation (Fischer in Mesolella et al., 1974). Ferric oxide, normally deposited in most marine sediment is commonly slowly reduced during the decay of organic matter entrapped in the sediment. In areas of slow sedimentation bacterial decay on the sea floor can, before burial, oxidize all organic matter whose absence within the sediment after burial prevents the ultimate reduction of the ferric iron. Pink encrinites, such as the Moccasin (Ordovician) of the Appalachians, Niagaran of the Michigan basin, Early Cretaceous of the Oman geosyncline, and the Hierlatzkalk (Lias) of the Austroalpine Alps are all basinal sediments, mostly in geosynclines. Nonferruginous sediments are naturally a light color in an oxidizing environment.

The geosynclinal rock types outlined above are normally cherty, the chert being in beds and nodules crudely parallel to the bedding. The silica is derived at early diagenetic stages from solution of the opaline silica of microplankton and spicules entombed in the sediment. It precipitates in replacement and void fillings, first as crystobalite which later alters to chalcedony. Petrographic study often shows much mutual replacements of silica and calcite, inferring multiple stages of varying pH conditions. The replacement and cavity filling begins very early in diagenesis and continues over a long time-period.

Sedimentary structures are simple, the strata away from the slopes are thin-bedded and commonly laminate; bioturbation is rare or absent. In places thick beds of breccia occur, channelling down into the even, planar beds and shovelling them up, crumpling and contorting them. Syngenetic conglomerates and breccias formed by seafloor solution and concretion growth are unique. Unlike terrigenous clastic sediments, even the thin lime mudstones show some early consolidation. Flame structures, load casts, and convolute bedding are rare to absent. Flaser (ball and flow) structure and evidence of resedimentation is known in slope areas. Sedimentary structures have been described in detail by Schlager and Schlager (1973).

The biota of geosynclinal strata is chiefly pelagic and identical with that described below under starved basin limestones.

Belt 1 B. Cratonic Basin Carbonates (Starved and Mostly Euxinic) (Fondothem)

These sediments are formed in centers of deep intracratonic and marginal cratonic (miogeosynclinal) basins, areas removed from coastlines or from the influence of carbonate-producing shelf areas. The deposition is dependent on the amount of influx of fine argillaceous and siliceous material and the rain of decaying plankton, for the water is too deep and dark for benthonic production of carbonate. Contributions from the land are slight in a typical carbonate-producing shelf or basin and come mainly from wind-blown material; consequently a starved and somewhat deep basin commonly results.

Sediment is formed generally in euxinic conditions developed below oxygenation level and below wave base. The water depth is at least 30 m and generally several 100 m. The bottom water flowing off of the surrounding shelves may become hypersaline and dense, preventing easy turnover. This situation causes oxygen deficiency which, coupled with the constant rain of decaying plankton, increases the tendency toward stagnant, reducing environments. Few burrowers can exist here and fine laminated sediment results.

a) *Prevailing rock types:* Dark thin beds of limestone with dark shale or silt and some thin-bedded anhydrite. In some evaporite basins thick halite may constitute a later fill-in deposit.

b) *Colors:* Dark brown or black (up to several percent bituminous organic matter). In some basins reddish colors prevail (see preceding discussion under leptogeosynclinal facies).

c) *Grain types and depositional texture:* Lime mudstones and calcisiltites, micropeloids and microbioclasts. Crinoidal accumulations are also present.

d) *Bedding and sedimentary structures:* Very even planar mm lamination, ripple cross-lamination, small scale rhythmic bedding consisting of even beds of limestone intercalated with thin shales.

e) *Terrigenous clastics:* Somewhat admixed with carbonates and also interbedded in thin layers; quartz silt and shale. This material is windblown as well as water-carried. Chert is very common, probably derived during early diagenesis from opaline organisms and later from solution and replacement of quartz silt by carbonate.

f) *Biota:* Exclusively nektonic-pelagic fauna preserved in local abundance on the bedding planes. Mass mortality of pelagic organisms is believed responsible for these accumulations. Megafauna includes graptolites, planktonic bivalves, ammonites, and sponge spicules. Microfauna is thoroughly admixed with fine sediment. It includes calcareous calpionellids, tintinnids and calcispheres, and siliceous radiolarians and diatoms.

Belt 2. Shelf Facies (Deep Undathem)

The water is tens of meters or even a hundred meters deep, generally oxygenated and of normal marine salinity, with good current circulation. The depth is sufficient to be below normal wave base but intermittent storms affect bottom sediments. Such shelves are generally wide and sedimentation is quite uniform. This is the typical realm of deeper neritic sedimentation and may consist of carbonate or shale. Whereas the stratigraphic record is replete with interpreted examples of widespread neritic shelf deposits, there is no modern model for this type of sedimentation. Interpretations are based mainly on the rock record. Further, this facies belt is very similar to Belt 7 which represents an open circulation shelf existing "inside" a shelf margin barrier. Distinction between Belts 2 and 7 is important but at present far from clear.

a) *Prevailing rock types:* Very fossiliferous limestone interbedded with marl. Well segregated beds.

b) *Colors:* Gray, green, red, and brown due to variable oxidizing and reducing conditions.

c) *Grain types and depositional texture:* Bioclastic and whole fossil wackestone. Occasional beds of winnowed bioclastic grainstone and coquina. Much pelleting of micrite matrix. Some calcisiltite.

d) *Bedding and sedimentary structures:* Sediment thoroughly burrowed, beds homogenized. Thin to medium, wavy to nodular beds. Ball and flow structures

are common in argillaceous limestones. Bedding surfaces ordinarily show diastems and lag concentrates of fossils. Mud mounds and pinnacle reefs occur.

e) *Terrigenous clastics:* Quartz silt, siltstone, and shale commonly interbedded
with limestones in clearly segregated layers.

f) *Biota:* Very diverse shelly fauna indicating normal marine salinity; preservation of both infauna and epifauna. Fauna may not be very abundant in places
but is generally present. Notable presence of stenohaline forms such as brachiopods, corals, cephalopods, and echinoderms.

Belt 3. Basin Margin or Deep Shelf Margin Facies (Clinothem)

This facies is formed at the toe of slope of a carbonate-producing shelf. Carbonate
sediments consist of contribution from pelagic organisms plus fine detritus
moved off from adjacent shallow shelves. The water is at least the same depth as
Facies Belt 2 and perhaps 200-300 m deep. The basin is situated generally below
wave base and barely at oxygen level. The strata are chiefly thin, well-segregated
carbonate beds with minor interbeds or mere partings of clayey and siliceous
material, much of the fine terrigenous matter having drifted or blown farther out
into the basin. These rocks may resemble basinal sediments but are less argillaceous and somewhat thicker. Some of these rhythmic, thin-bedded limestones are
hundreds of meters thick.

a) *Prevailing rock type:* Fine-grained limestone, in some places cherty.

b) *Color:* Dark to light.

c) *Grain types and depositional texture:* Mostly lime mudstone with some
calcisiltite and including some microbreccia beds, and coarser bioclastic-lithoclastic packstones.

d) *Bedding and sedimentary structures:* Some beds are laminated lime mudstones; in even rhythmic (flysch-like) beds. Other thicker-bedded units are of
massive unlaminated lime mudstone. Some beds are graded. Megaslump features
within evenly bedded limestone cause major discontinuities in the bedding. It has
been suggested that large scale but shallow channeling due to density currents
causes the discontinuities in these even-bedded lime mudstones. Occasionally
micrite bioherms occur.

Some beds are laminated lying above micrograded lithoclastic and bioclastic
debris from upslope, i.e., beds of allodapic limestone. Sporadic turbidites, exotic
blocks, and debris flows are found in sequences relatively close to shelves in
unstable areas.

e) *Terrigenous clastics:* Rarely present except as fine shale partings. Chert is
common.

f) *Biota:* The bioclastic detritus derived principally from upslope. The fauna is
open-shelf and normal marine but may be a mixture of older forms derived from
the shelf, benthonic organisms living on the slope and some pelagic forms.

Belt 4. Foreslope Facies of Carbonate Platform (Marine Talus, Clinothem)

The slope is generally located above the lower limit of oxygenated water and
stretches from above the wave base to below it. The material is debris deposited

on an incline formed seaward as the ramp grows. This may be as steep as 30 degrees. The sediment is somewhat unstable and varies greatly in size and shape. There may be bedded, fine-grained layers with megaslumps; some foreset bedded and wedge-shaped strata consist principally of lime sand and of mound- or lens-shaped masses of trapped and stabilized, fine-grained carbonate sediment.

a) *Prevailing rock types:* Variable types of limestone depending on the water energy upslope; lime muds and sands, boundstone, and sedimentary breccia.

b) *Color:* Dark to light.

c) *Grain type and depositional texture:* Lime silts and bioclastic wackestone-packstone. Lithoclasts of varying shapes and sizes are derived from cemented strata upslope. Much reworked material with locally derived organic debris, reef rudstone.

d) *Bedding and sedimentary structures:* Megaslumps in thinly bedded strata; large scale foreset bedding (wedges); large exotic blocks interrupting bedding; slope mounds of fine-grained sediment; syngenetic slumps, pull-aparts, and breccias; clastic injection dikes and fissure in-fills.

e) *Terrigenous clastics:* Mostly pure carbonate but some shale, silt, fine sand drifted downslope from above and mixed with carbonate or filling cavities.

f) *Biota:* Mostly bioclastic debris from upslope but also colonies of in place encrusting organisms. The facies may be very fossiliferous, its fauna varied and open marine.

Belt 5. Organic Reef of Platform Margin

The ecologic character varies dependent on the water energy, steepness of the slope, organic productivity, amount of frame construction, binding, or trapping, frequency of subaerial exposure, and consequent cementation. Three types of profiles with linear shelf-margin organic buildups may be discerned:

Type I: Downslope accumulations of carbonate mud and organic detritus.
Type II: Ramps of knoll reefs with intervening bioclastic sands.
Type III: Frame-constructed reef rims.

These types of shelf margins are discussed in the following section.

a) *Prevailing rock type:* Massive limestone and dolomite in places consisting solely of organisms. Also much bioclastic debris.

b) *Color:* Light.

c) *Grain types and depositional texture:* Masses and patches of organic boundstone. Interstices may be filled with lime mudstone in downslope reefs or banks and with grainstone and packstone in upslope accumulations. Some mounds formed by clumps of organisms growing in upslope position have only mud matrix owing to the protection from winnowing by reef frame. Interstices in boundstone of higher energy reefs are filled with lime sand and gravel. Inter-mound areas consist commonly of grainstone and packstone.

d) *Bedding and sedimentary structures:* Massive organic framework with constructed (roofed) cavities. Lamination is caused by organic growth; it swells and thickens upward. In mounds with considerable lime-mud matrix, stromatactis-like structure is common. Brecciation and fissuring of massive buildups may occur and injection dikes are present.

e) *Terrigenous clastics:* Essentially absent.

f) *Biota:* Colonies of sessile framebuilding organisms may or may not dominate. Growth form is determined by the water energy. The forms may also be low-
lying and encrusting. Ramose or dendritic forms exist in more protected places.
Communities of organisms, dwelling in various ecologic niches, may form beds of
abundant accessory organisms (e.g., layers of brachiopods, molluscs and crinoids).

Belt 6. Winnowed Platform Edge Sands

These take the form of shoals, beaches, offshore or tidal bars in fans or belts, or
eolianite dune islands. Sites of such marginal sands range from well above sea
level to 5 or 10 m deep. Much clean sand is winnowed and deposited by waves,
tidal, or longshore currents of 1-2 knots. The salinity is normal marine because of
good circulation. The environment is well oxygenated but not hospitable to marine life because of the shifting substrate.

a) *Prevailing rock type:* Cross-bedded calcareous or dolomitic lime sand.

b) *Color:* Light.

c) *Grain types and depositional texture:* Rounded and fairly well-sorted grainstones. Some are coated and oolitic. Others are merely rounded bioclasts.

d) *Bedding and sedimentary structures:* Marine sands with medium to small-
scale festoon cross-bedding. Eolianites have large-scale cross-bedding with dips of
more than 25 degrees. Surfaces representing stratigraphic hiatus are common in
both subenvironments. Eolianites have preserved old soil horizons and root casts.

e) *Terrigenous clastics:* Quartz sand may be present with the calcarenites.

f) *Biota:* Worn and abraided coquinas of benthonic animals living on reef and
foreslope are common. Few indigenous organisms occur because of the shifting
substrate. Large bivalves (megalodonts) or gastropods are common, as are fragmented remains of large dasycladacean algae and certain foraminifera. Such
forms are prevalent in these strata throughout the geologic column.

Belt 7. Open Marine Platform Facies (Shallow Undathem)

Geographically such environments are located in straits, open lagoons, and bays
behind the outer platform edge. The general term shelf lagoon is applicable.
Water is shallow, generally a few meters to tens of meters deep. Salinity varies
from essentially normal marine to somewhat higher; circulation is very moderate.
Water conditions are favorable for organisms but often the stenohaline forms are
excluded. The sediments are texturally varied but contain considerable amounts
of lime mud.

a) *Prevailing rock types:* Variable limestone and in some cases lenses and thin
beds of land-derived clastics.

b) *Color:* Light and dark.

c) *Grain types and depositional texture:* Great variety of textures, grainstone
to mudstone, e.g., lenses of lime sands commonly of shelly and angular fragments
or lumachelles (coquinas of essentially whole shells), beds of bioclastic wackestones; mounds and lenses of organically produced and trapped sediment; biostromes.

d) Bedding and sedimentary structures: Medium to platy bedded; burrowing and pelleting of sediment common. When clay is admixed, ball and flow compaction structures are present, as well as nodular and wavy bedding.

e) *Terrigenous clastics:* When present, generally in well-segregated beds intercalated with limestones.

f) *Biota:* Fauna may be abundant with mollusks, sponges, arthropods, foraminifera, and algae particularly common. Patch reefs are present; abundant marine grasses and trees play important roles in trapping and stabilizing fine sediment in shallow water. Organisms requiring normal marine salinity are present but may be rarer than in the open sea, e.g., brachiopods, cephalopods, echinoderms, and red algae.

Belt 8. Facies of Restricted Circulation on Marine Platforms

This facies includes mostly fine sediment in very shallow, cut-off ponds and lagoons which have restricted circulation and hypersaline water. Geographically the lagoons may be classified into those behind or between barrier reefs, those formed behind coastal spits, or those within atolls or faros. They are generally, but not exclusively, shallow. This belt also includes the highly diverse and well-known intertidal environments. Lime mud is characteristic of these deposits which form on levees, intertidal flats, ponds, and marshes. Coarser sediment exists in tidal channels and local beaches. The highly variable conditions here include fresh, salt, and even hypersaline water; areas of subaerial exposure; both reducing and oxygenated conditions, and vegetation of both marine and fresh-water swamp areas. Wind-blown clastics may contribute significantly in some places. All of these constitute a stress environment for organisms. Diagenetic effects are strongly marked in the resulting sediment.

a) *Prevailing rock types:* Generally lime muddy sediment with much dolomite.

b) *Color:* Light.

c) *Grain type and depositional texture:* Highly varied, most sediments consisting of lime mud; grainstone rare except for thoroughly pelleted sediment. Channels contain lithoclastic grainy sediment; clotted, pelleted mudstones and wackestones are most common.

d) *Bedding and sedimentary structures:* Much laminated mudstone, fenestral (birdseye) fabric, algal stromatolite, small-scale graded bedding, dolomite and caliche crusts. The channel sands show cross-bedding.

e) *Terrigenous clastics:* Rare except for wind-blown material; where present, usually in well-segregated layers.

f) *Biota:* Very limited fauna und flora, mainly gastropods, algae, foraminifera (miliolids), and ostracods. These organisms may occur locally in great abundance.

Belt 9. Platform Evaporite Facies

The supratidal and inland pond environment of the restricted marine platform developed in an evaporative climate—the areas of sabkha, salinas, salt flats. In-

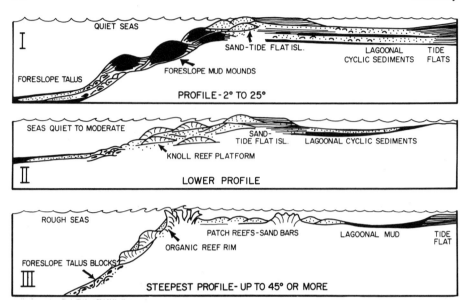

Fig. XII-3. Three types of carbonate shelf margins: I, downslope lime-mud accumulation; II, knoll reef ramp or platform; III, organic reef rim. From Wilson (1974, Fig. 1), with permission of American Association of Petroleum Geologists

tense heat and aridity is common, at least seasonally. Marine flooding is sporadic. The addition of gypsum or anhydrite, formed from evaporative concentration of sea water in the sediments, is both depositional and diagenetic. Precipitation in and replacement of the original sediment occurs as the dense brines move down through the sediment or are pulled to the surface through evaporation of interstitial waters. The precipitated sulfate is unstable and may deform through crystal growth, intake of water of crystallization, or compaction.

a) *Prevailing rock types:* Nodular and wavy anhydrite, or gypsum interlaminated with dolomite. Such rock types are commonly associated with redbeds.

b) *Color:* Highly variable, red, yellow, brown.

c) *Grain type and depositional texture:* Very fine grain carbonate sediment; gypsum/anhydrite crystals often form a felted mat of tiny lath-shaped crystals or, when secondary, are large, bladed, and poikilotopic.

d) *Bedding and sedimentary structures:* Laminate, both wispy and planar types, mud cracks, stromatolite and spongiostrome structure, gypsum rosettes and selenite blades (pseudomorphed by anhydrite), and syngenetic, diagenetic, and deformational structures such as nodular and chickenwire (flaser) structure, enterolithic (contorted) folding. Also diastem surfaces and caliche crusts are common.

e) *Terrigenous clastics:* May be very common, redbeds and windblown sediment.

f) *Biota:* Almost no indigenous fauna except for blue-green algal stromatolites, brine shrimp.

Three Major Types of Shelf Margin Profiles

The numerous examples of shelf margin profiles presented in Chapter IV through XI may be grouped into three classes (Wilson, 1974): I-Downslope carbonate mud accumulations, II-Ramps of knoll reefs, and III-Framebuilt reef rims (Fig. XII-3). The classification is based on the slope of the stratigraphic profile and the composition, shape, and disposition of subunits making up the marginal carbonate mass (e.g., mud mound cores, frontal knoll reefs, reef flats, flanking beds, patch reefs, capping beds, lime sand bars, eolianite dunes, aprons of slope sediments and talus conglomerates). The above grouping is empirical and its rationale is incomplete but the recognition of certain types of shelf margins can be a valuable tool for both outcrop mapping and subsurface petroleum exploration. It may aid in the prediction of the steepness of the slopes off the shelves and in understanding the character of the sediment expected along them.

Type I. Downslope Mud Accumulations

These are linear trends of bioclastic lime mud or belts of mounds located on the foreslope of the shelf margin with upslope sand beaches and islands. The downslope sediment contains variable amounts of somewhat limited and specialized sessile organisms which tend to trap and baffle sediment. Thus, the downslope muds may occur as superimposed (stacked-up) mounds, usually loaf-shaped bodies although such forms are not always observable. Locally, the slopes of such accumulations may be steep—from 25 to 30 degrees, about the angle of repose of constituent sediment, but the overall slope into the basin may vary from a degree or two, to as much as 25 degrees. This variation may be tectonically controlled because organic construction is not a factor in such buildups.

On steep slopes the carbonate mud may accumulate far enough downslope to be below the photic zone, and perhaps in water a hundred or so meters deep, although the sediment is derived from much shallower water. In such cases where clay is transported into the basin, well-defined mounds are separated by dark shales.

Where gentle foreslopes prevail, carbonate mud mounds will still develop but in shallow, photic zone water. The tops of such individual mounds may reach into the zone of wave action. When this occurs, organic growth on the tops may develop a reef framework and grade into Type II or III. But, even where this is observed, the uppermost part of the slope, which is generally well within the turbulent zone, is composed essentially of lime sand shoals, beaches, dunes, and islands, and normally contains few or no framebuilding and sediment trapping organisms. Examples are:

1. Capitan Formation of the Permian Reef Complex of the Guadalupe Mountains of New Mexico and West Texas (Achauer, 1969; Dunham, 1972).

2. Phylloid algal shelf-margin buildups of Pennsylvanian and Early Permian age in the southwestern U.S.A., (Plumely and Graves, 1953; Wray, 1962; Wilson, 1967a; Heckel and Cocke, 1969). Capped in places by tubular foraminiferal-*Tubiphytes* boundstone creating Type II.

3. Waulsortian (Lower Carboniferous) mounds in Belgium, England, Ireland, and North America (Hudson, 1930; Pray, 1958; Lees, 1961, 1964; Troell, 1962; Cotter, 1965).

4. Sponge-algal reefs of the Weiss Jura (Upper Jurassic) of Schwabia (Gwinner, 1962, 1971; Hiller, 1964).

5. Stromatoporoid-tabulate coral lime mud mounds of the Late Devonian (Frasnian) age in the Dinant basin, Belgium (Lecompte, 1956). Growth upward into wave base results in cap of stromatoporoid boundstone and development of Type II.

No exact modern analogs of these downslope mud mounds have been described, perhaps because we have not searched in the correct places for them. Such localities might exist north of the Great Pearl Bank on the gentle slope into the axis of the Persian Gulf.

Interestingly, dives in submersibles in the Florida straits off Bimini have encountered loaf-shaped current-oriented piles of fine and coarse lime sediment supporting growth of abundant organisms (Neumann et al., 1972). These are now at depths of about 700 m and may be analogous to some downslope bioherms seen in the geologic record. For example, small mounds of lime mud are seen at the foot of the Capitan depositional slope in McKittrick Canyon. The actual depth of the formation of these is dependent on the amount of sea-level drop during Late Permian time. Their position low on the slope would indicate a possibility of depth of at least 100-200 m.

Type II. Knoll Reef Ramps

These ramps consist of linear belts of ecologic (framebuilt) knoll reefs on gentle slopes at the outer edge of the shelf margin. These apparently began growth at the normal wave base or at positions a little farther downslope at depths of a few tens of meters. In the absence of strong waves or currents, there was little massive framebuilding but many sessile and encrusting organisms flourished. The framebuilders form chiefly branching and fasciculate colonies. Vertical ecologic zoning is common within the knolls, the growth form changing upward generally to massive encrusting forms.

The buildups are produced fully as much by organic productivity, binding, trapping, and encrusting, and by lack of removal of debris as by organic *in situ* frame construction. Interreef material may be volumetrically much greater than the patches of framebuilding organisms. The vast quantity of debris is mostly bioclastic and was derived from prolific growth on the tops of the reef knolls and not necessarily from destruction of the organic framework or previously lithified material. Water energy was sufficient to remove only the finest debris; hence mostly lime sand accumulated in interreef areas. Cores of the reef knolls commonly contain lime mud because in this realm of only moderate wave action, the framebuilders offered enough protection to prevent its removal. Moderate surge over the cavernous, vuggy knolls may have provided a suction or inward pumping which helps trap the finer sediment. In most of the examples studied the shallow ramp has a gentle seaward slope (from a few degrees up to about 15 degrees). From this one might infer that the ramps are a response to mild tectonic subsidence. Such gentle slopes effectively dampen the most violent wave action. Shoals and islands of laminated, tidal flat lime muds and sands generally lie behind such reef flats. Examples are:

1. Modern ledge-flat reefs of Bermuda, located atop the Bermuda platform (Verril, 1907).
2. Rudist reefs at the shelf margins in the Middle Cretaceous of Mexico, subsurface of south Texas, and the Middle East (Bonet, 1952; Nelson, 1959; Griffith et al., 1969; Harris et al., 1968.
3. Malm reefs of Swiss and French Jura Mountains developed in places into Type III frame-built rims (P. Ziegler, 1956; M. Ziegler, 1962; Bolliger and Burri, 1970).
4. *Thecosmilia*—sponge-spongiomorph Late Triassic reefs of the Northern Limestone Alps of Austria and Bavaria (Fabricius, 1966, Zankl, 1969). In places these developed into Type III framebuilt rims with slopes up to 20 degrees.
5. Stromatoporoid-tabulate coral reefs of Upper and Middle Devonian age in western Canada (Klovan, 1964; Dooge, 1966; Fischbuch, 1968).

Type III. Framebuilt Reef Rims

This is a linear belt of organic reef frame growing up to sea level or into the zone of turbulence. Submarine bars and shoals of lime sand exist behind the reef, filling the lagoon and in places even forming islands. These may be barrier or fringing reefs and are zoned ecologically in parallel belts according to their coral growth form. They are essentially due to the growth of hexacorals and associated stabilizing red algae and are chiefly Mesozoic to Holocene in age. Such reefs commonly possess steep slopes (more than 45 degrees or even vertical scarps) and much talus debris. Examples are known of transition from a complex of reef knolls in water of lower energy (Type II) to an active framebuilt reef (Type III).

Examples are:

1. Modern hexacoral reefs. Zonation exists from large rounded or sheety colonial corals in water from 70 to 10 m deep to *Acropora* species building into the zone of the wave-action, to a coralline algae flat of cemented rubble behind the reef front. Usually a sand shoal area exists behind the "*Lithothamnion*" flat.
2. Some Malm reefs of Swiss and French Jura Mountains developed from Type II.
3. Steinplatte, a Late Triassic *Thecosmilia* and sponge reef of the Northern Limestone Alps in Austria and Bavaria (Ohlen, 1959; Fabricius, 1966; Zankl, 1971). This reef growth developed in places over a complex of knoll reefs of Type II.
4. Kirkuk fringing reef of Middle Tertiary age in Iraq (Henson, 1950; Van Bellen, 1956; Dunnington, 1958).
5. Canning Basin-Late Devonian reef complexes, western Australia (Playford and Lowry, 1966). These reefs possess a 30-35 degree slope and well developed talus. In Alberta, Canada, beds of exactly the same age and bearing the same organisms develop shelf margins of very low angle and are generally of Type II.

Speculative Relationships between Types of Shelf Margins and Tectonics

Given a reasonable amount of subsidence, hydrologic and climatic controls appear more important than tectonics in determining which type of shelf margin develops. Buildups which are formed in areas of more or less major and continued subsidence, display all three types of shelf margins. These include margins of great offshore banks and edges of major platforms built out into miogeosynclines. The seaward slopes are steep in these cases, owing to basement faulting, and commonly such margins are fronted by open oceans. Theoretically, if tectonic down-warping causes steep enough slopes, an organic barrier might not develop at the shelf margin and a narrow ramp facies, with high energy belts along the littoral zone, might be formed (Ahr, 1973). Such narrow fringes at cratonic margins are rare but recognized as a special category in the classification of tectonic settings which follows. In most instances, high wave energy and upwelling in turn produce optimum growth of organic framework even on steep slopes in areas of consider-able subsidence. Under these circumstances Type III profiles develop, particularly in the Holocene when the coral, *Acropora* dominates. When factors inhibit or-ganic growth at the shelf margins (e.g., overflooding of saline brines from shelf lagoons), Type I profiles may develop below the narrow fringes at the shelf margin.

Conversely, it appears that in many cases where tectonic downwarping has been slight, shelf margins display ramps or platforms with Type II knoll reefs. In areas of moderate subsidence in shelf lagoons the standard belts of carbonate facies develop ramps with Type II knoll reefs built out in irregular patterns from margins of intracratonic basins (category 3 of the tectonic classification) or as fringes around mildly positive areas (category 4).

Carbonate Mounds and Associated Ecologic Reefs

Origin and Orientation of Mounds

Mounds and banks are typical forms of carbonate accumulation in a quiet-water environment. They usually consist of detrital, poorly sorted, bioclastic micrite with only minor amounts of organic boundstone. These accumulations formed either well below wave base or in water so shallow that effective wave action does not exist. In many places such bodies form the base upon which ecologic reefs grow. The numerous examples described in Chapters IV-XI show three preferred geographical positions for such carbonate bodies: distributed in various patterns in deeper basins, arranged just downslope along Type I shelf margins, and spread widely in shelf lagoons or shallow basinal areas. The latter occurrence is most common. The mound shape results from accretion of locally produced carbonate sediment up to the wave base and under conditions of stable or rising sea level. These shapes vary from flat lenses to steep, conical piles with slopes of 30-40 degrees. Studies of Holocene mounds and banks in Florida Bay, the Florida Straits, Yucatan and British Honduras lagoons, and in western Australia have been instructive in understanding the processes originating and perpetuating the mound phenomenon.

Modern mounds occur in several orientations—even when distributed across vast shelf areas. These mounds may occur at random or in crude lineations encircling shallow ponds or bays. These forms are believed to be responses to very sluggish currents or gyres set up in shallow water bodies by persistent moderate winds. Large, elongate, irregular banks are known at the open ends of large shallow bays in several modern areas: western Florida Bay, Cayo Sucio at the northern end of Isla Blanca Bay, northeast Yucatan, and Bulkhead shoal at the southern end of Chetumal Bay, British Honduras. These are caused by dampening-out of tidal currents and swells from the open sea and the mechanical accumulation of fine lime-mud sediment in slack water.

More sharply defined, bread-loaf, elongate mounds are possibly shaped by longshore currents. These chains of mounds parallel the coast lines and may form in multiple rows. Analogs from the Holocene include Rodriguez and Tavernier Keys off the Florida Keys (Turmel and Swanson, 1972). These same features may occur in groups across the shelf areas, many with axes parallel to the depositional strike and hence to the ancient coast lines. Detailed studies of some areas show that orientations of ellipsoidal mounds may also be normal to the shoreline. Such trends may be tidal deltas at the mouths of passes through the Pleistocene rock barriers or reefs which partically enclose the shelf lagoons. Tidal bars which may occur within wider passes are more stream-lined than the delta accumulations. These can vary in composition from clean sand to accumulations of lime mud and bioclastic debris stabilized by veneers of organisms. Fine examples of Holocene tidal deltas and tidal bars in passes may be observed between some of the middle and lower Florida Keys. The 10 km wide fringe of lime mud banks lining the mainline shore of Shark Bay, Western Australia display transverse tidal channels oriented normal to the coast (Davies in Logan et al., 1970). Accumulation results from interaction of tides and longshore drift with organic carbonate production on and trapping by mats of marine grass. Such transverse trends of micritic buildups are not common in the geologic record but are known in the Pennsylvanian strata of the Paradox basin and midcontinent.

Even mounds developed below wave base and out of reach of the surface currents may have well-defined trends. The mounds described by Neuman et al. (1972) off Bimini in the Florida channel are apparently current-formed bodies. Currents are observed here at depths of 700 m; the elongate shape of the bodies is distinctive. In contrast, the Lower Carboniferous Waulsortian mounds, developed downslope of the preexisting scarps, are more or less circular. In addition, they often occur across the basins in lines whose trends may be structurally controlled. Many deeper water mounds, as well as those located in shallow basins, lack apparent orientation. Swarms of pinnacle reefs appear in complex patches on platforms in some basins (Klovan, 1974). Purdy (1974a) has proposed that such a pimply or labyrinthine pattern is caused by growth on submerged karst surfaces. He notes such patterns within the Holocene-Pleistocene large Alacran atoll on the Campeche bank and on the wide shelf behind the British Honduras barrier reef. Numerous examples from the geologic record show that such patches of pinnacle reefs cluster on platforms or terraces which may be marginal to basins and later drowned by submergence. These include the West Irian Miocene (New Guinea); the north central Texas Canyon buildups; the Cooking Lake platform with Late Devonian Leduc reefs of Alberta, Canada; the Devonian Presqu'ile reefs of northern Alberta, and the Michigan basin Silurian.

At present very little data exist on the origin and orientation of basinal buildups. The above discussion of mound genesis and orientation emphasized their origin as hydrologic accumulations, drawing heavily on recent sediment analogies. In geological occurrences it is always difficult to explain the localization of

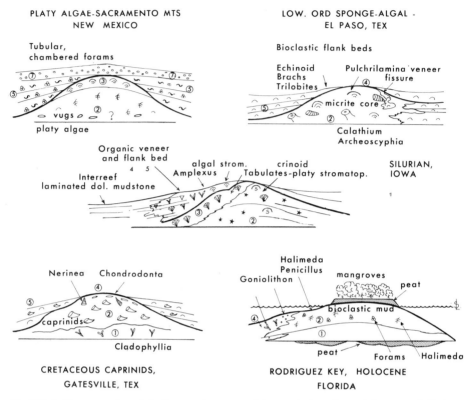

Fig.XII-4. Five examples of shelf mounds, summarizing variations in evolving growth facies as diagrammed in Fig.IV-9, IV-27, VI-25, and XI-4. Facies 1–7 are characterized in Fig.XII-5. For general legend see Fig.III-1

particular buildups. For practical field reasons it is usually impossible to observe outcropping beds just under a mound. In oil fields, production is generally from the tops of the buildups and few wells penetrate the mound base. Our best clue for determining the probable origin is perhaps the relationships of trends to regional paleogeography.

Mound Facies Sequence

Whatever their origins, many mounds developed on shelves and in shallow basins consistently show a vertical and lateral sequence of textural and organic facies. This is because the processes which create them are sufficiently rapid to cause shelf mounds normally to grow into wave base from deeper and quieter water. These processes may be summarized as follows:

1. Mechanical accumulation of both fine and coarse sediment through current and wave action. Probably the most important process localizing mound growth.

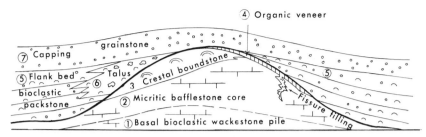

Fig. XII-5. Ideal carbonate mound with seven commonly developed facies. See also Fig. IV-9 and VI-25. The sequence of facies develops when the accumulation grows into wave base and is controlled by varying rates of sediment production, rates of subsidence, and hydrographic factors

2. Trapping and baffling of carbonate sediment produced locally at higher than normal rates. Probably the most important process contributing to the growth of the mound.

3. Stabilization of sediment by surface encrustation so that normal processes of marine erosion do not remove it.

4. Protection by a veneer or wall of frame-building organisms at a late stage in its development.

5. Protection by cementation. In lime mud deposited and remaining in the marine environment, cementation is very slow. In shallow-water banks, where chances of subaerial exposure are better, lithification of lime mud is more effective.

Although extrinsic, mainly hydrological, controls operated to cause a succession of changes in mound sediment as the structure builds, additionally intrinsic changes operated within the organic communities inhabiting the mounds. These were occasioned by creation of and competition for substrates by the mound dwellers (Alberstadt and Walker, 1973).

Examination of Paleozoic, Cretaceous, and Holocene sedimentary bodies referred to in previous Chapters permits generalization of the sequence which is summarized in diagrams on Figs. XII-4, XII-5, VI-25, and IV-9. Not all mounds progress through the whole sequence—only those in areas of moderate subsidence. But mounds commonly begin growth below wave base and build up into it, remaining there for sufficient time to develop crestal boundstone, flanking beds, talus, and perhaps a capping bed. Figs. IV-27a and b compare an arrested early stage of mound formation with a more complete development. Seven distinctive facies are commonly repeated in the examples despite vast differences in geologic age. They are defined by both textural and biological differences and are determined by organic and sediment reaction mainly to external factors.

1. Basal bioclastic wackestone pile: Most mounds begin with a very micritic sediment with much bioclastic debris. In Early and Medial Paleozoic strata this commonly contains considerable echinoderm fragments with less bryozoans, and brachiopod shells. In several instances piles of shell debris are seen at the base. The origin of these piles cannot be generally ascertained; presumably they are heaped up by gentle currents. No pronounced baffling or binding organisms are common in these beds.

2. Micritic bafflestone core: The thickest part of the mound generally consists of a micritic matrix replete with organisms capable of trapping or baffling fine lime sediment. These are commonly delicate or dendroid forms with upright growth habits. Each geologic age has its special biota which serves in this role. Often one form dominates almost to the exclusion of others:
a) Sponges and algae in Cambro-Ordovician
b) Bryozoans in Middle and Late Ordovician, Silurian and Early Carboniferous
c) Platy algae in Late Carboniferous (Pennsylvanian)
d) Large fasciculate-dendroid corals in Late Triassic shelf areas
e) Lithistid sponges in Late Jurassic
f) Rudists in Cretaceous shelf areas
g) Marine grasses at present time

Commonly this main core facies is extremely brecciated because it was originally a mixture of gelatinous, non-resistent mud, and rigid, brittle bioclasts. It probably slumped under its own weight.

Many mounds, particularly those in very shallow water or those swamped by argillaceous influx, never reached beyond this early stage of a micritic mound core, developed few flank beds, and no crestal boundstone.

3. Crestal boundstone: When a bafflestone mound reached into wave base, the topographic eminence formed by the stabilized soft sediment may have served as a seat for organic boundstone. This may take two forms: (a) An ecologic reef may have developed through colonization by large, sessile, massive invertebrates such as corals, hydrozoans (stromatoporoids and spongiomorphs), sponges, some rudists, richtofenid brachiopods, and red algae. Protected cavities showing vertically oriented fabrics (geopetals) were formed. (b) Quieter water boundstone developed with more lamellar forms and with protected cavities trending parallel to the outer slope of the mound. As pointed out by Alberstadt and Walker (1973), initial colonization may proceed to a high degree of biological diversification. In this intrinsic process the development of a mesh-work of complexly shaped organisms gives rise to a number of habitats and encourages a multitude of species—additional major framebuilders and binders as well as niche dwellers. The ecological succession is similar in many buildups, despite great differences in taxa through long geological periods. One logical explanation for the similar successions is a prevailing substrate control of growth forms.

4. Organic veneers and fissure filling: Another possibility exists for colonization of an earlier formed bafflestone mound. If conditions permitted only slow deposition on the mound top, and no extensive growth of frame-producing organisms occurred, the upper surface of the mound may have become coated with a thin veneer of a variety of encrusting or adnate organisms. Examples of these include *Chondrodonta* and oysters in the Cretaceous and certain sponges and stromatoporoids in Pennsylvanian beds. Specialized forms growing in tidal pools can be seen in certain Silurian buildups in Illinois. A second type of veneer occurs. Laminated or stromatolitic beds are common at the top of many Ordovician and Silurian mounds, noted by Alberstadt and Walker as a "Domination phase", where one organism dominates all others by successful adaptation to stress conditions. The same phenomenon has been described by Ahr (1971) in Late Cambrian mounds in Texas. This type of veneer may be a response to very shallow water at the tops of mounds in which wave action is essentially ineffective but where tides merely carry sediment across the flat top to be trapped and stabilized by blue-green algal growth. A third type of veneer is known in some mounds which originated in deeper water, and accumulated just to wave base. Here gradual winnowing may have concentrated worn and resistent bioclasts at the top of the bafflestone unit. In some mounds extensive vertical fissures are filled with the products of this winnowing, accumulated as blackened, worn, and coated particles washed in from above.

5. *Flanking beds:* If the mound top remained for a long time, just at wave base, and became colonized by more or less fragile, stalked organisms, moderate water movement and normal organic decay may have resulted in extensive flank beds. These include echinoderms, fenestrate bryozoans, small rudists, dendroid corals or stromatoporoids, branching red algae, and tubular foraminifera encrusting noncalcareous organisms. The flank beds lap up on the sides of the mound and are composed exclusively of bioclastic debris. Under conditions of only slight subsidence, the original mound core may be almost buried in flank beds which accreted out from it on all sides. Volumentrically such beds may be greater than the core itself. This is to be observed in some Silurian reefs in the Midcontinent as well as in Pennsylvanian algal plate mounds in New Mexico.

6. *Talus:* A rarer, but widespread flank facies, is that of marine talus composed of lithoclastic and bioclastic debris, the former being the key component. The lithoclasts represent chunks of partly or wholly lithified micrite torn from the surface of the mound by collapse or wave action and slumped or carried down by currents along the sloping sides of the mound into normal flank beds. Since carbonate mud mounds are generally found in areas of low wave energy, this deposit is often missing. Exactly what processes formed these local lithoclastic conglomerates still constitutes a mystery. The smooth, even shapes of most mounds do not indicate much erosion off the tops.

7. *Capping grainstones:* When sea level remained stable and the intermound areas filled in with sediment, often a shelf deposit formed across the top of the mounds. Commonly this is a single stratum of cross-bedded grainstone over the whole area. In many places this high energy, post-mound sediment usually contains a specialized fauna consisting of robust gastropods and dasycladacean algae.

Interrelations of Parameters Controlling and Modifying Carbonate Accumulations

Clear, well-lit, warm, marine tropical water producing carbonate sediments sustains a very ancient system of chemical and biochemical processes. It is as old as 2.5 billion years. This system, discussed in Chapter II and diagrammed in Fig. II-12, is responsible for thousands of meters of mostly shallow-water limestones and dolomite typical of the geological record. Carbonate production operates most efficiently in marine areas adjacent to tectonically stable land and in provinces where there is little fresh-water runoff. In such areas limpid shallow marine water borders and covers extensive shelves around land areas.

A variety of processes act upon the biochemical system. These effect the various carbonate facies and bodies of sediment discussed in this volume. The following section outlines the inter-relationships of the basic sedimentary controls of these processes: tectonics, climate, eustatic sea-level fluctuations, hydrology, organic composition, and character of substrate. Table XII-1 is an attempt to organize these in a meaningful way. Combinations of these parameters control degrees of depth, water agitation, circulation, salinity and substrate character and frequency of subaerial exposure. In turn these act upon sediments within the general tectonic settings represented by (1) wide shallow-shelf deposits, (2) carbonate shelf margins, and (3) basins and geosynclines. Table XII-2 outlines many of these relationships within the framework of the tectonic settings of prominent carbonate depositional patterns. A glance at the tables shows that the relationships may be complex, and that, unlike the case of terrigenous sediments, a tectonic classification of carbonate sedimentary patterns is not so useful in generalizing and arranging natural groups of rock. Some of the interrelationships are discussed below. Others are charted on the Tables.

Tectonism

Tectonic activity operates to control carbonate sedimentation at various levels. Its basic control is in the rate and continuity of subsidence without which sedimentation is negligible. Rapid and continuous subsidence results in extraordinarily thick sections in certain carbonate producing areas. As pointed out in Chapters I and II, tectonic subsidence when balanced by or slightly overcome by rate of sedimentation may result in carbonate upbuilding or sedimentary progradation just as impressive as in clastic deltas.

The tectonic framework of a region may directly affect the water circulation along its coasts. The complex morphology of the Arabian side of the Persian Gulf attests to this. Orientation of contemporaneous fold belts and fault trends oblique to the edges of basement blocks control the regularity of coastlines and embayments and hence exert control on the degree of restriction of circulation of marine waters. On the other hand, structural orientation normal to prevailing wind and wave direction encourages rapid organic growth causing barrier and fringing reefs. Local structural uplift, when drowned, may form patch or pinnacle reefs as is happening today in the Persian Gulf and in the Gulf of Mexico.

Moreover, slight subsidence of a basin may influence diagenesis by permitting the buildup of carbonate platforms even with the level of the sea.

Eustatic Sea-Level Changes and Substrate Effects

Both world-wide glaciation and megatectonic shifting of oceanic and cratonic plates may induce eustatic sea level changes, a process operating either with or in opposition to local tectonic subsidence. Where uncompensated by tectonic subsidence, eustatic sea-level lowering may bring about an unusual drop in the water table, improved subsurface drainage, and stronger diagenesis in carbonate shelf

Table XII-1. Interrelationships of parameters controlling and modifying carbonate accumulations

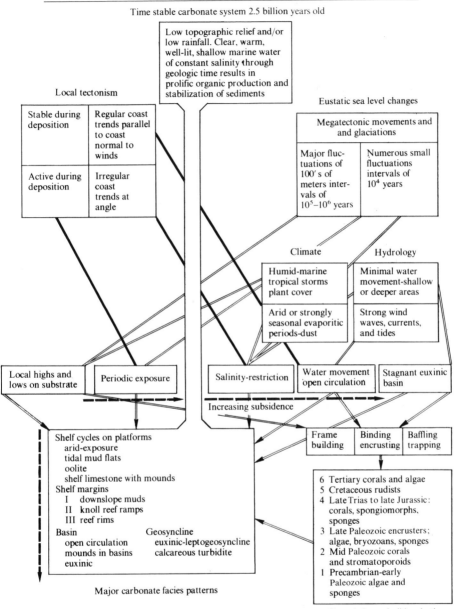

Time stable carbonate system 2.5 billion years old

Major carbonate facies patterns

Organic frame builders in time

strata. This happens particularly within climatic regimes of strongly seasonal rainfall and evaporitic conditions.

On shelves widely exposed by lowered eustatic sea level, tropical conditions bring about several stages of karst development which in turn control subsequent carbonate growth and facies patterns when resubmerged (Purdy, 1974a and b).

Table XII-2. Parameters influencing carbonate accumulations plotted against tectonic-

Increasing subsidence →

General controlling parameters	Specific controls	Geosynclines		Basins		
		Turbidite filled	Lepto	Leptoeuxinic	Mounds	Open circulation shelf, shallow basin
		1	1	1	2	2
Eustatic sea level lowering	Greater than subsidence					
Local tectonic	Active during deposition	Fills basin with shelf and land debris			Doming and faulting controls of trends of mounds	
	Trends oblique or parallel to coast					
Climate	Tropical-humid					
	Arid-seasonal		Hypersaline brines may induce lack of turnover, euxinic conditions encouraged			
Hydrographic — Enclosed	Shallow restricted circulation					
Hydrographic — Open shelf	Wind wave shallow currents fetch impt.					Storm wave base affects bottom
	Tides					
Hydrographic — Offshore deep	No deep currents		No currents — stagnant H_2S reducing conditions			
	Deep currents upwelling	Turbidity density flows	Faint currents — Ripple marks		Shapes and originates mounds by piling	Only strong enough to winnow fine sediments
Organisms	Frame					
	Binding encrusting					
	Baffle stone				Forms mounds at or below photic zone	Small mounds

environmental classification

Much less subsidence

Shelf margins of offshore banks and major platforms			Cyclic shelf on Platforms			
I Downslope muds 3	II Knoll reef ramps 5 and 6	III Reef rim 5	Widespread uniform sheet limestone 7	Cycles with oolite 6	Cycles with intertidal muds dominating 8	Cycles with arid exposure 9
Possible diagenesis	Diagenesis, karst control of substrate & reef form	Diagenesis, karst control of substrate & reef form	High stand of sea level, transgressive deposits	Control of cyclicity	Control of cyclicity	Important control on diagenesis
	Forms bases for individual reef bodies?	Forms bases for individual reef bodies?				
		Parallel to coast encourages reef growth		Parallel to coast causes oolite to form	Oblique to coast causes restricted circulation	
	Solution karst	Solution karst			Variable salinities, brackish marshes	
Extensive cementation of lime sands on barrier island in splash zones ⟶					Evaporite sabkha	Evaporite sabkha deposits
Induces salinity currents downslope	Restriction possible close to reef ramp very shallow water			Tops of some cycles	Restricted circulation hypersaline lagoons & tidal flats	Evaporitic tendency
Moderate surge	Moderate surge, positive effects	Strong surge	Causes shelly sand lenses in lime mud sediments			
	Tidal currents circulate water	Tidal currents have positive effects	Tidal currents aid in open shelf circulation	Production of tidal bar, well-formed ooids	Tidal effects minimal, only in channel	Tidal effects important if range is great
Lack of deep current down slope permits mud to accumulate						
		Rapid growth of organic frame enhanced by upwelling				
	Prominent in patches on ramp	Dominates modern reefs prominent in Mesozoic & Siluro-Dev.				
Mounds are encrusted	Prominent in all geologic ages knolls often micritic					
Dendroid lacy forms encourage trapping	Dendroid forms encourage trapping		Small mounds in normal marine parts of cycles ⟵⟶			

Solution rims developed at the edges of karst platforms form the substrate for barrier reefs, chains of islands develop over old karst drainage divides, and the atoll and faro forms may result from submergence and carbonate growth around isolated blocks or towers whose steep sides develop their own solution rims. Spurs and grooves and major passes between the reefs may result from a rejuvenated carbonate growth on the edges of former lagoon channels and solution rills.

Hydrology and Climate

The importance of hydrologic factors in causing and modifying buildups has been stressed by Stanton (1967) and Wilson (1974). Except for tides, climate and hydrologic factors are closely interrelated. A particularly good analysis is given by Stanton. Examples of how some of these controls interrelate follows:

1. Winds and waves produce several effects to encourage carbonate buildups as well as to destroy them. The bringing of fresh nutrient-laden water causes organisms to flourish. The loss of CO_2 in agitated water encourages $CaCO_3$ production with resultant cementation of debris. The piling effect of waves furnishes higher areas on which organisms may thrive. Even onshore winds along the coasts may encourage carbonate production by inducing upwelling water to bring nutrients. Of course, severe storms destroy carbonate accumulations and make difficulties for the growth of certain more delicate calcareous organisms, but the positive effects of water energy generally outweigh the destructive tendencies.

2. Combinations of climate and hydrology control salinity. Arid climate and continental winds result in an increased rate of evaporation; this, in turn with low tide range and sluggish circulation caused by complicated coastal configuration, results in evaporative conditions limiting the abundance and types of organisms capable of forming carbonate. The same combination enhances precipitation of carbonate cement in the splash zone and in shallow marine bays.

3. Proper combination of tectonic stabilization, eustatic sea-level fluctuations, and climate offer important controls over carbonate diagenesis. Lithification of carbonate may be brought about by subaerial exposure caused by long-term sea-level lowering. Seasonal rainfall and periods of aridity cause intense cementation of the carbonate mass. Conversely, under conditions of high rainfall and tropical plant cover, considerable solution of carbonate formations takes place.

Organic Composition

Controls partly dependent and partly independent of physical processes are afforded by variations in growth potential and framework construction of various organisms during geologic time.

Heckel (1974) presents a diagram showing the importance of various groups of organisms as producers of carbonate mass and framework throughout geologic time (Fig. XII-6). Table XII-1 shows six major phases of this evolution as well as the grouping of organisms into those with sediment-trapping, sediment-binding, and framework-building potential. Calcareous algae have continuity through the Phanerozoic, holding about the same ecologic positions and affecting carbonate sedimentation the same way, whereas waxing and waning of coelenterate groups such as corals and stromatoporoids have occurred, with special culminations of framebuilders in the Middle Paleozoic and from the Jurassic to the Holocene.

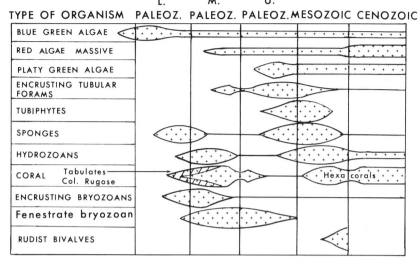

Fig. XII-6. Dominance of major organic groups in carbonate buildups through geologic time. After Heckel (1974)

In addition, special groups important in sediment binding and trapping dominate at certain periods in the geologic record. For example, colonial and encrusting foraminifera from Devonian to Jurassic, bryozoans in the Ordovician and Early Carboniferous, and rudists in the Cretaceous.

Review of Tectonic Settings for Carbonate Buildups and Cycles

Major amounts of carbonate generally do not accumulate in areas of extensive tectonic activity except for some isolated reefs in geosynclines. This is because orogenic uplift tends to cause intense erosion and produces in humid regions a considerable influx of fresh water and terrigenous clastics which suppress general carbonate production. Inasmuch as tectonic activity does not control source area and rate of deposition in the carbonate realm as in terrigenous clastic sedimentation, a lack of precise correlation between tectonics and carbonate deposition might be expected. Other controlling parameters are of equal or greater importance in shaping the major types of carbonate sediments. Nevertheless, megatectonic elements such as geosynclines, basins, and cratonic areas may serve as a useful framework for grouping carbonate facies patterns even though such groups may not be as sedimentologically distinctive as one would wish.

The relationships which exist between the outlined tectonic settings and the major groupings of carbonate patterns used in this volume are expressed on Tables XII-1 and XII-2.

Subcategories in the classification below are based on form and trend of the buildups as well as their facies, parameters which are related to the degree of subsidence of the major tectonic elements. Classifications presented by Krebs (1971) and Krebs and Montjoy (1972) and Wilson (1974) have more or less been expanded for wider application. A related classification has been presented also by Heckel (1974). As in Chapter II, the numerals on the following outline are chosen to agree with those of Krebs and Montjoy.

1. Basinal buildups in areas of great subsidence.
   a) In marginal cratonic basins or miogeosynclines:
      (1) Major offshore banks.
          E.g., Central Basin platform of the West Texas Permian basin, Cretaceous of central Mexico, Dolomites of south Tyrol.
      (2) Linear trends along basement faults.
          E.g., Leduc-Rimbey Late Devonian buildups, central Alberta basin, Canada.
      (3) Pinnacle reefs either in belts or random patches.
          E.g., Michigan basin Silurian; pinnacle buildups paralleling the edges of the basin; the Williston basin and Zama area in northern Alberta have mostly random patches; Pennsylvanian of the Midland basin, north central Texas, banks parallel to eastern shelf margin; Frasnian (Late Devonian) buildups of Dinant basin, Belgium.
   b) Within geosynclinal troughs on or around volcanic uplifts. E.g., Rhenish trough of Rhenish Shale Mountains, Germany.
   c) Major offshore banks under oceanic influence. E.g., Great Bahama Banks and related complexes.
2. Buildups at edges of major platforms and ramps developed off cratonic blocks; areas of great subsidence.
   a) Linear buildups along shelf margins of these platforms. E.g., the Mosul Cretaceous platform of northern Iraq; Permian Reef Complex surrounding the Delaware basin and built out from the eastern side of the Pedernal uplift in New Mexico; Devonian trends such as the Pres Qu'ile and around the Peace River high of northern Alberta; and the Lennard shelf fringing the Canning basin, Western Australia. Shelf margin Types I, II, and III.
   b) Complexes of individual banks and reefs widely spread along and upon such shelf margins. Shelf margin Type II. For example, Middle and Late Devonian buildups within the external shelf of the Rhenish trough, southeast of the Old Red Continent, northwestern Germany; the eastern edge of the western Morroccan shelf (Devonian); Silurian shelf (low clastic belt) north of the Illinois basin, U.S.A.
   c) Major but narrow fringes off faulted edges of cratonic blocks or orogenic ridges. E.g., the Smackover Late Jurassic trend around northern Gulf of Mexico; British Honduras (Belize) coast.
3. Buildups on ramps or platforms in areas of moderate subsidence; built out into shallow intracratonic basins; on shelf areas with scattered mounds and patch reefs. Shelf margin Type II most common.
   a) Micritic mounds and masses with very slight relief and in shallow water, ordinarily with only sponges or bryozoans and an algal stromatolite cap. E.g., Lower Ordovician of southwestern U.S.A., Middle Ordovician (Chazyan) of Adirondack shelf.
   b) Simple mud-mound accumulations with calcarenitic flank beds. E.g., Waulsortian mounds in Early Mississippian strata, Sacramento Mountains, New Mexico developed below shelf margins at edges of basins.
   c) Boundstone reefs atop banks or mud-mound accumulations. Silurian of northern Mid-continent, U.S.A; Spanish Sahara Devonian patch reefs.
   d) Faros or patch reefs of pure boundstone. E.g., Silurian and Devonian over wide areas of the U.S.A. and Canada; Devonian of Eifel west of Rhine River; Silurian of Gotland; modern patch reefs of Florida strait and British Honduras.
   e) Upward shoaling cycles across ramps from the shelf to the basin. E.g., the Mississippian Madison Group of the Williston basin.
4 and 5. Low fringing buildups or halos of cyclic character around local positive areas on shelves. E.g., Callovian bank, southeast Paris basin, north of Morvan massif and Edwards (Cretaceous) limestone around Llano uplift, central Texas.

There should be modern analogs to the categories defined by the above classification because the same descriptive parameters of megatectonic setting, trend and shape of constituent sedimentary bodies are applicable to Holocene carbonates. The problems of recognizing such analogs are greater than might be supposed, however, for several reasons: (1) Our knowledge of Holocene carbonate sedimentation is apt to be merely two-dimensional; in only a few places do we have sufficient borehole data to enable us to determine thickness and facies and Holocene-Pleistocene geological history. Geographic setting is more easily studied than tectonic framework. Of the half-dozen Holocene carbonate areas known today, only in the Florida and British Honduras regions is a partial three dimensional picture available. This prohibits the determination of rate of subsidence and an evaluation of recent tectonic activity. (2) Because of Tertiary worldwide tectonic activity and the relatively high stand of continents at the present day, only a few small, shallow, epicontinental seas exist. These are floored with only a thin veneer of sediment deposited within the last 5000 years or with no sediment at all. (3) Indeed, no place has been discovered on earth where normal neritic carbonate sedimentation has proceeded uninterrupted from Pleistocene to Holocene. This is because the drastic Wisconsin-Würm glacial lowering of the sea level (more than 100 m only 15000–18000 years ago) laid dry all of the stable shelves so characteristic of the carbonate producing regime. Additionally, continental shelves drowned by the recent rise of sea level have received hardly any sediment since the stabilization of sea level at its present level 5000–7000 years ago. Carbonate buildups at the margins of most of the shelves have not had sufficient time in which to reestablish.

In summary, there are hardly any continental seas or intracratonic basins to serve as modern tectonic models and modern shelves are strongly influenced by the recent sea-level change and by purely oceanic hydrology and climate. With recognition of these complications and limitations, the following attempt is made to analyze modern carbonate accumulations in terms of geographic, as well as tectonic, setting and to apply to them the tectonic classification presented above for ancient buildups. This is done to emphasize the problems of using recent analogs and models in our interpretative thinking about ancient carbonates.

1. Basinal buildups; areas of great subsidence.
   a) Offshore banks with oceanic influence. (No modern banks of large size exist in marginal cratonic basins.) E.g., Bahama Banks.
   b) Geosynclinal and oceanic volcanic areas. E.g., some Pacific atolls.
2. Major platforms and ramps developed off cratonic blocks or as fringes off the same, areas of great subsidence; all under oceanic influences.
   a) Reef-rimmed coastlines with lagoons (barriers) or reef-rims directly against coasts (fringes). E.g., British Honduras and Great Barrier Reef, Australia.
   b) Barrier island rimmed coastline with remnant Pleistocene barriers as well as those formed of Holocene sands. E.g., northeastern Yucatan coast; Shark Bay, Western Australia; Florida Keys and Straits.
3. Buildups within platforms; areas of moderate subsidence.
   a) Shelves presumably on ramps built out into shallow marginal cratonic or foredeep basins with scattered patch or pinnacle reefs. E.g., Persian Gulf (specifically the Great Pearl Bank). This is the single example of a land-locked epicontinental sea or foredeep basin existing in a carbonate-producing realm.
   b) Shelves on continental margins drowned by recent sea-level rise with only relict sediment. Some scattered outer reef knolls. E.g., Sahul, Campeche, West Florida, Nicaragua

Table XII-3. Use of facies models

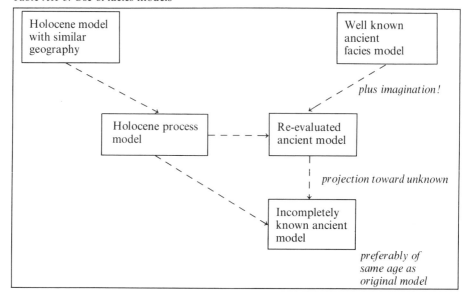

shelves, and Sunda shelf which is flooded with terrigenous clastics and has only marginal carbonate buildups.

c) Shelf seas across major cratonic blocks do not exist at the present time. Comparisons with ancient areas must be made with relatively shallow bays and lagoons along the present coasts. These commonly contain banks or mounds of bioclastic lime mud. E.g., the Florida Bay; lagoons on northeastern Yucatan coast; Shark Bay, Western Australia.

The above outline relates the geographic settings of modern depositional models to tectonic framework so that models will be more applicable (see Malek-Aslani, 1973). As pointed out above this does not work entirely satisfactorily. It *is* possible in many places to use parts of a geographic model regardless of the exact match of the whole. The Persian Gulf Trucial Coast model comes closest to a complete display of the standard carbonate facies progression but only when applied to an arid climate. Similarly, Florida Bay is a useful place to study the products of restricted circulation but is part of a larger geographic setting which is complicated by Pleistocene history. The outer barrier to the Bay is triple, consisting of the Florida Reef tract, the shallow Florida strait, and the Pleistocene rim of the Keys.

Perhaps the most valuable use of Recent carbonate deposition is to study the processes and to use this information to pique our imagination when transferring attention to an ancient geological example. This is usually more productive than merely matching facies patterns on models of different geologic age and perhaps of different geographic settings. This procedure is diagrammed in Table XII-3. A particularly thorough application of a Recent model to geological interpretations is found in the work of Griffith et al. (1969) on the Cretaceous of Mexico and Texas.

The exploration geologist normally begins with stratigraphic knowledge of a few wells or sections and a general understanding of the basin tectonics. As diagrammed on Table XII-3, a specific example existing in the stratigraphic record is observed and an attempt is made to find a geographically comparable Holocene model to be studied. From the Holocene study comes an understanding of sedimentary processes which can be measured and evaluated from marine investigation. The process model may then be used, with a little imagination, for a reevaluation of the well-known geological example and ultimately for projection to an incompletely known carbonate facies complex and for more accurate prediction of trends and facies in it.

As an example of the valuable interplay of Holocene models with ancient examples, consider the evolution of thought about the Permian Reef Complex. Newell et al. (1953) after extensive study of the Bahama Banks and following post World War II studies of the Pacific reefs, used the only known Holocene sedimentary models to help interpret the Guadalupian strata of New Mexico and West Texas as a major barrier reef trend. Size and geography of the Recent and Ancient models agreed quite well. But subsequent petrographic study of the Permian "reef" showed no framework capable of organic stabilization of the abrupt shelf edge and much discussion ensued in the literature about its origin. Beginning in 1960, processes of diagenesis in a strongly arid environment and in a landlocked sea were studied in the Persian Gulf. This was an area whose geographic setting did not serve as a very apt model for comparison with the Permian Reef Complex. Nonetheless, many diagenetic features along this arid coastline were recognized as identical to those of the Permian: evaporites on sabkha, splash zone drusy cementation, rapid hardening of peloidal sands on tidal flats, extensive development of bulbous algal stromatolites, desiccation polygons, and tidal flat dolomitization. Dunham (1972), recognizing the similarity of processes based on similarity of diagenetic features, proposed that lithification during a time of extreme climate and periodic drops of sea level was responsible for stabilization of the sharp shelf margin. At the present time, investigation of the classic Triassic banks of The Dolomites is progressing rapidly and these structures are being reinterpreted, aided both by an understanding of Recent processes of vadose zone diagenesis and by an improved interpretation of the Permian Reef Complex, a model not far removed in time from the Ladinian strata of the Alps (Bosellini and Rossi, 1974). Further investigation on the interiors of modern reefs has revealed the importance of submarine cementation in maintaining them. The role of this process now needs to be evaluated in the Permian buildups of the Guadalupe Mountains.

Thus geologic thought directed toward interpretation of carbonate buildups evolves as continuously as do the organisms which generate such masses. We have still much to learn. The didactic method employed in this volume, the numerous outlines, and classifications should not obscure this fact. Concepts will almost certainly alter as more data appear and are analyzed by experienced stratigraphers and petrographers. The author hopes that this volume has laid a beginning foundation for the description of carbonate depositional facies—a base from which more accurate interpretations of environments and prediction of vertical sequences and geographic patterns is possible. The study of present-day processes and development of Holocene models is vital to our interpretations of ancient strata, but the recognition and classification of what actually exists in the geologic record must form the primary basis for our concepts.

References

Achauer, C. W.: Origin of Capitan Formation Guadalupe Mountains, New Mexico and Texas. Am. Ass. Petrol. Geologists Bull. **53**, 2314–2323 (1969).

Adams, J. E., Rhodes, M. L.: Dolomitization by seepage refluxion. Am. Ass. Petrol. Geologists Bull. **44**, 1912–1920 (1960).

Ahr, W. M.: Paleoenvironment, algal structures, and fossil algae in the Upper Cambrian of central Texas. J. Sediment. Petrol. **41**, 205–216 (1971).

Ahr, W. M.: The Carbonate ramp: An alternative to the shelf model. Trans. Gulf Coast Ass. Geol. Soc. **23**, 221–225 (1973).

Aitken, J.: Middle Cambrian to Middle Ordovician cyclic sedimentation, southern Rocky Mountains of Alberta. Can. Petrol Geol. Bull. **14**, 404–441 (1966).

Alberstadt, L. P., Walker, K. R.: Stages of ecological succession in lower Paleozoic reefs of North America. Abstracts with Programs, Geol. Soc. America, Boulder, Colorado. **5**, No. 7. 530–532 (1973).

Alberstadt, L. P., Walker, K. R., Zurawski, R. P.: Patch reefs in the Carters Limestone (Middle Ordovician) in Tennessee and vertical zonation in Ordovician reefs. Geol. Soc. Am. Bull. **85**. 1171–1182 (1974).

Alderman, A. R., Skinner, H. C. W.: Dolomite sedimentation in the southeast of South Australia. Am. J. Sci. **255** 561–567 (1957).

Aldinger, H.: Ecology of the algal-sponge reefs in the Upper Jurassic of the Schwäbische Alb. In: Müller, G., Friedman, G. M. (Eds.): Recent Developments in Carbonate Sedimentology in Central Europe, p. 250–253. Berlin-Heidelberg-New York: Springer 1968.

Alling, H. L., Briggs, L. I.: Stratigraphy of Upper Siluriau Cayugan evaporites. Am. Ass. Petrol. Geol. Bull. **45** 515–547 (1961).

American Geological Institute: In: Gary, M., McAfee, R., Woof, C. L. (Eds.): Glossary of Geology, 805 pp. Washington, D. C. 1973.

Anderson, F. J.: Environmental models for Paleozoic communities. Lethaia **4**, 287–302 (1971).

Anderson, E. J.: Principles of benthic community analysis (notes for a short course), Community Patterns. In: Sedimenta IV, Comparative Sedimentology Laboratory, Fisher Island, Miami, Florida. 1974.

Arkell, W. J.: Jurassic of the World. New York: Oliver and Boyd Ltd., Hafner Pub. Company 1956.

Armstrong, A. K.: Stratigraphy and paleontology of the Mississippian in southwestern New Mexico and adjacent southeast Arizona. New Mexico State Bur. Mines Mineral Res. Mem. **8**, 95 pp. (1962).

Armstrong, A. K.: Carboniferous carbonate depositional models, preliminary lithofacies and paleotectonic maps, arctic Alaska. Am. Ass. Petrol. Geologists Bull. **58**, 621–645 (1974).

Assereto, R., Kendall, C. G. St. C.: Megapolygons in Ladinian limestones of Triassic of southern Alps: evidence of deformation by penecontemporaneous desiccation and cementation. J. Sediment. Petrol. **43**, 715–723 (1971).

Baars, D. L.: Petrology of carbonate rocks. In: Bass, O., Sharps, S. L. (Eds.): Shelf Carbonates of the Paradox Basin, p. 101–129. Four Corners Geol. Soc. 4th Field Conf. Symposium, Durango, Colorado 1963.

Badiozamani, K.: The Dorag dolomitization model-application to the Middle Ordovician of Wisconsin. J. Sediment. Petrol. **43**, 965–984 (1973).

Ball, M. M.: Tectonic control of the configuration of the Bahama banks. Trans. Gulf Coast Ass. Geol. Soc. **17**, 265–267 (1967a).

Ball, M. M.: Carbonate sand bodies of Florida and the Bahamas. J. Sediment. Petrol. **37**, 556–591 (1967b).

Barnetche, A., Illing, L. V.: The Tamabra limestone of the Poza Rica oilfield, 38 pp. XX Congreso Geológico Internacional, Mexico 1956.

Barthel, K. W.: Die obertithonische, regressive Flachwasser-Phase der Neuburger Folge in Bayern. 174 pp. Bay. Akad. Wiss. Abhandl. New Ser., 142 (1969).

Barthel, K. W.: On the deposition of the Solnhofen lithographic limestone (Lower Tithonian, Bavaria, Germany). Neues Jahrb. Geol. Paläontol. Abhandl. 135, 1–18 (1970).

Barthel, K. W., Janicke, V., Schairer, G.: Untersuchungen am Korallen-Riffkomplex von Laisacker bei Neuburg an der Donau (Unteres Untertithon, Bayern). Neues Jahrb. Geol. Paläont. Monatsh. 1, 4–23 (1971).

Bathurst, R. G. C.: Carbonate sediments and their diagenesis (Developments in Sedimentology 12), 620 pp. Amsterdam-London-New York: Elsevier 1971.

Bausch, W. M.: Der Obere Malm an der unteren Altmuhl; nebst einer Studie über das Riff-Problem 38 pp. Erlanger Geol. Abhandl. 49 (1963).

Bausch, W. M., Zeiss, A.: Zur Zusammensetzung des Kehlheimer Riffkalk. Geol. Gl. No. Bayern 16, 240–242 (1966).

Beales, F. W.: Ancient sediments of Bahaman type. Am. Ass. Petrol. Geologists Bull. 45, 1845–1880 (1958).

Beales, F. W.: Limestone Peels. J. Alta Soc. Petrol. Geologists 8, 132–135 (1960).

Beaudoin, B.: Sedimentation détritique d'une série carbonatée reputée "Pélagique" (Exemple du Jurrassique Supérieur-Crétacé Inférieur de sud-est de la France). Sedimentary Geology, p. 135–151. Amsterdam: Elsevier 1970.

Bebout, D. G., Maiklem, W. R.: Ancient anhydrite facies and environments, Middle Devonian Elk Point basin, Alberta. Can. Petrol. Geol. Bull. 21, 287–343 (1973).

Becerra, H. A.: Estudio bioestratigrafico de la Formacion Tamabra del Cretácico en el distrito de Poza Rica. Inst. Mex. Petróleo Rev. 2 No. 3, 21–25 (1970).

Behrens, E. W., Land, L. S.: Subtidal Holocene dolomite Baffin Bay, Texas. J. Sediment. Petrol. 42 155–161 (1972).

Bernoulli, D.: North Atlantic and Mediterranean Mesozoic facies: a comparison, p. 801–871. Initial Rept. Deep Sea Drilling Proj. 11, Washington D. C. 1972.

Bernoulli, D., Jenkyns, H. C.: A Jurassic basin: the Glasenbach Gorge, Salzburg, Austria. Verhandl. Geol. Bundesanstalt 4, 504–531 (1970).

Bernoulli, D., Jenkyns, H. C.: Alpine, Mediterranean and central Atlantic Mesozoic facies in relation to the early evolution of the Tethys: In: Modern and Ancient Geosynclinal Sedimentation. Soc. Econ. Paleontologists Mineralogists Spec. Publ. 19. 129–160 (1974).

Berry, W. B. N., Boucot, A. J.: Correlation of the North American Silurian rocks, 289 pp. Geol. Soc. Am. Spec. Paper 102 (1971).

Bishop, W. F.: Petrology of upper Smackover limestone in north Haynesville field. Claiborne Parish, Louisiana. Am. Ass. Petrol. Geologists Bull. 52, 92–128 (1968).

Bishop, W. F.: Environmental control of porosity in the upper Smackover limestone, North Haynesville field, Claiborne Parish, Louisiana. Trans. Gulf Coast Ass. Geol. Soc. 19, 155–169 (1969).

Black, M.: Great Bahama Bank—a modern shelf lagoon (abst.) Geol. Soc. Am. Bull. 41, 109 (1930).

Blount, D. M., Moore, C. H.: Depositional and nondepositional breccias, Chiantla quadrangle, Guatemala. Geol. Soc. Am. Bull. 80, 429–442 (1969).

Bolliger, W., Burri, P.: Sedimentologie von Shelf-Carbonaten und Beckenablagerungen in Oxfordien des centralen Schweizer Jura, 96 pp. Beiträge Geol. Karte Schweiz, Ser. 140 (1970).

Bonet, F.: La Facies Urgoniana del Cretácico Medio de la Region de Tampico. Asoc. Mex. Geológos Petroleros Bol. 4, 153–262 (1952).

Bonet, F.: Biostratigraphic notes on the Cretaceous of eastern Mexico. In: Geology of the Peregrina Canyon and Sierra de El Abra, Mexico. Corpus Christi Geol. Soc. Ann. Field Trip Guidebook, 36–48 1963.

Bornhauser, M.: Marine sedimentary cycles of Tertiary in Mississippi embayment and central Gulf Coast area. Am. Ass. Petrol. Geologists Bull. 31, 698–712 (1947).

Bosellini, A., Rossi, D.: Triassic carbonate buildups of the Dolomites, northern Italy. Soc. Econ. Paleontologists Mineralogists Spec. Publ. 18, 209–233 (1974).

Bott, M. H. P., Johnson, G. A. L.: The controlling mechanism of Carboniferous cyclic sedimentation. J. Geol. Soc. London **122**, 421–441 (1967).

Bouma, A. H.: Sedimentology of some Flysch deposits; a graphic approach to facies interpretation, 168 pp. Amsterdam: Elsevier Publ. 1962.

Brady, M. J.: Sedimentology and diagenesis of carbonate muds in coastal lagoons of N. E. Yucatan. Ph. D. Dissertation, 288 pp. Houston, Texas: Rice Univ. 1971.

Bricker, O. P. (Ed.): Carbonate Cements. The Johns Hopkins Univ. Studies in Geology, Vol. 19, 376 pp. Baltimore, Maryland: Johns Hopkins Press 1971.

Briggs, L. I., Gill, D.: Silurian shelf reefs and evaporites, Michigan basin. 8th Inter. Sed. Congress, Heidelberg, (Abstr.), 13 (1971).

Brown, L. F.: Virgil-Lower Wolfcamp repetitive depositional environments in north-central Texas. In: Elam, J. C., Chuber, S. (Eds.): Cyclic Sedimentation in the Permian Basin, 2nd Edition, West Texas Geol. Soc., 41–54. 1972.

Bryant, W. R., Meyerhoff, A. A., Brown, N. P., Furrer, M. A., Pyle, T. E., Antoine, J. W.: Escarpments, reef trends, and diapiric structures, eastern Gulf of Mexico. Am. Ass. Petrol. Geologists Bull. **53**, 2506–2542 (1969).

Carbone, F., Pratullon, A., Sirna, G.: The Cenomanian shelf-edge facies of Rocca Di Cave (Prenestini Mountains, Latium). Geologica Romana **10**, 131–198 (1971).

Carozzi, A.: Nouvelles observations microscopiques sur les dépôts de courants de turbidité du Malm de la Nappe de Morcle en Haute-Savoie, 30 pp. Bull. Inst. Nat. Genevòis **57**, (1955).

Carozzi, A., Zadnik, V. E.: Microfacies of Wabash Reef, Wabash, Indiana. J. Sediment, Petrol. **29**, 164–171 (1959).

Carozzi, A., Bouroullec, J., Deloffre, R., Rumeau, J-L.: Microfacies of the Jurassic of Aquitaine, petrography-diagenesis-geochemistry-petrophysics. Bull. Centre Rech. Pau-SNPA, **1**, 594 pp. (1972).

Carrasco, B.: La Formación El Abra (Formación el Doctor) en la Plataforma Valles-San Luis. Potosí. Inst. Mex. Petróleo, Rev. **2**, 97–99 (1970).

Carrasco, B.: Litofacies de la formación El Abra en plataforma de Actopan, Hgo. Inst. Mex. Petróleo. Rev. **3**, 5–26 (1971).

Cayeux, L.: Les roches sédimentaire de France: Roches carbonatées (calcaires et dolomies), 463 pp. Paris: Masson et Cie, 1935.

Chave, K. E.: Recent carbonate sediments — an unconventional view. J. Geol. Education **15**, 200–204 (1967).

Chilingar, G. V.: Relationship between Ca/Mg ratio and geologic age. Am Ass. Petrol. Geologists Bull. **40**, 2256–2266 (1956).

Chilingar, G. V., Bissel, H. J., Fairbridge, R. W. (Eds.): Carbonate rocks (Developments in Sedimentology 9). Amsterdam, N.Y.C.: Elsevier 1967.

Choquette, P. W., Pray, L. C.: Geological nomenclature and classification of porosity in sedimentary carbonates. Am. Ass. Petrol. Geologists Bull. **54**, 207–250 (1970).

Choquette, P. W., Traut, J. D.: Pennsylvanian carbonate reservoirs, Ismay field, Utah and Colorado. In: Bass, R. O., Sharp, S. L. (Eds.): Shelf Carbonates of the Paradox Basin, pp. 157–184. Four Corners Geol. Soc. 4th Field Conf. Symposium, Durango, Colorado 1963.

Chuber, S., Pusey, W. C.: Cyclic San Andres facies and their relationship to diagenesis, porosity and permeability in the Reeves field, Yoakum County, Texas. In: Elam, J. C., Chuber, S. (Eds.): Cyclic Sedimentation in the Permian Basin, 2nd Ed., pp. 135–150. Midland, Texas: West Texas Geol. Soc. 1972.

Church, S. B.: Sponge-algal reefs in the lower Ordovician Pogonip group, western Utah. (Abstr.) Cordilleran Section, Geol. Soc. Am. 155 (1974).

Clark, T. H., Stearn, C. W.: The Geological evolution of North America, 2nd Ed. 570 pp. New York: Ronald Press 1968.

Clayton, R. N., Jones, B. F., Berner, R. A.: Isotope studies of dolomite formation under sedimentary conditions. Geochim. Cosmochim. Acta **32**, 415–432 (1968).

Colter, V.: The Palaéoecology of the Wenlock limestone. Ph. D. Thesis, 315 pp. England· Univ. of Cambridge 1957.

Comparative Sedimentology Laboratory Symposium: Tidal deposits, a compilation of examples, 240 pp. Univ. of Miami 1973.

Coogan, A. H.: Recent and ancient carbonate cyclic sequences. In: Elam, J. C., Chuber, S. (Eds.): Cyclic Sedimentation in the Permian Basin, 2nd Ed., pp. 5–16, Midland, Texas: West Texas Geol. Society 1972.

Coogan, A. H., Bebout, D. G., Maggio, C.: Depositional environments and geologic history of Golden Lane and Poza Rica Trend, Mexico, an alternative view. Am. Ass. Petrol. Geologists Bull. **56**, 1419–1447 (1972).

Cook, H. E.: Miette platform evolution and relation to overlying bank ("reef") localization, Upper Devonian, Alberta. Can. Petrol. Geol. Bull. **20**, 375A–411 (1972).

Cook, H. E., McDaniels, P. M., Montjoy, E. W., Pray, L. C.: Allochtnonous carbonate debris flows at Devonian bank ("reef") margins, Alberta, Canada. Can. Petrol. Geol. Bull. **20**, 439–497 (1972).

Cooper, G. A.: Paleoecology of Middle Devonian of eastern and central United States. Geol. Soc. Am. Mem. **67**, 249–277 (1957).

Cotter, E.: Waulsortian-type carbonate banks in the Mississippian Lodgepole Formation of central Montana. J. Geology **73**, 881–888 (1965).

Cummings, E. R., Shrock, R. R.: Niagaran coral reefs of Indiana and adjacent states and their stratigraphic relations. Geol. Soc. Am. Bull. **39**, 579–620 (1928).

Cuvillier, J., Schürmann, H. M. E. (Eds.): International Sedimentary Petrographical Series, I–XV, Leiden: J. Brill and Co. 1951–1969.

Davies, G. R.: A Permian hydrozoan mound, Yukon, Territory. Can. J. Earth Sci. **8**, 973–988 (1971).

Davies, G. R., Ludlam, S. D.: Origin of laminated and graded sediments, Middle Devonian of western Canada. Geol. Soc. Am. Bull. **84**, 3527–3546 (1973).

Davies, G. R., Nassichuk, W. W.: The Hydrozoan? *Palaeoaplysina* from the Upper Paleozoic of Ellesmere Island, Arctic Canada. J. Paleontology **47**, 251–265 (1973).

DeBuisonje, P. H.: Recurrent red tides, a possible origin of the Solnhofen limestone. Koninkl. Ned. Akad. Wetenschap., Proc. Ser. B, **75**, No. 2, 152–177 (1972).

Deffeyes, K. S., Lucia, F. J., Weyl, P. K.: Dolomitization of Recent and Plio-Pleistocene Sediments by marine evaporite waters on Bonaire, Netherlands Antilles. Soc. Econ. Paleontologists Mineralogists Spec. Publ. **13**, 71–88 (1965).

Dickinson, K. A.: Upper Jurassic carbonate rocks in northeastern Texas and adjoining parts of Arkansas and Louisiana. Trans. Gulf Coast Ass. Geol. Soc. **19**, 175–188 (1969).

Dolphin, D. R., Klovan, J. E.: Stratigraphy and paleoecology of an Upper Devonian carbonate bank, Saskatchewan River crossing, Alberta. Can. Petrol. Geol. Bull. **18**, 289–331 (1970).

Dooge, J.: The stratigraphy of an Upper Devonian carbonate-shale transition between the north and south Ram Rivers of the Canadian Rocky Mountains. Leidse Geol. Med. **39**, 53 pp. Leiden: J. J. Groen en Zoon (1966).

Dott, R. H., Batten, R. L.: Evolution of the Earth 649 pp. New York: McGraw-Hill Co. (1971).

Duff, P. M. D., Hallam, A., Walton, E. K.: Cyclic Sedimentation (Developments in Sedimentology 10). 280 pp. Amsterdam: Elsevier 1967.

Dunham, R. J.: Classification of carbonate rocks according to depositional texture. In: Ham, W. E. (Ed.): Classification of Carbonate Rocks. Am Ass. Petrol. Geol. Mem. **1**, 108–121 (1962).

Dunham, R. J.: Vadose pisolite in the Capitan reef (Permian), New Mexico and Texas. In: Friedman, G. M. (Ed.): Depositional Environments in Carbonate Rocks. Soc. Econ. Paleontologists Mineralogists Spec. Publ. **14**, 182–191 (1969a).

Dunham, R. J.: Early vadose silt in Townsend mound (reef), New Mexico. In: Friedman, G. M. (Ed.): Depositional Environments in Carbonate Rocks. Soc. Econ. Paleontologists Mineralogists Spec. Publ. **14**, 139–181 (1969b).

Dunham, R. J.: Stratigraphic reefs versus ecologic reefs. Am. Ass. Petrol. Geologists Bull. **54**, 1931–1932 (1970).

Dunham, R. J.: Guide for study and discussion for individual reinterpretation of the sedimentation and diagenesis of the Permian Capitan geologic reef and associated rocks, New Mexico and Texas. In: Permian Basin Section, Soc. Econ. Paleontologists Mineralogists, 235 pp. Pub. **72–14** (1972).

Dunnington, H. V.: Generation migration, accumulation, and dissipation of oil in northern Iraq. In: Weeks, L. G. (Ed.): Habitat of Oil, A Symposium, pp. 1194–1251. Tulsa, Okla: Am. Ass. Petrol. Geologists 1958.

Dupont, H.: Contribution à l'étude des Faciès du Waulsortien de Waulsort. Mem. Inst. Geol. Louvain 24, 94–164 (1969).

Dzulvnski, S. T., Sanders, J. E.: Bottom marks on firm lutite substratum underlying turbidite beds, 1594. Geol. Soc. Am. Bull. 70 (1959).

Dzulynski, S. T., Smith, A. J.: Convolute lamination, its origin, preservation, and directional significance. J. Sediment. Petrol. 33, 616–627 (1963).

Edie, R.: Mississippian sedimentation and oil fields in southeastern Saskatchewan. Am. Ass. Petrol. Geologists Bull. 42, 94–126 (1958).

Elloy, R.: Réflexions sur quelques environements récifaux du paléozoïque. Bull. Centre Rech. Pau-SNPA, 6, 1–105 (1972).

Embry, A. F., Klovan, J. E.: A Late Devonian reef tract on northeastern Banks Island, Northwest Territories. Can. Petrol. Geology Bull. 19, 730–781 (1971).

Enos, P.: Reefs, Platforms, and Basins of Middle Cretaceous in northeast Mexico. Am. Ass. Petrol. Geologists Bull. 58, 800–809 (1974).

Evamy, B. D.: The application of a chemical staining technique to a study of dedolomitization. Sedimentology 2, 164–170 (1963).

Evans, J. K.: Depositional environment of a Pennsylvanian black shale (Heebner) in Kansas and adjacent states. Ph. D. Dissertation, 131 pp. Houston, Texas: Rice Univ. 1966.

Fabricius, F.: Beckensedimentation und Riffbildung an der Wende Trias/Jura in den Bayerisch-Tiroler Kalkalpen. Intern. Sed. Petr. Series IX, 144 pp. Leiden: E. J. Brill 1966.

Fairbridge, R. W.: Stratigraphic correlation by microfacies. Am. J. Sci. 252, 683–694 (1954).

Fichter, H. J.: Geologie der Bauen-Brisen-Kette am Vierwaldstättersee u. die zyklische Gliederung der Kreide u. des Malm der Helvetischen Decken, 129 pp. Dissertation, Bern, Switzerland: Stampfi and Cie 1934.

Field, R. M.: Paleo-oceanography of limestone seas (abst.). Geol. Soc. Am. Bull. 41, 110 (1930).

Fischbuch, N. R.: Stratigraphy, Devonian Swan Hills reef complexes of central Alberta. Can. Petrol. Geology Bull. 16, 444–587 (1968).

Fischer, A. G.: Stratigraphic record of transgressing seas in light of sedimentation on Atlantic coast of New Jersey. Am. Ass. Petrol. Geologists Bull. 45, 1656–1666 (1961).

Fischer, A. G.: The Lofer cyclothem of the Alpine Triassic. Kansas Geol. Survey Bull. 169, 107–149 (1964).

Fisher, W. L.: Sedimentary patterns in Eocene cyclic deposits, Northern Gulf Coast region. Kansas Geol. Survey Bull. 169, 151–170 (1964).

Fisher, W. L., Rodda, P. U.: Edwards formation (Lower Cretaceous), Texas: Dolomitization in a carbonate platform system. Am. Ass. Petrol. Geologists Bull. 53, 55–72 (1969).

Flügel, E.: Mikrofazielle Untersuchungen in der Alpinen Triassic—Methoden und Probleme. Mitt. Ges. Geol. Bergbaustud. 21, 9–64 (1972).

Flügel, E., Franz, H. E.: Electronen mikroskopischer Nachweis von Coccolithen im Solnhofener Plattenkalk (Ober Jura). Neues Jahrb. Geol. Paläont. Abhandl. 127, 245–263 (1967).

Folk, R. L.: Practical petrographic classification of limestones. Am. Ass. Petrol. Geologists Bull. 43, 1–38 (1959).

Folk, R. L.: Spectral subdivision of limestone types. In: Ham, W. E. (Ed.), Classification of Carbonate Rocks. Am. Ass. Petrol. Geologists Mem. 1, 62–84 (1962).

Folk, R. L.: The natural history of crystalline calcium carbonate: effect of magnesium content and salinity. J. Sediment. Petrol. 44, 40–53 (1974).

Freyburg, B.: Übersicht über den Malm der Altmühl Alb, 40 pp. Erlanger Geol. Abhandl. 70, 1968.

Friedman, G. M.: Identification of carbonate minerals by staining methods. J. Sediment. Petrol. 29, 87–97 (1959).

Friedman, G. M.: Significance of Red Sea in problem of evaporites and basinal limestones. Am. Ass. Petrol. Geologists Bull. 56, 1072–1086 (1972).

Galley, J. E.: Oil and Geology in the Permian basin of Texas and New Mexico. In: Weeks, L. G. (Ed.): Habitat of Oil, a Symposium, pp. 395–446. Tulsa, Okla: Am. Ass. Petrol. Geologists 1958.

Galloway, W. E., Brown, L. F.: Depositionai systems and shelf-slope relations on cratonic basin margin, uppermost Pennsylvania of north central Texas. Am. Ass. Petrol. Geologists Bull. **57**, 1185–1218 (1973).

Garrison, R. E.: Pelagic limestones of the Oberalm beds (Upper Jurassic-Lower Cretaceous), Austrian Alps. Can. Petrol. Geology Bull. **15**, 21–49 (1967).

Garrison, R. E., Fischer, A. G.: Deep Water limestones and radiolarites of the Alpine Jurrassic. In: Friedman, G. M. (Ed.) Depositional Environments in Carbonate Rocks. Soc. Econ. Paleontologists Mineralogists Spec. Publ. **14**, 20–56 (1969).

Ginsburg, R. N.: Environmental relationships of grain size and constituent particles in some south Florida carbonate sediments. Am. Ass. Petrol. Geologists Bull. **40**, 2384–2427 (1956).

Ginsburg, R. N., James, N. P.: Holocene reef wall and forereef lithofacies Belize, British Honduras. (abs.). Am. Ass. Petrol. Geologists and Soc. Econ. Paleontologists Mineralogists Annual Meeting Abstracts, 39 (1974).

Ginsburg, R. N., Marszalek, D. S., Schneidermann, N.: Ultrastructure of carbonate cements in a Holocene algal reef of Bermuda. J. Sediment. Petrol. **41**, 472–482 (1971).

Ginsburg, R. N., Rezak, R., Wray, J. L · Geology of calcareous algae (notes for a short course). Sedimenta I 61 pp. Comparative Sedimentology Lab., Univ. of Miami 1971.

Goodell, H. G., Garman, R. K.: Carbonate geochemistry of superior deep test well, Andros Island, Bahamas. Am. Ass. Petrol. Geologists Bull. **53**, 513–536 (1969).

Grabau, A. W.: A textbook of geology part I, General Geology, 83 pp. Boston-New York-Chicago: D. C. Heath and Co. 1920.

Griffith, L. S., Pitcher, M. G., Rice, G. W.: Quantitative environmental analysis of a Lower Cretaceous reef complex. In: Friedman, G. M. (Ed.): Depositional Environments in Carbonate Rocks. Soc. Econ. Paleontologists Mineralogists Spec. Publ. **14**, 120–138 (1969).

Grim, R. E.: Concept of diagenesis in argillaceous sediments. Am. Ass. Petrol. Geologists Bull. **42**, 246–253 (1958)

Guzman, E. J.: Reef-type stratigraphic traps in Mexico. Proc. 7th World Petrol. Congress, Vol. 2, pp. 461–470. London: Elsevier 1967.

Gwinner, M. P.: Geologie des weißen Jura der Albhochfläche (Württemberg). Neues Jahrb. Geol. Paläon. Abhandl. **115**, 137–221 (1962).

Gwinner, M. P.: Carbonate rocks of the Upper Jurassic in southwest Germany. In: Müller, G. (Ed.): Sedimentology of Parts of Central Europe, Guidebook, 8th Inter. Sed. Congress, pp. 193–208. Heidelberg 1971.

Gygi, R.: Zur Stratigraphie der Oxford-Stufe (oberes Jura-System) der Nordschweiz und des süddeutschen Grenzgebietes, 123 pp. Beitr. Geol. Karte Schweiz, New Ser. **136**, No. 11 (1969).

Hadding, A.: Silurian algal limestones of Gotland; indicators of shallow waters and elevation of land; some reflections on their lithological character and origin 26 pp. Lund Univ. Inst. Mineral. Paleon. and Quat. Geology, Pub. 70; Arsskr. ave. 2, **57** (1959).

Hallam, A.: Origin of the limestone-shale rhythm in the Blue Lias of England: a composite theory. J. Geology **72**, 157–169 (1964).

Hallam, A.: A pyritized limestone hardground in the Lower Jurassic of Dorset (England). Sedimentology **12**, 231–240 (1969).

Ham, W. E. (Ed.): Classification of Carbonate Rocks, a Symposium, 279 pp. Am. Ass. Petrol. Geologists, Mem. **1**, Tulsa, Okla 1962.

Hanshaw, B. B., Back, W., Dieke, R. G.: A geochemical hypothesis for dolomitization by ground water. Econ. Geology **66**, 710–724 (1971).

Harbaugh, J. W.: Mississippian bioherms in northeast Oklakoma. Am. Ass. Petrol. Geologists Bull. **41**, 2530–2544 (1957).

Harbaugh, J. W.: Significance of marine banks in southeast Kansas in interpreting cyclic Pennsylvanian sediments. Kansas Geol. Survey Bull. **168**, 199–204 (1964).

Harris, T. J., Hay, J. T. C., Twombley, B. N.: Contrasting limestone reservoirs in the Murban Field, Abu Dhabi. In: 2nd Regional Tech. Symposium, Soc. Petrol. Engineers of AIME, pp. 149–182. Dhahran, Saudi Arabia 1968.

Hay, W. W.: Probablistic stratigraphy. Eclogae Geol. Helvetiae **65**, 255–266 (1972).

Heckel, P. H.: Possible inorganic origin for stromatactis in calcilutite mounds in the Tully limestone. Devonian of New York. J. Sediment. Petrol. **42**, 7–18 (1972a).

Heckel, P. H.: Recognition of ancient shallow marine environments. In: Recognition of Ancient Sedimentary Environments, In: (Rigby, J. K., Hamblin, W. K. Eds.). Soc. Econ. Paleontologists Mineralogists Spec. Publ. **16**, 226–286 (1972b).

Heckel, P. H.: Pennsylvanian stratigraphic reefs in Kansas, some modern comparisons and implications. Geol. Rundschau **61**, pt. 2, 584–598 (1972c).

Heckel, P. H.: Carbonate buildups in the geologic record: a review. In: Laporte, L. F. (Ed.). Reefs in Time and Space, Selected Examples from the Recent and Ancient. Soc. Econ. Paleontologists Mineralogists Spec. Publ. **18**, 90–154 (1974).

Heckel, P. H., Cocke, J. M.: Phylloid algal mound complexes in outcropping Upper Pennsylvanian rocks of mid-continent. Am. Ass. Petrol. Geologists Bull. **53**, 1058–1074 (1969).

Henson, F. R. S.: Cretaceous and Tertiary reef formations and associated sediments in the Middle East. Am. Ass. Petrol. Geologists Bull. **34**, 215–238 (1950).

Hiller, K.: Über die Bank und Schwammfazies des Weißen Jura der Schwäbischen Alb (Württemberg), 189 pp. Geol.-Paläontologischen Institut Tech. Hochschule Stuttgart, New Ser., **40** (1964).

Hills, J. M.: Late Paleozoic sedimentation in West Texas Permian basin. Am. Ass. Petrol. Geologists Bull. **56**, 2303–2322 (1972).

Hoffman, P.: Shallow and deepwater stromatolites in Lower Proterozoic platform-basin facies change, Great Slave Lake, Canada. Am. Ass. Petrol. Geologists Bull. **58**, 856–867 (1974).

Holder, H.: Jurassic stratigraphische Geologie, Band IV, 603 pp. Stuttgart: Ferdinand Enke Verlag 1964.

Hollman, R.: Über Subsolution und die "Knollenkalke" des calcare Ammonitico Rosso Superiore im Monte Baldo (Malm; Norditalien). Neues Jahrb. Geol. Paläont. Monath. **4**, 163–179 (1962).

Horowitz, A. S., Potter, P. E.: Introductory Petrography of Fossils, 302 pp. Berlin-Heidelberg-New York: Springer

Houbolt, J. J. H. C.: Surface sediments of the Persian Gulf near the Qatar peninsula. Dissertation, Univ. Utrecht, 113 pp. The Hague: Mouton and Co. 1957.

Hsu, K. J., Siegenthaler, C.: Preliminary experiments on hydrodynamic movement induced by evaporation and their bearing on the dolomite problem. Sedimentology **12**, 11–25 (1969).

Hudson, R. G. S.: Carboniferous of the Craven reef belt Namurian unconformity at Scaleber, near Settle. Proc. Geologists Ass., 303–305 (1930).

Hudson, R. G. S., Cotton, G.: The Lower Carboniferous in a boring at Alport, Derbyshire. Proc. Yorkshire Geol. Soc. **55**, 192–215 (1945).

Illing, L. V., Wells, A. J., Taylor, J. C. M.: Penecontemporary dolomite in the Persian Gulf. Soc. Econ. Paleon. Mineralogists Spec. Publ. **13**, 89–111 (1965).

Ingels, J. J. C.: Geometry, paleontology, and petrography of Thornton reef complex, Silurian of northeastern Illinois. Am. Ass. Petrol. Geologists Bull. **47**, 405–440 (1963).

Ireland, H. A.: Terminology for insoluble residues. Am. Ass. Petrol. Geologists Bull. **31**, 1479–1490 (1947).

Irwin, M. L.: General theory of epeiric clear water sedimentation. Am. Ass. Petrol. Geologists Bull. **49**, 445–459 (1965).

Jacka, A. D., Thomas, C. M., Beck, R. H., Williams, K. W., Harrison, S. C.: Guadalupian depositional cycles of the Delaware basin and northwest shelf. In: Elam, J. C., Chuber, S. (Eds.): Cyclic Sedimentation in the Permian Basin, 2nd Ed., pp. 151–195. West Texas Geol. Soc. Publ. 1972.

James, G. W.: Stratigraphic geochemistry of a Pennsylvanian black shale (Excello) in the Mid-continent and Illonois basin. Ph. D. dissertation, 92 pp. Houston, Texas: Rice University 1970.

Jenik, A. J., Lerbekmo, J. F.: Facies and geometry of Swan Hills reef member of Beaverhill Lake formation (Upper Devonian), Goose River field, Alberta, Canada. Am. Ass. Petrol. Geologists Bull. **52**, 21–56 (1968).

Jenks, S.: Deposition and diagenesis of the Mississippian Lodgepole Formation, central Montana. M. A. Theis, 50 pp. Houston, Texas: Rice University.

Jodry,R.L.: Growth and dolomitization of Silurian Reefs, St. Clair County, Michigan. Am. Assoc. Petrol. Geologists Bull. **53**, 957–981 (1969).

Johnson,G.A.L.: Lateral variations of marine and deltaic sedments in cyclothemic deposits with particular reference to the Visean and Namurian of northern England. Compte rendu 4th Congress pour l'Avancement des Études de Stratigraphique et de Geologie du Carbonifere. **II**, 323–330 (1961).

Johnson,J.H.: Bibliography of fossil algae, 1942–1955, 92 pp. Quarterly of the Colorado School of Mines, Golden, Colorado, **52**, (1957).

Johnson,J.H.: Limestone-building algae and algal limestones. 297 pp. Colorado School of Mines Publ. Boulder, Colorado: Johnson Publ. Co. 1961.

Johnson,J.H.: Bibliography of fossil algae algal limestones, and the geological work of algae, 1956–1965, 148 pp. Quarterly of the Colorado School of Mines, Golden, Colorado, **62**, (1967).

Jurgan,H.: Sedimentologie des Lias der Berchtesgadener Kalkalpen. Geol. Rundschau **58**, 464–501 (1969).

Jux,v.U.: Die Riffe Gotlands u. ihre angrenzenden Sedimentationsräume. Contrib. Geology **I–4**, pp.41–90. Stockholm: Acta Univ. 1957.

Jux,v.U.: Die Devonischen Riffe im Rheinischen Schiefergebirge. Neues Jahrb. Geol. Paläontol, Abhandl. **110**, 186–258 (1960).

Kay,M.: Geology of the Utica quadrangle, New York. N. Y. State Museum Bull. **347**, 126 pp. (1953).

Keith,J.W.: Environmental interpretation of subsurface Washita-Fredericksburg limestones, northern Live Oak County, south Texas, pp. 72–78. In: Peregrina Canyon and Sierra de El Abra, Guidebook Corpus Christi Geol. Soc. Field Trip, Corpus Christi, Texas 1963.

Kendall,C.G.St.C.: An environmental re-interpretation of the Permian evaporite/carbonate shelf sediments of the Guadalupe Mountains. Geol. Soc. Am. Bull. **80**, 2503–2526 (1969).

Kendall,C.G.St.C., Skipwith,P.A.d'E.: Recent algal mats of a Persian Gulf lagoon. J. Sediment. Petrol. **38**, 1040–1058 (1968).

Keulegan,G.H., Krumbein,W.C.: Stable configuration of bottom slope in a shallow sea and its bearing on geological processes. Am. Geophysics Union Trans. **30**, 855–861 (1950).

King,P.B.: Permian stratigraphy of trans-Pecos, Texas. Geol. Soc. Am. Bull. **45**, 697–798 (1934).

King,P.B.: Permian of West Texas and southeastern New Mexico: West Texas-New Mexico Symposium, Part II. Am. Ass. Petrol. Geologists Bull. **26**, 535–763 (1942).

King,P.B.: Geology of the southern Guadalupe Mountains, Texas, 183 pp. U. S. Geol. Survey Prof. Paper **215** (1948).

King,P.B.: The Evolution of North America, 190 pp. Princeton, New Jersey: Princeton Univ. Press 1959.

Kinsman,D.J.J.: Modes of formation, sedimentary associations and diagnostic features of shallow-water and supratidal evaporites. Am. Ass. Petrol. Geologists Bull. **53**, 830–840 (1969).

Klovan,J.E:: Facies analysis of the Redwater reef complex, Alberta, Canada. Can. Petrol. Geology Bull. **12**, 1–100 (1964).

Klovan,J.E.: Development of western Canadian Devonian reefs and comparison with Holocene analogues. Am. Ass. Petrol. Geologists Bull. **58**, 787–799 (1974).

Klüpfel,W.: Über die Sedimente der Flachseen im Lothringer Jura. Geol. Rundschau **7**, 97–109 (1917).

Konishi,K., Wray,J.L.: *Eugonophyllum*, a new Pennsylvanian and Permian algal genus. J. Paleontology **35**, 659–666 (1961).

Krebs,W.: Devonian reef limestones in the eastern Rhenish Schiefergebirge, pp.45–81. In: Müller,G. (Ed.): Sedimentology of Parts of Central Europe, Guidebook, 8th Inter. Sed. Congress, Heidelberg 1971.

Krebs,W.: Devonian carbonate complexes of central Europe. In: Laporte,L.F. (Ed.): Reefs in Time and Space, Selected Examples from the Recent and Ancient, Soc. Econ. Paleontologists Mineralogists Spec. Pub. **18**, 155–208 (1974).

Krebs,W., Mountjoy,E.W.: Comparison of central European and western Canadian Devonian reef complexes. 24th. Inter. Geol. Congress, Sec.**6**, 295–309 (1972).

Krumbein,W.: Criteria for subsurface recognition of unconformities. Am. Ass. Petrol. Geologists Bull. **26**, 36–62 (1942).

Krumbein,W., Sloss,L.L.: Stratigrapny and Sedimentation, 660 pp. San Francisco-London: W.H.Freeman Co. 1963.

Kuenen,P.H.: Paleogeographic significance of graded bedding and associated features. Koninkl. Ned. Akad. Wetenschap, Proc. Ser. B, **55**, 28–36 (1952).

Kuenen,P.H.: Significant features of graded bedding. Am. Ass. Petrol. Geologists Bull. **37**, 1044–1066 (1953).

Kukal,Z., Saadallah,A.: Aeolian admixtures in the sediments of the northern Persian Gulf. In: Purser,B.H. (Ed.): The Persian Gulf-Holocene Carbonate Sedimentation and Diagenesis in a Shallow Epicontinental Sea, pp. 115–121. New York-Heidelberg-Berlin: Springer 1973.

Land,L.S.: Submarine lithification of Jamaican reefs. In: Bricker,O.P. (Ed.): Carbonate Cements. Johns Hopkins Univ. Studies in Geology, Vol. 19, 59–62. 1971.

Land,L.S.: Contemporaneous dolomitization of Middle Pleistocene reefs by meteoric water, North Jamaica. Bull. Marine Sci. **23**, 64–92 (1973a).

Land,L.S.: Holocene meteoric dolomitization of Pleistocene limestone, North Jamaica. Sedimentology **20**, 411–424 (1973b).

Laporte,L.F.: Recent carbonate environments and their paleoecologic implications, pp.231–258. Evolution and Environment. New Haven Conn.: Yale University Press 1968.

Laporte,L.F.: Recognition of a transgressive carbonate sequence within an epeiric sea: Helderberg Group (Lower Devonian) of New York State. In: Friedman,G.M. (Ed.): Depositional Environments in Carbonate Rocks. Soc. Econ. Paleontologists Mineralogists Spec. Publ. **14**, 98–119 (1969).

Laudon,L.R., Bowsher,A.L.: Mississippian formations of Sacramento Mountains, New Mexico. Am. Ass. Petrol. Geologists Bull. **25**, 2107–2160 (1941).

Lecompte,M.: Quelques précisions sur le phénomène récifal dans le Devonien de l'Ardenne et sur le rythme sédimentaire dans lequel il s'intègre. Bull. Inst. Roy. Sci. Nat. Belg. **32**, (21), 1–39 (1956).

Lecompte,M.: Die Riffe in Devon der Ardennen und ihre Bildungsbedingungen. Geologica et Palaeontologica (Marburg) **4**, 25–71 (1970).

Lees,A.: The Waulsortian "Reefs" of Eire: a carbonate mudbank complex of Lower Carboniferous age. J. Geology **69**, 101–109 (1961).

Lees,A.: The structure and origin of the Waulsortian (Lower Carboniferous) "Reefs" of west-central Eire. Phil. Trans. Roy. Soc. London, Ser. B, No. 740, **247**, 483–531 (1964).

Lees,A.: Les Dépots carbonates de plate-forme. Bull. Centre Recherches Pau-SNPA **7**, 177–192 (1973).

Leighton,M.W., Penedexter,C.: Carbonate rock types. In: W.E.Ham (Ed.): Classification of Carbonate Rocks. Am. Ass. Petrol. Geologists Mem. **1**, 33–61 (1962).

Leonardi,P., Le Dolomiti: Geologic dei monti tra Isarco e Piave, Vol. 1 and 2, 1019 pp. Rome: Nat. Res. Council 1967.

Lineback,J.A.: Illinois basin-sediment starved during Mississippian. Am. Ass. Petrol. Geologists Bull. **53**, 112–126 (1969).

Lloyd,E.R.: Capitan limestone and associated formation of New Mexico and Texas. Am. Ass. Petrol. Geologists Bull. **13**, 645–658 (1929).

Logan,B.W., Davies,G.R., Read,J.D., Cebulski,D.E.: Carbonate sedimentation and environments, Shark Bay, Western Australia, 223 pp. Am. Ass. Petrol. Geologists Mem. **13** (1970).

Logan,B.W., Rezak,R., Ginsburg,R.N.: Classification and environmental significance of algal stromatolites. J. Geology **72**, 68–83 (1964).

Lombard,A.: Géologie sédimentaire, pp.508–522. Les Séries Marines. Paris: Masson et Cie. 1956.

Lopez-Ramos,E.: Distribucion de la porosidad en las calizas del Cretacio medio. In: Eguia-Huerta,A. (Ed.): La exploracion gravimetirica y seismologica en estructuras arrecifales tipo Faja de Oro. Asoc. Mex. Geólogos Petroleros Bol. **7**, 241–311 (1955).

Lowenstam,H.A.: Niagaran reefs in the Great Lakes area. J. Geology **58**, 430–487 (1950).

Lowenstam, H. A.: Niagaran reefs in the Great Lakes area. In: Ladd, H. S. (Ed.): Treatise on Marine Ecology and Paleoecology, 2, Paleoecology. Mem. Geol. Soc. Am. **67**, 215–248 (1957).

Lucia, F. J.: Recognition of evaporite-carbonate shoreline sedimentation. In: Rigby, J. K., Hamblin, W. K. (Eds): Soc. Econ. Paleontologists Mineralogists Spec. Publ. **16**, 160–191 (1972).

Machielse, S.: Devonian algae and their contribution to the western Canadian sedimentary basin. Can. Petrol. Geology Bull. **20**, 187–237 (1972).

Maiklem, W. R., Bebout, D. G., Glaister, R. P.: Classification of anhydrite—a practical approach. Can. Petrol. Geology Bull. **17**, 194–233 (1969).

Majewske, O. P.: Recognition of Invertebrate fossil fragments in rocks and thin sections, 101 pp. 106 pls. Inter. Sed. Pet. Series XI, Leiden: Brill, E. J. 1969.

Malek-Aslani, M.: Lower Wolfcampian reef in Kemnitz Field, Lea County, New Mexico. Am. Ass. Petrol. Geologists Bull. **54**, 2317–2335 (1970):

Malek-Aslani, M.: Environmental modeling: A useful exploration tool in carbonates Trans. Gulf Coast Ass. Geol. Soc. **23**, 239–244 (1973).

Mamet, B.: Remarque sur le micro-fauna de foraminifers du Dinantian. Soc. Géologie Belgique Bull. **70**, 166–173 (1962).

Manten, A. A.: Silurian reefs of Gotland (Developments in Sedimentology 13), 539 pp. Amsterdam: Elsevier Pub. Co. 1971.

Maslov, V. P.: Fossil calcareous algae of the U.S.S.R. Akad. Nauk S.S.R., Trudy, Inst. Geol. Nauk **160**, 301 pp. (1956).

Matthews, R. K.: Diagenetic environments of possible importance to the explanation of cementation fabric in subaerially exposed carbonate sediments. In: Bricker, O. P. (Ed.): Carbonate Cements, The Johns Hopkins University Studies in Geology Vol. 19, 127–132, 1971.

Matthews, R. K.: Dynamic stratigraphy, 370 pp. New Jersey: Prentice-Hall, Inc. 1974.

McBride, E. F.: Stratigraphy and sedimentology of the Woods Hollow Formation (Middle Ordovician), Trans-Pecos, Texas. Geol. Soc. Am. Bull. **80**, 2287–2301 (1969).

McBride, E. F.: Stratigraphy and origin of Maravillas Formation (Upper Ordovician), West Texas. Am. Ass. Petrol. Geologists Bull. **54**, 1719–1745 (1970).

McCrossan, R. G.: Sedimentary boudinage structures in the Upper Devonian Ireton Formation of Alberta. J. Sediment. Petrol. **28**, 316–320 (1958).

McKee, E. D.: Cambrian history of the Grand Canyon region. Part I, Stratigraphy and ecology of the Grand Canyon Cambrian, 168 pp. Carnegie Inst. Washington Publ. **563** 1945.

McKee, E. D., Weir, G. W.: Terminology of stratification and cross stratification. Geol. Soc. Am. Bull. **64**, 381–390 (1953).

McKee, E. D., Chronic, J., Leopold, E. B.: Sedimentary belts in lagoon of Kapingamarangi Atoll. Am. Ass. Petrol. Geologists Bull. **43**, 501–562 (1959).

McQueen, H. S.: Insoluble residues as a guide in stratigraphic studies. Missouri Bureau Geology and Mines, 56th Biennial Rept. for 1929–1930, p. 102–131 (1931).

Meischner, K. D.: Allodapische Kalke, turbidite in Riff-Nahen Sedimentations-Becken. In: Bouma, A., Brouwer, A. (Eds.): Turbidites, pp. 156–191. Amsterdam: Elsevier Publ. Co. 1965.

Meissner, F. F.: Cyclic sedimentation in Middle Permian strata of the Permian basin, West Texas and New Mexico. In: Elam, J. C., Chuber, S. (Eds.): Cyclic Sedimentation in the Permian Basin, 2nd Ed., pp. 203–232. West Texas Geol. Society Midland, Texas, 1972.

Merriam, D. F. (Ed.): Symposium on cyclic Sedimentation, 636 pp. Kansas Geol. Survey Bull. **169**, Vol. 1 and 2 (1964).

Mesolella, K. J., Robinson, J. D., Ormiston, A. R.: Cyclic deposition of Silurian carbonates and evaporites in Michigan basin. Am. Assoc. Petrol. Geologists Bull. **58**, 34–62 (1974).

Meyers, W. J.: Chertification and carbonate cementation in the Mississippian Lake Valley formation, Sacramento Mountains, New Mexico. Ph. D. Thesis, 353 pp. Houston, Texas: Rice Univ. 1974.

Milliman, J. D.: Marine Carbonates, 375 pp. Berlin-Heidelberg-New York: Springer 1974.

Mitterer, R. M.: Influence of organic matrix on skeletal and non-skeletal $CaCO_3$ precipitation, p. 68. 8th Inter. Sed. Congress, Heidelberg, (Abstr.) (1971).

Mojsisovics, E. M.: Die Dolomit-Riffe von Südtirol und Venetien. 551 pp. Wien: A. Holder Co. 1879.

Moore, D.: Role of deltas in the formation of some British Lower Carboniferous cyclothems. J. Geology **67**, 522–539 (1959).

Moore, W. E.: Ecology of Recent foraminifera in northern Florida Keys. Am. Ass. Petrol. Geologists Bull. **41**, 727–741 (1957).

Morgenstein, M.: Sedimentary diagenesis and rates of manganese accretion on the Waho Shelf, Kauai Channel, Hawaii, pp. 121–136. Interuniversity program of research on Ferro-manganese deposits of the ocean floor (unpublished). Seabed Assessment Program Inter. Dec. Ocean Exploration, Nat. Sci. Foundation, Washington D.C. (1973).

Motts, W. S.: Geology and paleoenvironments of the northern segment, Capitan shelf, New Mexico and West Texas. Geol. Soc. Am. Bull. **83**, 701–722 (1972).

Mountjov, E. W.: Factors governing the development of the Frasnian Miette and Ancient Wall reef complexes (banks and biostromes), Alberta. First Internat. Symp. Devonian System. Alberta Geol. Soc. II, 387–408, 1968.

Mountjoy, E. W., Cook, H. E., Pray, L. C., McDaniel, P. N.: Allochthonous carbonate debris flows-worldwide indicators of reef complexes, banks or shelf margins, pp. 172–189. 24th Inter. Geol. Congress, sect. **6**, 1972.

Müller-Jungbluth, W.-V.: Sedimentary petrologic investigation of the Upper Triassic "Hauptdolomit" of the Lechthaler Alps, Tyrol, Austria. In: Müller, G., Friedman, G. M. (Eds.): Recent Developments in Carbonate Sedimentology in Central Europe, 228–239, 1968.

Murray, J. W.: An oil-producing reef-fringed carbonate bank in the Upper Devonian Swan Hills Member, Judy Creek, Alberta. Can. Petrol. Geology Bull. **14**, 1–103 (1966).

Murray, R. C.: Origin of porosity in carbonate rocks. J. Sediment. Petrol. **30**, 59–84 (1960).

Natland, M. L., Kuenen, P. H.: Sedimentary history of the Ventura basin, California and the action of turbidity currents. Soc. Econ. Paleontologists Mineralogists Spec. Publ. **2**, 76–107 (1951).

Nelson, H. F.: Deposition and alteration of the Edwards Limestone, central Texas. Symposium on Edwards limestone in central Texas. Univ. of Texas Publ. **5905**, 21–85 (1959).

Neumann, A. C., Gebelein, C. D., Scoffin, T. P.: The composition, structure, and erodability of subtidal mats Abaco, Bahamas. J. Sediment. Petrol. **40**, 274–297 (1970).

Neumann, A. C., Keller, G. H., Kofoed, J. W.: "Lithoherms" in the straits of Florida: Abstracts with Programs. Geol. Soc. Am. **4**, no. 7, 611 (1972).

Newell, N. D.: Bahamian platforms. In: The Crust of the Earth, a Symposium. Geol. Soc. Am. Spec. Paper **62**, 303–315 (1955).

Newell, N. D., Imbrie, J., Purdy, E. G., Thurber, D. L.: Organism communities and bottom facies, Great Bahama Bank. Bull. Am. Mus. Nat. Hist. **117**, 177–228 (1959).

Newell, N. D., Rigby, J. K., Fischer, A. G., Whiteman, A. J., Hickox, J. E., Bradley, J. S.: The Permian Reef Complex of the Guadalupe Mountains Region, Texas and New Mexico, 236 pp. San Francisco: W. H. Freeman and Co. 1953.

Noble, J. P. A.: Biofacies analysis, Cairn formation of Miette reef complex (Upper Devonian) Jasper National Park, Alberta. Can. Petrol. Geology Bull. **18**, 493–543 (1970).

Ohlen, H. R.: The Steinplatte reef complex of the Alpine Triassic (Rhaetian) of Austria, 122 pp. 20 pls. Ph. D. Dissertation. Princeton Univ. 1959.

Oliver, T. A., Cowper, N. W.: Depositional environments of the Ireton formation, central Alberta. Can. Petrol. Geology Bull. **2**, 183–202 (1963).

Ott, E.: Segmentierte Kalkschwämme (Sphinctozoa) aus der Alpinen Mitteltrias u. ihre Bedeutung als Riffbilder im Wettersteinkalk 96 pp. Bayer. Akad. Wiss. Abhandl. N. F., **131**, 1967.

Parkinson, D.: Lower Carboniferous reefs of northern England. Am. Ass. Petrol. Geologists Bull. **41**, 511–537 (1957).

Parks, J. M.: Reef-building biota from Late Pennsylvanian reefs, Sacramento Mountains, New Mexico (abst.). Am. Ass. Petrol. Geologists Bull. **46**, 274 (1962).

Perkins, B. F. (Ed.): Trace fossils, a field guide to selected localities in Pennsylvanian, Permian, Cretaceous and Tertiary rocks of Texas and related papers, 147 pp. S.E.P.M. Field trip, April 1–3, La. State Univ. Misc. Pub. 71–1 (1971).

Peterson, J.A., Ohlen, H.R.: Pennsylvanian shelf carbonates, Paradox basin. In: Bass, R.O., Sharp, S.L. (Eds.): Shelf Carbonates of the Paradise Basin, Four Corners Geol. Soc. 4th Field Conf. Symposium, Durango, Colorado, 65–79 (1963).

Pettijohn, F.J., Potter, P.E.: Atlas and Glossary of Primary Sedimentary structures, 370 pp. Berlin-Heidelberg-New York: Springer 1964.

Philcox, M.E.: Banded calcite mudstone in the Lower Carboniferous "reef" knolls of the Dublin basin, Ireland. J. Sediment. Petrol. **33**, 904–913 (1963).

Philcox, M.E.: A Waulsortian bryozoan reef ("cumulative biostrome") and the off-reef equivalents, Ballybeg, Ireland Compte Rendu, Sixth Inter. Congress Strat. and Geology of the Carboniferous, Sheffield, Eng. 1359–72 (1967).

Philcox, M.E.: Coral bioherms in the Hopkinton formation (Silurian), Iowa. Geol. Soc. Am. Bull. **81**, 969–974 (1970).

Pitcher, M.: Evolution of Chazyan (Ordovician) reefs of eastern U. S. and Canada. Can. Petrol. Geology Bull. **12**, 632–691 (1964).

Playford, P.E., Lowry, D.C.: Devonian reef complexes of the Canning basin, Western Australia, 50 pp. Geol. Survey of Western Australia Bull. **118** (1966).

Plumley, W.J., Graves, R.W.: Virgilian reefs of the Sacramento mountains. J. Geology **61**, 1–16 (1953).

Poborski, S.J.: Virgin Formation (Triassic) of the St. George, Utah, area. Geol. Soc. Am. Bull. **65**, 971–1006 (1954).

Polsak, A., Praturlon, A., Sirna, G.: Corrélation stratigraphique du facies biostromal dans le Cenomanien des Dinarides externes et de l'Apennin central, pp. 37–40. 7th Geol. Congress S. F. R. J. (Sadrzaji Referata pub.) Zagreb 1970.

Porter, J.W., Fuller, J.G.C.M.: Lower Paleozoic rocks of northern Williston basin and adjacent areas. Am. Ass. Petrol. Geologists Bull. **43**, 124–189 (1959).

Potter, P.E., Pettijohn, F.J.: Paleocurrents and basin analysis, 296 pp. Berlin-Göttingen-Heidelberg: Springer 1963.

Powers, R.W.: Arabian Upper Jurassic carbonate reservoir rocks. In: Ham, W.E. (Ed.): Classification of Carbonate Rocks, a Symposium, Am. Ass. Petrol. Geologists. Mem. **1**, 122–192 (1962).

Pray, L.C.: Fenestrate bryozoan core facies, Mississippian bioherms, southwestern United States. J. Sediment. Petrol. **28**, 261–273 (1958).

Pray, L.C.: Stratigraphic and structural features of the Sacramento Mountains escarpment, New Mexico, 250 pp. Roswell Geol. Soc. and Permian Basin Sec. Soc. Econ. Paleontologists Mineralogists, Guidebook of the Sacramento Mountains. 1959.

Pray, L.C.: Geology of the Sacramento Escarpment, Otero County, New Mexico, 144 pp. New Mex. Bur. Mines and Min. Resources, Bull. **35**, Socorro, New Mex. 1961.

Pray, L.C., Murray, R.C. (Eds.): Dolomitization and limestone diagenesis, 180 pp. Symposium, Soc. Econ. Paleontologists Mineralogists Spec. Publ. **13** (1965).

Price, N.B.: Some geochemical observations on manganese-iron oxide nodules from different depth environments. Marine Geology **5**, 511–538 (1967).

Pümpin, V.F.: Riffsedimentologische Untersuchungen im Rauracien von St. Ursanne und Umgebung (Zentraler Schweizer Jura). Eclogae Geol. Helvetiae **58**, No. 2, 799–976 (1965).

Purdy, E.G.: Bahamian oolite shoals. Am. Ass. Petrol. Geologists Bull. **45**, 53–62 (1961).

Purdy, E.G.: Carbonate Diagenesis: an Environmental Survey. Geol. Romana **7**, 183–228 (1968).

Purdy, E.G.: Reef configurations: Cause and effect. Soc. Econ. Paleontologists Mineralogists Spec. Publ. **18**, 9–76 (1974a).

Purdy, E.G.: Karst-determined facies patterns in British Honduras: Holocene carbonate sedimentation model. Am. Ass. Petrol. Geologists Bull. **58**, 825–855 (1974b).

Purser, B.H.: Syn-sedimentary marine lithification of Middle Jurassic limestones in the Paris basin. Sedimentology **12**, 205–230 (1969).

Purser, B.H.: Subdivision et interprétation des séquences carbonatées. Mem. B. R. G. M. **77**, 679–698 (1972).

Purser, B.H. (Ed): The Persian Gulf, Holocene carbonate sedimentation and diagenesis in a shallow epicontinental sea, 471 pp. Berlin-Heidelberg-New York: Springer 1973.

Purser, B. H., Loreau, J. P.: Aragonitic Supratidal encrustations on the Trucial Coast of the Persian Gulf, pp. 343–376. In: Purser, B. H. (Ed.): The Persian Gulf, 471 pp. Berlin-Heidelberg-New York: Springer 1973.

Pusey, W. C.: Recent calcium carbonate sedimentation in northern British Honduras, 247 pp. Ph. D. Dissertation. Houston, Texas: Rice University 1964.

Read, J. F.: Paleo-environments and paleography in Pillara formation (Devonian), Western Australia. Can. Petrol. Geology Bull. 21, 344–394 (1973).

Rich, J. L.: Three critical environments of deposition, and criteria for recognition of rocks deposited in each of them. Geol. Soc. Am. Bull. 62, 1–20 (1951).

Riding, R., Toomey, D. F.: The sedimentological role of *Epiphyton* and *Renalcis* in Lower Ordovician mounds, southern Oklahoma. J. Paleontology 46, 509–519 (1972).

Rigby, K.: Two new Upper Paleozoic Hydrozoans. J. Paleontology 32, 583–586 (1958).

Roehl, P. O.: Stony Mountain (Ordovician) and Interlake (Silurian) facies analogs of recent low-energy marine and subaerial carbonates, Bahamas. Am. Ass. Petrol. Geologists Bull. 51, 1979–2032 (1967).

Rose, P. R.: Comparison of type El Abra of Mexico with "Edwards Reef Trend" of south-central Texas, pp. 57–64. In: Geology of Peregrina Canyon and Sierra de El Abra, Guidebook Corpus Christi Geol. Soc. Field Trip, Corpus Christi, Texas 1963.

Rose, P. R.: Edwards group, surface and subsurface, central Texas, 198 pp. Rept. Invest. 74, Bur. Econ. Geology, Univ. of Texas 1972.

Ross, C. A.: Stratigraphy and depositional history of the Gaptank Formation (Pennsylvanian), West Texas. Geol. Soc. Am. Bull. 78, 369–384 (1967).

Ross, R. J., Cornwall, H. R.: Bioherms in the upper part of the Pogonip in southern Nevada. U. S. G. S. Prof. Paper 424-B, Art. 97, B 231–233 (1961).

Runnels, D. D.: Diagenesis, chemical sediments and the mixing of natural waters. J. Sediment. Petrol. 39, 1188–1201 (1969).

Rutten, M. D., Jansonius, J.: The Jurassic reefs on the Yonne (southeastern Paris Basin). Am. J. Sci. 254, 361–371 (1956).

Sander, B.: Beiträge zur Kenntnis der Anlagerungsgefüge (rhythmische Kalke und Dolomite aus der Trias). Mineral. Petrog. Mitt. 48, 27–139 (1936).

Sander, B.: Contributions to the study of depositional fabrics (rhythmically deposited Triassic limestones and dolomites), 209 pp. Transl. E. B. Knopf, Am. Ass. Petrol. Geologists 1951.

Sangree, J. B.: Silurian of northern Indiana. Ph. D. dissertation, Northwestern University. 1960.

Sarntheim, M.: Versuch einer Rekonstruktion der Mitteltriadischen Paläogeographie u Innsbruck, Österreich. Geol. Rundsch. 56, 116–127 (1967).

Schlager, W., Schlager, M.: Clastic sediments associated with radiolarites (Tauglboden-Schichten, Upper Jurassic, eastern Alps). Sedimentology 20, 65–89 (1973).

Schmidt, V.: Facies, diagenesis, and related reservoir properties in the Gigas beds (Upper Jurassic), northwestern Germany. Soc. Econ. Paleontologists Mineralogists Spec. Publ. 13, 124–168 (1965).

Schmidt, V., Klement, K. W.: Early diagenetic origin of reef framework in the Permian Capitan reef complex, Guadalupe Mountains, Texas and New Mexico. 8th Inter. Sed. Congress, Heidelberg, (Abstr.), 89 (1971).

Scholle, P. A.: Sedimentology of fine-grained deep-water carbonate turbidites, Monte Antola flysch (Upper Cretaceous), northern Apennines, Italy. Geol. Soc. Am. Bull. 82, 629–658 (1971).

Schwarzsacher, W.: Über die sedimentare Rhythmik des Dachsteinkalkes von Lofer. Geol. Bundesanstalt, Wien, Verhandl. 1947, 10–12, 175–188 (1948).

Schwarzsacher, W.: Petrology and structure of some Lower Carboniferous reefs in northwestern Ireland. Am. Ass. Petrol. Geologists Bull. 45, 1481–1503 (1961).

Scoffin, T. P.: The conditions of growth of the Wenlock reefs of Shropshire, England. Sedimentology 17, 173–219 (1971).

Seilacher, A.: Biogenic sedimentary structures. In: Imbrie, J., Newell, N. D. (Eds.): Approaches to Paleoecology, 296–316. New York: John Wiley and Sons 1964.

Seilacher, A.: Bathymetry of trace fossils. Marine Geology 5, 413–428 (1967).

Selley, R. C.: Nearshore marine and continental sediments of the Sirte basin, Libya. Quart. J. Geol. Soc. London 124, 419–460 (1970).

Shaver, R. H.: Silurian reefs of northern Indiana: reef and interreef macrofaunas. Am. Ass. Petrol. Geologists Bull. **58**, 934–956 (1974).

Shaw, A. B.: Time in Stratigraphy, 353 pp. New York: McGraw-Hill 1964.

Shearman, D. J., Fuller, J. G. C. M.: Anhydrite diagenesis, calcitization, and organic laminites, Winnipegosis Formation, Middle Devonian, Saskatchewan. Can. Petrol. Geology Bull. **17**, 496–525 (1969).

Shinn, E. A.: Practical significance of birdseye structures in carbonate rocks. J. Sediment. Petrol. **38**, 215–223 (1968 a).

Shinn, E. A.: Burrowing in recent lime sediments of Florida and the Bahamas. J. Paleontology **42**, 879–894 (1968 b).

Shinn, E. A.: Submarine lithification of Holocene carbonate sediments in the Persian Gulf. Sedimentology **12**, 109–144 (1969).

Shinn, E. A.: Sedimentary accretion along the leeward, southeast coast of Qatar Peninsula, Persian Gulf. In: Purser, B. H. (Ed.): The Persian Gulf, Holocene carbonate sedimentation and diagenesis in a shallow epicontinental sea, pp. 199–209. Berlin-Heidelberg-New York: Springer 1973.

Shinn, E. A., Ball, M. M., Stockman, K. W.: The geologic effects of hurricane Donna on south Florida. J. Geology **75**, 583–597 (1967).

Shinn, E. A., Ginsburg, R. N., Lloyd, R. M.: Recent supratidal dolomite from Andros Island, Bahamas. Soc. Econ. Paleontologists Mineralogists Spec. Publ. **13**, 112–123 (1965).

Shinn, E. A., Lloyd, R. M., Ginsburg, R. N.: Anatomy of a modern carbonate tidal flat, Andros Island, Bahamas. J. Sediment. Petrol. **39**, 1202–1228 (1969).

Shrock, R. R.: Sequence in layered rocks, 507 pp. New York: McGraw-Hill 1948.

Sieber, R.: Neue Untersuchungen über die Stratigraphie u. Ökologie der Alpinen Triasfaunen. Neues Jahrb. Mineral. Geol. Paläont. Beil. **78**, pt. B, 123–188 (1937).

Silver, B. A., Todd, R. G.: Permian cyclic strata, northern Midland and Delaware basins, west Texas and southeastern New Mexico. Am. Ass. Petrol. Geologists Bull. **53**, 2223–2251 (1969).

Sipple, R. F., Glover, E. D.: Structures in carbonate rocks made visible by luminescence petrography. Science **150**, No. 3701, 1283–1287 (1965).

Sloss, L. L., Dapples, E. C., Krumbein, W. C.: Litho-Facies Maps., 108 pp. New York: John Wiley and Sons 1960.

Smith, D. B.: Geometry and correlation along Permian Capitan escarpment, New Mexico and Texas: discussion. Am. Ass. Petrol. Geologists Bull. **57**, 940–945 (1973).

Smith, D. B.: Origin of tepees in Upper Permian shelf carbonate rocks of Guadalupe Mountains, New Mexico. Am. Ass. Petr. Geologists Bull. **58**, 63–67 (1974 a).

Smith, D. B.: Sediments of upper Artesia (Guadalupian) cycle-shelf deposits of northern Guadalupe Mountains, New Mexico. Am. Ass. Petrol. Geologists Bull. **58**, 1699–1730 (1974 b).

Smith, D. L.: Stratigraphy and carbonate petrology of the Mississippian Lodgepole Formation in central Montana, 143 pp. Ph. D. dissertation, Univ. of Montana 1972.

Soderman, J. W., Carozzi, A. V.: Petrography of algal bioherms in Burnt Bluff Group (Silurian), Wisconsin. Am. Ass. Petrol. Geologists Bull. **47**, 1682–1708 (1963).

Sorby, H. C.: The structure and origin of limestones. Proc. Geol. Soc. London **35**, 56–95 (1879).

Stanton, R. J., Jr.: Factors controlling shape and internal facies distribution of organic carbonate buildups. Am. Ass. Petrol. Geologists Bull. **51**, 2462–2467 (1967).

Stauffer, K. W.: Silurian-Devonian reef complex near Nowshera, West Pakistan. Geol. Soc. Am. Bull. **79**, 1331–1350 (1968).

Steidtmann, E.: The evolution of limestone and dolomite I: J. Geology **19**, 323–345 (1911).

Steidtmann, E.: Origin of dolomite as disclosed by stains and other methods. Geol. Soc. Am. Bull. **28**, 431–450 (1917).

Steineke, M., Bramkamp, R. A., Sander, N. J.: Stratigraphic relations of Arabian Jurassic oil. In: L. G. Weeks (Ed.): Habitat of Oil, A Symposium, Am. Ass. Petrol. Geologists, Tulsa, Okla., 1294–1329 (1958).

Stockman, K. W., Ginsburg, R. N., Shinn, E. A.: The production of lime mud by algae in South Florida. J. Sediment. Petrol. **37**, 633–648 (1967).

Stone,R.A.: Waulsortian-type bioherms of Mississippian age, central Bridger Range, Montana (Abstr.). In: Ann. Meeting Geol. Soc. Am.. Washington, D. C. 1971.

Stricklin,F.L., Smith,C.I.: Environmental reconstruction of a carbonate beach complex: Cow Creek (Lower Cretaceous) Formation of central Texas. Geol. Soc. Am. Bull. **84**, 1349–1368 (1973).

Stricklin,F.L., Smith,C.I., Lozo,F.E.: Stratigraphy of Lower Cretaceous Trinity deposits of central Texas. Rept. Invest. 71, Bur. Econ. Geol. Univ. Texas. Austin, Texas 1971.

Tebbutt,G.E., Conley,C.D., Boyd,D.W.: Lithogenesis of a distinctive carbonate rock fabric. Univ. Wyoming Contr. Geology **4**, 1–13 (1965).

Textoris,D.A.: Algal cap for a Niagaran (Silurian) carbonate mud mound of Indiana. J. Sediment. Petrol. **36**, 455–461 (1966).

Textoris,D.A., Carozzi,A.V.: Petrography and evolution of Niagaran (Silurian) reefs, Indiana. Am. Ass. Petrol. Geologists Bull. **48**, 397–426 (1964).

Textoris,D.A., Carozzi,A.V.: Petrography of a Cayugan (Silurian) stromatolite mound and associated facies, Ohio. Am. Ass. Petrol. Geologists Bull. **50**, 1375–1388 (1966).

Thomas,C.M.: Origin of Pisolites. Am. Ass. Petrol. Geologists Bull. **49**, (Abstr.), 360 (1965).

Thomas,G.E.: Grouping of carbonate rocks into textural and porosity units for mapping purposes. In: Ham,W.E. (Ed.): Classification of Carbonate Rocks. Am. Ass. Petrol. Geologists Mem. **1**, 193–223 (1962).

Thomson,A.F., Thomasson,M.R.: Shallow to deep water facies development in the Dimple Limestone (Lower Pennsylvanian), Marathon region, Texas. Soc. Econ. Paleontologists Mineralogists Spec. Publ. **14**, 57–78 (1969).

Toomey,D.F.: Note on a supposed "algal-foraminiferal consortium" from the Permian of West Texas. Contrib. Cushman Found. Foraminiferal Res. **13**, 52–54 (1962).

Toomey,D.F.: The biota of the Pennsylvanian (Virgilian) Leavenworth Limestone, midcontinent region. Part I: Stratigraphy, paleogeography, and sediment facies relationships. J. Paleontology **43**, 1001–1018 (1969a).

Toomey,D.F.: The biota of the Pennsylvanian (Virgilian) Leavenworth Limestone, midcontinent region. Part 2: Distribution of Algae. J. Paleontology **43**, 1313–1330 (1969b).

Toomey,D.F.: An unhurried look at a lower Ordovician mound horizon, southern Franklin mountains, West Texas. J. Sediment. Petrol. **40**, 1318–1334 (1970).

Toomey,D.F.: The biota of the Pennsylvanian (Virgilian) Leavenworth Limestone, midcontinent region. Part 3: Distribution of calcareous foraminifera. J. Paleontology **46**, 276–298 (1972).

Toomey,D.F., Croneis,C.: Gunsight (Virgilian) Wewokellid sponges and their depositional environment. J. Paleontology **39**, 1–16 (1965).

Toomey,D.F., Finks,R.M.: The Paleoecology of Chazyan (lower Middle Ordovician) "reefs" or "mounds" and Middle Ordovician (Chazyan) mounds, southern Quebec, Canada, a summary report. pp. 93–134, New York State Geol. Ass. Guidebook to field excursions. College Arts and Sciences, Plattsburgh, N. Y. 1969.

Toomey,D.F., Ham,W.E.: *Pulchrilamina*, a new mound-building organism from Lower Ordovician rocks of west Texas and southern Oklahoma, J. Paleontology **41**, 981–987, (1967).

Toomey,D.F., Johnson,J.H.: *Ungdarella americana* a new red alga from the Pennsylvanian of southeastern New Mexico. J. Paleontology **42**, 556–560 (1968).

Toomey,D.F., Mountjoy,E.W., Mackenzie,W.S.: Upper Devonian (Frasnian) algae and foraminifera from the Ancient Wall carbonate complex, Jasper National Park, Alberta, Canada. Can. J. Earth Sci. **7**, 946–981 (1970).

Toomey,D.F., Winland,H.D.: Rock and biotic facies associated with Middle Pennsylvanian (Desmoinesian) algal buildup, Nena Lucia field, Nolan County, Texas. Am. Ass. Petrol Geologists Bull. **57**, 1053–1074 (1973).

Toschek,P.H.: Sedimentological investigation of the Ladinian "Wettersteinkalk" of the "Kaiser Gebirge" (Austria). In: Müller,G., Friedman,G.M. (Eds.): Carbonate Sedimentology in Central Europe, pp.219–227, Berlin-Heidelberg-New York: Springer 1968.

Troell,A.R.: Lower Mississippian bioherms of southwestern Missouri and northwestern Arkansas. J. Sediment. Petrol. **32**, 629–644 (1962).

Troell, A. R.: Sedimentary facies of the Toronto Limestone, lower limestone member of the Oread megacyclothem (Virgilian) of Kansas. Ph. D. Thesis, 213 pp. Houston, Texas: Rice University 1965.

Trümpy, R.: Paleotectonic evolution of the Central and Western Alps. Geol. Soc. Am. Bull. **71**, 843–908 (1960).

Tsien, H. H.: The Middle and Upper Devonian reef-complexes of Belgium. Petrol. Geology Taiwan **8**, 119–173 (1971).

Turmel, R., Swanson, R. G.: Rodriguez Bank. South Florida carbonate sediments, Sedimenta 2, Laboratory Comparative Sedimentology, Univ. of Miami, Florida, pp. 26–31. (Reprint Guidebook 1, Geol. Soc. Am. Field Trip, 1964). 1972.

Turner, G.: Paleozoic stratigraphy of the Fort Worth Basin. Ft. Worth Basin Field Trip Guidebook, Ft. Worth and Abilene Geol. Societies, 61 (1957).

Tyrrell, W. W., Jr.: Criteria useful in interpreting environments of unlike but time-equivalent carbonate units (Tansill-Capitan-Lamar), Capitan Reef Complex, west Texas and New Mexico. In: Friedman, G. M. (Ed.): Depositional Environments in Carbonate Rocks, a Symposium. Soc. Econ. Paleontologists Mineralogists Spec. Publ. **14**, 80–97 (1969).

Udden, J. A.: Geology and mineral resources of the Peoria quadrangle, Illionois, 103 pp. U. S. Geol. Survey Bull. 506 (1912).

V. d. Borsch, C. C.: Source of ions for Coorong dolomite formation. Am. J. Sci. **263**, 684–688 (1965).

Van Bellen, R. C.: The Stratigraphy of the "Main Limestone" of the Kirkuk, Bai Hassan, and Qarah Chauq Dagh structures in north Iraq. J. Inst. Petroleum **42**, 233–263 (1956).

Van Siclen, D. C.: Depositional topography—examples and theory. Am. Ass. Petrol. Geologists Bull. **42**, 1897–1913 (1958).

Van Siclen, D. C.: A depositional model of late Paleozoic cycles on the eastern shelf. In. J. C. Elam, S. Chuber (Eds.): Cyclic Sedimentation in the Permian Basin 2nd, edition. West Texas Geol. Soc. 17–27 (1972).

Van Straatan, L. M. J. U.: Origin of Solnhofen Limestone. Geol. Mijnbouw **50**, 3–8 (1971).

Verrill, A. E.: The Bermuda Islands. Connecticut Acad. Arts and Sci. Trans. **12**, 45–348 (1907).

Van Tuyl, F. M.: The origin of Dolomite. Iowa Geol. Survey Ann. Rept. for 1914, **25**, 251–422 (1916).

Vest, E. L.: Oil Fields of Pennsylvanian-Permian Horseshoe Atoll, west Texas. In: Halbouty, M. (Ed.): Geology of Giant Petroleum Fields. Am. Ass. Petrol. Geology Mem. **14**, 185–203 (1970).

Waage, K. M.: Origin of repeated fossiliferous concretion layers in the Fox Hills Formation. In: D. Merriam (Ed.): Symposium on Cyclic Sedimentation. Kansas Geol. Survey Bull. **169**, 541–563 (1964).

Walther, J.: Einleitung in die Geologie als historische Wissenschaft — Beobachtungen über die Bildung der Gesteine und ihrer organischen Einschlüsse, Vol. 1, Jena: Fischer 1893.

Walther, J.: Die Fauna der Solnhofen Plattenkalk. Denkschrift 11 (Haeckel Festschrift), Jena 1904.

Wanless, H. R.: Local and regional factors in Pennsylvanian cyclic sedimentation. In: D. F. Merriam, (Ed.): Cyclic Sedimentation. Kansas Geol. Survey Bull. **169**, 593–606 (1964).

Wanless, H. R.: Eustatic shifts in sea level during the deposition of late Paleozoic sediments in the central U. S. In: Elam, J. C., Chuber, S. (Eds.) Cyclic Sedimentation in the Permian Basin, 2nd Ed., pp. 41–54. West Texas Geol. Soc. 1972.

Wanless, H. R., Shepherd, F. P.: Sea level and climatic changes related to Late Paleozoic cycles. Geol. Soc. Am. Bull. **47**, 1177–1206 (1936).

Wanless, H. R., Tubb, J. B., Gednetz, D. E., Weiner, J. L.: Mapping sedimentary environments of Pennsylvanian cycles. Geol. Soc. Am. Bull. **74**, 437–486 (1963).

Ward, W. C.: Diagenesis of Quaternary eolianites of northeast Quintana Roo, Mexico, 207 pp. Ph. D. Dissertation Houston, Texas: Rice University 1970.

Ward, W. C., Folk, R. L., Wilson, J. L.: Blackening of eolianite and caliche adjacent to saline lakes, Isla Mujeres, Quintana Roo, Mexico. J. Sediment. Petrol. **40**, 548–555 (1970).

Weaver, C. E.: Geologic interpretation of argillaceous sediments, Part I, Origin and significance of clay minerals in sedimentary rocks. Am. Ass. Petrol. Geologists Bull. **42**, 254–271 (1958).

Weeks, L. G.: Origin of carbonate concretions in shales, Magdalena Valley, Colombia. Geol. Soc. Am. Bull. **68**, 95–102 (1957).

Weller, M. J.: Stratigraphic Principles and Practice, 725 pp. New York: Harper and Brothers 1960.

Wells, A. J.: Cyclical Sedimentation: a review. Geol. Mag. **97**, 389–403 (1960).

Wengerd, S. A.: Reef limestones of Hermosa Formation, San Juan Canyon, Utah. Am. Ass. Petrol. Geologists Bull. **35**, 1038–1051 (1951).

Wengerd, S. A.: Biohermal trends in Pennsylvania strata of San Juan Canyon, Utah, pp. 70–77. In: Four Corners Field Conf. Guidebook, Geology of Parts of the Paradox, Black Mesa. and San Juan basins. Four Corners Geol. Soc. 1955.

Wengerd, S. A.: Stratigraphic section at Honaker Trail, San Juan Canyon, San Juan County, Utah, pp. 235–243. In: R. O. Bass, S. L. Sharps (Eds.): Shelf Carbonates of Paradox Basin. 4th Field Conf., Four Corners Geol. Soc. Symposium 1963.

Westoll, T. S.: The standard model cyclothem of the Visean and Namurian sequence of northern England. Congr. Avan. Etudes Stratigraph. Géol. Carbonifère, Compte Rendu **4**, Heerlen, 1958, **3**, 767–773 (1962).

Wilson, B. W., Hernandez, J. P., Meave, E.: Un banco calizo del Cretácico en la parte oriental del estado de Querétaro, Mexico. Bol. Soc. Geol. Mexicana **18**, 1–10 (1955).

Wilson, E. C.: No new *Ungdarella* (Rhodaphycophyta) in New Mexico. J. Paleontology **43**, 1245–1247 (1969).

Wilson, E. C., Waines, R. H., Coogan, A. H.: A new species of *Komia* Korde and the systemic position of the genus. Paleontology **6**, pt. 2, 246–253 (1963).

Wilson, J. L.: Upper Cambrian stratigraphy in the central Appalachians. Geol. Soc. Am. Bull. **63**, 275–322 (1952).

Wilson, J. L.: Late Cambrian and Early Ordovician trilobites from the Marathon uplift. Paleontology **28**, 249–285 (1954).

Wilson, J. L.: Cyclic and reciprocal sedimentation in Virgilian strata of southern New Mexico. Geol. Soc. Am. Bull. **78**, 805–818 (1967a).

Wilson, J. L.: Carbonate-evaporite cycles in lower Duperow Formation of Williston basin. Can. Petrol. Geology Bull. **15**, 230–312 (1967b).

Wilson, J. L.: Microfacies and sedimentary structures in "deeper water" lime mudstones. Soc. Econ. Paleontologists Mineralogists Spec. Publ. **14**, 4–19 (1969).

Wilson, J. L.: Depositional facies across carbonate shelf margins. Trans. Gulf Coast Ass. Geol. Soc. **20**, 229–233 (1970).

Wilson, J. L.: Influence of local structure in sedimentary cycles of Beeman and Holder formations, Sacramento Mountains, Otero County, New Mexico 41–54. In: Elam, J. C., Chuber, S. (Eds.): Cyclic Sedimentation in the Permian Basin, 2nd Ed. West Texas Geol. Soc. 1972.

Wilson, J. L.: Characteristics of carbonate platform margins. Am. Ass. Petrol. Geologists Bull. **58**, 810–824 (1974).

Winland, H. D.: Nonskeletal deposition of high Mg calcite in the marine environment and its role in the retention of textures. In: Bricker, O. P. (Ed.): Carbonate Cements, Johns Hopkins Univ. Studies Geology **19**, 278–284 (1971)

Winterer, E. L., Murphy, M. A.: Silurian reef complex and associated facies, central Nevada. J. Geology **68**, 117–139 (1960).

Wood, G. V., Wolfe, M. J.: Sabkha cycles in the Arab-Darb formation off the Trucial coast of Arabia. Sedimentology **12**, 165–191 (1969).

Wray, J. L.: Pennsylvanian algal banks, Sacramento Mountains, New Mexico. In: Kansas Geol. Soc. 27th Annual Field Conf., Guidebook, 158 pp. Wichita Kansas Geol. Soc. 1962.

Wray, J. L.: Late Paleozoic phylloid algal limestones in the United States. 23rd. Inter. Geol. Congress Proc. **8**, Prague, 113–119 (1968).

Young, L. M.: Early Ordovician sediments and history of Marathon geosyncline, Trans-Pecos, Texas. Am. Ass. Petrol. Geologists Bull. **54**, 2303–2316 (1970).

Young, R. G.: Sedimentary Facies and intertonguing in the Upper Cretaceous of the Book Cliffs, Utah-Colorado. Geol. Soc. Am. Bull. **65**, 177–202 (1955).

Zangrl, R., Richardson, E. S.: The paleoecological history of two Pennsylvanian black shales, 352 pp. Fieldiana Geol. Mem. **4**, (1963).

Zankl, H.: Die Karbonatsedimente der Obertrias in den nördlichen Kalkalpen. Geol. Rundschau **56**, 128–139 (1967).

Zankl, H.: Der Hohe Göll. Aufbau und Lebensbild eines Dachsteinkalk-Riffes in der Obertrias der nördlichen Kalkalpen, 123 pp. Abhandl. Senckenberg Naturforsch. Ges. **519** (1969).

Zankl, H.: Upper Triassic carbonate facies in the Northern Limestone Alps, pp. 147–185. In: Müller, G. (Ed.): Sedimentology of Parts of Central Europe. Guidebook, 8th Inter. Sed. Congres, Heidelberg 1971.

Zeiss, A.: Untersuchungen der Paläontologie der Cephalopoden des Unter-Tithon der südlichen Franken Alb. Burg. Akadem. Wiss. Abhandl. NF **132**, (1968).

Zenger, D. H.: Significance of supratidal dolomitization in the geologic record. Geol. Soc. Am. Bull. **83**, 1–12 (1972a).

Zenger, D. H.: Dolomitization and Uniformitarianism. Jour. Geol. Education (Council Educ. Geol. Sciences no. 19) **20**, 107–124 (1972b).

Ziegler, M. A.: Beiträge zur Kenntnis des untern Malm im Zentralen Schweizer Jura. 55 pp. A. G. Winterthur (1962).

Ziegler, M. A.: A study of the lower Cretaceous facies developments in the Helvetic border chain, north of the Lake of Thun (Switzerland). Eclogae Geol. Helvetiae **60**, No. 2, 509–527 (1967).

Ziegler, P. A.: Geologische Beschreibung des Blattes Courtelary (Berner Jura) und zur Stratigraphie des Sequanien im zentralen Schweizer Jura, 102 pp. Beiträge Geol. Karte Schweiz (N. F.) 1956.

Subject Index

Acid etching   57
Adnet Triassic buildup   248
Adneterkalk (Jurassic)   90, 253
Aggregated lumps   12, 62, 242
Alberta basin (Late Devonian)   43, 128, 129, 317
Algae   1, 3, 68, 74, 96, 97, 103, 112, 115, 117, 118, 143, 153, 165, 171, 239, 263, 264, 297, 345, 359, 368, 374
Algae (blue-green, Schizophytes)   3, 5, 62, 89, 143, 172, 173, 232, 241, 271, 279, 360, 369
Algae (codiacean green)   3, 5, 61, 87, 172, 173, 176, 186, 199, 210, 241, 242, 245, 271, 296, 322, 345
Algae (dasycladacean)   65, 97, 103, 110, 137, 143, 174, 183, 193, 209, 210, 223, 232, 237, 240–242, 253, 255, 264, 293, 295, 296, 304–306, 329, 342, 345, 358
Algae (diplopore)   240
Algae, phylloid (see platy algae)
Algae, (red or coralline)   5, 61, 121, 143, 172, 176, 185, 188, 197, 199, 210, 242, 255, 270, 299, 322, 329, 359, 363, 368, 369
Algal balls (onkoids)   69, 239
Algal borings   69, 296
Algal coatings   66, 269
Algal filaments   62
Algal mat blisters   83, 270
Algal mat fenestral fabric   121, 245
Algal plate mud mounds and facies   42, 112, 178, 185, 186, 196, 199, 200, 206, 210, 271, 368, 369
Algal plates (phylloid algae)   87, 88, 167, 172, 174, 176, 179, 185, 193, 197, 200, 209–211, 361
Algal-sponge buildups   98, 108, 232
Algal stromatolites   68, 82, 102, 108, 109, 112, 143, 165, 172, 173, 189, 198, 210, 229, 232, 241, 265, 272, 302, 320, 376, 379
Allochthonous debris   61
Allodapic limestone   64, 133, 264, 353
Alpine Muschelkalk   241
Alveolinids   347
Ammonites   150, 243, 254, 255, 257, 258, 264, 265, 267, 274, 277, 278, 324, 355
Ammonitico Rosso   90, 257, 265
Amphipora   121, 128, 138, 143, 146, 173
Ancient Wall bank (Alberta)   133
Andros Islands (Bahamas)   6, 47, 302

Aneth field (Utah)   178, 184
Anhydrite-gypsum   4, 19, 26, 85, 290, 294, 295, 299, 302, 310, 314, 332, 360
Aptychus limestone   257, 264
Arab Zones or Formation   289, 295, 298, 315, 316
Arabian Shield   47, 258, 288, 315, 339
Aragonite   63, 69, 146, 147, 264, 265, 296, 325
Aragonite, fibrous crystals   4, 17, 69, 232
Aragonite needles   5, 8, 17, 69
Arbuckle Formation   301
Archaeocyathids   96, 145
Atascosa Formation   337
Atesino High (Dolomites)   237
Atoll   22, 134, 240, 359, 374, 377
Aurora limestone   328
Autochthonous sedimentation   7, 24

Backreef   112, 128, 138, 242, 250, 254, 255, 270, 298, 299, 321, 324, 339, 346
Bafflestone   14, 65, 121, 198, 223, 244, 368
Baffling process   165–167, 367, 368
Bahama Banks (platform)   302, 379
Bahama Islands   41, 47, 254
Bahamite particles   67
Ball and flow   78, 195, 345, 354, 359
Banff Formation (Alberta)   150
Bagestan Limestone (Iran)   346
Banks (major offshore and on shelves)   21, 39, 119, 185, 200, 237, 238, 242, 254, 255, 281, 317, 319, 327, 329, 332, 339, 363, 364, 376
Banks (organic)   23, 128, 213, 287
Banks Island (Arctic Canada)   120, 141
Barmstein Limestone   264
Barrier island or bars   233, 377
Barrier Reef   22, 119, 133, 137, 139, 229, 245, 297, 359, 363, 365, 379
Bars   199, 363
Basal bioclastic pile   367
Basin   100, 128, 185, 201, 208, 216, 223, 237, 245, 250, 261, 276, 327, 338, 354, 370
Basin (starved)   25, 42, 100, 112, 131, 217, 238, 245, 255, 258, 289, 338
Basin buildups   28, 365, 376
Basin margin   25, 174, 226, 328, 356
Basinal facies   25, 120, 162, 195, 241, 336, 353
Basinal sediments and strata   202, 211, 241, 331, 338, 344, 346

Basrah basin   288, 290, 293, 340, 341
Baugh field (New Mexico)   191
Beach deposits   84, 208, 353, 359
Beach rock   69, 296
Beaverhill Lake Formation   129
Belemnites   267
Bentonites   91, 349
Bermuda (ledge-flat reefs)   363
Bernal Formation   222
Biancone Formation   264
Big Blue Series (Kansas)   171
Big Hatchet Mountains (New Mexico)   174,
   177–179, 199
Big Snowy Range (Montana)   148, 159–161,
   282
Bimini mounds (Florida Strait)   168
Bindstone   14, 65, 121, 223, 244
Bioclastic debris or detritus   50, 121, 172,
   193, 199, 241, 244, 267, 337, 338, 344, 356,
   361, 362, 364, 365, 369
Bioclastic grainstone   189, 250, 263, 264,
   274, 297, 346, 353
Bioclastic packstone   223, 263, 267, 287,
   356, 357
Bioclastic wackestone   121, 355, 357, 358
Bioclasts   5, 9, 12, 61, 62, 87, 137, 195, 199,
   210, 314
Biogenic particles   9
Bioherm   23, 139, 173, 175, 209, 279, 356
Biolithite   5, 14
Biomicrite   14, 244
Biomicrudite   14
Biosparite   244
Biostratigraphic zones   150, 342
Biostromes   239, 266, 299, 320, 322, 325, 345,
   358
Bioturbation   62, 68, 78, 209, 244, 354
Birdseye fabric (see fenestral fabric)
Birmenstorf beds   264, 267
Biscayne Bay (Florida)   199
Bivalves   241, 253, 264, 265, 267, 272, 273,
   296, 329, 355, 358
Black shale   208, 213, 215
Blackened pebbles   81, 270, 287, 297, 369
Blastoids   110, 143
Blockriffe   142
Blocks and boulders in foreslope   356
Blocky mosaic calcite cement   62, 70, 264
Boone Formation (Oklahoma)   148
Boquillas-San Felipe Formation   92
Bored surfaces   80, 274, 296, 345
Bouma sequences   75
Boundstone   1, 5, 12, 14, 63, 65, 121, 129, 146,
   173, 174, 176, 192, 193, 195, 196, 201, 223,
   232, 255, 357, 362, 368, 376
Brabant massif   126, 127, 150

Brachiopods   5, 61, 87, 98, 100, 103, 110, 119,
   121, 126, 138, 139, 144, 160, 162, 190, 195,
   204, 208–211, 233, 239, 241, 254, 263–265,
   267, 271, 272, 273, 285, 286, 290, 293, 299,
   305, 307, 319, 324, 356, 358, 359, 367, 368
Brecciation-breccia beds   81, 112, 115, 146,
   163, 172, 179, 183, 185, 191, 195, 200, 231,
   245, 253, 265, 297, 305–307, 329–331, 350,
   353, 354, 357
Bridger Range (Montana)   148, 160, 284
British Honduras shelf   41, 200, 364, 365,
   376, 377
Bryozoan mounds   5, 108, 233, 239, 273,
   367–369, 375, 376
Bryozoans   5, 71, 87, 100, 102, 103, 109, 115,
   126, 137, 145, 162, 164, 165, 167, 179, 185,
   190, 195, 204, 209, 241, 250, 263, 269,
   272, 285
Buckner Formation   294, 295
Buildups   173, 193, 200, 341, 344, 345, 361,
   362
Burgess shale   96
Burrows   63, 78, 84, 88, 89, 209, 211, 265,
   296, 299, 304, 306, 329, 330, 345, 349, 353,
   355, 359

Cairn Formation (Alberta)   121
Calcarenite   13, 121, 290
Calcarenite-calcisiltite debris   250, 253
Calcilutite   13
Calcirudite   13
Calcisiltite   13, 63, 109, 223, 233, 285, 287,
   355, 356
Calcispheres   160, 223, 232, 299, 355
Calcite   153, 162, 163, 264
Calcite (Mg, high Mg)   4, 5, 17, 69, 70, 265,
   310
Calcite (low Mg)   4, 17
Calcitornellid forams   172
Caliche   70, 85, 297, 303, 307, 360
Calpionellids   264, 353, 355
Campeche Bank   6, 28, 313, 365, 377
Canning Basin (Australia)   120, 133, 317,
   363, 376,
Capitan Formation   222, 361
Capping bed or capping grainstone   178,
   183, 196, 199, 210, 361, 367, 369
Caprinid rudists   319, 321, 322, 329, 330,
   333, 337
Caprotinid rudists   322
Carbonate banks   242, 253, 254, 287
Carbonate buildup (trends)   20, 40, 313
Carbonate ramps, platforms, mounds   34,
   126, 134, 258, 262, 302
Carlsbad Formation   222
Carters Formation   103

Cathod luminescense 59
Cavity linings (in reefs and mounds) 88, 231, 234
Cementation (meteoric) 231
Cementation (submarine) 231, 264
Cements (calcite mosaic) 17, 62, 70, 264
Cements (cloudy) 62
Cements, coarse druse (lining reef cavities) 70, 137, 146, 232, 307
Cements, (dogtooth spar) 17, 62
Cements, (druse) 183, 243, 264, 370
Central Basin platform (west Texas) 42, 217, 301
Central Colorado Basin 316
Central Montana High 32, 33, 52, 159, 283, 284
Central Texas Cretaceous 33
Cephalopods 61, 100, 102, 112, 307, 356, 359
Chalk (chalky) 71, 95, 270, 341, 345, 346
Chalk Bluff Formation 222, 229
Channels 139, 186, 188–190, 208, 211, 213, 215, 231, 289, 329, 330, 346, 356, 359, 374
Channels (tidal) 68, 69, 98, 102, 365
Charles Formation 283
Charophyte 270
Chazyan (Ordovician of Lake Champlain) 102, 376
Chert (cherty limestone) 151, 155, 160, 176, 196, 204, 238, 241, 245, 267, 314, 328, 353, 356
Cipit exotix blocks 239, 243, 317
Clay "contaminants" 93
Clay particles 3
Clay and silt (horizons, breaks, markers) 1, 3, 202
Clearfork Formation 301
Climate 51, 171, 221, 232, 254, 257, 279, 282, 289, 300, 303, 307, 308, 313, 327, 363, 370, 371, 374, 376
Clinoform 36, 349
Clinothem 25, 348
Coal 171, 203, 204, 213, 215, 260, 324
Coated bioclasts 65, 269, 297, 345
Coated grains 87, 223, 240, 295, 304, 305, 329, 337, 358
Coated peloids 295
Coccoliths 8, 17, 262, 263, 264, 271, 276
Collapse brecciation 70, 79, 196
Collapsed mud matrix 163, 183
Color 77, 279
Comanche Shelf 315, 336, 338
Compaction 63, 69, 178, 205, 213, 266, 277, 345, 359
Concretions 79, 204, 208, 349
Conglomerate (channel fill) 76
Conglomerate (lag) 196

Connate water diagenesis 71, 164, 308, 313, 316–318
Conodonts 137, 150, 208
Constructed cavities 193, 357
Cooking Lake Platform (Alberta) 298, 365
Coquina 65, 209, 358
Corals (Anthozoan) 5, 71, 74, 87, 111, 118, 126, 137–139, 147, 150, 153, 160, 199, 204, 209, 210, 232, 241, 243–245, 250, 254, 255, 263, 273, 278, 299, 322, 325, 333, 335, 337, 356, 363, 374
Corals (dendroid) 109, 121, 329, 368, 369
Corals (domical) 103
Corals (encrusting) 102, 263, 273
Corals (fasciculate) 49, 121, 142, 368
Corals (lamellar, dish-shaped) 142, 267
Corals (massive growth forms) 142
Corals (reefy) 115, 257, 266, 273
Corals (Rugosa or Tetracorala) 5, 74, 100, 117, 119, 141, 145
Corals (solitary or horn) 160, 162, 195, 273
Corals (tabulate) 100, 102, 103, 109, 110, 112, 117, 138, 141, 145, 362
Corals (thickets, mats) 115, 145
Cornuspirid foraminifera 196
Corrasion zones 80
Correlation 44, 150
Cratonic uplifts or blocks 171, 216
Craven faults (England) 155
Crestal boundstone 198, 367
Crinkly laminae 83, 210, 243, 297, 307
Crinoidal bioclastic debris 115, 119, 243, 305
Crinoidal-bryozoan association 112, 153
Crinoidal flank beds 146, 197
Crinoidal limestone 151, 155, 156, 185
Crinoids 72, 93, 109, 110, 112, 115, 117, 119, 121, 137, 138, 143, 160, 162, 164, 165, 195, 199, 204, 209, 210, 211, 233, 239, 241, 253, 255, 264, 265, 269, 271, 273, 277, 278, 285, 286, 293, 294, 297, 299, 305, 307, 355, 357
Cross-bedding 86, 112, 203, 204, 210, 269, 273, 285, 286, 329, 355, 358, 359, 369
Cuesta del Cura Formation 328
Cuesta de El Abra 44, 329, 330, 336, 337
Culm facies 150, 153
Currents 6, 168, 199, 241, 358
Currents (tidal) 47, 358
Currents (turbidity) 276
Cycles (sedimentary) 47, 126, 191, 202, 240, 242, 293
Cycles (shelf or backreef) 211, 250, 281
Cyclic patterns 46, 50, 302, 349
Cyclic-reciprocal sedimentation 42, 208, 216, 229, 330

Cyclothems (Carboniferous)   49, 52, 202, 206, 208, 211

Dachstein limestone (Austria-Bavaria)   52, 244, 250, 305
Daonella (Pelagic bivalves)   238, 243
Dark shale beds   327, 340
Dasycladacean grainstones   65, 68, 183, 223
Debris mud flows   133, 134, 176, 350, 353, 356
Dedolomitization   70, 318
"Deep Edwards"-Stuart City reef trend   322, 336, 339
"Deeper water" facies   42, 160, 238, 241, 266, 285, 350
Delaware basin   192, 217, 219, 221
Deltas (deltaic sediments)   202, 204, 206, 208, 211, 213, 214
Denver-Julesburg (north Denver) Basin   316
Depositional relief or slopes   36, 179, 210, 218, 221, 233, 238, 254
Derbyshire, England   148
Detrital sand grains   91
Diablo Platform   159, 219
Diagenesis   69, 70, 79, 114, 146, 162, 183, 200, 231, 308, 312, 313, 325
Diatoms   355
Diceras   263, 319, 320
Differential subsidence   37
Dinant Basin (Belgium)   120, 125, 150, 151, 376
Dogtooth spar (see cements)
Dolomite   30, 57, 58, 62, 71, 171, 196, 242, 290, 293, 297, 299, 310, 330, 359
Dolomite crusts   70, 83, 306, 312, 320, 330
Dolomite (hydrothermal)   310
Dolomite (stratigraphic)   171, 196, 299, 310
Dolomite (sucrose texture)   71, 133, 314
Dolomites (Triassic of Italy)   39, 137, 164, 232, 233, 304, 307, 316, 317, 335
Dolomitization   19, 112, 131, 147, 151, 184, 201, 231, 239, 243, 274, 293, 296, 299, 300, 303, 307, 315, 333, 337, 339, 342, 379
Dorag dolomitization   313
Dorp (shelf-bioherm facies in Eifel)   127
Druse (see cement)
Druse veins   84, 231, 296
Dunes   86, 171
Duperow Formation (Williston Basin)   91, 298, 316, 317
Duvernay Formation (Alberta)   120, 131

Earp Formation (New Mexico)   171, 201
Eastern shelf, Midland Basin   43, 171, 197
Echinoderms (Echinodermata)   5, 61, 87, 98, 110, 190, 239, 263, 267, 290, 353, 356, 359, 367, 369

Echinoids   87, 209, 211, 233, 239, 241, 264, 271, 273, 296, 324, 347
Edgewise conglomerate (see Intraformational conglomerate)
Edwards Formation   315, 321, 376
Effinger Marl   267
Eifel (Devonian reef)   146, 376
El Abra Formation   39, 328–331, 333
El Paso Formation   301
Elk Point evaporites   130
Ellenburger Formation   98, 301, 317
Encrinites (see also crinoidal limestone)   65, 90, 126, 155, 164, 186, 203, 204, 353, 354
Encrusting Mn and Fe   70
Enfacial angles   62, 88
Engleboro-Settle area (England)   203
Enterolithic structure   85, 360
Eolianite   86, 302, 358, 361
Escabrosa shelf   159
Eustatic sea level change   51, 202, 213, 216, 300, 370
Euxinic conditions   208, 211, 217, 255, 354
Evaporites   4, 7, 46, 69, 77, 129, 130, 171, 175, 201, 202, 217, 222, 243, 244, 254, 257, 258, 279, 280, 283, 289, 293, 297, 304, 306, 315–317, 326, 327, 342, 355, 371, 374, 379
"Evinospongia"   239
Excello shale (Kansas)   208
Exotic blocks and boulders   243, 255, 317, 350, 357
Extrinsic (hydrologic) controls in mound formation   367

Facies (bank interior)   234, 297, 327, 329, 330, 332
Facies (basinal)   25, 120, 162, 195, 223, 241, 336, 354
Facies ("deeper water")   160, 222, 238, 241, 266, 285, 350, 354, 356
Facies (marine talus, foreslope, forereef debris)   25, 128, 193, 199, 222, 239, 245, 356
Facies (open marine platform)   25, 358
Facies (organic buildups)   25, 65, 222, 357
Facies (platform edge sand)   25, 222, 358
Facies (platform evaporite)   26, 222, 359
Facies (restricted circulation)   26, 222, 359
Facies (shelf)   25, 153, 329, 355
Facies (shelf margin)   25, 174, 330, 357
Facies (starved basin)   25, 100, 114, 217, 258, 354
Facies Belts (standard)   37, 221, 348, 350
Faro (shallow atoll)   22, 134, 138, 146, 240, 359, 374, 376
Fasciculate (growth form)   49, 72, 142, 241, 263, 270, 274

Fault control (of buildups) 155, 234, 254, 330, 333, 363, 376
Fecal pellets 5, 6, 61, 67, 88, 242, 263, 294
Fenestral fabric 63, 67, 68, 70, 82, 128, 132, 138, 163, 223, 231, 242, 245, 262, 297, 304, 307, 309, 330, 359
Fenestrate (growth form) 72, 160, 162–164, 190, 304, 369
Ferric iron 90, 353
Ferro-manganese crusts-nodules 70, 265, 353
Ferroan cement 62
Festoon cross-bedding 79
Fibrous aragonite 4, 17, 69, 232
"Filamentous limestone" 265, 353
Fill-in carbonate-evaporite cycle 240, 281, 313, 318
Fillings of cavities and fissures 200, 231, 368
Fish 277, 278, 306
Fissures in mounds 126, 254, 262, 357, 368, 369
Flank beds 78, 116, 118, 119, 146, 173, 178, 182, 186, 193, 199, 210, 238, 325, 361, 367, 369
Flank debris 176, 197
Flaser bedding (see also ball and flow) 78, 85, 354, 360
Flat pebble conglomerate (see intraformational conglomerate)
Floatstone 13, 14, 65, 69, 114, 134, 137, 138, 241, 329
Florida Bay 6, 47, 364, 378
Florida Keys 199, 365, 377, 378
Florida shelf, straits, and reef tract 7, 14, 197, 199, 313, 336, 364, 376–378
Flute casts 75, 265
Flysch bedding 75, 153, 202, 243
Flysch Calcaire (Alps) 350
Fondoform 36
Fondothem 25, 36, 348, 354
Foraminifera 5, 7, 61, 68, 71, 87, 150, 151, 160, 171, 172, 183, 185, 197, 199, 209, 210, 211, 232, 239, 241, 245, 250, 253, 263, 264, 271, 272, 278, 290, 293, 304, 306, 329, 342, 358, 359, 361, 375
Forereef debris 134, 238, 344
Forereef shoals 333
Foreset bedding 78, 210, 221, 286, 329, 357
Foreslope facies 25, 193, 226, 239, 245
Fractures 200, 305, 308, 318, 346
Framestone 14, 65
Framework (organic, framebuilt) 137, 146, 357, 358, 362–364, 367
Freshwater sediments 257, 260, 267, 270
Fringe (carbonate halo) 28, 31, 231, 266, 364, 376, 377

Fungi borings 69
Fusulinids 46, 55, 87, 185, 186, 195, 199, 209, 211, 223, 232, 255

Gamma ray markers 91, 299
Gastropods (snails) 61, 68, 86, 87, 103, 110, 119, 138, 183, 210, 223, 232, 239, 255, 264, 267, 270, 273, 290, 304, 305, 329, 353, 358, 359, 369
Gatesburg-Conococheague Formation 315
Geopetal structure 63, 65, 70, 80, 183, 193, 210, 242, 296
Georgetown Formation 337
Gigas beds (Late Jurassic of Germany) 315
Girvanella 211, 307
Glaciations (eustatic sea level change) 52, 202, 377
Glauconite zones 296, 349
Glen Rose Formation 301
Glide surfaces 76
Globigerinid marls or lime mudstone 342, 344, 347
Globigerinid ooze 70, 76
Globigerinids 64
Golden Lane Bank or Platform (Faja del Oro) 30, 44, 46, 327, 328, 331, 333, 337, 338, 339
Goniatites 137
Goniolithon (red alga) 197
Gotland (Silurian reefs) 104, 115, 117, 141, 376
Graded-bedding 238, 265, 273, 336, 359
Grain contacts 62
Grain or particle types 12, 240, 304, 277, 305, 329
Grain size (size sorting of particles) 13
Grainstone 12, 155, 174, 177, 183, 185, 186, 189, 193, 196, 197, 204, 210, 223, 241, 245, 253, 261, 264, 283, 287, 290, 293, 297, 330, 337, 342, 344, 355, 357, 358
Grain support 11, 62
Grapestone 1, 5, 6, 12, 62, 67, 287, 304
Graptolites 64, 105, 115, 355
Great Bahama Bank 14, 16, 30, 376
Great Scar Limestone 155, 159, 203
Great Slave Lake 97
Groove and tool markings 76
Grumelous fabric-matrix 63, 70, 89
Guadalupe Mountains 217, 221, 229, 361, 379
Gulf of Mexico 31, 47, 325, 327, 336, 370
Gypsum (see anhydrite)

Hallstatt Formation 245, 250
Halobia (pelagic bivalve) 245
Halos (fringes or rims) of limestone 29, 31, 170, 237, 325, 340, 376

Hard grounds 70, 80, 262, 270, 287, 296, 297, 324, 345
Hauptdolomit Formation 243, 244, 250, 305, 316
Hawar shale (Arabia) 344
Hay River basin 129
Haynesville A. Formation 295
Heebner shale (Kansas) 208
Helderberg Group (New York) 39
Hemicycles 281
Hermosa Formation 175
Hierlatzkalk (Eastern Alps) 90
Hohe Göll 250
Holston Formation 103
Homogenization by burrowing (bioturbated) 65, 209
Horsehoe Atoll (North Central Texas) 30, 186, 200
Horseshoe crab *(Limulus)* 278
Hueco Mountains 176
Hydrologic-hydraulic processes 312, 370, 374, 377
Hydrozoans 5, 74, 174, 232, 239, 241, 254, 255, 295, 322, 335

Illinois (Bainbridge) Basin 105, 376
Indiana-Illinois Silurian Shelf 112
Infilling (see also filling of cavities) 195, 200, 357
Infiltering of sediment 62, 63, 88
Injection dikes 80, 357
Insoluble residues 90, 91
Internal sediment, fillings 63, 126, 162, 163, 183, 200, 243, 307
Intertidal muds 163, 299, 338
Intraclasts 12, 67
Intraformational conglomerate 69, 82, 112, 245, 320, 330
Intramicrite 9
Intrasparite 9, 245
Intrinsic (substrate) controls in mound formation 367, 368
Inundative phase (cycles) 179
Ireton Shale, Alberta 120, 131
Ismay field 177, 199
Isopachous cement rims 62, 69

Jameson bank (north central Texas) 185
Jawan Formation 342
Jefferson Formation 299
Jubaila Formation 290

Kapingamarangi Atoll 7
Karst (karstic collapse) 81, 254, 305, 307, 308, 331, 333, 365, 371, 374

Karwendel bank 241
Key beds (horizons-correlation markers) 44, 54, 349
Kirkuk anticline 342, 363
Knoll reefs (Knollenriffe) 22, 128, 142, 156, 330
Komia 173, 197, 199, 201
Kössen Formation 245, 248, 250

Lag sediment 66, 69, 81, 196, 297
Lagoons 68, 89, 138, 151, 193, 200, 213, 217, 229, 240, 242, 243, 274, 279, 297, 298, 299, 306, 315, 320, 358, 359, 363, 364, 377
Lake Valley Formation 156, 159
Laminae 67, 77, 96, 151, 153, 360
Lamination 63, 68, 195, 238, 293, 297, 299, 306, 329, 330, 354, 355, 359, 360, 363, 368
Laminite (fenestral) 128, 132, 147, 242, 304, 306, 330
Latimar (Dolomites) 304
LaValle-Wengen Formation 236, 243
Leached fractures 183
Leduc (Devonian) reefs 317
Leffe Formation (Belgium) 151
Lennard shelf (Australia) 133, 146
Leptogeosynclinal sediments 350
Lime mud matrix 10, 172, 346
Lime mud mounds 23, 126
Lime mud and mudstone 1, 3, 4, 10, 69, 77, 146, 148, 163, 199, 262, 264, 290, 325, 327, 330, 337, 338, 342, 344, 346, 347, 353, 356, 359, 365
Lime mud-sabkha cycles 115, 282, 297, 301
Linear reef trends 41, 42, 156
Lithoclastic packstone (talus) 151, 223, 267
Lithoclasts 12, 61, 63, 87, 98, 133, 137, 195, 199, 210, 245, 265, 299, 304, 329, 346, 357, 359, 369
Livinallongo beds 237
Llano Uplift 170, 376
Load casts 76
Lodgepole Formation 160, 283
Lofer facies 52, 305, 307
Loferite 67, 68, 82, 253, 304, 309
Longshore currents 199, 358
Lower Ordovician carbonate platform (Ellenburger-Arbuckle) 16, 91, 376

Madison Group 283, 314, 316, 376
Magnesium (Mg) 310
Maiolica Formation 264
Malm 261, 265, 271, 363
Manganese and ferric crusts-nodules 70, 76, 265, 373
Marathon foldbelt 317
Maravillas Limestone 350
Marbre Noir Formation (Belgium) 152

Marine (submarine) cementation 164, 232, 250
Marine talus 25, 199, 126
Mass of Carbonate 21, 116, 155
Maverick Basin 338
Mediterranean Tethyan trough 258
Megabreccia 133, 134
Megacyclothems 213, 216
Megalodonts 253, 305, 306, 358
Megaslumps 356, 357
Meniscus cement 70, 296
Meteoric water (Diagenesis) 70, 79, 146, 162, 183, 200, 231, 308, 312, 313, 325
Mexican geosyncline 16, 327
Mexico 39, 327, 363, 378
Michigan basin 30, 105, 112, 314, 315, 365
Micrite 10, 13, 17, 50, 62, 69, 165, 209, 223, 232, 245, 296, 302, 364
Micrite (brecciated) 178
Micrite matrix 11, 63, 65, 115, 174, 178, 193, 195, 209, 210, 233, 239, 243, 244, 253, 272, 274, 314, 329, 355
Micritic mounds 100, 139, 148, 162, 172, 198, 271, 356, 359, 376
Micritic rinds 61, 69, 273, 337
Micritization 61, 87, 293, 296, 297
Microbioclastic 63, 267, 355
Microbreccia 64, 238, 265, 350, 356
Microfacies 60, 63, 171, 209, 223, 244, 262, 336
Micrograded layers 82
Micropeloid 353, 355
Microplankton 265
Microspar 88, 208, 209
Microstalactic cement 70
Microstylolites 265
Midcontinent 47, 206, 209, 213
Midland basin 185, 200, 217, 376
Midlands, England 152, 203
Miette Bank 133, 317
Miliolid foraminifera 5, 330, 337, 346, 359
Millimeter laminae 82
Minturn Formation (central Colorado) 176
Mission Canyon Formation 283
Moccasin Formation (Tennessee) 90
Mollusks (Mollusca) 5, 7, 110, 199, 208, 209, 211, 213, 223, 232, 239, 240, 241, 243, 245, 250, 255, 263–265, 270, 271, 273, 274, 290, 293, 330, 345, 347, 358, 359
Monopleurid rudists 319, 321, 345, 346
Montsech, Spain 261, 276
Monument upwarp 174
Morocco (Devonian) 139
Mosul block (uplift) 262, 340, 342, 376
Mottling 63, 78, 306
Mound core or reef core 119, 151, 174, 178, 182

Mounds or mud mounds (see also micritic mounds) 21, 126, 154, 156, 186, 199, 219, 253, 329, 333, 342, 356–358, 361, 362, 364–366, 369, 376
Mud balls 82
Mud chips 83
Mud cracks 83, 304, 360
Mud supported fabric 62
Mudflows (debris flows) 75, 231
Mudstone 12, 133, 146, 211, 270, 285, 287, 294, 296, 299, 307, 337, 346, 357
Muleshoe mound or bioherm 156
Muschelkalk 254

Namur-Dinant basin 151
Nautiloids 49, 110, 119, 137, 144, 160, 162, 233, 241
Needle fiber cement 70
Nena Lucia bank 185
Neomorphism 60, 71
Neptunian dikes 80, 307
Niagaran Formation 105, 354
Nodular anhydrite 85, 299, 345
Nodular bedding 78
North German basin 261
North Pennine block 203
North Texas-Tyler basin 338
Northern Limestone Alps 234, 239, 243, 363
Nowshera, Pakistan (Devonian reefs) 119, 141
Nusplingen 276, 277

Oberalm Formation 264
Onkoids 12, 61, 69, 79, 121, 138, 204, 209–211, 213, 245, 263, 269–271, 287, 293, 304, 305, 307
Ooids 1, 6, 7, 9, 12, 61
Oolite (oolitic) 6, 66, 110, 138, 174, 190, 204, 213, 257, 258, 261, 267, 269, 270, 280, 282, 286, 290, 293, 296, 297, 358
Oolite-grainstone cycles 186, 204, 282, 287, 298
Oosparite 13, 245
Ophiuroids 264
Opthalmid forams 172, 173
Orbitolina 342, 344, 347
Organic slime 89
Oro Grande basin 176, 177, 199, 216
Ostracods 5, 7, 61, 66, 68, 87, 88, 160, 162, 211, 223, 232, 260, 270, 299, 304, 330, 359
Overpacking 62, 66, 285
Oxidation zones 80, 287, 296
Oysters 61, 80, 278, 287, 296, 322, 347, 368
Ozark dome 170, 189

Packing concept 11
Packstone 12, 63, 87, 88, 102, 109, 121, 138, 151, 161, 193, 208–210, 212, 223, 242, 273, 285, 330, 345, 346, 356, 357
Paine member (Lodgepole Formation) 160, 284
Paleoaplysina 174
Paleontologic zonation (biostratigraphy) 55, 257, 349
Paleotectonic settings 28
Palisade cement 62, 69, 296
Paradox basin (Utah, New Mexico) 43, 173, 174, 177, 178, 184, 199, 216, 316, 365
Paris basin 261, 376
Partnach beds 241, 242
Peace River high (Alberta, Canada) 376
Pebble conglomerate 69, 114
Pedernal uplift 159, 178, 216, 219, 231, 376
Pedregosa basin 171, 179, 199
Peels 58, 59
Pelagic bivalves 245, 262, 264, 267, 333, 336, 353, 355
Pelagic lime mudstone 64, 257, 262, 264, 328, 333, 337, 338, 340, 344, 353
Pelagites 350
Pelleting of mud (pelleted mud) 13, 67, 184, 210, 267, 285, 286, 295, 359
Pelmatozoans 103, 143
Peloids or pelletoids (peloidal) 1, 6, 9, 12, 61, 68, 87, 133, 151, 153, 185, 223, 239, 241, 244, 274, 283, 285, 287, 293, 295, 296, 299, 304, 307, 314, 329, 330, 337, 342, 345, 379
Pelsparite 67, 245
Pembrokeshire (England) 148
Pendent (microstalactitic) cement 70
Pennines (England) 148, 154, 155
Permian Reef Complex 28, 37, 44, 164, 221, 232, 240, 255, 307, 317, 361, 376, 379
Persian Gulf 6, 14, 31, 33, 47, 91, 288, 293, 296, 298, 302, 339, 340, 362, 370, 377, 379
Phosphate zones and nodules 296
Photosynthesis process 389
Phreatic zone diagenesis 70, 312, 318
Piles (of sediment) 23, 165, 166, 198, 199, 367
Pinnacle reefs 22, 30, 39, 112, 316, 349, 356, 365, 369, 370, 376
Pir-i-Mugrun ridge (Iraq) 342
Pisoids (including concretionary vadose pisolite) 70, 85, 138, 223, 231, 239, 240, 269, 297, 304, 307
Platform edge sands 358
Platform interior cycles 282
Platforms 21, 24, 34, 130, 140, 182, 199, 202, 221, 226, 237, 241, 250, 255, 258, 261, 262, 281, 319, 322, 325, 327, 338, 348, 349, 357, 363, 370, 376, 377

Platy algae 87, 88, 172, 174, 176, 179, 193, 198, 199, 210
Platy (phylloid) algal mounds 185, 196, 199, 200, 206
Pleistocene eolianite 28, 302, 377, 378
Pogonip Formation 98
Polished plaquettes 56
Porosity 17, 58, 70, 71, 93, 95, 112, 147, 176, 183–186, 193, 195–198, 325, 333, 345
Poza Rica 328, 331, 333
Presqùile Formation 376
Pressure solution 71, 277
Profiles of shelf margins 223, 229, 238, 242, 254
Progradation or outbuilding 37, 202, 206, 213, 215, 297, 300, 313, 348, 349
Ptygmatic structure 85
Pumping of evaporative waters (evaporative pumping) 312

Qatar peninsula 341
Qatar-Surmeh positive block 262, 288, 290, 341
Queen sands 231
Quintnerkalk 262, 264, 267, 273

Radiolarians 64, 238, 239, 278, 347, 355
Radiolarite 90, 258, 262, 265, 340, 344, 353
Radiolitid rudists 319–321, 329, 330, 333
Raibl Formation 243, 316
Raindrop prints 83
Ramp 21, 24, 28, 34, 134, 182, 199, 221, 223, 268, 302, 325, 327, 348, 361, 362, 376, 377
Rancheria Formation 156
Rasenriffe 128, 141
Rate of carbonate sedimentation 16
Receptaculites 49, 137
Reciprocal sedimentation 177, 229, 349
Red beds 171, 254, 360
Red encrinites 353, 354
Red nodular limestone 70, 257, 262, 265, 353
Red River-Viola-Montoya 48, 314
Red soils 297, 305, 307
Red stained zones 80, 86, 183, 191, 201, 296, 355
Redeposited (retextured) sediment 77, 262
Reducing conditions 89
Redwater bank 131, 132
Reef 6, 72, 128, 193, 254, 264, 266, 297, 375
Reef (barrier) 22, 119, 139, 245, 297, 359, 363, 365, 379
Reef (blocky) 128
Reef crest 137
Reef definition 23
Reef (ecologic or organic framework) 5, 23, 25, 109, 222, 229, 357, 363, 364

Reef flank   115, 117, 361
Reef (fringing)   22, 134, 363
Reef (geologic or stratigraphic)   21, 193
Reef (knolls)   128, 250, 253, 273, 361–363
Reef (patch)   22, 78, 119, 245, 261, 270, 273, 299, 345, 359, 361, 370, 376
Reef (rudist)   345, 363
Reef rudstone   65
Reef (transgressive)   39
"Reef tufa"   162
Reflux of brines   310, 312, 315, 317, 318
Regressive sedimentation   50, 51, 54, 179, 189, 191, 202, 206, 213, 215, 219, 258, 267, 270, 274, 281, 285, 338, 344, 348
Relief, depositional   186, 188, 196, 200, 221, 233, 253, 271, 273, 326, 330, 345, 348
Renalcis   137, 138, 143, 146, 173
Requienid rudists   319, 320, 329
Residemented clasts   77, 250
Restricted marine circulation   6, 46, 132, 237, 282, 293, 297, 299, 300, 359, 370
Restricted marine faunas   133, 245, 320
Retextured sediments   77
Rhaetic limestone   244, 245, 250
Rhenish trough, Eifel region   376
Rhizocretions   86
Rhythmic-thin-bedding   356
"Ribbing"   86
Rill marks   83
Rim (buildup of carbonate-halo-fringe)   217
Rim cement   70, 71
Ripple marks   77, 78
Rodriquez Bank (Florida)   197, 199, 365
Root hair casts   86, 297
Roots   83
Rötelwand mound   248, 253
Rub al Kali basin   289, 293, 341, 344
Rubenriffe   128
Rudists   319, 324, 325, 329, 333, 335, 338, 342, 344, 345, 368, 369, 375
Rudites   195
Rudstone   13, 69, 241, 329, 357

Sabkha   293, 295, 297–300, 302, 307, 312, 315, 318, 359, 379
Sacramento Mountains   55, 86, 150, 155, 165, 173, 176–179, 216
Salina Formation   112, 114
Saline waters (brines)   310, 364
Salt   262, 289, 293
Salt hoppers   85
San Andres Formation   92, 254, 300, 316
San Andres Mountains, New Mexico   176
San Felipe Formation   92
San Marcos Arch   338
St. Cassian Formation   243
St. Georges Land (England)   154

Satellite mounds or reefs   111, 115, 117, 139, 145, 173
Schwabian Alb   271
Schwelm (interreef-basinal facies)   127
Scurry County fields (Horseshoe atoll)   185
Sea level drops   137, 179, 195, 196, 201, 202, 212, 214, 215, 221, 229, 231, 282, 309, 333, 338, 370
Sea mounts   262
Sedimentary boundinage (see "ball and flow")   78
Sedimentary topography   36
Selenite crystals   69, 85
Serla Dolomite platform   237, 243
Serpulids   81, 269, 271, 272
Sessile benthos   71
Shale-silt "breaks" or markers   196, 355
Sheet cracks   84, 304, 307
Shelf   21, 202, 217, 221, 223, 261, 267, 281, 302, 322, 325, 338, 370, 377
Shelf cycles   49, 179, 211, 349
Shelf lagoon   21, 358, 364, 365
Shelf margin   21, 42, 44, 177, 192, 193, 200, 202, 214, 217, 219, 221, 229, 254, 257, 258, 266, 267, 269, 297, 317, 322, 329, 337, 339, 350, 361, 363, 370
Shelf margin profiles I, II, III   25, 155, 168, 200, 229, 242, 245, 255, 274, 357, 361, 363, 364, 376
Shelf mounds   322
Shoaling cycles (see upward shoaling cycles)
Shuaiba Formation   315, 344, 345
Sierra Madre (Mexico)   327, 330
Size-sorting   13, 62
Sligo-Cupido Formation   338
Slopes (depositional, steep)   36, 133, 218, 250, 316, 326, 330, 357, 361, 362, 363
Slumps (soft sediment)   76, 79, 126, 147, 155, 161, 163, 179, 238, 262, 263, 265, 273, 277, 329, 335, 369
Smackover Jurassic   28, 31, 280, 293, 294
Snails (see gastropods)
Soil zones   86
Sole markings   76
Solenopora (red alga)   109, 137, 143, 232, 241, 250, 253
Solnhofen   257, 261, 263, 273, 276, 278
Solution brecciation   85, 172
Solution compaction   62, 70
Solution porosity   200
South Alps   233, 238, 242
Southesk Formation   133
Sparry cement (see also cements)   13, 146, 153, 241
Sphaerocodium   118, 239, 240, 243, 263, 307
Sphinctozoans (beaded sponges)   72, 241
Spicules   211, 264, 271

Spiculite 64, 174, 192, 195, 196, 223, 241, 264, 267, 286, 353
Spirorbids 62, 211
Sponge-alga (algal-sponge) reef or mound 5, 266, 279, 317, 362
Sponges 62, 71, 72, 102, 109, 110, 137, 145, 167, 174, 190, 197, 211, 223, 229, 232, 241, 244, 250, 253, 254, 257, 261, 263, 267, 271–273, 322, 325, 329, 355, 359, 368, 376
Spongiomorph hydrozoans 245, 250, 253, 263, 273, 322, 329, 333, 337, 368
Spongiostrome 69, 82, 108, 118, 250, 360
Stabilization of sediment (in banks, reefs) 367, 369
Stachyoides (dendroid stromatoporoid) 138, 173
Staining 58
Starved basins 100, 112, 217, 255, 258, 289, 340, 348, 354
Steinplatte reef (Sonnenwände) 245, 253
Still-stand of deposition 49, 296, 345, 349
Stoney Mountain (Maquoqueta-Sylvan formations) 48
Stromatactoid structures 70, 79, 82, 96, 108, 109, 126, 137–139, 146, 153, 160, 162, 165, 167, 183, 238, 253, 357
Stromatolites (stromatolite mounds) 97, 102, 108, 109, 118, 137, 145, 165, 172, 174, 198, 210, 227, 229, 239, 304, 306, 360, 368
Stromatoporoids 110–112, 115, 117–119, 126, 137, 139, 141, 142, 145, 147, 174, 198, 201, 239, 244, 263, 362, 363, 368, 369, 374
Stromatoporoids (bulbous, globular, cabbage head) 72, 96, 109, 128, 138, 143
Stromatoporoids (dendroid) 138
Stromatoporoids (massive, encrusting) 62, 71, 74, 100, 102, 121, 143, 250, 253, 299
Stromatoporoids (plate-like) 109
Stromatoporoids (tabular-lamellar) 71, 102, 103, 109, 112, 117, 119, 121, 126, 138, 143, 167
Structural trends (multilinear, control on carbonate buildups) 129, 192, 333, 338, 345
Stuart City-Edwards reef 328, 336
Stylolites (stylolitization) 62, 71, 93, 204, 318, 328, 339, 345
Subaerial exposure surfaces 49, 70, 147, 196, 201, 231, 240, 254, 296, 297, 345, 357, 367, 374
Submarine cementation 287
Subsidence, differential 37, 238, 254, 335, 370
Subsidence control on sedimentation 37, 51, 214, 363
Subsolution of carbonate 70, 76, 265
Sun River dolomite 316

Supratidal 84, 299
Surface exposure 183, 218
Surmeh shelf (see Qatar-Surmeh block)
Swan Hills Formation 130
Swiss Jura reef facies 44, 260, 261, 266–271

Talus (marine-see also Facies, marine talus) 25, 151, 195, 231, 245, 267
Tamabra Formation 329, 333, 335
Tamaulipas Formation 328, 335
Tectonic activity 202, 213, 216, 281, 288, 289, 362, 370
Tectonic controls 51, 150, 219, 235, 293, 361, 363
Tectonic setting 170, 174, 198, 200, 221, 349, 369, 375, 377
Tepee structures 84, 231, 304
Terrigenous (clastic) influx 202, 204, 205, 216, 340, 342
Tethys (Tethyan geosyncline) 254, 258, 261, 319
Texas craton 301
Textural inversion 62, 65
Thamama Group 315, 344
"Thecosmilia" (organ-pipe coral) 167, 239, 241, 244, 250, 253, 363
Thickness variations 40, 44
Thin sections (use of) 57
Thin-shelled bivalves 265, 353
Thin subsurface correlation markers 43
Tidal bars 6, 7, 66, 68, 199, 219, 358, 365
Tidal channels 98, 306
Tidal currents 47, 358, 374
Tidal deltas 6, 365
Tidal flats 47, 96, 171, 217, 229, 240, 242, 267, 270, 276, 290, 293, 295, 298, 299, 302, 307, 310, 312, 315, 317, 363, 379
Tidal variations 282
Topography (control on carbonate growth) 42
Topography (see slopes and depositional relief)
Townsend-Kemnitz field 173, 192, 195, 196
Trace fossils 78, 353
Transgressive reef 39
Transgressive sedimentation 39, 50, 51, 54, 179, 185, 189, 196, 202, 204, 213, 215, 216, 258, 289, 304, 338, 348
Trenton limestone 91
Trilobites 5, 98, 100, 110, 112, 139, 144, 162, 211
Tubiphytes 71, 173, 174, 193, 196, 198, 201, 223, 229, 232, 239, 241, 243
Tubular foraminifera 172, 173, 176, 193, 196, 198, 199, 201, 369
Tubules and pipes 241, 304
"Tufa mounds" 152

Turbidites   105, 245, 261, 264, 265, 350, 353, 356, 361

Um Shaif field (Persian Gulf)   298
Umbrella effect   62, 65
Undaform   36
Undathem   25, 348, 355
Uniform lithosomes   47
Upward shoaling cycle   241, 281, 283, 297, 304, 313, 314, 318, 330, 333, 376
Upwelling   6, 364, 374

Vadose pisolites   239, 240, 297, 304, 307
Vadose silt   70, 88, 304
Vadose zone diagenesis   70, 112, 164, 177, 195, 196, 201, 229, 231, 250, 296, 307
Val Seriana quarries, Bergamasc Alps   304
Valles Platform   30, 329, 330, 337
Veneer (reef and mound coatings)   365, 367, 368, 377
Voids (constructed)   250
Vorbourg beds   270
Vugs   95, 176, 243

Wackestone   12, 14, 109, 121, 138, 151, 160, 176, 185, 186, 193, 196, 209–211, 223, 229, 241, 267, 270, 271, 287, 289, 293, 299, 306, 330, 337, 342, 345, 347, 355

Waterways Formation   130
Waulsortian facies mounds   148, 157, 162, 165, 167, 285, 362, 376
Wave action   6, 47, 199, 358, 364, 369, 374
Wavy bedding   78, 209
Wenlock (Silurian reefs)   104
West Texas–New Mexico (Tobosa) basin   105, 137
Wetterstein limestone   240–242
Williston basin   32, 47, 52, 55, 129, 159, 283, 285, 289, 298, 299, 301, 314, 316, 376
Wind (onshore)   6, 241, 364
Wind action   354, 355, 359, 374
Winnowing   65, 167, 184, 185, 197, 209, 210, 222, 268, 337, 355, 357, 369
Woodbend Formation   129, 130, 132

Yates Formation   231
Yoredale cycles   153, 155, 202, 214
Yucatan (northeast coast)   6, 41, 200, 313, 364, 365, 377, 378
Yvoir Formation (Belgium)   151

Zama Rainbow area (northern Alberta)   30, 112, 316, 376
Zebra rock   84, 163, 307
Zechstein Formation   112
Zuloaga Formation   293

Plate I. Pennsylvanian Mound Flank Sediment, Standard Microfacies 5

Uppermost bed of foreset flanking strata of Late Pennsylvanian phylloid algal lime mud mound in the Sacramento Mountains, north wall of Dry Canyon above State Highway 83, 6 km northeast of Alamogordo, New Mexico. Sample AHP; see Chapter III for a detailed discussion of this rock. The location of the sample is at the left end and at the top of the main biohermal mass of flank beds in the Frontispiece photo. Facies P-7, Chapter VII, X 15

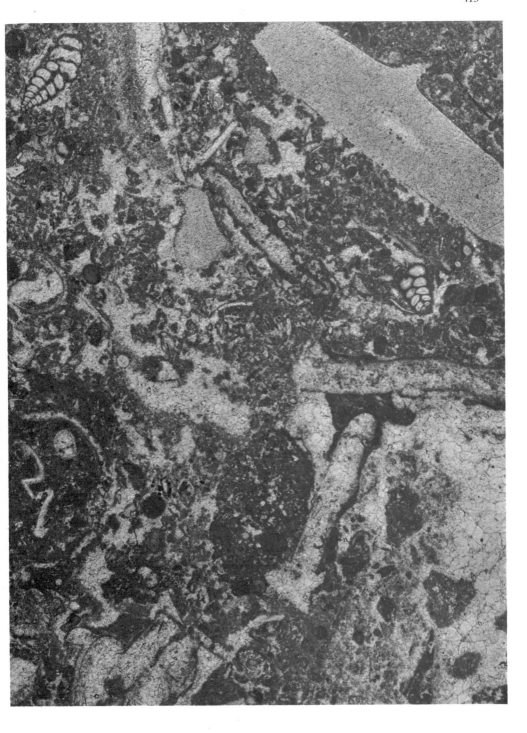

Plate II. Basinal and Offbank Facies

(A) Spiculitic wackestone basinal strata composed of very light and delicate mon-axon spicules (25 microns across), now partly calcified in black, organic-rich, siliceous micrite, standard microfacies 1; facies P-11, Chapter VII. The sediment accumulated below wave base in water more than 30 m deep, quiet, with sufficient current to orient the spicules but not to winnow the micrite and calcisilt. The facies occurs on the basinal side of the Wolfcampian Townsend-Kemnitz reef in Lea County, New Mexico. See Figs. VI-19 to VI-22. Sample: Shell ETA-5, 10337ft; thin section, × 36

(B) Microbioclastic peloidal calcisilt, standard microfacies 2. The sediment is composed of off-bank debris partly indigenous and partly drifted off a bank of very low relief. Currents which deposited the silt grains also winnowed the lime mud. Note the mm laminae grading from tiny echinodermal bioclasts up to dark peloids or fine-grained lithoclasts. Cure section, Middle Jurassic, southeast Paris basin, Burgundy, France. Photograph courtesy of Bruce Purser and Koninklijke Shell Exploration and Production Laboratory, The Netherlands. See Figs. X-7 and X-8 for location of the section and calcisilt facies. Thin section, × 36

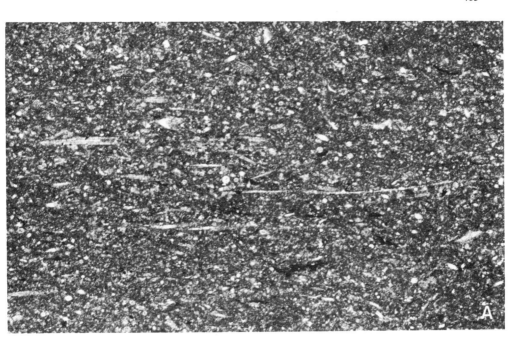

Plate III. Mesozoic Pelagic Lime Mudstones, Standard Microfacies 3

(A) Dense, dark micrite with some microbioclasts. The photograph contains a pelagic crinoid or ophiuroid plate. Sample is from basinal, off-bank strata of the Mexican miogeosyncline, basal Tamaulipas limestone (Lower Cretaceous), in the Peregrina Canyon. Tamaulipas, Mexico, (Chapter XI). Thin section, × 36

(B) Calpionellids in dove gray, laminated pelagic micrite. Sample Jur-16 is from type Oberalmkalk (Maiolica or Biancone facies) of the Austroalpine nappes. Such sediment is common in the latest Jurassic (Tithonian) troughs within the Tethyan geosyncline, facies J-14. Thin section, × 90

(C) *Oligostegina* (calcispheres) and globigerinid foraminifera in dark micrite (wackestone). Typical off-reef facies of Middle Cretaceous basins and basin edges. Cenomanian to upper Albian "Georgetown" overlying the Deep Edwards Reef. Gulf Tartt No. 1 well, 10410-15H feet, Fashing Field, Atascosa County, Texas (Chapter XI). See also Oligocene globigerinids, Horowitz and Potter, 1971, Pl. 67. Thin section, × 90

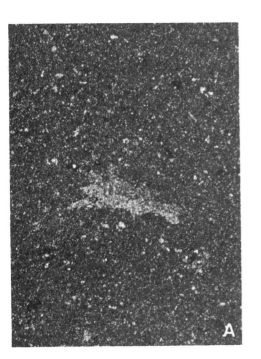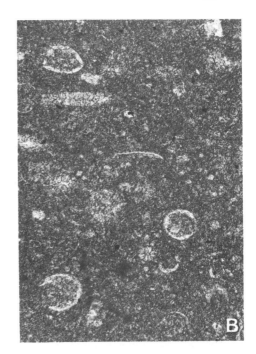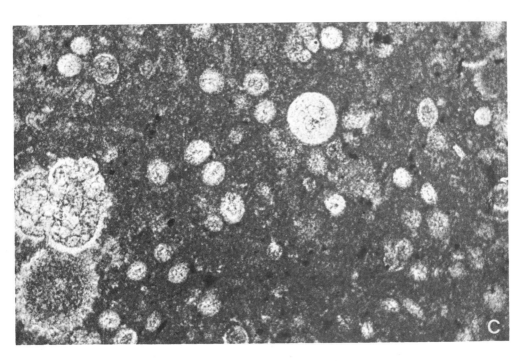

Plate IV. Reef Flank and Reef Boundstone of Virgilian (Late Pennsylvanian)

(A) Standard microfacies 5 is bioclastic, lithoclastic packstone composed of debris on the mound slope. Sample DC-88 is from an off-reef position at the eastern base of the lower mound at the mouth of Dry Canyon, section 29, Sacramento Mountains, New Mexico (see Frontispiece) and consists of rotted debris of algal plates, tubular foraminifera, and *Tubiphytes* (lower left corner of photo). The depositional environment and interpretation of this Late Pennsylvanian sediment is very similar to that of AHP shown on Plate I. The mixture of ill-sorted, partly worn and partly angular bioclasts and lithoclasts is typical of reef flank deposits, facies P-7, Chapter VII. Thin section, × 15

(B) Encrusting bindstone, standard microfacies 7. Sample consists of altered encrusting codiacean? algal laminae with masses of dark, microporous, tubular calcitornellid foraminifera. Cores of Middle Pennsylvanian mounds in the Paradox basin, Utah contain organic structures similar in configuration to the coarsely crystalline laminae shown here from Virgilian strata. The former have a cortical tubular structure much like that of codiacean algae. This type of boundstone caps phylloid algal micrite mounds located on the gently sloping outer margin of the narrow shelf on the west side of the Sacramento Mountains. Sample AHT is from the top of the western (left) side of the main mound, north wall of Dry Canyon (see Frontispiece). See also tubular foraminiferal bindstone, Plate XX. See description in Chapter VI. Thin section, × 15

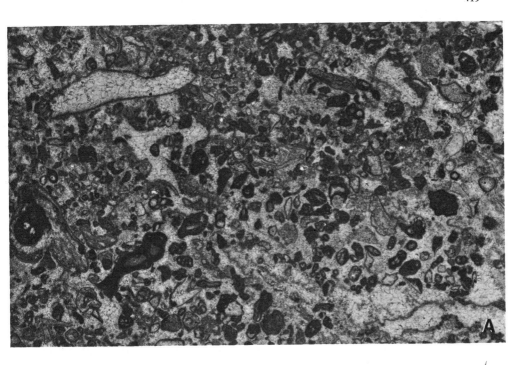

Plate V. Mound Bafflestone and Whole Fossil Wackestones

(A) Bafflestone, standard microfacies 7. Tubular-cylindrical shaped sponges and sponge-like organisms occur in micritic mounds in the Cambro-Ordovician strata in North America. Such mounds also may have stromatolitic structures. Following Embry and Klovan's (1971) interpretation, the trapping and baffling action of such a loose framework results in accumulation of lime mud and mound buildups like those seen around dendroid corals in the Devonian and Mesozoic. Fenestrate bryozoans and crinoids serve this function in the Middle Paleozoic. This sample and photograph, courtesy of D.F. Toomey, University of Texas Permian Basin, is from the Lower Ordovician Arbuckle group, Arbuckle Mountains, on U.S. Highway 77 between Anadarko and Davis in southern Oklahoma. Thin section, × 3

(B) Whole fossil wackestone, standard microfacies 8. Sample is from a biostrome of *Monopleura* in the middle Glen Rose formation (Albian, Cretaceous) at the upper end of Medina Lake, Medina County, Texas. The bed is a light-colored argillaceous micrite and contains abundant whole monopleurid rudists which have been somewhat crushed by compaction. Note the preservation of the uncompacted pelloidal fabric within the protection of the shells compared with the micrite texture between the fossils. The *Monopleura* also formed a bafflestone texture, being always in small mounds or biostromes. The occurrence of well-preserved whole fossils in micrite matrix indicates generally *in situ* burial and a very quiet water environment. Thin section, × 10

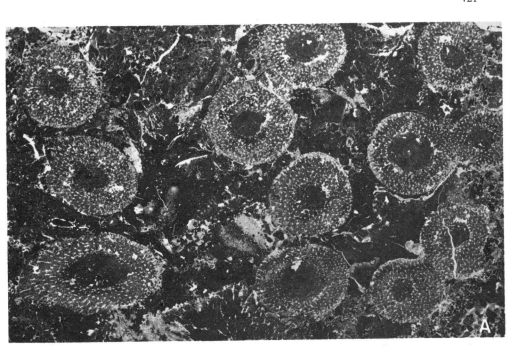

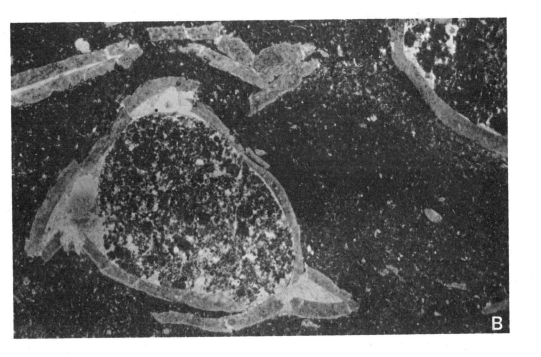

Plate VI. Restricted and Open Marine Facies

(A) Coquina or grainstone of dasycladacean alga, *Clypeina jurassica*, standard microfacies 18; facies J11, Chapter IX. Broken pieces of *Clypeina* and rotted, round pieces of indistinguishable bioclasts represent sediments on shoals of shelf areas of Arab cycles on the Arabian Shield. Moderate water energy is indicated. Sample DK-51 (40) is from 6935–7050 feet in the Iraq Petroleum Company Dukhan 51 well, from Arab D zone, Late Jurassic, southern edge of Qatar Peninsula, Arabia. Acetate peel, × 18

(B) Normal marine bioclastic, lithoclastic, thoroughly homogenized wackestone, standard microfacies 9, facies P3, Chapter VII. Sample is from the Leavenworth limestone of central Kansas, number two limestone of Oread megacyclothem (see Fig. VII-6) and contains bioclasts of gastropods, coated crinoid pieces, echinoid spines, platy fragments of algae or mollusks, numerous fusulinids, and a few dark lithoclasts. Photograph is courtesy of D. F. Toomey, University of Texas Permian Basin. Thin section, × 3

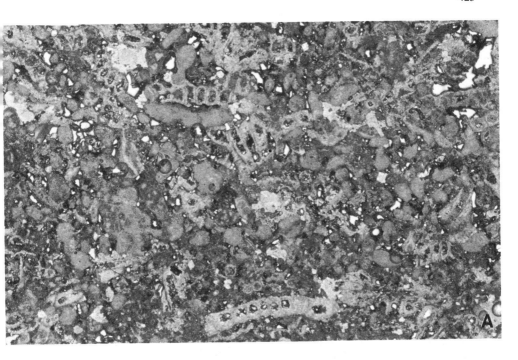

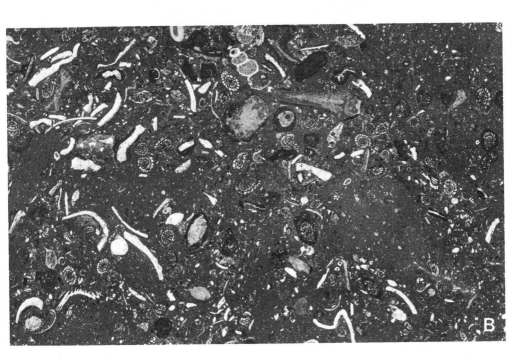

Plate VII. Coated Rounded Bioclastic Grainstones, Standard Microfacies 11

(A) Rounded clam shells and gastropod tests mixed with smaller ooids and black peloids, coarse grainstone. Some larger particles have micrite rinds caused by boring filamentous blue-green algae and indicating particle movement in shallow sunlit water less than 10m deep. Particles are completely and evenly lined with fibrous druse but much original pore space is unfilled (white). Large blocky crystals on lower left are replacement selenite. Sample DK-51 (37) from shoal in Late Jurassic (Arab D Formation) on the Arabian Shield. Facies J 10, Chapter IX. Peel, × 18

(B) Shoal grainstone-packstone, coated and rounded. Sample MS-13-A is from the Mississippian Salem limestone, Cleveland quarry, Harrodsville, Indiana. Particle types are worn bioclasts of bivalves, brachiopods, gastropods, and endothyrids. Many particles have dark micrite rinds (see above). There is no drusy lining (marine cement?). The original void space was partly filled by peloidal calcisilt and remaining portion with coarse crystalline calcite; upper side of the sample is to left. Photograph is courtesy of D.F.Toomey, University of Texas Permian Basin. Thin section, approximately × 9

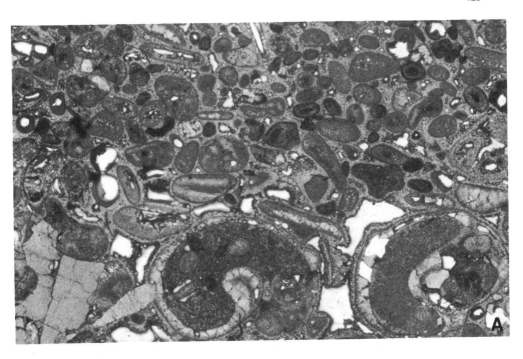

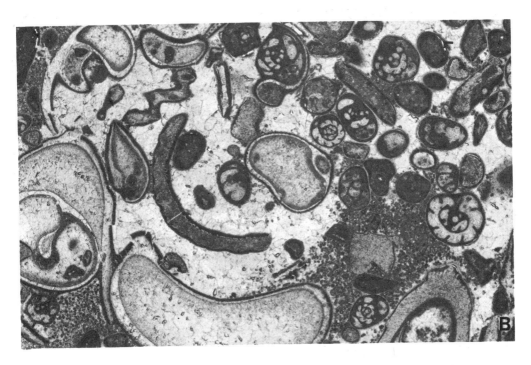

Plate VIII. Bivalve Coquinas, Bioclastic Packestones, Standard Microfacies 12

(A) Oyster shell hash, packstone with pelleted calcisilt matrix is represented by sample No. 64, from the upper 100 m of the Cupido Formation (Lower Cretaceous), Arteaga Canyon, near Saltillo, Coahuila, Mexico. Sediment is of shallow water, bank-interior facies. Note the umbrella effects, patches of white coarse crystalline calcite under rounded shell fragments. Such biomodal sorting occurs in the sediment where the larger shell fragments were abraded and worn by wave action on the shoals and ultimately deposited with pelleted mud in quieter areas. Note borings in the largest piece of oyster shell. Thin section, $\times 12$

(B) Shell hash, packstone with abundant micritic matrix. Shell debris is exclusively of bivalves, somewhat worn and with micrite rinds. Umbrella effects are seen beneath shell where finer sediment has settled down leaving calcite-filled original void space. Top is to upper left. The sediment represented by sample SP-2 was deposited in a shallow basin and forms a limestone bed in a dark shale sequence in the Kössen basin below the Steinplatte reef in Austria (Fig. VIII-22), Upper Triassic in age. Thin section, $\times 9$

428

Plate IX. Grainstones

(A) Grainstone (pelsparite), standard microfacies 16. Peloids composed partly of micritized and rounded, abraded, tubular foraminifera. Top of a cycle in the ·Pennsylvanian Virgilian Holder Formation in the Sacramento Mountains (see Chapter VI). Sample 23–39 from the west end of a ridge north of and above Alamogordo, Indian Wells Canyon, New Mexico. Thin section, × 15

(B) Onkoidal grainstone, standard microfacies 13; MF-9, Chapter VIII. Sample HG-6 is Late Triassic (Norian) near backreef sediment behind the Hohe Göll coral ramp (Chapter VIII). Particles are dasycladaceans and coral debris coated with micrite and nubicularid foraminifera and blue-green algae. The coarse sediment is cross-bedded and represents very shallow shoal sands probably at or close to the reef crest in the splash zone. Note the coarse, centripetally oriented crystals of cement between the grains. Thin section, × 10

(C) Lag grainstone, standard microfacies 14. A composite of mixed resistant particles of blackened, phosphatized and iron-stained lithoclasts and peloids and badly worn, resistant echinoderm and foraminiferal pieces. Other "lags" commonly possess bones and teeth. This sample 23–38 is from a thin persistent bed at the contact of two sedimentary cycles in the Upper Pennsylvanian (Virgilian) Holder Formation on a ridge of Indian Wells canyon above Alamogordo, Sacramento Mountains, New Mexico. The bed persists across the shelf area, transgresses older cycles and represents a long period of non-deposition in shallow marine reducing environment. Its grains have been compacted by solution possibly owing to occasional saturation by meteoric water. Thin section, × 15

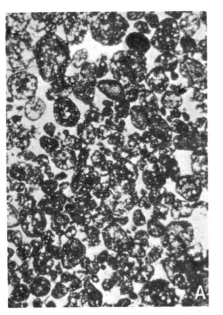

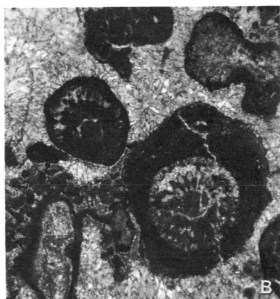

Plate X. Ooid Grainstones, Oolite, Standard Microfacies 15

(A) Over-packed oolite with both superficially coated grains and well-formed ooids. Packing, to the extent of fitted and planar contacts between the grains, is probably the result of early solution. The grains are evenly lined with a fine druse interpreted as probable marine cement. Such history indicates deposition and lithification in and out of the marine environment on shoals. The presence of foraminifera and coated bivalve or brachiopod fragments indicates possibly open marine circulation. Sample is from the Upper Jurassic Arab C Formation in Qatar Petroleum Company, Fuweirat well, 5230 ft depth, northeast coast of Qatar, Arabia. Peel, × 18

(B) Normally packed oolite with mostly well-formed, multiple-coated, spherical grains. The nuclei are composed of echinodermal, foraminiferal, and molluscan fragments. Shoal deposition is indicated in open marine water. Today such well-formed ooids are found mainly in tidal bars. Higher magnification shows that the cement between grains is a mosaic of blocky calcite and perhaps is late diagenetic, possibly from meteoric water in the phreatic zone. The sample is from the Late Jurassic Smackover Formation from a Skelly Oil Company well in northeast Texas at 8813–8834 ft. See Fig. X-11 for comparable facies in another well. Thin section, × 16

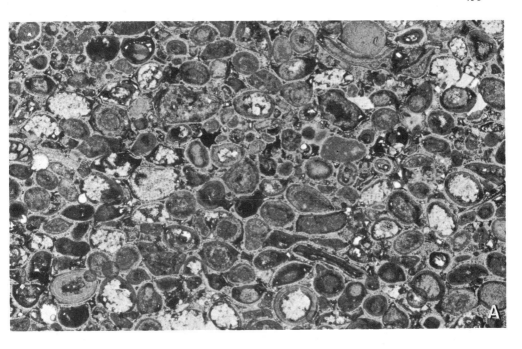

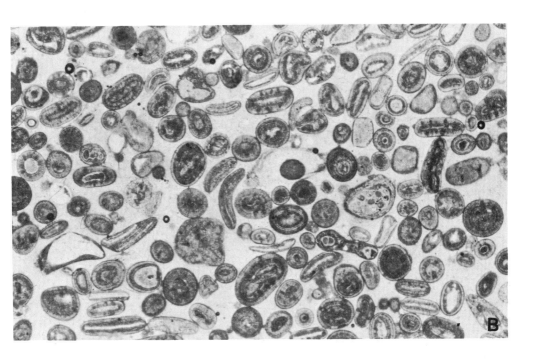

Plate XI. Spongiostrome Structure in Pelleted Micrite
and Pelsparite

(A) Spongiostrome laminite, standard microfacies 21, composed of peloids originally bound together by filamentous blue-green algae existing on tidal flats. Note
the irregular anastomosing fabric between the grains. The laminae are caused by
micrograding or at least alternation of coarse and fine layers. The finer, dark
micrite layers are formed at times of minor marine flooding when only the most
minute particles can be carried across the algal mats. Sample SP-50, Hauptdolomit (Upper Triassic-Norian) below the Steinplatte Haus, above Waidring, Austria. Thin section, × 18

(B) Pelsparite with three, centimeter-thick, micrograded laminae with small-
scale ripples, standard microfacies 16. Peloids are discrete and well-rounded,
shaped and deposited by moving water. Sample SJ-1 is from the San Juan Canyon, Utah. Middle Pennsylvanian sediment at the top of a sedimentary cycle. Just
above the thin micritic layers, large reworked lithoclasts and a few bioclasts occur
with the peloids. Thin section, × 11

See also Plate IX for higher magnification of standard microfacies 16. Also
Horowitz and Potter (1971), Plate 100 for pelsparite with lithoclasts from the
Palliser Formation, Devonian of Alberta, Canada

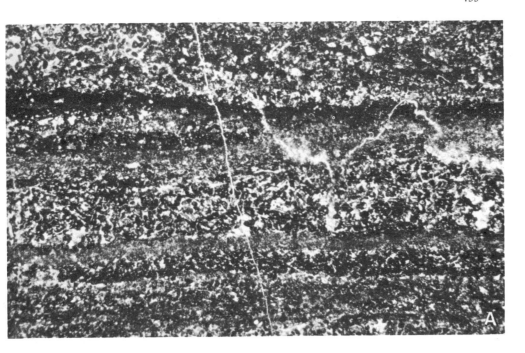

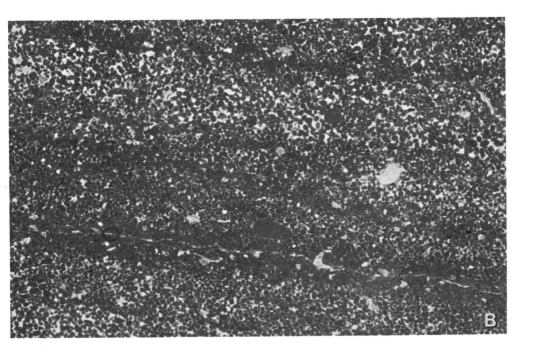

Plate XII. Laminites, Standard Microfacies 19

(A) Laminated dolomite-lime mudstone without bioclasts. Planar, rather even-bedded mm laminite from the evaporitic top of a Duperow (Devonian) cycle. Lighter layers are cryptocrystalline dolomite and larger scattered white specks are dolomite rhombs, 25 microns in size. Note the small scale truncation of laminae. Close interlamination of lime mudstone and dolomite indicates the very delicate control of dolomitization in the tidal flat environment. Such lamination may be induced by algal mats although no evidence of stromatolitic doming is seen. Quartz silt grains are commonly scattered along certain laminae. Such a microfacies occurs interbedded with sabkha anhydrite in the upper part of a regressive, fill-in, carbonate-evaporite cycle (see Chapter X). Sample is from the Duperow Formation in the subsurface of the Williston Basin, Shell Northern Pacific Richey No. 1 well, 8916 ft. Thin section, × 9

(B) Laminated pelleted mudstone-wackestone. Sample K-1 is from unit B, of the typical Lofer cycle of A. G. Fischer (1964). See Chapter X and facies MF-8, Chapter VIII. Laminae are only a mm or two thick. The coarser layers show peloids. The fabric was originally porous. Note the internal micritic sediment in the thin desiccation sheet-cracks (fenestral fabric) at the top and even in the tiny calcite cemented holes in the lower laminae (root or small burrows). The large calcite blotch in the lower right is probably a horizontal burrow. Dachsteinkalk on the Kehlstein highway below entrance to Kehlsteinhaus (Eagle's Nest) above Berchtesgaden, West Germany. Thin section, × 18

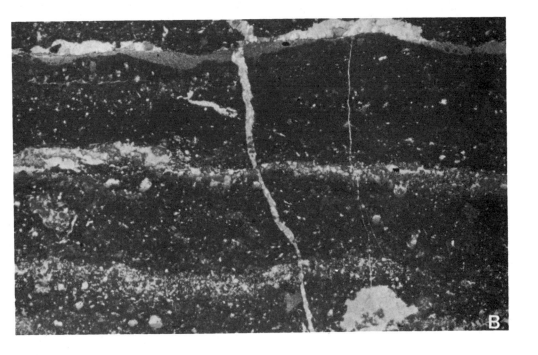

Plate XIII. Restricted Marine Microfacies

(A) Peloidal packstone with fenestral fabric due to desiccation of pelleted lime mudstone, also standard microfacies 16 and facies PM-12, Chapter VIII. Calcispheres (probably fruiting cases of algae) are scattered in the sediment. The fenestrules contain crystalline calcite with a void-filling pattern. Aggregation of such semi-hardened peloids with coating and micritization of the outside of such composite masses results in creation of grapestone lumps. The abundance of the latter in peloidal sparite sediment defines standard microfacies 17. Indeed, a spectrum exists from pelleted mudstone, to pellet packstone-grainstone, to grapestone pelsparite (grainstone). This sample PR-11 shows an intermediate stage and is from the backreef of the Permian Reef Complex, Guadalupe Mountains, Tansill Formation, in Rocky Arroyo, 30 km northwest of Carlsbad, New Mexico. Thin section, × 13

(B) Foraminiferal, dasycladacean grainstone-packstone. See also Plate VI-A, standard microfacies 18. This sample is a packstone of *Mizzia* (?) a dasycladacean, and miliolid-like tubular foraminifera, and a few shell fragments; from the Permian of Yugoslavia. Photograph courtesy of D.F. Toomey, University of Texas Permian Basin. Such coarse grainy sediment is commonly found in bars and shoals heaped up by tidal currents in shallow lagoons and bays and is common throughout the geologic record. Thin section, × 15

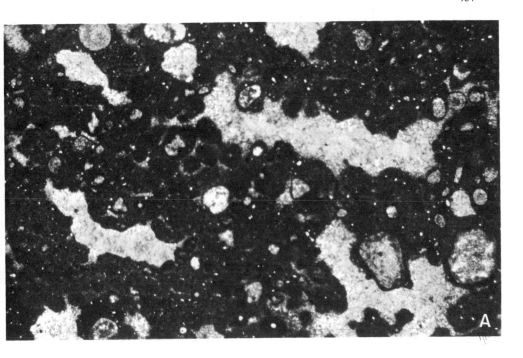

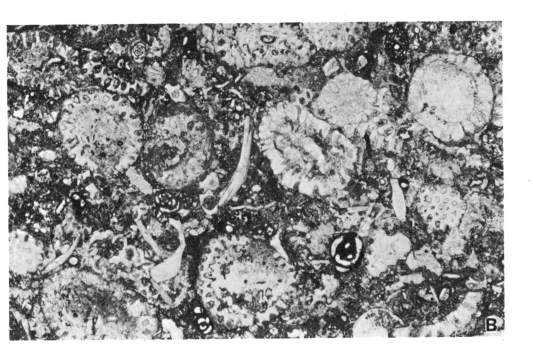

Plate XIV. Stromatolites

(A) Algal stromatolites (laterally linked hemispheroids); plan view of an eroded surface. Late Cambrian, upper Hoyt Limestone at Saratoga Springs, New York. Photograph courtesy of D. F. Toomey, University of Texas Permian Basin. The white scale is about 10 cm long

(B) Algal stromatolite in cross-section, standard microfacies 20. The sample includes some spongiostrome layers representing standard microfacies 21 and facies MF-8, Chapter VIII. See also Plate XI-A. Note the upward increasing relief of the convex bulges caused by increased algal growth and more trapping of sediment on high areas. This anti-gravity thickening of layers permits easy recognition and orientation of stromatolites in contrast to purely detrital laminae. Sample is from the Mendola Dolomite (Ladinian, Middle Triassic) 12 km southwest of Bolzano, Italy. Photograph is courtesy of A. J. Wells and Koninklijke Shell Exploration and Production Laboratory, The Netherlands. Polished surface, × 5

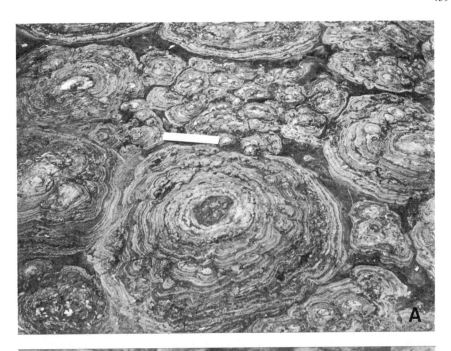

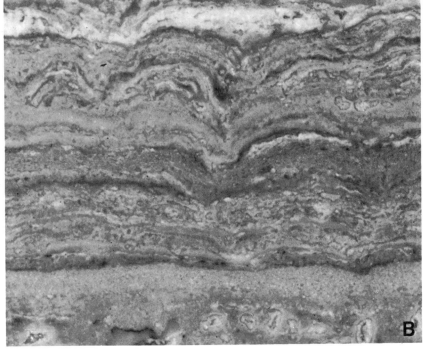

Plate XV. Onkoids, Pisolite and Mudstone Backreef Facies

(A) Onkoidal wackestone-packstone, Mumienkalk, standard microfacies 10 and 22 and facies J3, Chapter IX. Large *Girvanella*-bored algal biscuits encrust shell, coral, and snail fragments. The matrix consists of a wackestone with fine shell debris. The white spots are tiny onkoids. The sediment represents a deposit of shallow water with somewhat restricted circulation. Such sediment may not represent the original depositional environment of the algal nodules which require moving water to turn them over at intervals. This condition is not consistent with the presence of micritic matrix. The nodules may have grown on sides of a channel and have been buried in deep or shallower water in a muddy facies. The sample is from Les Sagnettes south of La Chaux-de-Fonds and from the main algal bed in the Malm (Upper Jurassic) sequence, Jura Mountains of northern Switzerland (see Chapter IX, Fig. IX-3). Photograph is courtesy of Koninklijke Shell Exploration and Production Laboratory and Martin Ziegler. Polished surface, × 2

(B) The well-known pisolite facies of the Upper Permian Capitan Limestone of Guadalupe Mountains, New Mexico. The exact origin of these particles, which were originally considered to be onkoids is still debated. The sample probably represents reworked cave pearls, or vadose pisoid concretions, formed either by meteoric or marine splash-zone water in porous carbonate sediment subjected alternately to wetting and extreme drying (see Chapter III and VIII). Note the infiltered crystal silt, of possible vadose origin, between the pisoids (Dunham, 1969b). Sample PRC-6 from switchback curve on the road up Walnut Canyon to Carlsbad Cavern. Thin section, × 15. (pisoid illustrated is 0.5 cm long)

(C) Homogeneous pure lime mudstone with blades of replacement selenite crystals, standard microfacies 23. Sample DK-35 is from the evaporitic top of an Arab D cycle and consists of chalky lime mudstone with early replacement gypsum, probably a sabkha sediment like that on present littoral salt flats. Iraq Petroleum Company, Dukhan 51 well, about 6900ft depth, southern Qatar. Peel, × 18

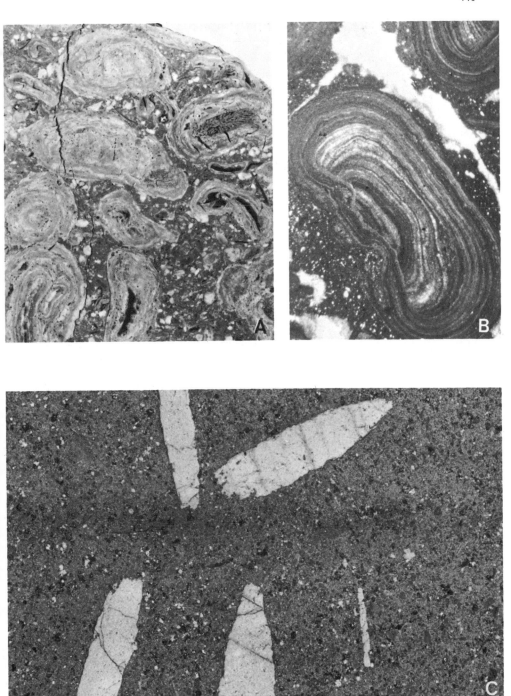

Plate XVI. Types of Intraformational Pebble Conglomerates

(A) Lithoclastic packstone, standard microfacies 24. Calcisilt laminated pebbles a cm or so long, incorporated in a fine calcisilt matrix with little or no micrite. The tiny grains in the matrix include a few bioclasts. The lithoclasts are locally derived intraclasts; such sediment occurs typically in tidal channels. Sample AGH is from El Paso Limestone (Lower Ordovician) at Skyline Drive, Franklin Mountains, El Paso, Texas. Thin section, × 15

(B) Lithoclastic grainstone, standard microfacies 24. Rounded aggregates and locally eroded lithoclasts (intraclasts of Folk) are of bioclastic peloidal wackestone. Some are peripherally micritized and all are thoroughly bored by blue-green filamentous algae (*Girvanella* structure) although this is too fine to be seen in the photograph. The cement is blocky calcite mosaic. Infiltered silt-size grains caught against larger clasts show that the top is to the right side of illustration. This sample AGE is also from a tidal channel in the El Paso limestone (Lower Ordovician) at Skyline Drive, Franklin Mountains, El Paso, Texas. Thin section, × 15

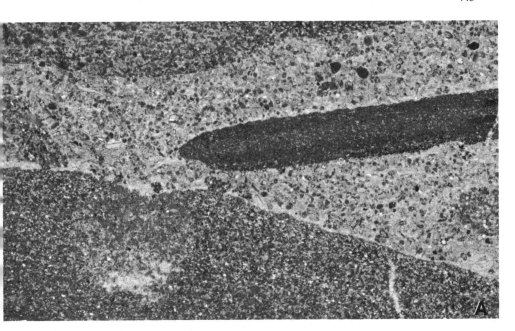

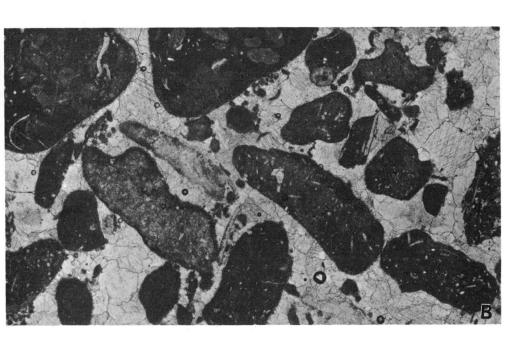

Plate XVII. Devonian Stromatoporoid Boundstone

(A) Globular to massive encrusting stromatoporoid *Clathrodictyon* interlaminated with the dendroid tabulate *Thamnopora* (in both upper left and right hand portion of photograph). The dark band represents a micrite (lime mudstone) layer which halted framestone growth. Standard microfacies 7, framestone or bindstone of Embry and Klovan (1971) and facies 8D of Chapter IV. Sample is from a small bioherm or biostrome in the subsurface Duperow Formation, (lower part of an upward shoaling cycle) Phillips Petroleum Company Hoehn No. 1 well, at 11 563 feet, Williams County, North Dakota. Thin section, × 14

(B) Globular, cabbage-head stromatoporoid in wackestone matrix with red alga *(Solenopora)* and brachiopod fragments composes standard microfacies 8 and facies 4D of Chapter IV, a normal marine bioclastic wackestone with some whole fossils. Phillips Petroleum Company, Hoehn No. 1 well at 11 561 feet, Williams County, North Dakota. Duperow Formation, (lower part of the typical cycle). See Fig. X-12 in text. Thin section, × 14

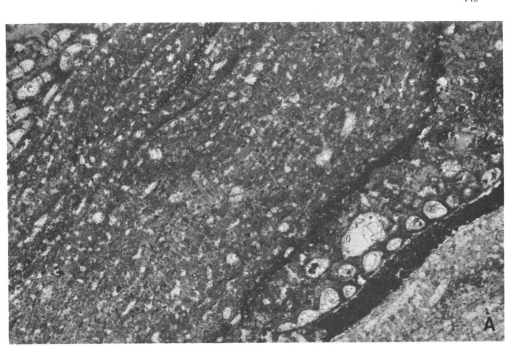

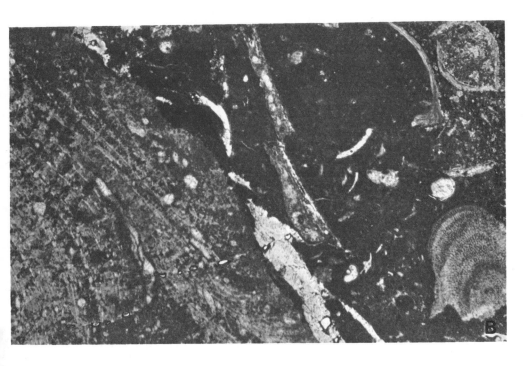

Plate XVIII. Devonian Stromatoporoid and Stromatactoid Structure

(A) Tabular stromatoporoid in micrite from a mound (bioherm) in Frasnian strata of the Dinant basin, Belgium: bindstone of Embry and Klovan (1971), standard microfacies 7 and facies 6D of Chapter IV. Such forms occur intergrown with sheet-like masses of the tabulate coral *Alveolites* and are more common in the lower, quieter water portion of the mound. Photograph and preparation courtesy of F.R. Allcorn, Tenneco, Houston, Texas. Polished slab with acetate peel affixed, × 1

(B) Stromatactoid structure. White areas are of void-filling coarse calcite. The digitate top and even base are well displayed. Such cavities in micrite matrix are common in lime mud mounds and are regularly developed in en echelon sheet-like patterns in Lower Carboniferous and Devonian bioherms in Europe (see Chapter V). This sample is from the Frasnian Marbre Rouge, Carriere de Croisettes, Dinant Basin, Belgium. Preparation and photograph courtesy of F.R. Allcorn of Tenneco, Houston, Texas. Polished slab with acetate peel affixed, × 1½

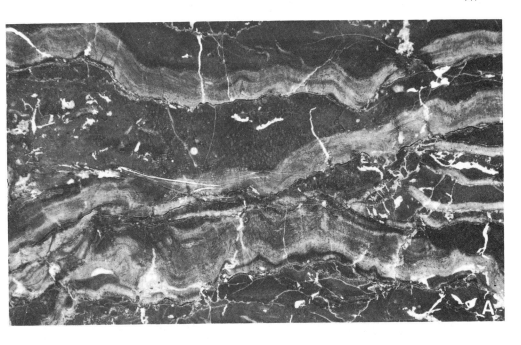

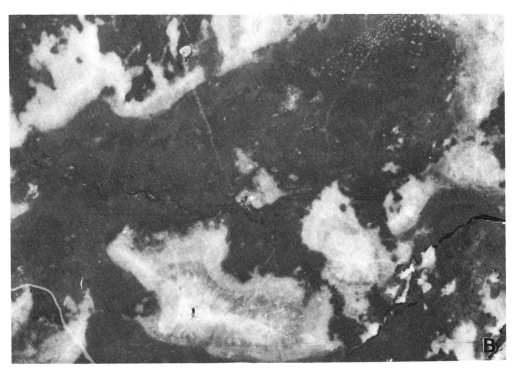

Plate XIX. Devonian Backreef Organisms

(A) Backreef calcisphere wackestone, facies 12D, Chapter IV. The radiosphaerids are probably the fruiting cases of algae, perhaps dasycladaceans, which thrived in the quiet, shallow waters of lagoons commonly in interiors of carbonate banks. Note the grumelous, vaguely pelleted matrix and small-scale fenestral fabric. Photograph is courtesy of J.E. Klovan, University of Calgary, Alberta, from Late Devonian beds of western Canada. Thin section, ×65

(B) Tubular stromatoporoid, *Amphipora*, in micrite from Devonian of Alberta, Canada; facies 11D, Chapter IV. Southesk Formation, Burnt Timber Creek, northwest of Calgary. Photograph is courtesy of A.J. Wells and Koninklijke Shell Exploration and Production Laboratory, The Netherlands. Polished slab, ×2¼

(C) Cross section of *Amphipora* in micrite from Alberta basin. Photograph is courtesy of J.E. Klovan, University of Calgary, Alberta, Canada. Thin section, ×25. Specimen about 4.0 mm across

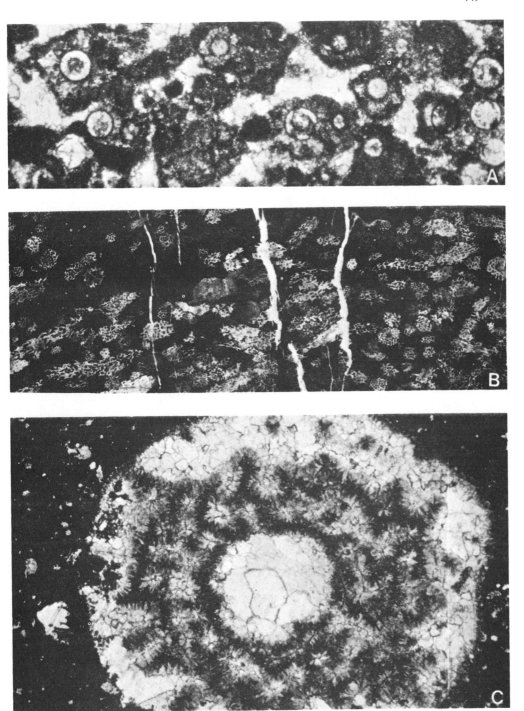

Plate XX. Encrusting Bindstone

(A) Masses of tubular calcitornellid foraminifera from capping rock of a lower mound stage buried in the interior of a Late Pennsylvanian Virgilian algal plate accumulation. See Chapter VI. Sample C-10 of Yucca mound comes from north of State Highway 83, at the mouth of Dry Canyon, Sacramento Mountains, New Mexico (see Fig. VI-10). Photograph is courtesy of D.F. Toomey, University of Texas Permian Basin. Thin section, × 15

(B) Masses of probable foraminifera *Renalcis* interlaminated with stromatolite from Late Devonian of Alberta. This encrusting organism is common in the Late Devonian of Australia (see Chapter IV) and in Cambrian beds in several parts of the world. Its biological affinity is unknown but its similarity to tubular encrusting forams of Late Paleozoic to Triassic age is considerable; and it occupies a similar habitat, characteristically capping mounds and reefs and forming an encrusting type of boundstone. Photograph is courtesy J.E. Klovan, University of Calgary, Alberta, Canada. Thin section, × 25

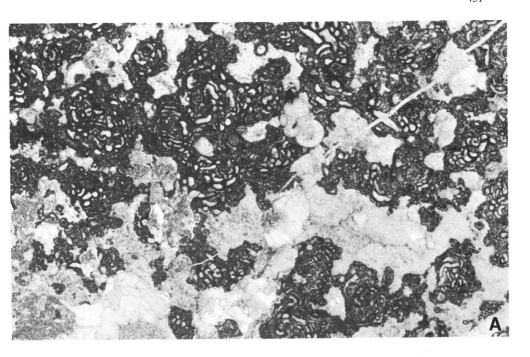

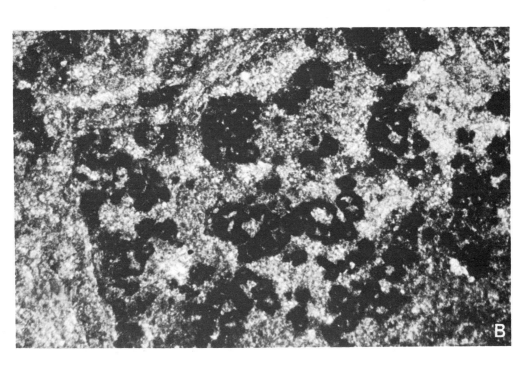

Plate XXI. Mississippian Facies

(A) Lime mudstone-wackestone of Waulsortian mound. Sample AGR is from the top of the mound at the north side of the mouth of Alamo Canyon, Sacramento Mountains, New Mexico and is composed of fine bioclastic debris of crinoids, ostracods, and shreds of bryozoan. The round discs with holes and arrowhead-shaped particles are crinoid stems and arm plates respectively. This is the typical composition of the all Waulsortian mounds in the Sacramento Mountains (see Chapter V). Thin section, × 15

(B) Encrinite-crinoidal bryozoan packstone. Sample AGN typifies flank facies off the same Waulsortian mound of which the above sample AGR is typical. Note sutured contacts of crinoid particles owing to microstylolitization during early compaction (see Chapter V). Other encrinites representing this special kind of standard microfacies 12 are seen in Horowitz and Potter 1971, Plates 83, 89, 91, 97. Encrinites are very common flanking beds of Paleozoic and early Mesozoic mud mounds developed in water of varying depths. Thin section, × 15

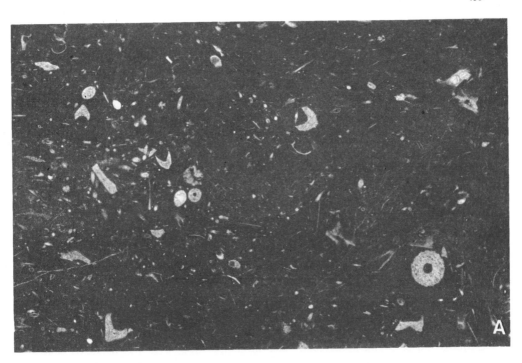

Plate XXII. Late Paleozoic Platy Algal Mound Rock

(A) Platy (phylloid) algal segments occur in typical micrite matrix; facies P6 of Chapter VII, also see Chapter VI. Cortex shows as dark, poorly preserved, short columns along the outer edges of crystalline plates. This is a typical codiacean structure. Tubular foraminifera encrust some plates as well as the tiny chambered foraminifer *Tuberatina* and the multi-chambered bell-shaped *Tetrataxis*. A fusulinid is seen in the lower left corner. This "plate rock" is common in Pennsylvanian to middle Permian strata in the southwestern United States. Sample 23-42 is from the Virgilian Holder Formation on the north side of Indian Wells Canyon on the ridge above Alamogordo, Sacramento Mountains, New Mexico. Thin section, × 10

(B) Brecciated platy algal lime mud wackestone common in mud mounds with abundant bioclastic particles; facies P6 of Chapter VII. Crinoids, brachiopods, and bryozoans are also common in this sample (see Chapter VI). The brecciation is an early slumping phenomenon. The hardened clasts and fossil fragments are fractured but the fracturing dissipates within the matrix between the grains. Sample DCWR-2 is from the lower mound at the mouth and on the north side of Dry Canyon, Sacramento Mountains, New Mexico. See Frontispiece for location. Thin section, × 4

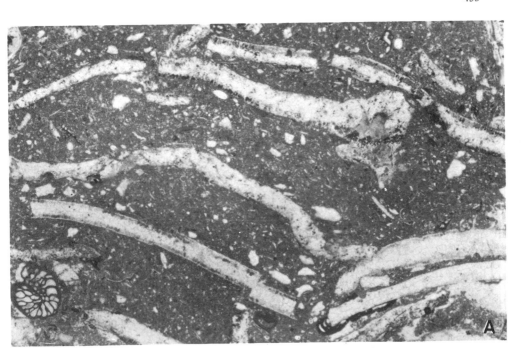

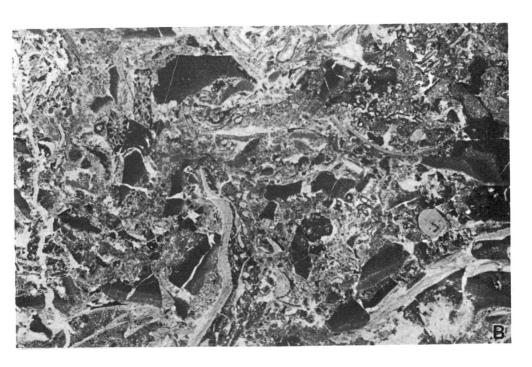

Plate XXIII. Special Pennsylvanian Organisms

(A) A fusulinid packstone-wackestone with interstitial matrix of silt to fine sand size bioclastic debris—striking bimodal sorting; facies P5 of Chapter VII and standard microfacies 10. Note the orientation of the spindle-shaped fusulinids is such that most are transverse sections (the lower right corner displays a perfect axial-longitudinal section). Most of the fusulinids are worn and thoroughly micritized showing transport for some time or distance before deposition. Seven lithoclasts and a transverse section of a crinoid stem are seen. Leavenworth Limestone (No. 2 limestone) from Oread megacyclothem of Kansas, Locality 3 of D.F. Toomey (1969a) who also kindly furnished the photograph. Thin section, × 7

(B) *Komia* packstone, thick-bedded detrital Mid Pennsylvanian limestone. Finer particles have in-filtered the coarse grain-supported fabric of *Komia* with lithoclasts. *Komia* is most probably a small dendroid, branching stromatoporoid and is found in detrital accumulations representing shoal environments in Middle Pennsylvanian strata of the southwestern United States (see Fig. VI-24). The sample EDCW-26, from the Bug Scuffle Limestone, is from a Desmoinesian carbonate bank in the Sacramento Mountains of New Mexico. Thin section, × 15

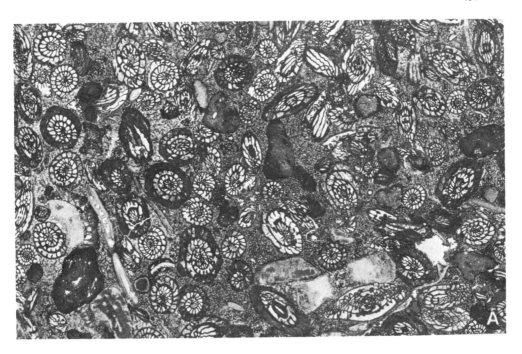

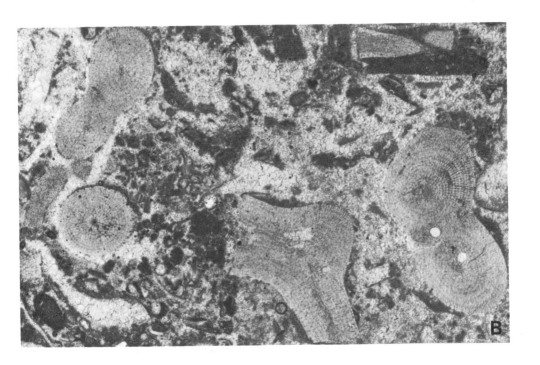

Plate XXIV. Triassic Boundstones

(A) Late Triassic coralline reef with encrusting algal fabric. The framestone is of coral stalks and bulbous, irregularly laminated forms, coarser than *Tubiphytes*; facies MF-2 of Chapter VIII. Interstices are lined with the same stromatolitic crinkly laminae with tiny tubules. Voids ultimately have been filled with coarse centripetally oriented calcite. Sample SP3 is from the Steinplatte reef front near Waidring, Austria (see Fig. VIII-22). Thin section, × 11

(B) *Tubiphytes* encrusting boundstone (bindstone) from the reefy front of the Dolomite Middle Triassic banks; facies MF-1 of Chapter VIII. Sample MS-4 is from a Cipit (exotic) limestone block in the Rosszähne at the Seiseralm in north Italy. Fabric is mainly encrusting laminated growths indistinguishable from much of the *Tubiphytes* in the Permian of southwestern United States. Several cavities are filled with drusy calcite but others are filled with silt-size detrital, contemporaneous internal sediment. This is probably a typical "reef-margin" sediment of the Dolomite carbonate banks. Thin section, × 9

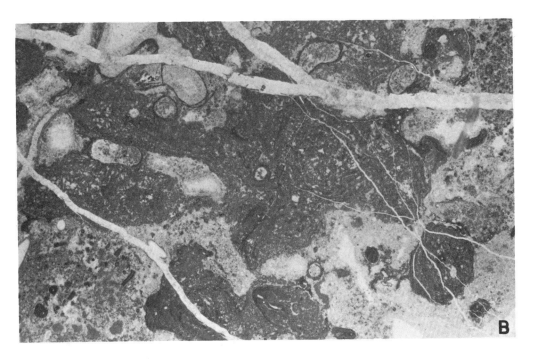

Plate XXV. Permo-Triassic Cavity Fillings in Micrite and Boundstone

(A) Cavity in "reef fabric" in Triassic (Ladinian) Dolomite showing laminate encrusting boundstone (dark) with several stages of cavity filling: (1) light-colored, irregular and open laminae, (2) centripetally oriented sparry calcite, and (3) second generation of lighter calcite. Sample MS-17 is from a Cipit (exotic) block on the northeast Rosszähne trail at the edge of Seiseralm, northern Italy. Thin section, × 10

(B) Cavity in Permian Reef Complex of the Guadalupe Mountains, New Mexico; facies Pm-5 of Chapter VIII. A micritic matrix on the right and left edges of the photograph is typical of much of the "reef". This matrix is covered by encrusting bryozoan, tubular foraminifera and finally an irregular lamina of possible algal origin. This encrusting growth has surrounded a few cavities in the boundstone, the largest of which contains some internal sediment. The geopetal sediment indicates that the top is to the right. The major cavity is lined with dark micrite on both sides. The cavity is filled with several generations of fine, gray druse. A later generation of coarser druse appears at the lower left corner. Sample AAF-12 from Pinery Trail above Guadalupe Pass, Culberson County, Texas. Thin section, × 16

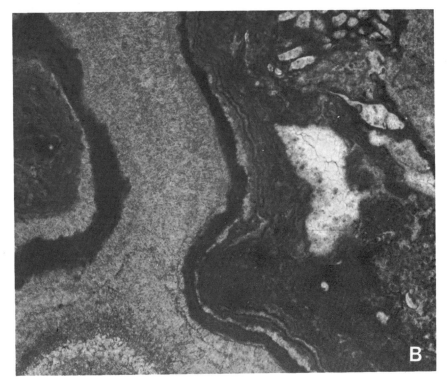

Plate XXVI. Triassic Reef Fabric

(A) Sphinctozoan segmented sponge with encrusting algae; facies MF-2 of Chapter VIII. The sample MS-15 is from a Cipit block on the Rosszähne trail in the Dolomites, at the Seiseralm, northern Italy. Similarity of the biota of these undolomitized exotic blocks with that of the Ladinian Wetterstein Limestone of the Northern Limestone Alps indicates that the blocks are derived from a reefy rim of encrusting boundstone. This is similar to that seen around the rim of banks which have been described in Austria and Bavaria by E. Ott (1966) and illustrated by Fig. VIII-16. Thin section, × 11

(B) Spongiomorph hydrozoan from low in the Rhaetic Rötelwand bioherm, south of Hallein, Austria. The bioherm occurs as an isolated mass in Kössen shale basin but is capped by impressive organic encrusting boundstone. Sample RK-15 contains a framestone of *Spongiomorpha* encrusted with a fine tubular organism; facies MF-1 of Chapter VIII. Such spongiomorph fabric, along with corals, sponges, and red algae, makes up major boundstone reefs in the Triassic. Thin section, × 6

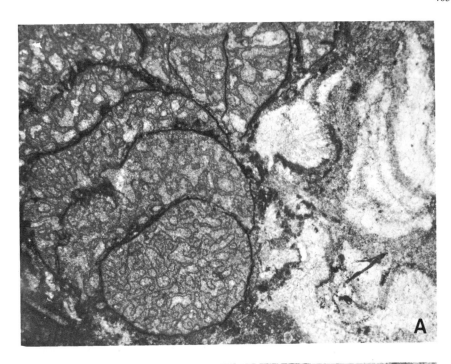

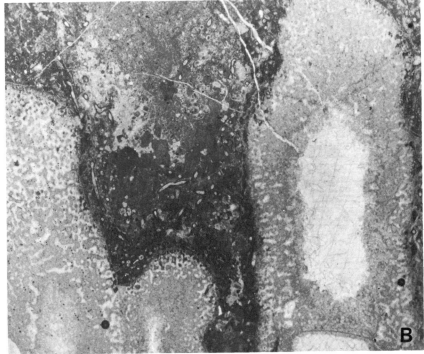

Plate XXVII. Solnhofen Limestone

(A) Krummelagen or soft sediment slump structure less than 2 m high in otherwise regularly bedded, platy, deep lagoonal sediment typical of the formation. Small quarry near Solnhofen, Franconia, West German Federal Republic

(B) Typical thin to platy, even-bedded Solnhofen Limestone of Kimmeridgian (Late Jurassic) age from the Bergér Quarry near Eichstätt, Franconia, West German Federal Republic. Note hammer for scale

Plate XXVIII. Solnhofen Limestone Biota

(A) *Saccocoma* (pelagic swimming crinoid) typical of the Jurassic and Cretaceous open marine basinal limestones, on a block purchased from the Museum Bergér Harthof near Eichstätt, Franconia. Polished slab with affixed acetate, × 2. (Photograph and preparation by F. R. Allcorn, Tenneco, Houston, Texas)

(B) Solnhofen micrite with tiny echinoderm plate; facies J9 of Chapter IX. Sample SB is from the Bergér Quarry near Eichstätt, Franconia. This remarkably fine-grained matrix under a scanning electron microscope shows abundant pieces of coccolithophores as well as tiny 3 to 4 micron rhombs. Thin section, × 90

(C) Late Jurassic Solnhofen shrimp about natural size from a photograph purchased from the Museum Berger Harthof near Eichstätt, Franconia, West German Federal Republic. The remarkably faithful reproduction of the completely flattened soft-bodied organisms is typical of the famous Solnhofen fauna

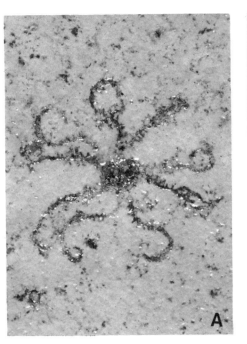

Plate XXIX. Jurassic Basinal Sediments of Austroalpine Nappes in the Northern Limestone Alps

(A) Radiolarite of Upper Jurassic age; facies J16 of Chapter IX. Sample Jur-13 is red, laminated, siliceous micritic pelagic sediment. It is from the unit 3 of the Glasenbach section, south of Salzburg, Austria. Such thin strata, with abundant radiolaria, must represent sediment of very deep water (thousands of meters) in certain troughs on the outer continental slope into the Tethys geosyncline. Thin section, × 127

(B) Red nodular limestone, Ammonitico Rosso, or rote Knollenkalk. This sample, Jur-11, is Lower Jurassic (Lias) of unit 2 of the Glasenbach section just south of Salzburg, Austria. It consists of red nodular conglomeratic sediment in which the nodules are of pelagic lime mud and are set in a micrite matrix also with exclusively pelagic microfauna and ammonites. Nodules are surrounded by microstylolites concentrating clay, iron, and manganese oxides. This is considered to be an early diagenetic fabric developed principally on sea mounts within the geosyncline and not a true detrital conglomerate. The sediment is widespread in the Austroalpine area. It is facies J17 of Chapter IX. Thin section, × 16

(C) Filamentous bivalve micrite, standard microfacies 3; facies J 15 of Chapter IX. The sample Jur-17 is of green, siliceous, Middle Jurassic limestone, unit 2E of the Glasenbach section, just south of Salzburg, Austria. Such accumulations of very thin-shelled pelagic bivalves are common in Triassic to Jurassic offshore, basinal, somewhat argillaceous dark limestone. The fauna represents sudden extinctions of pelagic or nektonic fauna with resulting sea-bottom deposition of tremendous numbers of shells. These form thin layers in otherwise very fine grain sediment. Thin section, × 14

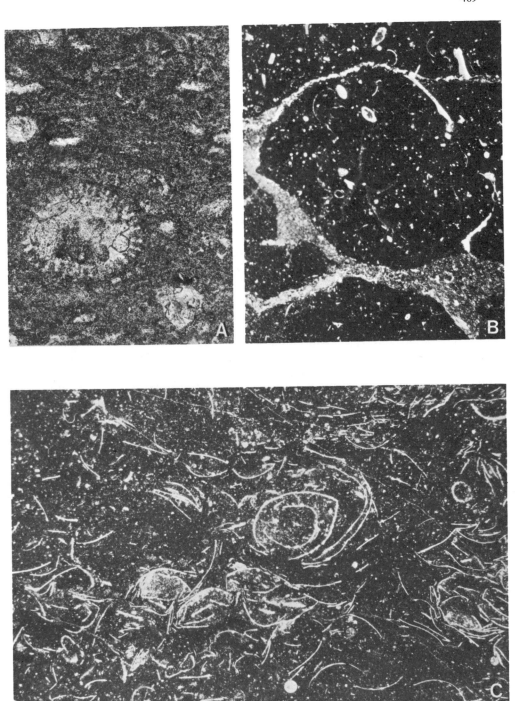

Plate XXX. Foraminifera and Rudist from the Cretaceous

(A) Alveolinid limestone. *Praealveolina* wackestone shoal sediment, characteristic of bank interiors with moderately open circulation. Sample Wso-25, Upper Wasia Formation, Cenomanian from Wadi Mi'aidin, Oman Range, Arabia. Thin section, × 14

(B) *Orbitolina* packstone. Sample H is of Aptian age and is from the Schrattenkalk of the Niederhorn area in the Helvetics, south of the Thunersee, central Switzerland. The sediment includes small foraminifera and dasycladacean fragments and is characteristic of standard microfacies 18. *Orbitolina* beds are common in the Cretaceous in slightly argillaceous strata of open marine circulation and salinity and they may form a facies belt exterior to the bank center occupied by an alveolinid or miliolid rich limestone. Thin section, × 14

(C) Miliolid-lithoclastic grainstone with grain aggregates, standard microfacies 17–18. This is a typical and widespread Cretaceous and Tertiary facies of the most restricted marine environment of the bank interior. Only certain rudists such as *Toucasia* and *Requienia* and dasycladacean algae occur with the miliolids. The sample is from the Christy Mitchell No. 1 Laskowski well at a depth of 10980-10990 ft, Fashing Field, Karnes County, Texas and from the Edwards (Albian) backreef, behind the Deep Edwards reef trend. The above three foraminiferal facies are displayed in Fig. XI-14 from a Middle Eastern field. Thin section, × 15

(D) Large foraminiferal packstone, *Choffatella* from the Lower Cretaceous, standard microfacies 18. Sample Sumac-71 is from the Cúpido Formation in Arteaga Canyon near Saltillo, Coahuila, Mexico. The sample shows probable current orientation of the grains and an early solution compaction. A thin coating of drusy calcite around most grains may well represent marine cement pseudomorphed after aragonite. The sediment probably thus represents a shoal facies of normal marine salinity with intermittent exposure and marine cementation. Grainstone and packstone with such foraminifera are also present in the Middle East. Thin section, ×18

(E) Caprinid of Albian age from the lowest beds of a shelf margin mound on the eastern edge of the Valles platform. Sample K-Mex 22 is from the lower part of Taninul Quarry, Cuesta El Abra, 16 km east of Valles, San Luis Potosi, Mexico. See Fig. XI-4. The vesicular structure of the shell wall is typically caprinid. Note that the fossil was originally buried in lime mud and is now part of a lithoclast incorporated in a calcisilt matrix composed of small bioclastic fragments. Thin section, × 6

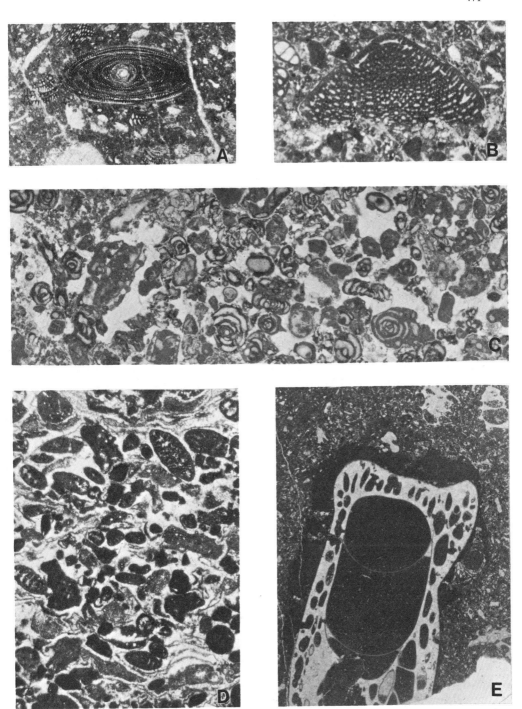

Depositional Sedimentary Environments
With Reference to Terrigenous Clastics
Second, Revised Edition

H.E. Reineck, Institut für Meeresgeologie und Meeresbiologie,
Senckenberg-Institut, Wilhelmshaven, FRG
I.B. Singh, Department of Geology, Lucknow University, Lucknow,
India

This book offers up-to-date information on primary sedimentary structures
throughout the world and provides comprehensive accounts of the various
sedimentation structures, describing their genesis and distribution in
modern environments.

Contents

PRIMARY STRUCTURES AND TEXTURES Depositional Environments •
Physical Parameters • Current and Wave Ripples • Surface Markings and
Imprints • Scour Marks • Tool Marks • Penecontemporaneous
Deformation Structures • Bedding • Sediment Grain Parameters •
Chemical and Mineralogical Parameters • Biological Parameters •
Environmental Reconstructions
MODERN ENVIRONMENTS Importance of Sequence in Environmental
Reconstruction • Glacial Environment • Desert Environment • Lake
Environment • Fluvial Environment • Estuarine Environment • Deltaic
Environment • The Coast • The Shelf • Examples of Beach-Shelf Profiles
from Modern Environments • Coastal Lagoons • Tidal Flats • Continental
Margin, Slope, and Ocean Basin

1980/549 pp./683 illus./paper/ISBN 0-387-10189-6

Microfacies Analysis of Limestones
Revised and Expanded Edition

E. Flügel, University of Erlangen-Nürnberg

Designed for exploration geologists who need rapid and intensive training
in the modern methods of microfacies analysis, this volume is a practical
introduction to a discipline of geology crucial in the investigation of
depositional environments and the diagenesis of carbonate rocks.

Contents
Introduction to Facies Analysis • Recent Carbonate Sedimentation •
Carbonate Diagenesis • Microfacies Characteristics • Fossils in Thin
Sections • Classification of Carbonate Rocks • Microfacies Types •
Standard Microfacies Types • Complementary Methods • Facies Diagnosis
and Facies Models • Case Histories

1982/630 pp./78 illus./53 plates/cloth/ISBN 0-387-11269-3